W9-DCW-382

LESKO'S NEW TECH SOURCEBOOK

Also by Matthew Lesko

Something for Nothing

Getting Yours

Information USA

How to Get Free Tax Help

The Maternity Source Book (with Wendy Lesko)

The Computer Data and Data Base Source Book

Lesko's New Tech Sourcebook

A Directory to Finding Answers in Today's Technology-Oriented World

Matthew Lesko

SHARON ZAROZNY, *Research Director*

HARPER & ROW, PUBLISHERS, New York
Cambridge, Philadelphia, San Francisco, London
Mexico City, São Paulo, Singapore, Sydney

1817

LESKO'S NEW TECH SOURCEBOOK. Copyright © 1986 by Matthew Lesko. All rights reserved. Printed in the United States of America. No part of this book may be used or reproduced in any manner whatsoever without written permission except in the case of brief quotations embodied in critical articles and reviews. For information address Harper & Row, Publishers, Inc., 10 East 53rd Street, New York, N.Y. 10022. Published simultaneously in Canada by Fitzhenry & Whiteside Limited, Toronto.

FIRST EDITION

Designer: Sidney Feinberg

Library of Congress Cataloging in Publication Data

Lesko, Matthew.
  Lesko's new tech sourcebook.

  Includes index.
  1. High technology—Information services—United States—Directories. 2. High technology—United States—Directories. I. Zarozny, Sharon. II. Title.
T10.63.A1L47 1986   607.2073.   85–45209
ISBN 0–06–181509–8   86 87 88 89 90 RRD 10 9 8 7 6 5 4 3 2 1
ISBN 0–06–096036–1 (pbk.)   86 87 88 89 90 RRD 10 9 8 7 6 5 4 3 2 1

To the free flow of information, the life blood of a prosperous society

Contents

Acknowledgments

I would like to thank Monica Horner and Nadia Schuman for their special interest and dedication to this book, and I would like to express my gratitude to Amy Summers, Barbara Cannon, Natalie Hartman, and Maria Manolatos for their special research contribution. A special thank-you goes to the Institute of Electrical and Electronics Engineers, the Association for Computing Machinery, and the National Bureau of Standards. Our researchers found the staff and members of these organizations extremely helpful and cooperative in the preparation of this book. And finally, I would like to thank each of the high-tech experts who agreed to be listed as a resource in this book.

Introduction

Once considered simply the dream of science fiction writers, "high tech" is now an exciting and dynamic reality. Scientific and technological advances have changed every aspect of our life, so much so that in 1982 *Time* magazine selected the computer as its "Man of the Year," and in 1984 President Reagan signed into law a Congressional Resolution designating September 30 through October 6 as National High Tech Week.

Today computers and microprocessors outnumber people in the United States. Our development of robotics and artificial intelligence has resulted in "smart machines" running factories without people—and far more efficient factories, as robots don't call in sick, tire of mundane tasks, or need time off to sleep. The United States Department of Defense is funding what's been called "Star Wars" projects, and NASA is planning the first American-manned space station. Human "spare parts" in the form of artificial organs, limbs, and joints have brought us closer to building a bionic man or woman, and other medical advancements have led to moral dilemmas over human cloning, test-tube babies, and the definition of death. Electronic banking is a convenience many of us enjoy, and stores are now beginning to design sales floors with electronic shoppers in mind. Telecommuting and teleconferencing are changing the work environment. The way we use our leisure time, and even dating, has been altered by our use of home computers and videotaping. New words and phrases have entered our vocabulary: biotechnology, humanoids, bytes, paperless office, technophobia, technoshock, sports biomechanics, and the electronic age. Other words have taken on new meaning: bits, chips, cottage industries, entrepreneurship, incubators, and venture capital.

Since the 1950s we have seen an explosion of technology in which, as incredible as it may seem, 90 percent of all scientific knowledge has been generated. And, according to the United States Department of Commerce, this knowledge is expected to double again in the next ten to fifteen years.

Once recognized as the leader in the technology explosion, the United States is rapidly losing this position. In the mid-1970s our country, which has 5 percent of the world population, was generating 75 percent of the world's technology. Today, ten years later, it is estimated that we are producing only 50 percent of it, and it is expected that that rate will drop to 30 percent in another decade. Why? Most countries now see technology as essential to both the economy and the quality of life;

therefore, an international race is on as countries battle for worldwide high-tech dominance.

In the United States, research spending for industry as a whole has grown faster than either sales or profits over the last five years. The United States spent an estimated $84 billion on research and development in 1983—more than Japan and Europe combined. Yet, Japan is proving a strong technological challenge to the United States. It has set a national goal of winning 18 percent of the computer business in the United States and a 30-percent share of global sales by 1990.

Japan is pouring at least half a billion dollars into developing a "supercomputer" that thinks like a human by 1990. An international sense of urgency has developed over research leading to this new family of computers, dubbed "fifth generation," which will be able to understand normal speech, accept spoken requests, and make decisions about stored information. It is commonly believed that the country that first markets this product will become the world leader in high technology.

While Japan has spent fewer dollars on research and development, a recent study by the United States Department of Commerce reports that it is taking the lead in important aspects of computer hardware and software, semiconductors, fiber optics, biotechnology, pharmaceuticals, and robotics. One explanation for this is that the Japanese have perfected the art of intelligence gathering, so much so that some American executives are convinced there is a Japanese CIA-type network tracking worldwide technological developments. The Japanese talk to our experts; read our journals, high-tech papers, conference proceedings, newspapers, and magazines; and report back to industry in their homeland. The United States, on the other hand, is considered to be losing ground because we don't even know how to take advantage of existing scientific knowledge produced in our own country. Recognizing this problem, American competitors are now banning together and forming cooperatives with universities and government—mainly for the purpose of sharing information and expertise.

The individual, too, needs to learn about new tech. While it enhances our lifestyle, some sociologists claim that technology has the potential of creating a greater gap between the rich and the poor, the educated and the uneducated. Therefore, in order to make wise decisions about your career, education, or retraining, you should become familiar with what new tech offers and requires.

As progress in science and technology depends upon the utilization of existing knowledge, access to technological information is becoming increasingly important. Industry needs the latest technical and business information available to make product innovations that will sustain America's leadership position in the world marketplace. Scientists and engineers need to keep up with the rapidly changing developments in their fields. And, since advancements affect every aspect of our life—from job opportunities to health—the average individual needs to learn about technological achievements in order to make wise choices about his or her future, and the future of humanity.

Computer storage and retrieval capabilities have ushered in a new challenge frequently called the "Information Age." Current estimates indicate that the amount of information at least doubles every ten years. We are now faced with information overload. It is estimated, for example, that in mid-1984 there were almost 3,000 books available about personal computing alone, and that figure has probably doubled by now. Therefore, it has become increasingly important to know where to turn for the best and most timely information about a particular subject or problem.

To help you determine where to begin your search, we have interviewed hundreds of experts in the major areas of high technology. We asked them which organizations they belonged to, what journals and periodicals they read, and where they recommend others—businesspeople, students, professionals, and the general public—go for information. The result is this Directory, which leads you to scientific and technical organizations, as well as government agencies, pertinent literature, data bases, and experts.

Whether you are investing in high-technology research and development, thinking about a career, or responsible for your company's strategic planning, you will find these information leads helpful.

Good luck as you explore this relatively new, exciting, and growing field, and welcome to a unique information sourcebook for the high-tech age.

How to Use *Lesko's New Tech Sourcebook*

This Directory leads you to a wide range of information sources in the major areas of high technology. Chapters describing resources for high-tech careers, films, and funding opportunities are also included. The Appendix presents our recommendations for information sources covering the overall field of new technology, and resources you can consult to obtain definitions of new tech terms and fields.

People are an important information resource, and your success in gathering what you need often depends upon how you approach them; therefore, we have included a section below entitled "The Art of Obtaining Information." It contains ten techniques to keep in mind when calling someone for information.

For each subject included in the Directory, we have provided a variety of information sources organized under the following categories:

Organizations and Government Agencies

This section lists professional societies, trade associations, government offices, businesses, research centers, and other places to go for information. A description including purpose, services, and publications is provided for each entry.

Professional societies and trade associations often are an important resource. Usually the staff can refer you to their members who are experts in the field, and many of these groups publish membership directories that are an excellent tool for identifying people, companies, and so on, in a particular area. Most sponsor conferences and workshops, the proceedings of which are usually published. Technical papers, journals, bibliographies, audiovisuals, career information, statistics, and materials designed to educate the public are usually available from these groups. On-line data bases are also maintained by some societies.

Government agencies are another good starting point for information. Often you can locate a staff member who is an expert in a particular aspect of the field you are researching. He or she can give you an overview, guide you to government and nongovernment organizations supporting research, supply statistics, and recommend literature. The government produces many publications, most of which are available free of charge or for a nominal fee. Frequently, government offices also operate in-house, on-line data bases that staff members will search upon request.

Companies, research centers, and educational institutions are also mentioned in this section. Public affairs offices in these companies will often send you brochures, product literature, and sometimes technical papers. Depending upon company policy, you may even be able to speak with a department responsible for researching the area in which you are interested. Educational institutions and research centers have specialists who can tell you about current developments in their field as well as technical literature.

Publications

In this section are descriptions of magazines, journals, information source directories, and other printed materials. Given the rapidly changing nature of new technology, we have focused on frequently updated publications. This section should be used as a guide to the most widely read publications, not as an exhaustive listing of materials available. Free publications are noted. As professional associations produce about 40 percent of all scientific literature, you should also scan the publications we have listed in entries under "Organizations and Government Agencies."

Data Bases

Described here are on-line data bases recommended by people we interviewed. If you are trying to identify data bases, you should also check entries under "Organizations and Government Agencies."

Additional Experts and Resources

This category contains listings of companies, educational institutions, and miscellaneous sources that can be consulted for information. It also includes people we contacted who were willing to answer questions from the public. These individuals are either experts in a particular high-tech field or knowledgeable about where to go for information.

Limitations

This Directory is intended to help you identify useful information sources pertaining to high technology. It is not to be considered an exhaustive listing of available high-tech resources.

Unfortunately some of the addresses and telephone numbers we have listed may change by the time you use this book. We regret the inconvenience and suggest you try to track down new contact information for the sources identified.

Also, as we received no guarantee on the prices listed for publications, we suggest you verify costs before ordering.

Finally, we recommend you exercise consideration when contacting the resources listed, especially the individuals. Many are working scientists and engineers, and while they have agreed to help the public, many would prefer to be contacted only when you have exhausted other resources or at least done some preliminary research. Government and association staffs are sometimes small and quite busy; therefore, they should be used wisely.

The Art of Obtaining Information

Information requests must be made carefully. Your approach may be the sole reason why a source is worthwhile as opposed to unhelpful. Following are ten techniques developed by the author when he established Washington Researchers, a company specializing in gathering information from federal bureaucrats:

1. *Introduce yourself cheerfully.* The way you open the conversation sets the tone for the entire interview. Your greeting and initial comments should be cordial and cheery. They should give the feeling that yours is not going to be just another anonymous telephone call, but a pleasant interlude in your source's day.
2. *Be open.* You should be as candid as possible with your source, since you are asking the same of him or her. If you are evasive or deceitful in explaining your needs or motives, your source may be reluctant to provide you with information. If there are certain facts you cannot reveal because of client confidentiality, for example, explain why. Most people will understand.
3. *Be optimistic.* Throughout your entire conversation, you should exude a sense of confidence. If you call a source and say, for example, "You don't have information that can help me, do you?," it makes it easy for your source to say, "You're right. I can't help you." A positive attitude encourages your sources to stretch their minds to see the various ways they can be of assistance.
4. *Be humble and courteous.* You can be optimistic and still be humble. Remember the old adage that you can catch more flies with honey than you can with vinegar. People, in general, and experts, in particular, love to tell others what they know, as long as their position of authority is not questioned or threatened.
5. *Be complimentary.* This goes hand in hand with being humble. A well-placed compliment about your source's expertise or insight into a particular topic will serve you well. In searching for information in the government, you are apt to talk to people who are colleagues of your source. It would be reassuring to your source to know he is respected by his peers.
6. *Be concise.* State your problem in the simplest possible manner. If you bore your source with a long-winded request, your chances for a thorough response are diminished.
7. *Don't be a "gimme."* A "gimme" is someone who says "give me this" or "give me that" and has no consideration for the other person's feelings.
8. *Talk about other things.* Do not spend the entire time talking about the informa-

tion you need. If you can, discuss briefly a few irrelevant topics, such as the weather, the Washington Redskins, or this year's pennant race. The more social you are, the more likely it is that your source will respond favorably.

9. *Return the favor.* You might share with your source some bits of information, or even gossip, you have learned from other sources. However, be certain not to betray the trust of your client or another source. If you do not have any information to share at the moment, keep your source in mind and call him or her back if something develops when you are further along in your research.

10. *Send thank-you notes.* A short note, typed or handwritten, helps ensure that your source will be just as cooperative in the future.

New Tech Sourcebook

Acoustics

Organizations and Government Agencies

Acoustical Society of America (ASA)

335 East 45th Street
New York, NY 10017

212-661-9404

ASA aims to increase and disseminate acoustics knowledge and promote its practical applications. Members are physicists, engineers, psychologists, and audiologists in various fields of acoustics. Areas that the Society covers include architectural acoustics, underwater sound, musical acoustics, speech communication, bioacoustics, shock and vibration, psychological and physiological acoustics, and noise. The organization has several committees that produce standards in these areas. It holds two meetings each year, at which technical papers are presented. A limited amount of information is available from the staff, including referrals to members and consultants. The Society publishes:

• *Journal of the Acoustical Society of America*—a monthly research journal directed to professionals in the field. ASA members receive a complimentary subscription, and nonmembers can subscribe for $240 a year.

• *Catalog of Acoustical Standards*—is available from the ASA Standards Secretariat. Also, one can purchase individual acoustic standards through the Standards Secretariat.

Bolt, Beranek and Newman, Inc.

10 Moulton Street
Cambridge, MA 02138

617-491-1850—Erich Bender

A consulting firm, the company has several hundred professionals working in all aspects of acoustics, including architectural acoustics, industrial noise control, product noise control, sound system development, sound masking system development, underwater acoustics, environmental studies, aircraft noise control, and speech processing. Dr. Bender can answer questions or refer you to specific experts. Exploratory telephone conversations are free; however, there is a fee for consulting services. The company also has offices at 1300 North 17th Street, Arlington, VA 22209, 703-524-4870, and at 21120 Vanowen Street, Canoga Park, CA 91303, 213-347-8360.

Building Acoustics

Room A105, Sound Building
National Bureau of Standards
Gaithersburg, MD 20899

301-921-3783

This group is involved in research in the following areas: architectural acoustics, performance of acoustical materials, noise generation, and propagation in and around buildings. The staff can answer questions in their areas of expertise and make referrals.

Central Institute for the Deaf
818 South Euclid Avenue   314-652-3200
St. Louis, MO 63110

This private, nonprofit institution conducts research in the following areas: hearing aids and digital processing of sound, effects of noise on hearing, auditory perception, the auditory nervous system, and sound-to-tactile aids for the hearing impaired. The staff can answer questions and/or make referrals in their areas of expertise.

Mechanical Production Metrology Division
Room A149, Sound Building   301-921-3607—Dr. Victor Nedzelnitsky
National Bureau of Standards
Gaithersburg, MD 20899

This government division performs research and provides services to both industry and government in the following areas: microphone calibrations; testing and evaluation of acoustical instruments—for example, sound level meters, noise dosimeters, and hearing aids. Their special laboratory facilities, called anechoic chambers, are available for cooperative research. Staff members have written technical papers, which are available to scientists and other professionals in the field. Dr. Nedzelnitsky can answer questions and/or make referrals to other experts.

Bruel and Kjaer Instruments, Inc.
15944 Shady Grove Road   301-948-0494
Gaithersburg, MD 20877

This company manufactures sound and vibration test equipment. The salespeople, who are application engineers, can answer questions and make referrals. The company's free application notes show how equipment was used in certain types of tests. It also has a local service center for the equipment it sells.

Additional Experts and Resources

Mead Killion, Ph.D., President
ETYMOTIC Research, Inc.   312-228-0006
61 Martin Lane
Elk Grove Village, IL 60007

Dr. Killion has an M.S. in mathematics, and a Ph.D. in audiology. He has published nine U.S. patents on subminiature microphones, earphones and amplifiers, and has written and lectured extensively on hearing aids and earmolds. Dr. Killion has also written papers in the field of electroacoustics, psychoacoustics, physiological acoustics, and audiology. His company's research focuses on devices for hearing and hearing testing.

ADA

See: Programming Languages (Computer)

Aeronautics

Organizations and Government Agencies

Aerospace Education Foundation (AEF)
1750 Pennsylvania Avenue, NW
Washington, DC 20006

202-637-3370

This association is dedicated to educating people about different aspects of aerospace. It sponsors educational symposia, seminars, workshops, and achievement programs, and it maintains a library that is open to the public by appointment. It publishes the following:

● *Air Force*—a monthly magazine of the Air Force Association covering different aspects of the aerospace industry. Annual subscription is $15 a year.

● *Crusade for Air Power*—an educational book on recent aerospace history. It is available for $16.45.

Aerospace Industries Association of America (AIA)
1725 DeSales Street, NW
Washington, DC 20036

202-429-4600

A national trade association of U.S. companies, AIA is engaged in research, development, and manufacture of aerospace systems. Staff members, in the Research Centers, perform research and economic data services. The library of aerospace literature is open to the public. The Office of Public Affairs can provide you with information about the aerospace industry. It publishes the following:

● *Aerospace*—a quarterly nontechnical magazine that explores the role of the aerospace industry in high technology, international trade, national defense, and space exploration. The 16-page publication is available free of charge.

● *AIA Annual Report*—a yearly publication, issued each January, which covers the previous year. It reports AIA activities and lists members of each of the association's councils, including procurement and finance, the technical council, the operations council, the international council, and the public affairs council. It also reports activities of the legislative counsel. The publication is available free of charge.

• *Aerospace Facts and Figures*—an annual reference book. Its eight chapters provide statistical and analytical information about the U.S. aerospace industry. Topics include aircraft production, missile programs, space programs, foreign trade, and employment. The 175-page publication is available for $10.95 from McGraw-Hill, Inc., 1221 Avenue of the Americas, New York, NY 10020, 212-512-2123.

American Astronautical Society

6060 Duke Street   703-751-7721
Alexandria, VA 22304

This scientific and technical society is devoted to space activities, and to the advancement of the astronautical sciences and space flight engineering. It publishes a wide variety of books dealing with space flight and the astronautical sciences, including:

• *Journal of the Astronautical Sciences*—a quarterly archival journal that contains articles dealing with astrodynamics, celestial mechanics, guidance, and control. The annual subscription price is $75.

• *AAS Directory*—a membership directory published every two years. It lists officers, technical committees, and awards. The price is $30.

• *Science and Technology Series*—annual publications that cover a variety of topics and serve as a supplement to the meetings proceedings. Prices range from $20 to $60.

• *Index to All AAS Society Papers and Articles (1954-1978)*—a publication covering the years 1954 through the end of 1978. The price is $30.

American Institute of Aeronautics and Astronautics (AIAA)

557 West 57th Street   212-247-6500
New York, NY 10019

The Institute is devoted to science and engineering in aviation, as well as space technology and systems. Part of its services include publishing books, journals, and technical papers. It publishes the following:

• *International Aerospace Abstract*—a semimonthly compendium of worldwide aerospace literature. This data base includes more than 690,000 items, such as journal articles, conference papers, publications of learned academies and professional societies, books, and other types of aerospace literature. An Annual Cummulative Index is also available. The annual subscription is $525 a year, for 24 issues.

• *Astronautics and Aeronautics*—a monthly interdisciplinary magazine devoted to technical and scientific coverage of the aerospace field. Recent issues have included articles on turbine engine systems, ramjet missiles, tactical air power in land warfare, goals in planetary exploration, advanced robotics, advanced composites in commercial aircraft, and computer-aided design. The *AIAA Bulletin*, included in each issue, provides announcements of all AIAA technical programs and papers. It is available for $42 a year.

● *Aerospace America*—a monthly magazine that provides general information about technology applied to aerospace. Subscription is $51 a year.

● *Journal of Aircraft*—a monthly publication containing technical articles about aircraft. Subscription is $165 a year.

● *Journal of Spacecraft and Rockets*—a monthly publication reporting on the latest technology used in spacecraft and rockets. Subscription is $85 a year.

● *The AIAA Student Journal*—a quarterly containing articles of general interest to students and young professionals in the aerospace disciplines. Subscription is $12 a year.

American Society of Mechanical Engineers

Aerospace Division   212-705-7745
345 East 47th Street
New York, NY 10017

The Aerospace Division, one of 32 technical divisions of the Society, is concerned primarily with the mechanical engineering aspects of aircraft, spacecraft, and missile design and operation. The primary functions of the Division include the organization and sponsorship of technical conferences, publication of technical volumes, and technical support to the Society and government organizations.

National Aeronautics and Space Administration (NASA)

NASA Information Center   202-453-1000
NAS-2
Washington, DC 20546

Photographs, speakers, and aeronautical information of interest to the scientist as well as the general public can be obtained from NASA. The office above can direct you to the appropriate resource. Listed below are NASA offices that also can be of assistance.

Public Affairs Officer
Office of Aeronautics and Space Technology
NASA Headquarters
Washington, DC 20546
202-453-2754

Free pamphlets, brochures, and lithographs about NASA aeronautics programs are available from this office. Written requests are preferred.

NASA Headquarters Library
600 Independence Avenue, SW
Room A-39
Washington, DC 20547
202-453-8545

The Library is open to the public. Librarians will help you find scientific and technical materials, recommend other sources of information, and answer questions of a general nature.

Speakers Bureau
Public Services Branch
NASA Headquarters
Washington, DC 20546
202-453-8315

NASA maintains a Speakers Bureau. If you are interested in the availability of a speaker for your group, send this office a written request.

Space Photographs

PO Box 486
Bladensburg, MD 20710
Written requests only

You can purchase NASA photographs from Space Photographs, a company under contract to NASA to take photographs and make them available to the general public. Upon receiving your request, the company will send you a photo index and a price/order blank.

Space Flight Participant Program

Mail Code ME
NASA Headquarters
Washington, DC 20546

202-453-2556

While NASA is not yet taking reservations for the space shuttle, it does have an office coordinating opportunities for private citizens to fly on space shuttle missions. As opportunities develop, such as the first teacher in space, the above office will announce them through the media and provide information packets and applications. The following is currently available:

● *Space Flight Participant Program*—a free fact sheet describing the program and where to learn about it.

Space Services, Inc. (SSI)

7015 Gulf Freeway, Suite 140
Houston, TX 77087

713-649-1716

SSI will arrange to have the cremated remains of your loved ones sent into space to orbit for 64 million years. The company has reserved places for 10,333 $2'' \times \frac{5}{8}''$ containers to travel in a space vehicle which people on earth can view with a telescope. Free information packets are available about this service.

Publications

Anti Jet Lag Diet

Office of Public Affairs
Argonne National Laboratory
9700 South Cass Avenue
Argonne, IL 60439

Information about reducing the adverse side effects of air travel by altering your diet is provided in this free publication. It is available free of charge by sending a stamped self-addressed envelope to the above address.

Aviation Week and Space Technology

McGraw-Hill, Inc.
1221 Avenue of the Americas
New York, NY 10020

212-512-2000

This monthly magazine provides complete coverage of scientific, technical, operational, financial, political, and legislation developments, both domestic and international, in all major segments of the aerospace industry. Topics covered include air transport, aeronautical engineering, space technology, business flying, satellite communications, management, missile engineering, manufacturing technology, and avionics. It is available for $45 a year from *Aviation Week and Space Technology*, P.O. Box 1505, Neptune, NJ 07753, 800-922-0793.

Transactions on Aerospace and Electronic Systems

IEEE Headquarters   212-705-7900
345 East 47th Street
New York, NY 10017

This bimonthly publication covers equipment, procedures, and techniques applicable to the organization, installation, and operation of functional systems designed to meet high-performance requirements of earth and space systems. It is available for $108 a year.

Additional Experts and Resources

International Trade Administration

Aerospace Division   202-377-0678—Randy Meyers or Gene Kingsbury
Room 4047—Department of Commerce
Washington, DC 20006

Both experts are responsible for preparing studies on the aerospace industry. Their work includes forecasting the market and describing what shape it is in, keeping track of aerospace trade, and making statistical analyses. They will answer questions about all aspects of the aerospace industry, and refer you to other sources of information.

Agriculture

Organizations and Government Agencies

Agricultural Research Service (ARS)

U.S. Department of Agriculture (USDA)   301-344-2264
Bldg. 005, BARC-West, Room 307
Beltsville, MD 20705

The principal scientific research agency of USDA, the Agricultural Research Service focuses on crops, animals, post-harvest technology, resource conservation, human nutrition, and systems technology. A large part of current ARS research is fundamental, attacking problems needing long-term and high-risk research approaches. A growing proportion of the research falls under the broad heading of biotechnology. ARS laboratories exist at 150 locations around the country, including the campuses of all land-grant colleges and universities. Laboratories are also maintained in eight foreign countries. More than 8,500 people are employed by ARS: one third are scientists; the rest are technicians and support personnel. Many ARS scientists have published technical reports and papers in their field of expertise. These publications are available, often free of charge, from the scientists themselves. Listed below are the center directors of several ARS field offices. As their job entails keeping up to date on all ARS research performed in their respective geographical areas, they are a good source for information and referrals. Also listed are the four ARS chemistry labs, which also have experts in agriculture-related fields. If your regional field office cannot assist you, the staff in the main office listed above are a good secondary source. They can refer you to an appropriate expert, lab, or center involved in the previously mentioned subjects.

Dr. Herbert L. Rothbart, Area Director
North Atlantic Area
USDA-ARS
600 East Mermaid Lane
Philadelphia, PA 19118
215-233-6593

Dr. Waldemar Klassen, Area Director
Beltsville Area
USDA-ARS
Room 227, Bldg. 003, BARC-West
Beltsville, MD 20705
301-344-3078

Dr. Ernest L. Corley, Area Director
South Atlantic Area
USDA-ARS
PO Box 5677
College Station Road
Athens, GA 30613
404-546-3311

Dr. Carleton D. Ranney, Area Director
Mid-South Area
USDA-ARS
PO Box 225
Stoneville, MS 38776
601-686-2311, ext. 265

Dr. Joseph R. Johnston, Area Director
Southern Plains Area
USDA-ARS
VTERL Complex, F&B Road
PO Drawer EC
College Station, TX 77841
409-260-9346

Dr. Robert A. Rhodes, Area Director
Midwest Area
USDA-ARS
1815 North University Street
Peoria, IL 61604
309-682-4011, ext. 602

Dr. Paul A. Putnam, Area Director
Central Plains Area
USDA-ARS
PO Box 70
Ames, IA 50010
515-239-8205

Dr. Kenneth L. Lebsock, Area Director
Northern State Area
USDA-ARS
University of Minnesota
Room 266, North Hall
St. Paul, MN 55108
612-349-3268

Dr. William G. Chace, Jr., Area Director
Northwest Area
USDA-ARS
809 Northeast 6th Avenue, Room 204
Portland, OR 97232
503-231-2240

Dr. Lambertus H. Princen
Center Director, NRRC
USDA-ARS
1815 North University Street
Peoria, IL 61604
309-685-4011, ext. 542

Dr. Jan van Schilfgaarde, Area Director
Mountain States Area
USDA-ARS
2625 Redwing Road, Suite 350
Fort Collins, CO 80526
303-223-3459

Dr. H.P. Binger
Center Director, WRRC
USDA-ARS
800 Buchanan Street
Albany, CA 94710
415-486-3421

Dr. John P. Cherry
Center Director, ERRC
USDA-ARS
600 East Mermaid Lane
Philadelphia, PA 19118
215-233-6593

Dr. Ivan W. Kirk
Center Director, SRRC
USDA-ARS
1100 Robert E. Lee Boulevard
PO Box 19227
New Orleans, LA 70179-0227
504-589-7511

American Society of Agricultural Engineers (ASAE)
2950 Niles Road   616-429-0300
St. Joseph, MI 49085

A nonprofit technical, scientific, and educational Society, ASAE is dedicated to the betterment of agriculture through the improved application of engineering principles. The Society coordinates the efforts of educators, USDA researchers, state agricultural experimentation researchers, and engineers in regulatory agencies and agriculture-related manufacturing and processing industries. ASAE is divided into five major technical areas, each producing periodicals, conference proceedings, technical papers, and so on. The divisions are Power and Machinery, Soil and Water, Structures and Environment, Electric Power and Processing, and Food Engineering.

Through its Cooperative Standards Program (CSP), ASAE develops voluntary standards to serve the agricultural industry. ASAE's 52 local chapters nationwide sponsor programs for area members and promote agricultural engineering at the local level. The Society sponsors two national meetings annually, as well as special national and international conferences and symposia. ASAE also provides career guidance information regarding agricultural engineering opportunities.

ASAE publishes periodicals, standards, monographs, conference proceedings, and more than 800 technical papers yearly. A free publications catalog is available. Examples of the Society's publications include:

- *Transactions of the ASAE*—a bimonthly technical journal. Its peer-reviewed articles describe the application of engineering principles to the solution of agricultural problems. Each issue is approximately 300 pages, and the subscription is $75 a year.

• Divisional Editions of the *Transactions of the ASAE*—at the end of each year, each ASAE Division publishes its transactions and full-length papers in special editions according to subject matter.

American Society of Agronomy (ASA)

677 South Segoe Road   608-274-1212
Madison, WI 53711

This organization promotes the wide use of natural resources in the production of food, feed, and fiber crops. The staff can refer you to members who are scientists working in private industry, universities, colleges, and state and federal governments. ASA participates in the annual joint meeting held by the Agronomy Society of America, the Crop Science Society of America, and the Soil Science Society of America. The society's publications include:

• *Agronomy Journal*—a bimonthly technical publication with papers on all aspects of crop and soil sciences, including crop physiology, and production and management, along with the relationship to soil fertility and climatic conditions. Approximately 160 pages per issue, the publication is available for $65 a year or $13 an issue.

International Society for Plant Molecular Biology

Department of Biochemistry   404-542-2086
University of Georgia
Athens, GA 30602

This scientific Society provides a framework for bringing plant molecular biologists together. About one third of the members are involved in the cloning of genes. The Society sponsors several meetings; its first international congress was held in 1985. The staff can answer questions and/or make referrals. Publications include:

• *Plant Molecular Biology*—a bimonthly technical publication directed toward scientists working in the field. It has approximately 80 pages an issue; the subscription rate is $50. Write to Maratinus Nijhof/W.J. Junk, PO Box 566, 2501 C N The Hague, the Netherlands.

Soil Science Society of America

677 South Segoe Road   608-274-1212
Madison, WI 53711

This organization focuses on all fields of soil science, including physics, chemistry, microbiology, fertility, morphology, classification, and mineralogy. It participates in the annual joint meeting held by the American Society of Agronomy, the Crop Science Society of America, and the Soil Science Society of America. The staff can make referrals.

The organization publishes the following:

- *Soil Science Society of America Journal*—a bimonthly publication that contains papers on new developments in soil physics, mineralogy, chemistry, microbiology, fertility and plant nutrition, soil genesis and classification, soil and water management, forest and range soils, and fertilizer use and technology. Approximately 160 pages an issue, the publication is available for $65 a year or $13 an issue.

Crop Science Society of America
677 South Segoe Road   608-274-1212
Madison, WI 53711

This organization comprises scientists interested in all aspects of crop science, including the improvement, culture, management, and utilization of field crops. It participates in the annual joint meeting held by the American Society of Agronomy, the Crop Science Society of America, and the Soil Science Society of America. The staff can answer questions and make referrals. Publications include:

- *Crop Science*—a bimonthly technical journal that reports on recent developments in crop breeding and genetics, crop physiology and biochemistry, ecology, cytology, crop and seed production, statistics, and weed control. Approximately 160 pages an issue, the publication is available for $65 a year or $13 an issue.

Data Bases

AGNET
University of Nebraska   402-472-1892
Lincoln, NE 68583-0713

This data base, compiled by U.S. Department of Agriculture specialists at several land-grant universities, deals with agriculture and home economics. It contains information about programs related to livestock and crop production, grain handling, marketing data, financial analysis, and home economics. It enables the user to keep up to date on market conditions, trends, and USDA reports. It also has an electronic mail and conferencing service. Access is available via the National IN-WATS network or by direct-dial GTE Telenet.

AGRICOLA
Information Systems Division   301-344-3813
National Agricultural Library, Fifth Floor
U.S. Department of Agriculture
10301 Baltimore Boulevard
Beltsville, MD 20705

The data base covers agricultural data gathered from worldwide journals and monographic literature, as well as U.S. government reports. Topics covered are agriculture, agricultural economics, consumer protection, nutrition, rural sociology, animal science, veterinary medicine, forestry, plants, agricultural chemistry, natural resources, entomology, and agricultural engineering. In print, AGRICOLA is the *Bibliography of Agriculture*. It is available on BRS and DIALOG.

AGRICULTURE RESEARCH REVIEW
Lloyd Dinkins   901-274-9030
PO Box 22642
Memphis, TN 38122

    The data base, put together from researchers' scientific papers, covers research either in progress or recently completed. It focuses on public research carried out by the U.S. Department of Agriculture and state experiment stations. It contains the full text reports, research plans, and bibliographies of agricultural research. The entire scope of agricultural research is covered, including research on crops and livestock. Information is from October 1982 to the present, and it is available on NewsNet, Inc.

CRIS/USDA (Current Research Information System)
U.S. Department of Agriculture   301-344-3846
USDA/CRIS, NAL Building
Beltsville, MD 20745

    This system consists of information about current research in agriculture and related sciences, sponsored or conducted by USDA research agencies, state agricultural experiment stations, state forestry schools, and other cooperating state institutions. Also incorporated are active and recently completed projects within the last two years. Contents include the following: biological, physical, social, and behavioral sciences related to agriculture in its broadest applications, including natural resource conservation and management; marketing and economics; food and nutrition; consumer health and safety; family life, housing, and rural development; environmental protection; forestry; outdoor recreation; and community, area, and regional development. Information is provided for the past two years, and it is updated quarterly. It is available on DIALOG.

Additional Experts and Resources

Dr. Robert T. Fraley
Manager, Plant Molecular Biology Group   314-694-2178
Monsanto Company
800 North Lindbergh Boulevard
St. Louis, MO 63167

    Dr. Fraley conducts research in plant genetic engineering. He is involved in the development of methodology for introducing foreign genes into plants, with emphasis on plant gene expression and crop improvement. Dr. Fraley has several publications to his credit.

Dr. Gerald W. Isaacs
Chairman, Agricultural Engineering Department   904-392-2522
University of Florida
Frazier Rogers Hall
Gainesville, FL 32611

Dr. Isaacs and several of his staff members are researchers in the field of robotic application in agriculture. They can answer questions in this area. The First International Conference on Robotics and Intelligent Machines in Agriculture was cosponsored by the Agricultural Engineering Department and the American Society of Agricultural Engineers (ASAE). Proceedings are available from ASAE, 2950 Niles Road, St. Joseph, MI 49085, 616-429-0300.

Alternative Fuels

See: Fuels, Alternative

Animation

See: Computer Animation

Antimissile Systems

See: Star Wars: Antimissile Systems in Space

Apparel

Begins next page. See also: Textiles

Organizations and Government Agencies

American Apparel Manufacturers Association (AAMA)
1611 North Kent Street   703-524-1864
Arlington, VA 22209

This trade association with 450 members represents the major apparel manufacturers. The staff can answer questions and/or make referrals to helpful sources. Members serve on AAMA committees established to study topics and problems of concern to the industry. The committees have published more than 65 reports on their findings. These reports are available to the public for a fee; topics covered include economics, finance, marketing, planning, personnel, systems, and technical developments and application. A free publications catalog is available from AAMA.

Draper Laboratories Automation Project
U.S. Department of Commerce
Office of Trade Adjustment Assistance
14th and Constitution Avenue, NW
Washington, DC 20230

Under a project jointly funded by the government, private industry, and the garment union, Draper Laboratories has been studying automated sewing systems and computer robotics. The research and development helps the apparel industry develop new technologies, to regain a competitive edge in the marketplace. Although a large portion of the research is classified information, reports are being produced annually for the years 1983 through 1986. This office can provide information about the project and information about other apparel industry-related publications. The Draper report must be purchased from NTIS, 5285 Port Royal Road, Springfield, VA 22161, 703-487-4763. The First Year Report (#PB83-155895) is five volumes and costs $50.

The Management Services Group (MSG)
PO Box 1986   803-771-7500
Columbia, SC 29202   800-845-8820

MSG is a technical and training division of Bobbin International, Inc. It maintains an extensive network of information resources relating to apparel. Its staff will provide free consulting and referral services. The group conducts seminars nationwide on a variety of high-tech topics, including technical apparel plant engineering. MGS has published the following:

• *World-Wide Technology Update*—a one-time report, which appeared in the July 1984 edition of *Bobbin International Magazine*. The study profiles technology utilization in the apparel industry, addressing the current situation, what is lacking, and future trends. The report, which is also available in lecture format, can be purchased through the office at the above address.

Publications

Apparel World

386 Park Avenue   212-683-7520
New York, NY 10016

    Apparel World is a monthly publication covering cut and sewn apparel, the latest apparel fashion technology and manufacturing technology, management ideas, business trends, and machinery. A subscription to *Apparel World* includes a *Buyers Guide*, which is published in September and provides a detailed listing of equipment, products, and services supplied to the apparel trade, and a listing of companies and machinery. Along with the *Buyers Guide*, you receive a Yearbook with references for apparel manufacturers. The cost is $12 a year.

Bobbin International

PO Box 1986   803-771-7500
Columbia, SC 29202   800-845-8820

    Bobbin International publishes the following:

    • *The Bobbin International*—this publication covers all aspects of the apparel industry, such as technical advances, industrial engineering, new products, robotics, productivity, apparel, manufacturing, people in the news, linen, intimate apparel, and sportswear. The 250-page monthly is available for $18 a year.

    • *La Bobbina*—this bimonthly publication is a Spanish version of *Bobbin International* and is available free of charge.

    • *Textbooks for the Sewing Products Industry Publications Catalog*—describes more than 60 textbooks available from Bobbin International about topics such as apparel manufacturing, business, finance, industrial engineering, and quality control. The catalog is available free of charge.

Daily News Record

Fairchild Publications   212-741-4000
7 East 12th Street
New York, NY 10003

    Published Monday through Friday, this newspaper reports on men's apparel, textiles, and retail. The paper provides information about new technological developments, and it contains legislative and financial news. The annual subscription is $50.

Additional Experts and Resources

Joel Sternberg and Kurt Salmon, Chair—Apparel Management
Fred Fortress, Director of Apparel Research
Department of Textiles   215-951-2750
Philadelphia College of Textiles and Science
Schove House Lane and Henry Avenue
Philadelphia, PA 19144

The College faculty provides instruction in apparel management and fashion design. The school serves as a resource center for industry and houses the only accredited U.S. laboratory for testing wearing apparel, fibers, and fabric. Currently the College is the only U.S. institution with access to Japan's KES-FB system, which has been billed as the most revolutionary apparel technology in the past 25 years.

Joel Sternberg and Fred Fortress are experts about new technological developments in the apparel industry. They can provide assistance, answer questions, and make referrals. Their office sends out notices of apparel-related seminars, courses, and demonstrations. Call or write the above office to be added to the mailing list.

Appropriate Technology

See also: Renewable Energy Technology; Solar Technology; Transfer Technology

Organizations and Government Agencies

Institute for Local Self-Reliance
2425 18th Street, NW   202-232-4108
Washington, DC 20009

The Institute is a non-profit, private organization that helps cities and neighborhoods achieve economic development. The staff can provide technical assistance, information and referrals, public policy advice, and research. The Institute is involved in energy conservation, small-scale power production, resource recycling, and composting. A free listing of its publications is available. Publications range in price from $1.25 to $35.

National Appropriate Technology Assistance Service (NATAS)
U.S. Department of Energy   800-428-1718 (Montana residents)
PO Box 2525   800-428-2525 (Elsewhere in the U.S.)
Butte, MT 59702-2525

NATAS is an information and referral service funded by the Department of Energy. Its staff provides three primary services relating to appropriate technology: (1) general information; (2) engineering, scientific, and technical assistance to help callers solve problems related to energy conservation and renewable energy uses; and (3) commercial technical assistance to help entrepreneurs develop the business side of energy-related appropriate technology. NATAS works closely with federal, state, and local programs to coordinate these activities. The staff can assist with energy-conservation planning, design systems, product patent licensing, as well as business planning, organization and development. The staff can provide helpful publications, and they will do on-line searching of both commercial and private appropriate technology data bases.

National Center for Appropriate Technology (NCAT)

PO Box 3838   406-494-4572

Butte, MT 59702-3838

   NCAT is a nonprofit organization established to research, develop, and transfer small-scale technologies. Its goal is to promote applications of conservation and renewable energy technology. Energy-efficient home construction and home weatherization are major research areas for NCAT. The professional staff can provide technical assistance and training, publications, research, and referrals. NCAT has more than 60 publications in stock on topics ranging from the use of solar mobile homes to small-scale fuel alcohol research.

Volunteers in Technical Assistance, Inc. (VITA)

1815 North Lynn Street   703-276-1800

Suite 200

Arlington, VA 22209

   VITA is a nonprofit international development organization involved in accessing or transferring small-scale technology to Third World countries. VITA maintains an on-line data base of 4,500 volunteer consultants worldwide who are experts in the various appropriate technologies. The staff will search their data base for the general public. Searches can be done by geographical area, language, and skill level. VITA experts will answer questions, provide technical assistance, and make referrals. VITA has published more than 100 publications relating to appropriate technologies. A free catalog of the publications, which range in price from $1.25 to $35, is available.

Aquaculture

Organizations and Government Agencies

American Fisheries Society (AFS)

5410 Grosvenor Lane   301-897-8616

Bethesda, MD 20814

   This scientific group is dedicated to the advancement of fisheries science and the conservation of renewable aquatic resources. Members include fisheries, aquatic science professionals, and students. The staff can respond to written inquiries and make referrals. AFS holds an annual meeting, and the fish culture section of the society holds its own meetings. A free publications brochure lists books available. Additional fisheries science materials, including periodicals and journals, are also available from the Society. Publications include:

   • *Progressive Fish Culturist*—a quarterly publication containing papers on design and construction of culture systems; hatchery materials and equipment; growth, feeding, and nutrition; disease diagnosis and treatment; population studies, breeding, and rearing. The annual subscription price is $12.

Department of Fisheries and Allied Aquacultures

Auburn University   205-826-4786
Swingle Hall
Auburn, AL 36849

The Department's research focuses on many elements: hatchery management and the production of young fish that can be stocked in ponds and lakes; selective breeding and fish genetics to improve the strains of fish; fish nutrition with reference to providing diets for fish at minimum cost and for optimal growth; the study of fish parasites and diseases; economic analysis, and so on. The university has been involved in international aquaculture programs in about 12 different countries. It makes available a publications list that describes research reports, department reports, and the international research and development series. The staff will answer questions or make referrals.

National Fisheries Center—Leetown

Fish and Wildlife Service   304-725-8461
Department of the Interior
Box 700, Route 3
Kearneysville, WV 25430

The Center's research focuses on the production of healthy fish, including the development of new techniques to produce quality fish. One of its major operations is the Fish Health Research Laboratory, located at Leetown with backup from seven other laboratories and field stations in other parts of the country. The staff in the Technical Assistance Section can provide technical advice. The library, which participates in interlibrary loans, contains 20,000 volumes, reprints of leaflets, articles, and so on.

World Mariculture Society

Division of Continuing Education   504-388-3137
Louisiana State University
341 Pleasant Hall
Baton Rouge, LA 70803

This society serves as an international clearinghouse for information on the status, potential, and problems of cultivation of aquatic organisms. Members are scientists, fish farmers, and businessmen. The Society holds a major technical meeting annually and works continually to further the development of mariculture throughout the world. The staff can answer questions or make referrals. Publications include:

• *The Journal of the World Mariculture Society*—an annual publication consisting of papers presented at the yearly meeting. The cost is $30.

• Three publications pertaining to meetings that the Society has held are available for a fee.

Publications

Aquaculture

The Journal Information Center   212-867-9040, ext. 303
Elsevier Publishing Company
52 Vanderbilt Avenue
New York, NY 10017

This international journal is devoted to research about the exploration, improvement, and management of all aquatic food resources, both floristic and faunistic, related directly or indirectly to human consumption. It is published approximately every two and one half weeks, and is directed to aquaculturists, fish scientists, and marine biologists. Annual subscription price is $556.95.

Aquaculture Digest

9434 Kearney Mesa Road   619-271-0133
San Diego, CA 92126

A monthly report on fish and shellfish farming in the U.S., *Aquaculture Digest* has a heavy private-sector orientation. It covers business development, scientific research, and production for human consumption. The digest is directed toward fish and shellfish farmers, government agencies, and colleges and universities. The 24-page publication is available for $50 a year.

Achill River Corporation

23 Sunset Terrace   704-254-7334
PO Box 2329
Asheville, NC 28802

Achill River has two publications:

● *Aquaculture Magazine*—is a bimonthly periodical directed toward people working in the field of aquaculture. Its features and news articles describe what is happening internationally in the field. Articles cover products raised under controlled conditions, including all types of fish and shellfish, as well as specific fish farm operations. The 52-page publication is available for $15 a year.

● *The Buyer's Guide*—this 100-page annual publication contains the following information: aquaculture organizations and associations at the national and international levels; state aquaculture and fish farming associations; state extension specialists in the fish and wildlife field; soil conservation biologists working in the field; diagnostic services available; universities and institutions offering aquaculture courses; a directory of products and services, including consultants; a cross reference that lists suppliers to the industry. Subscribers to *Aquaculture Magazine* receive the *Guide* free of charge. Others can purchase it for $8.

Farm Pond Harvest

Professional Sportsman's Publishing Company   815-472-2686
Rural Route 2
PO Box AA
Memence, IL 60954

   Farm Pond Harvest is a quarterly that primarily covers small waters. It contains fundamental information on both the recreational and commercial raising of fish, frogs, shrimp, and so on in the U.S. and foreign countries. Approximately 36 pages long, the publication is available for $8 a year.

Additional Experts and Resources

James J. Davis

Department of Wildlife and Fisheries Science   409-845-7473
210 Nagle Hall
Texas A&M University
College Station, TX 77843

   James J. Davis provides leadership in the area of freshwater and marine aquaculture. Areas of expertise include catfish farming, either intensive or pond-culture; crawfish or crayfish; pilapia; freshwater shrimp; freshwater sport fish; marine shrimp; redfish; and other marine fin fish. Dr. Davis can furnish publications about specific topics as well as a publications list. He can answer general questions, and he will refer requests for more specific information to the appropriate specialist in the department.

Aquatic Farms

See: Aquaculture

Architectural Engineering

Organizations and Government Agencies

American Society of Mechanical Engineers (ASME)

345 East 47th Street   212-705-7440
New York, NY 10017

   ASME is a nonprofit technical and professional membership organization with over 111,000 individual members. There are 33 technical divisions covering various engineering disciplines, such as aerospace, rail transportation, product engineering, electronic packaging, heat transfer, biomechanical engineering, and computers. ASME sponsors about 30 national conferences a year, each devoted to a specific study of engineering, and also offers continuing education in the form of short study courses and various student competitions. There is a codes and standards division that establishes and publishes voluntary codes for engineering practice. ASME has a large publications catalog, available upon request, which includes technical transaction journals, books, reports and conference proceedings and papers. Membership fees begin at $25 a year, and include subscription to the Society's magazine. Publications include:

• *Mechanical Engineering Magazine*—a monthly general trade magazine with articles on all aspects of engineering. Regular features include information on new products and technologies, Washington news, technical papers, code and standard interpretations, and letters and comments from professional engineers. There is also a meetings and education calendar, and indexes of opportunities and consulting services. About 120 pages an issue; subscription is $30 a year for nonmembers.

• *Computers in Mechanical Engineering*—a bimonthly magazine covering the current emphasis on the use of computers in all aspects of the engineering field. Regular topics covered include automation research, computers in high level control, reliability verification, and computer calculations. Also included is a research supplement with various technical journal papers. Each issue contains a new products section, reviews of current literature, industry news, software discussions, and editorials. About 150 pages an issue; the annual subscription is $30.

Architecture & Engineering Performance Information Center (AEPIC)

University of Maryland   301-935-5544
3907 Metzerott Road
College Park, MD 20742

AEPIC, a self-supporting nonprofit center, is engaged in the collection, classification, and analysis of data describing problems, malfunctions, and failures in architectural and civil projects, their histories, and their outcomes. All data is arranged in computerized data files or in AEPIC's several libraries. The University of Maryland is the international center and national repository of AEPIC, with a number of international repositories being developed in other countries. AEPIC also offers custom services for the collection and analysis of performance data. An AEPIC newsletter covers trend analysis in the field of architecture and engineering, and the *Journal of Performance Analysis* emphasizes theoretical and conceptual issues of performance. Annual membership ranges from $100 to $800 a year. Services include:

• Computerized Citation Files—files referring to published information about performance problems, drawing from journals, newspapers, agency investigation reports, and so on.

• Dossier Library—documentation of performance data about the incidents and information in the case files maintained by AEPIC.

• Visual Materials Library—collection of photographs, slides, and other visual materials related to the AEPIC case files.

• Reference Library—current and historical codes, standards, and other technical references.

Association of Physical Plant Administrators of the Universities and Colleges (APPA)
1446 Duke Street   703-684-1446
Alexandria, VA 22314-3442

APPA is an international educational association of over 1,200 institutional members worldwide dedicated to developing a sense of professional responsibility for the administration, care, operation, planning, and development of facilities of higher education. APPA offers seminars, institutes, publications, and a network of facilities professionals to address daily management problems. Current emphasis is on computerized management of physical plant services. A publications bulletin lists numerous references and videotapes, including a management manual and a directory of consultants. Membership dues begin at $135 a year and include a subscription to the APPA official newsletter.

• *APPA Newsletter*—official monthly publication of APPA, written especially for the physical plant administrator in higher education. The newsletter emphasizes current developments in physical plant departments, federal agencies and higher education associations. It also includes news of APPA seminars and publications and an information exchange column. Other regular features include employment opportunities, book reviews, upcoming events, and papers on various technical and management procedures. Subscription for nonmembers is $40 a year.

National Society of Professional Engineers (NSPE)
Washington Engineering Center   703-684-2800
1420 King Street
Alexandria, VA 22314

NSPE is a nonprofit professional association of over 80,000 engineers, focusing on the nontechnical aspects of engineering: economics, education, legislation, and private practice. NSPE members work to influence and monitor legislation crucial to engineering, both on national and state levels. NSPE also conducts seminars, reference programs, readiness courses, and other educational programs. Its practice divisions (industry, construction, education, government, and private practice) unite engineers whose concerns are similar to provide solutions to problems in their specific areas. There is also a worldwide referral system called DPTSI, Design Professions Technical Specialties Index, to market products and services. NSPE also coordinates a liaison society program with some 30 engineering-related organizations participating. The staff will make information referrals to specialists and publications. Publications include:

• *Engineering Times*—a monthly newspaper covering the complete gamut of engineering news: what is going on in Washington, industry news, salary and employment trends, and engineering education. About 50 pages an issue; annual subscription is included with membership, and costs $14 for nonmembers.

Society for Computer Applications in Engineering, Planning and Architecture (CEPA)
15713 Crabbs Branch Way   301-926-7070
Rockville, MD 20855

CEPA is an independent, nonprofit professional association with approximately 400 member organizations, mostly engineering consultant firms, university engineering departments, public agencies from city to federal level, and data processing vendors. CEPA's work consists of collecting, compiling, and distributing information to members about the use of computers in engineering practice, using a referral method to connect members to others with computer software and hardware for specific results. CEPA staff will also refer nonmembers to a member who may be able to assist with very specific problems or information. CEPA maintains a library of software available to members, with abstracts of some 500 engineering programs available. CEPA holds two annual conferences open to the public, and will provide on request a publications listing that cites special reports, conference proceedings, and the newsletter.

• *CEPA Newsletter*—a quarterly publication of about 20 pages an issue. In addition to articles written by members on various hardware and software in the engineering field, it contains conference information, technical reports, and notices of related organizations' meetings and events. Included with the $225 CEPA membership, the newsletter costs $30 a year for nonmembers.

World Computer Graphics Association (WCGA)
2033 M Street, NW, Suite 399   202-775-9556
Washington, DC 20036

WCGA is a private, nonprofit association committed to the basic goal of improving international productivity and technological advancement through the promotion of the global computer graphics community. WCGA works with representatives from countries around the world to support user-vendor relationships through conferences, exhibitions, and seminars. Annual events include the following conferences: "Computer Graphics in the Building Process," "Computer Graphics Applications for Management and Productivity," "International Graphics," and "Defense Computer Graphics."

Publications

Design Graphics World
Communications Channels, Inc.   404-256-9800
6255 Barfield Road
Atlanta, GA 30328

This monthly trade magazine is available at no cost to qualified engineers, architects, managers, government workers, manufacturers, construction foremen, and those in other industries concerned with design graphics. Geared chiefly to architects and engineers, the magazine covers topics in the discipline of graphics, including computer graphics, micrographics, and reprographics. Monthly columns include systems drafting, design models, information management, engineering micrographics, and software. Approximately 50 pages an issue, the magazine is available to nonmembers at an annual subscription of $18.

Plan and Print

International Reprographics Association, Inc.   312-671-5356
9931 Franklin Avenue
Franklin Park, IL 60131

   The official publication of the American Institute for Design and Drafting (918-258-8651), *Plan and Print* is a monthly trade magazine for the design and reproduction manager. Strong emphasis is placed on computer-aided design (CAD), with frequent articles on both the hardware and software generating this process. A regular department, "CAD Angles," contains news and information from the National Computer Graphics Association (NCGA). Other regular features include new product and literature reviews, person and places highlights, a calendar of related events, design improvements, and classified ads. It is about 50 pages an issue, and the annual subscription is $15.

Progressive Architecture

Box 6142   203-348-7531
Cleveland, OH 44101

   This monthly magazine is produced for those in the fields of mechanical and architectural engineering. Each issue contains a large illustrated section devoted to showing the best solutions to building design problems, including interiors. Regular features include new product information, architectural law interpretations, government developments, book reviews, classified advertising, and an events calendar. There is also a section on designs and numerous articles on specification practice. It is about 250 pages an issue, and the annual subscription is $18.

Artificial Insemination

See: Reproductive Technology

Artificial Intelligence

Begins next page. See also: CAD/CAM; Computer Integrated Manufacturing; Robotics; Supercomputers

Organizations and Government Agencies

American Association for Artificial Intelligence (AAAI)
445 Burgess Drive   415-328-3123
Menlo Park, CA 94025

This scientific society is involved in the presentation of worldwide research in the field of artificial intelligence. It sponsors several conferences, and the papers presented at the conferences are published as proceedings. The Association annually publishes materials prepared for six or seven tutorials held each year during the National Conference on Artificial Intelligence. Examples of tutorial topics include programming technology, languages and machines, and expert systems. The publications are each available for $7 from the Association. AAAI also publishes:

• *AI Magazine*—the official publication of AAAI. It includes informative articles on the current state of the art in AI and keeps readers posted on AAAI-related matters. It also contains general survey articles, reports of research in progress, reviews of books and articles published elsewhere, announcements of meetings and conferences, and job openings in the field of AI. Annual subscription is $25, $15 for students.

• AAAI Conference Proceedings are available from: William Kaufman, Department A154, 95 First Street, Los Altos, CA 94022, 415-948-5810.

Publications

The Artificial Intelligence Report
AI Publications   415-949-2324
95 First Street
Los Altos, CA 94022

Published ten times a year, the *AI Report* keeps track of AI applications, new companies, growth potential of AI, AI research, U.S. and foreign research and development, and so on. It is available for $200 a year.

Artificial Intelligence Journal
Elsevier Science Publishing Co., Inc.   212-867-9040
Journal Information Center
52 Vanderbilt Avenue
New York, NY 10017

The *AI Journal* deals with the theory and practice of computer programs and other artifacts that manifest intelligent behavior. The journal tries to bring together related aspects of the subject of artificial intelligence and to publish accounts of current research. The annual subscription is $214.62.

William Kaufman, Inc.   415-948-5810
95 First Street
Los Altos, CA 94022

AI books available from this publisher include:

● *The Handbook of Artificial Intelligence*—a three-volume reference book which presents more than 200 short articles covering important ideas, techniques, and systems developed in 25 years of research in the AI field. All articles are written for the reader who has little or no familiarity with the AI field. Volume I costs $39.50; Volume II, $42.50; and Volume III, $59.50; or the three volume set for $120.

● *Principals of Artificial Intelligence*—This book, by Nils Nilsson of SRI International, explains the core ideas of AI, the fundamental principles of this rapidly expanding computer science. Exercises at the end of each chapter expand on subjects treated in the text, and pertinent references aid further study. It is available for $30.

Scientific DataLink   212-838-7200
850 Third Avenue, 21st Floor
New York, NY 10022

A microfiche library of reports on artificial intelligence gathered from labs and research centers. Available reports and research papers are from the following schools and research centers: Massachusetts Institute of Technology; Stanford University; Purdue University; SRI International; Carnegie-Mellon; University of Maryland; Bolt, Beraneck and Newman; Yale University; University of Illinois; and the University of Pennsylvania.

Tracer Bullets   202-287-5580
Reference Section   202-287-5639
Division of Science & Technology
Library of Congress
Washington, DC 20540

The Division of Science & Technology of the Library of Congress has put together a "tracer bullet" on artificial intelligence. It lists recommended information sources, including literature and organizations.

Additional Experts and Resources

Several universities and institutes across the country conduct research in the area of artificial intelligence, and many publish results of their research. Contact these research centers for specific technical questions, or for information about their publications.

MIT   **Stanford University**
Artificial Intelligence Lab   Artificial Intelligence Lab
545 Technical Square   Stanford, CA 94305
Cambridge, MA 02139   415-497-2800
617-253-6773

Carnegie-Mellon University
Computer Science Department
AI Research
Schenley Park
Pittsburgh, PA 15213
412-578-2597

University of Maryland
Computer Science Department
College Park, MD 20742
301-454-2002

Yale University
Artificial Intelligence Lab
330 Dunham Lab
10 Hillhouse Avenue
New Haven, CT 06520
203-436-0606

SRI International
Artificial Intelligence Center
Menlo Park, CA 94025
415-859-2311

Artificial Organs, Limbs, and Joints

See: Medical Technology; Transplantation Technology

Attitudes Toward High Technology

See: Labor and Technological Change; Social Impact of Technology; Technophobia

Audio Systems

See: Acoustics

Banking, Electronic

See: Electronic Funds Transfer

Bar Code Technology

See also: Recognition Technology

Organizations and Government Agencies

Automatic Identification Manufacturers (AIM)
1326 Freeport Road   412-782-1624
Pittsburgh, PA 15238

AIM, a section of the Material Handling Institute, specializes in bar code and other automating material-handling systems. It offers information on all automatic identification technology, including fixed-beam and moving-beam scanners, hand-held bar code readers, stand-alone and portable data collection terminals, photosensors, bar code printing devices, automatic data collection, and material-handling systems. Publications include:

- *Uniform Symbol Descriptions*—a number of Uniform Descriptions describing the various automatic identification symbols in popular use; costs $16.

- *Glossary of Automatic Identification Terms*—compilation of all known glossaries of terms in the industry; includes terms from the areas of OCR, fixed beam, moving beam, printing, code applications, presence sensing, and CCD array sensing. It is available free of charge.

- *Bibliography of Automatic Identification Articles*—a listing of articles that have appeared in leading business publications dealing with automatic identification equipment systems. Listing includes articles from 1962 to present. Available free of charge.

Society of Manufacturing Engineers (SME)
Computer Automation Systems   313-271-1500, ext. 389—Nancy McAlpine
One SME Drive   313-271-1090—Publications Office
PO Box 930
Dearborn, MI 48128

SME holds one or two seminars a year on bar coding, which are oriented toward the manufacturing aspects. The seminars are a good source of educational material. A binder containing information presented by speakers at each seminar is available from the Publications Office.

Publications

Bar Coding Reprints
American Production & Inventory Control Society (APICS)
500 West Annandale Road
Falls Church, VA 22046-4274

These reprints are compilations of material on bar coding that have appeared in APICS' journal, *Production and Inventory Management*, and its *Annual Conference Proceedings*, which cost $10.

Logistics Applications of Automated Marking and Reading Symbols **(LOGMARS)**
Superintendent of Documents   202-783-3238
U.S. Government Printing Office
Washington, DC 20402

A report in 14 volumes, LOGMARS has been prepared by the Office of the Assistant Secretary, Department of Defense. The objective of the study is to establish a standard machine-readable symbology to be marked by commercial vendors and DOD activities on items, unit packs, outer containers, and selected documentation, and to establish procedures for the use of the symbology. Cost is $20.

North American Technology, Inc. (NATI)
174 Concord Street   603-924-7136
Peterborough, NH 03458

This firm publishes the following:

• *Bar Code Manufacturers and Services Directory*—an annual directory with listings of every major bar code manufacturer and service. Each entry consists of a brief description of the product line or service; general information about the company; and name, address, and phone number of a knowledgeable representative. It is available for $29.95.

• *Bar Code News*—a bimonthly journal of bar code system applications, which covers topics such as new products, new applications, code formats, market analysis, investments, standards. Write or call the publisher to see if you qualify for a free subscription.

• *Contemporary Applications of Optical Bar Code Technology*—a comprehensive report and software package that shows how to use bar code technology with small computers. It details the techniques, application situations, and theory of optical bar code technology. It includes a complete bibliography of available source materials and costs $100.

Scan Newsletter
11 Middle Neck Road   516-487-6370
Great Neck, NY 11021

Scan is a monthly newsletter devoted to bar code scanning and related technologies. It is a good source for tracking the companies, products, and developments in this emerging technology. The annual subscription is $75 for 12 issues, or $110 for *Scan* plus the *International Edition*.

Additional Experts and Resources

American Bar Code Systems (ABCS)

91–31 Queens Boulevard   718-457-2400
Elmhurst, NY 11373

ABCS develops bar code data collection systems for the market, and also publishes *The Bar Code Banner*. This monthly newsletter keeps the reader informed about bar coding and its many applications, with articles written by experts in the field.

Thomas Sobczak

Little People's Productivity Center   516-623-6295
2580 Grand Avenue
Baldwin, NY 11510

Mr. Sobczak, a private consultant, is an expert on bar code technology. He will answer questions and put together information on the topic, either for a fee or free, depending on the amount of information needed.

Bioconversion

See: Biomass/Bioconversion

Bioelectricity

See also: Bioengineering

Organizations and Government Agencies

Bioelectromagnetics Society (BEMS)

1 Bank Street, Suite 307   301-948-5530
Gaithersburg, MD 20878

This organization's purpose is to promote scientific study of the interaction of electromagnetic energy (at frequencies ranging from 0 hertz through those of visible light) and acoustic energy with biological systems. Members with a Ph.D. degree in engineering, physics, and the biological sciences are doing research on the effects of electromagnetic and acoustic energies of biological systems. BEMS holds an annual scientific conference, at which approximately 200-250 technical papers are presented. Abstracts of these papers are available at $5 a book. BEMS usually holds another meeting each year, a joint meeting with a different organization doing related work. The staff can respond to written inquiries. Publications include:

• *Bioelectromagnetics*—a quarterly technical journal directed toward researchers in the field of bioelectromagnetics. The approximately 100-page publication is available to members directly from the Society for $25 a year. Nonmembers can order it from Alan R. Liss, Inc., 150 Fifth Avenue, New York, NY 10011, for $90.

Foundation for the Study of Bioelectricity
PO Box 82   617-527-1395
Boston, MA 02135

The Foundation's goal is to educate the scientific and medical community and the general public about the potential importance of the field. It provides scientific references to the literature on bioelectricity and also has information on the medical applications of electricity and the electromagnetic field. It has an advisory board of professors of medicine and physics.

Research Services Branch of the Intramural Research Division of the National Institute of Mental Health
Building 36, Room 2A03   301-496-4957
National Institutes of Health
Bethesda, MD 20205

This group of engineers, technicians, and computer scientists is involved in the design and development of instrumentation and computer systems for neurophysiological, neuropharmacological, and neuropsychological research. The staff can answer questions and/or make referrals.

Publications

Journal of Bioelectricity
Marcel Dekker   212-696-9000
270 Madison Avenue
New York, NY 10016

This technical publication is directed toward scientists who study the effects of electromagnetic energy on living systems. It serves as a forum for publishing research results in the field. The periodical runs approximately 200 pages, and the subscription price for two to three issues is $75 a year.

Additional Experts and Resources

Boguslaw Lipinski, Ph.D.
Foundation for the Study of Bioelectricity   617-527-1395
PO Box 82
Boston, MA 02135
　　Dr. Lipinski is affiliated with Tufts University School of Medicine and has done research in bioelectricity for the last ten years. He can deliver lectures on the principles of bioelectricity, specifically on the effect of environmental electricity (natural and man-made). He is associate editor of the *Journal of Bioelectricity*, as well as president of the Foundation for the Study of Bioelectricity.

Dr. Andrew A. Marino
Department of Orthopedic Surgery   318-674-6180
Louisiana State University Medical Center
1501 Kings Highway
Shreveport, LA 71130
　　Dr. Marino is a biophysicist whose research focuses on the effects of electromagnetic energy on living systems. Two specific aspects of his research are the use of electromagnetic therapy for treating diseases and the evaluation of health risks associated with electromagnetic pollution. He is editor of the *Journal of Bioelectricity*.

Bioengineering

See also: Biomechanics; Biotechnology

Organizations and Government Agencies

Alliance for Engineering in Medicine and Biology
4405 East-West Highway, Suite 402   301-657-4142
Bethesda, MD 20814
　　This federation of professional associations shares an active interest in the interaction between engineering and the physical sciences, and medicine and the biological sciences. There are 20 engineering and medical associations as members, which can help refer inquiries to their staff. The Alliance holds an annual conference on engineering in medicine and biology. Publications include:
　　● A free brochure on biomedical engineering—send a self-addressed stamped envelope.
　　● Proceedings of the annual conferences, dating back to 1969—cost ranges from $7.50 to $32.

Association for the Advancement of Medical Instrumentation

1901 North Fort Meyer Drive   703-525-4890
Arlington, VA 22209

This nonprofit association of health-care professionals has members from many disciplines and professions, which develop, use, or manage biomedical technology or instrumentation. The basic purpose is to assist health-care professionals to keep current with advancements in medical technology, to share their knowledge and experience, and to further the improvement and use of medical technology. They hold many conferences, seminars, and training meetings, as well as regional and national membership meetings. The association develops standards and recommends practices in the field of biomedical engineering. Publications include:

• *Medical Instrumentation*—a bimonthly journal containing original papers and technical articles that deal with developments and trends in applications of technology to patient-care standards, management techniques, safety, and maintenance in the medical instrumentation field. Regular features include original research, applications, technological assessment, clinical device notes, new products, abstracts, book reviews, a calendar of events, and classified advertising. The publication, which averages 50 pages, is available for $65 a year.

American Society for Testing and Materials

1916 Race Street   215-299-5521
Philadelphia, PA 19103

The main purpose of this nonprofit organization is the development of voluntary consensus standards, which originate in committees composed of producers, users, and general-interest organizations. Two committees issuing standards are the F4 Committee on Medical and Surgical Devices and the F29 Committee on Anesthesiology. F4 has about 89 standards; F29, a new committee, has about 15 standards. Standards vary in price from $8 to $20. Raymond Sansone, staff manager of both committees, can answer questions and make referrals. The Society has many publications, including journals, books, STPs (Standard Technical Publications) with respect to certain committees, and catalogs.

Biomedical Engineering and Instrumentation Branch

Division of Research Services   301-496-5771
National Institutes of Health (NIH)
Bldg. 13, Room 3W13
Bethesda, MD 20205

This organization provides engineering resources for the intramural research program of NIH. It also publishes an annual report in November. The staff can answer questions and/or make referrals.

Biomedical Engineering Society

PO Box 2399   213-206-6443
Culver City, CA 90231

The purpose of this nonprofit organization is to promote the advancement of biomedical engineering knowledge and its utilization. Members are researchers and professors in universities and hospitals. They hold one annual meeting, for which proceedings are published in the Society's journal, *Annals of Biomedical Engineering*. Publications include:

● *Annals of Biomedical Engineering*—a bimonthly journal directed toward professionals in biomedical engineering. It publishes original research in biomaterials, biomechanics, rehabilitation engineering, instrumentation, and physiological systems analysis. Annual subscription price is $20 for members and $140 for nonmembers. Available from Pergamon Press, Inc., Fairview Park, Elmsford, NY 10523.

● *The Biomedical Engineering Society Membership Directory*—an approximately 100-page guide published annually in the fall. It is primarily a directory of individual members, including name, degree, address, phone number, and areas of specialization. The cost is $25.

Biomedical Engineering Division

American Society for Engineering Education   202-293-7080
11 Dupont Circle, Suite 200
Washington, DC 20036

This Division is a good source for educational information. The staff can refer inquiries to the chairman of the Biomedical Engineering Division.

The Biomedical Engineering Division in the Technological Institute

Northwestern University   312-492-7672
Evanston, IL 60201

This Division is involved in research and teaching in the areas of rehabilitation, analysis of data, and artificial organs. The staff can answer questions, make referrals, and send available brochures upon request.

Division of Bioengineering

American Society of Mechanical Engineers   212-705-7740
United Engineering Center
345 East 47th Street
New York, NY 10017

This Division deals with an array of concerns, including: biomechanics; medical devices, sports equipment; heat transfer; rehabilitation engineering; skeletal muscular analysis; fluid dynamics, prosthesis. The Division's members are medical doctors, mechanical engineers, and others involved in the field of bioengineering. The Society's public relations staff can refer you to an appropriate member who can answer technical questions. The office can also provide you with information about the Society's books and journals relating to bioengineering, as well as other information sources. At present the collection includes:

• *Advances in Bioengineering*—a five-volume series of books covering developments from 1979 through 1983. Each book contains technical papers presented at the annual conference. The volumes run 200-300 pages and range from $12 to $20 for members, and $24 to $40 for nonmembers.

• *Journal of Applied Mechanics*—a quarterly journal containing technical papers with discussions. The costs is $18 a year for members, and $90 for nonmembers.

• *Journal of Biomechanical Engineering*—see description under "Biomechanics." Subscription price is $18 a year for members, and $72 a year for nonmembers.

Engineering in Medicine and Biology Society

Institute of Electrical and Electronics Engineers (IEEE)   212-705-7890—Technical Activities Board
345 East 47th Street
New York, NY 10017

The Society has members knowledgeable in the areas of bioengineering and biomedical engineering. The Technical Activities Board can refer you to such experts. Publications include:

• *Engineering in Medicine and Biology Magazine*—a quarterly publication containing both general and technical short articles on current technologies and methods used in biomedical and clinical engineering. Current news items, book reviews, patent descriptions, and a correspondence section are included. The publication is available for $40 a year.

• *Transaction on Biomedical Engineering*—a monthly publication covering concepts and methods of the physical and engineering sciences applied in biology and medicine, ranging from formalized mathematical theory through experimental science and technological development to practical clinical applications. The publication is available for $104 a year.

• Order the above publications from: IEEE Service Center, Publication Sales Department, 445 Hose Lane, Piscataway, NJ 08854, 201-981-1393.

Food, Pharmaceutical and Bioengineering Division

American Institute of Chemical Engineers (AICHE)   212-705-7335—Publications
345 East 47th Street   212-705-7315—Main Number
New York, NY 10017

This division of AICHE sponsors various symposia in its five programming areas: food, pharmaceuticals, biotechnology, biomedical, and fundamentals and life sciences. It also has a Research Institute in Food Engineering (RIFE), which is primarily a conduit for obtaining funds for research in food engineering and transferring them to appropriate academic researchers. The staff at AICHE can refer inquiries to the current chairman of the Food, Pharmaceutical and Bioengineering Division. Publications include:

• *Biotechnology Progress*—a quarterly magazine carrying news items and papers that have been presented at the various symposia. The cost of the approximately 120-page publication is included in the membership fees. For information on the nonmember subscription price, call 212-705-7663.

• *Chemical Engineering Progress*—a monthly technical journal directed toward chemical engineers. It contains papers presented at symposia sponsored by the American Institute of Chemical Engineers. The approximately 100-page publication is available at the nonmember subscription price of $35 a year.

National Rehabilitation Information Center

Catholic University of America   202-635-5826
4407 Eighth Street, NE
Washington, DC 20017

This rehabilitation information service and research library is funded by the Department of Education's National Institute of Handicapped Research. The Center maintains a collection of more than 10,000 documents, including research reports, periodicals, and reference documents. The staff can refer you to specialists in appropriate research reports, and government-funded research projects at universities, research centers, hospitals, and the private sector. The Center maintains two data bases on literature and projects in the field. The staff will search the data bases for the public; the minimum charge for this service is $10, which entitles you to complete citations on 100 documents. The data bases are available on BRS. The library is open to the public. The Center also publishes the following:

● *Periodical List*—an annotated listing of all periodicals collected by the Center. It includes over 200 publications and costs approximately $20.

● *NARIC Subject Catalog*—descriptions of all subjects covered by NARIC, as well as complete annotations. The cost is $25.

Office of Technical Transfer

Rehabilitation Research and Development Service   212-620-6670—Lilly Homs,
Veterans Administration   Technical Information Specialist;
252 Seventh Avenue   Ronald Coons, Assistant Managing Editor of the *Journal*
New York, NY 10001

The Office can supply you with bioengineering information as it relates to rehabilitation research and development. The library maintains an extensive research reference collection, and the staff can refer you to experts, research centers, laboratories, and literature in the field. A bibliographic listing of bioengineering publications is available from the library. The Office also publishes the following:

● *Journal of Rehabilitation Research and Development*—published three times yearly, with an additional issue entitled *Rehabilitation R&D Progress*. The *Journal* contains scientific articles about rehabilitation research and development, including papers on prosthetics, amputation, spinal cord injury and sensory aids. Ten to 15 abstracts of recent articles, a bibliography of bioengineering articles appearing in current literature, and a calendar of events also appear in each issue. The publication is available free of charge to institutions and organizations involved in rehabilitation research and development, as well as to educational institutions, hospitals, and libraries.

Orthopedic Research Society (ORS)
444 North Michigan Avenue, Suite 1500   312-644-3285
Chicago, IL 60611

A Society of orthopedic surgeons and other investigators in the field, ORS promotes orthopedic research and provides a meeting place for the presentation and discussion of research activities. Each year it publishes *Transactions of ORS' Annual Meeting*. This volume is a good resource for information about current research in bioengineering. The volume runs approximately 400 pages and costs $25.

Rehabilitation Engineering Society of North America
4405 East-West Highway, Suite 402   301-657-4142
Bethesda, MD 20814

This interdisciplinary society is concerned about the advancement of rehabilitation technology. It can refer inquiries to technical experts in the field. It also sponsors an annual conference, for which proceedings are available. A publications list, consisting of reports, books, and so on, is available. Publications include:

• *Wheelchair III*—a 1982 report of a workshop on specially adapted wheelchairs and sports wheelchairs. The cost is $10.

• *Technology for Independent Living Sourcebook*—a book containing extensive new material contributed by many people associated with rehabilitation engineering centers and hospitals around the country. Areas covered include information services and resources; equipment section; educational and vocational technology; the workplace; recreation and leisure; personal mobility; control communication and sensory aids; computer applications; funding, models, policy, and statistics.

Special Interest Group in Biomedical Computing
Association of Computing Machinery (ACM)   212-869-7440
11 West 42nd Street
New York, NY 10036

This ACM Group encourages and facilitates communication and exchange of information concerning problem areas, computer routines, techniques, and activities among individuals and laboratories involved in biomedical research applications and hospital or patient-care applications; it stimulates better understanding of potentialities of digital and analog computers in these areas. ACM can refer inquiries to the group chairman. Publications include:

• *SIGBIO Newsletter*—a quarterly publication reporting on applications and research in computers in medicine. The approximately 26-page publication is available for $21 a year for nonmembers, and $14 a year for members. The student member price is $5 a year.

Publications

CRC Press, Inc.
2700 Corporate Boulevard, NW   305-994-0555, ext. 342
Boca Raton, FL 33431

This press publishes:
- *CRC Critical Reviews in Biomedical Engineering*—a technical publication directed toward scientists and researchers. It covers engineering in medicine and biology, including biomedical instrumentation. Each volume consists of four issues. Usually, one full volume is published each year. About 80-128 pages, the publication is available for $104 a volume.

- *CRC Critical Reviews in Biocompatability*—this new journal covers the application of biomaterials to the surgical, dental, and veterinary fields, and the production of biomaterials. Usually one full volume is published each year. The publication, which runs about 100 pages, costs $104 a volume.

Data Bases

BIOSIS Previews
BioSciences Information Service   800-523-4806 (except in Pennsylvania)
2100 Arch Street   215-568-4016, ext. 244 or 241
Philadelphia, PA 19103

BIOSIS Previews contains citations from such periodicals as *Biological Abstracts*, *Biological Abstracts/Reports*, *Reviews*, *Meetings*, and *Bioresearch Index*. These publications provide comprehensive worldwide coverage of research in the life sciences, including bioengineering and biomedical engineering. Material scanned for *BIOSIS Previews* includes periodical literature, monographs, books, textbooks, technical reports, published theses, meetings, nomenclatural rules, notes, letters, annual reports, bibliographies, and guides. Available through DIALOG, 3460 Hillview Avenue, Palo Alto, CA 94304, 800-227-1960; or BRS, 1200 Route 7, Latham, NY 12110, 800-833-4707.

Additional Experts and Resources

Robert E. Mates, Ph.D.
Department of Mechanical & Aerospace Engineering   716-636-2726
SUNY at Buffalo
Buffalo, NY 14260

Dr. Mates is an expert in bioengineering, with a special interest in cardiovascular physiology. He is a good resource for information about what is happening in the field, and where to go for information. He can also answer technical inquiries.

Dr. Roy Rada
National Library of Medicine   301-496-2475
Bethesda, MD 20209

Dr. Rada can answer specific questions in the area of biomedical computing and can make referrals to other sources. He has been editor of the *SIGBIO Newsletter* of the Association for Computing Machinery.

Bioethics

See also: Biotechnology; Social Impact of Technology; Technology Assessment

Organizations and Government Agencies

American Society of Law and Medicine (ASLM)
765 Commonwealth Avenue   617-262-4990
Boston, MA 02215

This nonprofit educational organization sponsors 12 conferences a year on such topics as new reproductive technologies; nursing and the law; mental health law; malpractice; genetics, legal and ethical health care; the elderly; and bioethical issues. The Society publishes:

● *Law, Medicine, and Health Care*—a bimonthly journal with articles on a variety of topics, such as nursing and the law, hospital law, medical malpractice, genetics and the law, and bioethical issues related to law and health care.

● *American Journal of Law and Medicine*—a quarterly law review journal copublished with the Boston University School of Law, which covers the legal issues of medicine. The 150-page journal is available for $40.

Center for Bioethics
Kennedy Institute of Ethics   202-625-2383
Georgetown University
Washington, DC 20057

The Center deals with ethical problems arising in the areas of health care, and biomedical and behavior research. It maintains a collection of research materials, including books, journal articles, legal materials, regulations, government publications, and other documents pertaining to biomedical issues. The scope of the Center covers such fields as philosophy, medicine, law, religion, science and technology, and social sciences. The Center is an excellent information resource.

Dr. Goldstein, the director of the library, is an expert in the field of bioethics, and can answer questions and make referrals to other information sources.

Hastings Center

360 Broadway   914-478-0500
Hastings-on-Hudson, NY 10706

The Center deals with ethical issues in science, medicine, and other professions, such as health-policy concerns dealing with death and dying, doctor/patient relationships, and newborn intensive-care issues. The Center's staff is available to answer questions and make referrals to other information sources in the field of bioethics. Publications include:

• *Hastings Center Report*—a bimonthly publication dealing with ethical issues in science and medicine, covering organ transplants, prenatal diagnosis of genetic diseases, recombinant DNA research, and so on. The 48-page publication is available for $29 for individuals, $25 for students, and $41 for institutions.

• *Focus*—an internal newsletter that provides an overview of the Center. Available to members only.

• *IRB Review of Human Subjects Research*—a bimonthly publication for members of institutional research boards (IRB). It deals with ethical issues, and the design and conduct of human subject research; it covers such issues as ethics of research protocol, ethics of doing clinical trials, and various other ethical issues related to medical research. The 12-page publication is available for $30 for individuals, $40 for libraries, and $180 for institutions.

National Center for Institutional Ethics Committees

765 Commonwealth Avenue   617-262-4990
Boston, MA 02215

The Center is an information clearinghouse that covers guidelines on ethics committees. Bibliographies on ethics committees are available, and the Center maintains a consultant, a Speakers Bureau, and research projects information. Each year conferences are held, and the proceedings are published and available for sale. Also published is:

• *Ethics Committee Newsletter*—a quarterly publication that provides a vehicle for communication among individuals who serve on institutional ethics committees. Four to 20 pages long, the newsletter is free.

Publications

Bioethics: A Guide to Information Sources

Gale Research Company   313-961-2242
Book Tower   800-521-0707
Detroit, MI 48226

This book, by Doris Meuller Goldstein, is a completely annotated bibliography listing organizations, journals, and other resources in the field. The 350-page publication is available for $58.

Ethics Magazine
University of Chicago Press
PO Box 37005
Chicago, IL 60637

The quarterly publication is devoted to the study of ideas and principles that form the basis of ethical questions in medicine. It covers the philosophy, social and political theory, and economic analysis of the bioethical issues. The 100-page publication is available for $18 for students, $25 for individuals, and $45 for institutions.

Issues of Science and Technology
National Academy of Sciences   202-334-2000
2101 Constitution Avenue, NW
Washington, DC 20418

This quarterly journal deals with policy as it relates to science and technology, covering such topics as bioethics, health care, industrial policy, and military and defense issues. The 130-page publication is available for $24.

Journal of Bioethics
Human Sciences Press   212-243-6000
72 Fifth Avenue
New York, NY 10011

This quarterly journal is primarily concerned with helping doctors become more aware of ethical issues in the field of medicine. The 80-page publication is available for $22 for individuals, and $49 for institutions.

Science, Technology and Human Values
John Wiley and Sons, Inc.   212-692-6000
605 Third Avenue
New York, NY 10158

This quarterly journal contains analyses of the interaction of science technology and its social context. It emphasizes the effects of technology on human beings and human society, especially as it relates to social conditions, ethics, and values. The 100-page journal is available for $48.

Additional Experts and Resources

Dr. Alexander Morgan Capron
Professor of Law, Medicine, and Public Policy   213-743-6473
The Law Center
University of Southern California
Los Angeles, CA 90089-0071

Dr. Capron is a professor of bioethics who writes books and articles for scientific journals. He also has testified before Congress on bioethical issues, and was the Executive Director of the President's Commission for Study of Ethics and Biomedical Research. He is available to answer questions in his area of expertise as well as to refer you to other good sources of information.

Rachelle Hollander, Ph.D.
Ethics and Values in Science and Technology   202-357-7552
National Science Foundation
1800 G Street, NW
Washington, DC 20550

Dr. Hollander directs an NSF program supporting research and related activities that examine the ethical and value issues associated with current scientific, engineering, and technical activities. Funded projects involve scientists, engineers, scholars, consultants, and groups affected by specific technological developments. Dr. Hollander and her staff can provide information about ongoing bioethics research and experts in the field.

Biomass/Bioconversion

See also: Fuels, Alternative; Renewable Energy Technology; Solar Technology

Organizations and Government Agencies

Bio-Energy Council
1815 North Lynn Street, Suite 200   703-276-9411
PO Box 12807
Arlington, VA 22209

The Bio-Energy Council is a nonprofit international organization to promote the sensible use of biomass to energy without harming the environment. It receives financial support from industry and individual associates. Dr. Paul F. Bente, Jr., is available to answer questions and/or make referrals. He has a Ph.D. in chemistry, with industry, government, and nonprofit sector experience. The information center is a reference library of technical reports. Publications include a series of directory handbooks, which contain summaries of progress being made in the fields in which the Council is involved. *The International Bioenergy Directory and Handbook* is an annual publication that includes five to six essays on emerging technologies, and two on state-of-the-art. Cost is $95.

Biomass Energy Technology

Renewable Technology   202-252-6750—Dr. Beverly J. Berger, Director
Conservation and Renewable Energy, DOE
FORSTL, CE-321, Room 5F059
1000 Independence Avenue, SW
Washington, DC 20585

This office oversees long-term research and development for increasing the supply of biomass feedstocks and for developing conversion technologies that will produce liquid and gaseous fuels and petrochemical substitutes from a variety of biomass feedstocks. The office investigates aquatic, herbaceous, and woody crops with potential to increase biomass yields, and it investigates thermochemical, biochemical, and photobiological conversion systems.

Forest Service

Department of Agriculture   202-447-2702—Dr. F. Bryan Clark, Research
PO Box 2417   202-447-6677—J.B. Hilmon
12th Street and Independence Avenue, SW
Washington, DC 20013

This office conducts biomass energy-related research, and coordinates research activities with federal and state agencies, as well as universities. These activities include obtaining, analyzing, and projecting national statistics on woody biomass supply and use. The office also administers National Forest lands in a multiple-use approach that includes sales of industrial wood and personal-use fuel wood.

Biomechanics

See also: Bioengineering

Organizations and Government Agencies

American Alliance for Health, Physical Education, Recreation & Dance (AAHPER)

1900 Association Drive   703-476-3400
Reston, VA 22091

This professional association is composed of 3,600 members, from college professors to elementary school health teachers, many of whom are involved in research. The Alliance sponsors seminars and conferences. Publications include:

• *Basic Staff Series*—includes six books, four of which are geared toward biomechanics: (1)Exercise Physiology; (2) Kinesiology; (3) Motor Learning; and (4) Motor Development. The books run about 200 pages and are written for junior and senior high school teachers. The books cost $6.55 each, or $34.10 for the entire series.

● *Research Quarterly for Exercise and Sport*—a scholarly journal for research-oriented professionals in the fields of physical education, athletics, sports, and recreation. This quarterly frequently contains articles on biomechanics. The membership fee of $42 includes free subscription; distribution of the journal is restricted to members.

American Society of Biomechanics

1753 West Congress Parkway   312-942-5813
Chicago, IL 60612

This Society comprises 200 engineers, biologists, and others studying sports-related activities and injuries as well as ergonomics. Members will answer public inquiries and/or make referrals. Each year the Society sponsors an annual meeting. A bound volume of the meetings' transactions (including abstracts of papers presented) is available for $20.

Biomechanics Laboratory

Centinela Hospital Medical Center   213-673-4660
555 East Hardy Street
Inglewood, CA 90307

Centinela's Biomechanics Laboratory is a motion-analysis laboratory using high-speed film and electromyographic techniques to analyze body movement. The staff will respond to public inquiries and make referrals to other laboratories. The Laboratory has prepared numerous publications, ranging in complexity from highly technical to public-oriented guides, such as "Shoulder and Arm Exercises for Baseball Players." These publications are generally available free of charge.

Biomechanics Laboratory

U.S. Olympic Committee   303-578-4516
1750 East Boulder Street
Colorado Springs, CO 80909

This Laboratory is part of the Sports Medicine Program within the U.S. Olympic Committee. It is involved in improving the efficiency and technical capabilities of U.S. athletes in training for international competition. The staff can answer questions about biomechanics and direct you to experts in the field, ongoing research projects, companies producing biomechanics-related equipment, and current literature. The following materials are available free of charge:

● *Abstracts of Biomechanical Research Performed by the USA Olympic Committee*—contains abstracts of biomechanical projects conducted by 15 programs across the country to satisfy the needs of 22 sports.

● *Biomechanical Research Projects for Olympic Athletes*—this listing provides contact information for 30 biomechanical projects currently being conducted in the U.S. Each project covers a different sport.

● An information packet on the various types of equipment used in the Olympic Committee's Biomechanics Laboratory. It includes contact information on manufacturers supplying equipment. Brochures about all aspects of biomechanics are also available.

Division of Bioengineering

American Society of Mechanical Engineering (ASME)   202-705-7740—Public Relations Office
United Engineering Center
345 East 47th Street
New York, NY 10017

    Many members of the Division of Bioengineering are very involved in the various aspects of biomechanics. ASME's Public Relations Office can put you in contact with an appropriate member who will answer your questions and refer you to other information sources. The Bioengineering Division publishes the following journals related to biomechanics:

    • *Journal of Biomechanical Engineering*—a journal containing technical papers about the scientific results of biomechanical engineers engaged in research and development. The quarterly discusses practical applications and provides information about industry, books, and journals. Topics covered include: heat transfer rehabilitation and cancer therapy; biofluid mechanics; orthopedics; bio-solid mechanics; bone-remolding bio-materials; arteriosclerosis; artificial organs, and prostheses; lung disease; bio-instrumentation and measurement; and simulation of physiological systems. Issues run approximately 130 pages. The annual subscription is $18 for members and $72 for nonmembers.

    • *Journal of Applied Mechanics*—a quarterly containing technical papers with discussions; papers on biomechanics often appear. The annual subscription is $18 for members and $90 for nonmembers.

International Society of Biomechanics (ISB)

Richard Nelson, Ph.D.   814-865-3445
The Pennsylvania State University
Biomechanics Laboratory
University Park, PA 16802

    A professional research organization, ISB is composed of about 600 members, many of whom are sports biomechanists, engineers, and orthopedic surgeons. The Society's purpose is the interchange of information throughout the world, and it sponsors a conference for that purpose every other year. Most of its services are for members only, but the office will refer you to a biomechanics expert who can answer your questions. ISB publishes the proceedings of their international conferences. The volumes include selected monographs and texts dealing with various aspects of biomechanics.

Orthopedic Research Society

444 North Michigan Avenue, Suite 1500   312-644-3285
Chicago, IL 60611

    This Society of orthopedic surgeons and other investigators promotes research and provides a meeting place for the presentation and discussion of research activities. Each year the Society publishes *Transactions of ORS' Annual Meeting*. This volume is a good resource for information about current research in biomechanics. The yearly volume runs approximately 400 pages and costs $25.

Publications

International Journal of Sport Biomechanics

Human Kinetics Publisher, Inc.   217-351-5076
Box 5076
Champaign, IL 61820

This is a new scholarly journal, launched to serve as an international source for disseminating sport biomechanics research and scholarly inquiry. The quarterly includes original research reports, reviews, theoretical papers, abstracts of recent articles of interest for other publications, research notes, and comments. The annual subscription is $24 for individuals and $48 for institutions.

Journal of Biomechanics

Pergamon Press, Inc.   914-592-7700
Maxwell House
Fairview Park
Elmsford, NY 10523

This international journal contains fundamental studies of many aspects of biomechanics, including: muscularskeletal system orthopedics, cardiovascular and respiratory; dentistry; human injury; rehabilitation; and sports. The monthly journal is available free of charge to members of the American Society of Biomechanics. The annual subscription is $55 for individuals and $270 for institutions.

Journal of Orthopedic Research

Orthopedic Biomechanics Laboratory   617-735-2940
Beth Israel Hospital
330 Brookline
Boston, MA 02215

The quarterly journal is designed to serve both as an archive for orthopedic research papers and as a forum for current developments. It is geared toward both clinicians and researchers. It covers biomechanics engineering of surgeons pertaining to all aspects of the body. The annual subscription is $65 for individuals and $80 for institutions. Subscription requests should be directed to: Raven Press, 1140 Avenue of the Americas, New York, NY 10036, 212-575-0335.

Medicine and Science in Sports and Exercise

American College of Sports Medicine   317-637-9200
PO Box 1440
Indianapolis, IN 46206

This is a bimonthly medical journal containing articles written by exercise physiologists and other professionals involved in sports. Biomechanics is frequently covered in the journal. The annual subscription to the 90-page bimonthly is $50. Single copies are available for $7.

Additional Experts and Resources

Stephen Gordon, Ph.D.
Director of Muscularskeletal Diseases Program   301-496-7326
NIADDK/NIH
Westwood Building, Room 407
Bethesda, MD 20205

The program Dr. Gordon directs is the single largest funder of orthopedic biomechanics research. He and his staff can provide you with information about government funding in biomechanics and orthopedics, as well as information about ongoing research in the field. The staff will also discuss biomedical projects and provide advice and consultation. Dr. Gordon has done a lot of research on head injury, as well as kinetics and kinematics of body motion.

Jack L. Lewis, Ph.D.
Professor of Civil Engineering and Orthopedic Surgery   312-649-8560
Rehabilitation Engineering Program
Northwestern University
345 East Superior Street
Chicago, IL 60611

Dr. Lewis is a specialist in muscularskeletal biomechanics as it applies to orthopedic surgery and sports medicine.

Richard Skalak, Ph.D.
Director of the Institute of Bioengineering   212-280-2520
Columbia University
610 S.W. Mudd Building
New York, NY 10027

Dr. Skalak is an expert in biomechanical problems, especially blood flow. He is knowledgeable about research in the field, including: pulmonary mechanics, arteriosclerosis, hearing aids, growths, and urology. His background also includes work in fluid and solid-state mechanics and work with mechanical systems. He has published extensively and is technical editor of ASME's *Journal on Biomechanical Engineering*.

Biomedical Engineering

See: Bioengineering

Biomedical Instrumentation

See: Bioengineering; Medical Technology

Biotechnology

See also: Bioengineering

Organizations and Government Agencies

Association of Biotechnology Companies
PO Box 39221   301-977-0084
Washington, DC 20016

This nonprofit organization, with members from biotechnology companies and related industries, is concerned with potential roadblocks to the future of the industry. The staff will answer questions and provide referrals to appropriate Association members on subjects such as patents, government research programs-contracts, regulations, commerce, and other topics pertinent to the biotechnology industry. It also holds workshops, seminars, and educational seminars on subjects relevant to the industry.

Center for Advanced Research in Biotechnology
University of Maryland   301-853-3700
Wilson H. Elkins Building
3300 Metzerott Road
Adelphi, MD 20785

The newly established Center is a cooperative venture organized by the National Bureau of Standards and several government agencies involved in biotechnology: FDA, Department of Agriculture, EPA, NIH. The Center focuses on physical and chemical measurements and modeling. It coordinates seminars and workshops, and its principal products are publications resulting from collaborative projects with industry, universities, and a variety of government agencies.

Committee for Responsible Genetics
PO Box 1759
Cambridge, MA 02139

617-491-5655—Terri Goldberg

5 Doane Street, 4th Floor
Boston, MA 02109

617-227-8035

This national public-interest organization can provide you with information about the impact of genetic technologies on society. The Committee serves as a forum for people concerned about all issues of biotechnology, and it provides a link between the scientific community and the general public. Drawing from her membership list, Ms. Goldberg can refer you to experts active in the technical and social aspects of biotechnology. Membership is open to all, and the organization comprises scientists, religious leaders, environmental specialists, industry and labor representatives, civil rights advocates, and the general public. It is developing panels to examine the various problematic and controversial issues of biotechnology. The Committee publishes the following newsletter:

● *Gene Watch Newsletter*—a bimonthly publication that covers a variety of topics and issues of concern to the general public. Newsletter topics have included: prenatal screening, biotechnology in the Third World, biological weaponry, and the human, social, and industrial uses of biotechnology. The cost is $10.

Industrial Biotechnology Association (IBA)
2115 East Jefferson Street
Rockville, MD 20852

301-984-9598

IBA is the oldest and largest biotechnology association. Its membership consists of both domestic and foreign for-profit companies actively engaged in biotechnology. IBA's affairs are devoted exclusively to the promotion of biotechnology, and include: monitoring regulatory and legislative developments on federal and local levels; sponsoring seminars and workshops on topics such as industry-university cooperation, patents, social and ethical issues, public education, and publications. The staff can provide you with up-to-date information and opinions from an industrial perspective. While most of the Association's services are restricted to members, IBA does publish educational materials for student, community, government, and media audiences. IBA's recent publication *Primer on What Biotechnology Is* provides a good introduction to biotechnology for the general public.

Publications

Bio/Technology
National Publishing Company
PO Box 316
Martinsville, NJ 08836

800-824-7888
212-477-9600—Editorial Department

Bio/Technology is a monthly magazine containing general articles and information about new products, research papers, technology reports, conferences, meetings, and courses. The subscription is $112 a year.

Biotechnology
International Trade Administration   202-377-3808
U.S. Department of Commerce
Washington, DC 20230

This publication presents the proceedings of meetings betweeen representatives of the biotechnology industry and the federal government. It includes a summary of issues raised at the meetings and transcriptions of industry presentations. Topics covered include the impact of federal research and regulatory policies on the biotechnology industry; export control of biotechnology; and technology transfer as the key to the commercialization of biotechnology. A comparison of the development of biotechnology in the U.S. and other countries is provided. This comparative analysis includes an examination of the major foreign challenges to U.S. firms; an identification of actual or potential barriers to the successful commercialization of biotechnological innovations; and a discussion of key issues and options related to U.S. government policies affecting the international standing of American biotechnology firms. *Biotechnology* is available for $7 from the Superintendent of Documents, U.S. Government Printing Office, Washington, DC 20402, 202-783-3238. Refer to stock number 003-009-00430-6.

Biotechnology Investment Opportunities
High Tech Publishing Company   802-869-2833
PO Box 266
Brattleboro, VT 05301

This journal identifies and analyzes emerging investment opportunities in genetic engineering. The publication follows trends and conditions having significant impact on the development and commercial application of leading-edge biotechnology research. It emphasizes new enterprise formation and emerging markets and applications. The annual subscription is $195 for ten issues.

Biotechnology Newswatch
Newsletter Publishing Center   212-512-6090—Editorial
McGraw-Hill, Inc.   212-512-4462—Subscription
1221 Avenue of the Americas, 43rd Floor
New York, NY 10020

This eight-page biweekly newsletter contains inside information about what businesses are doing in the field. The newsletter covers recent discoveries, company news and patents, and other information on developments in the U.S. and abroad. Topics regularly covered include genetic engineering, biomass, patents, enzymes, energy, monoclonals, and recombinant DNA. The subscription is $417 a year for 26 issues.

Biotechnology Resources

Research Resources Information Center
301-881-4150
U.S. Department of Health and Human Services
Public Health Service
1776 East Jefferson Street
Rockville, MD 20852

This free 85-page directory lists biotechnology-related research centers nationwide. Information is provided about services available, and the centers' research, emphasis or application. Resources are described for biochemical materials, biological structure and function, biomedical engineering, and computers. A geographical index is included.

Commercial Biotechnology: An International Analysis

Office of Technology Assessment (OTA)
202-226-2155
U.S. Congress
Washington, DC 20510

This report by OTA, an analytical support agency of the Congress, assesses the competitive position of the U.S. in biotechnology with respect to Japan and four European countries. The report, published January 1984, analyzes commercial activity from three perspectives: (1) the number and kinds of companies; (2) markets targeted by industrial research and development; and (3) the interrelationships among companies and the overall organization of the commercial effort. The study also examines the factors potentially important to the commercialization of biotechnology. A free summary of the report is available from OTA. The publication in its entirety must be purchased from the Superintendent of Documents, U.S. Government Printing Office, Washington, DC 20402. The report costs $20, and its order number is 052-003-00939-1.

Genetic Engineering and Biotechnology Firms Worldwide

Technical Insights, Inc.
PO Box 1304
Fort Lee, NJ 07024

The report describes some 1,200 firms in 29 countries, with information on the holdings and activities of the largest companies, and names, addresses, and phone numbers for the smaller companies. It includes information on research lab locations, number of employees, research activity areas, commercial fermentation products available, and so on. Available for $177.

Genetic Engineering News

Mary Ann Liebert, Inc.
212-289-2300
157 East 86th Street
New York, NY 10028

An important information source on the biotechnology industry, this journal provides major coverage of significant issues, regulatory and scale-up guidelines, research and development financial news, funding and government news, corporate profiles, university news, meeting reports, foreign reports, new product and literature information, and critical in-depth articles on new research and technology. Focuses on recombinant DNA, hybridoma, and other advanced technologies in the health sciences, agriculture, energy, food processing, pharmaceuticals, minerals, and diagnostics. Available for $115 a year for eight issues.

Telegen

48 West 38th Street   212-944-8500
New York, NY 10018

Telegen publishes the following:

● *Telegen Reporter*—is a monthly publication which monitors all aspects of biotechnology, with over 300 summaries of articles, reports, and conference proceedings indexed by subject, source, author, and SIC code.

● *Telegen Alert*—is an electronic alerting service for up-to-date information on the latest developments in biotechnology.

● *Telegen Quicktext*—this document retrieval service is for full text-delivery (microfiche or hardcopy) of documents summarized in *Telegen Reporter*. A complete microfiche file is available retrospective to 1973.

● *Telegen Annual Review*—is a year-end, hard-cover two-volume edition. One volume cumulates the year's indexes, the other contains individual summaries for the year. The *Review* also has an annual overview of significant activities of the preceding 12 months.

● *Telegenline*—provides on-line access to the data base.

Additional Experts and Resources

Genex

Corporate Communications Department   301-770-0650
6110 Executive Boulevard
Rockville, MD 20852

The staff members at Genex are a good source of information regarding developments in the industry. They can answer questions and make referrals to other resources.

Dr. William Gartland and Dr. Elizabeth Milewski

Office of Recombinant DNA Activities
National Institutes of Health (NIH)
Building 31, Room 3B10
Bethesda, MD 20205

Doctors Gartland and Milewski are an excellent resource for finding out about past and prospective government activities regarding recombinant DNA. They can refer you to other experts, advise you about federal guidelines and policies, and provide you with information about seminars, workshops, and new publications. Dr. Milewski is the editor of *Recombinant DNA Technical Bulletin*, a quarterly newsletter designed to help industry understand what is happening inside the federal government regarding recombinant DNA. Examples of topics covered include: federal regulations, workshops, and publications.

Dr. Sheldon Krimsky
Department of Urban and Environmental Policy
Tufts University
Medford, MA 02155

Dr. Krimsky is an expert on the social impact of biotechnology. His publications include:

● *Genetic Alchemy: The Social History of Recombinant DNA Controversy*—published by MIT Press, the book is a comprehensive social history of recombinant DNA covering the years 1970 through 1981.

● *International Network on Social Impacts of Biotechnology*—published twice yearly by Tufts University, this resource guide is designed for writers and social scientists interested in the social impact of biotechnology on civilization. It includes names of resource people, what has been written in field research, research being conducted, and occasionally abstracts. Write to Dr. Krimsky to be placed on his mailing list.

Senator Albert Gore, Jr.
U.S. Senate Office   202-224-3121
Washington, DC 20510

Senator Gore has spent many years looking at the federal role in biotechnology. When a member of the House, he served as chairman of the House Science and Technology Subcommittee on Investigations and Oversight. He has held hearings and seminars, and prepared publications examining what the government should and should not be doing in this field.

Dr. Bruce Mackler
General Counsel
Association of Biotechnology Companies
4400 Jennifer Street, NW
Washington, DC 20015

A scientist and attorney, Dr. Mackler is an expert on the legal and scientific issues of biotechnology.

Henry I. Miller, M.D., Medical Officer
Center for Drugs and Biologics   301-443-4864
FDA HFN-823
5600 Fishers Lane
Rockville, MD 20857

Dr. Miller is involved in the social and ethical aspects of recombinant DNA–derived products and human genetic theory. He is also involved in the use of recombinant DNA and other genetic engineering techniques for the production of pharmaceuticals, and the chemistry pharmacological, toxicological and clinical aspect of recombinant DNA–derived pharmaceuticals. Dr. Miller is available on a limited basis for referrals to other experts in the field. He has written over 30 publications, a complete listing of which is available from his office.

Mark Segal
U.S. Environmental Protection Agency (EPA)   202-382-3502
Office of Toxic Substances TS 796
401 M Street, SW
Washington, DC 20460

Mr. Segal is an expert on data bases that deal with biotechnology. He is willing to answer questions and can refer you to other information sources.

Building and Construction

See: Architectural Engineering

Business Advisory Services

See: Small Business Incubators; State High-Tech Initiatives; Venture Capital

Business Incubators

See: Small Business Incubators

Cable Television

Organizations and Government Agencies

Cable Television Bureau

Federal Communications Commission (FCC)   202-632-7076—Regina Barrett, Reference Room
2025 M Street, NW
Washington, DC 20554

The Bureau maintains files on all operating systems in the U.S., as well as applications for licenses, annual reports, and history cards. The reference room, with all the files, is open to the public, and a public information bulletin with general information on the cable industry is available free of charge. The staff will answer questions and recommend other sources of information.

Cable Television Information Center

1800 North Kent Street, Suite 1007   703-845-1700
Arlington, VA 22209

A national clearinghouse for cable and communications information, the Center maintains one of the most complete cable television libraries. It is open to the public for use, but appointments must be made. Regional seminars are held throughout the year. Publications include:

- *CTIC Cable Reports*—a monthly newsletter available for $165 a year.
- *CTIC CableBooks*—handbooks providing a comprehensive guide to cable services, cable policy options, legal issues, and regulatory concerns. Two volumes are available for $27, or $15 each.
- *Technology of Cable Television*—$4.50.
- *Cable Bibliography*—$2.
- *Glossary of Cable Terms*—$2.

Community and Antenna Television Association

3977 Chain Bridge Road   703-691-8875
Fairfax, VA 22030

This association is composed of cable television operators. A seminar is held each year for cable operator members, and new technology and methods are discussed. The staff is willing to answer questions from the public. A monthly newsletter keeps members and associates up to date on the latest developments in the industry.

National Cable Television Association (NCTA)

1724 Massachusetts Avenue, NW   202-755-3622
Washington, DC 20036

A trade association for the cable television industry, NCTA sponsors conventions, seminars, workshops, and conferences. The library contains the latest research and technical information about the industry, but it is not open to the public. Librarians will answer questions and make copies of documents for free, depending on the quantity of research. Services are primarily for members, but they will help the general public. The Public Affairs office will also answer questions. Publications include:

   • *Interaction*—a monthly newsletter highlighting legislation and regulatory development on the state level. Available for $24 a year.

   • *Tech Line*—a monthly newsletter, with data on the latest technological developments. Available for $24 a year.

Publications

Cable Marketing

352 Park Avenue South   212-685-4848
New York, NY 10010

   This magazine is a monthly marketing/management publication. Areas covered include issues affecting sales and profit performance; news reports and short features emphasizing market principles; a special section on engineering and technology, including information on new equipment and products; focus on the growth of the programming side of the cable industry; and financial issues. Available for $38 a year.

Cable Television Business

Cardiff Publishers   303-694-1522
6530 South Yosemite Street
Englewood, CO 80111

   Published biweekly, the magazine contains articles on the business aspects of the cable television industry. Topics covered include new business developments, finance, marketing, programming, ad sales, and engineering management. Available for $27 a year.

CableNews

Phillips Publishing, Inc.   301-986-0666
7315 Wisconsin Avenue, Suite 1200N
Bethesda, MD 20814

   This weekly newsletter covers such cable topics as competing technologies, marketing, regulation, and new products. The newsletter is available for $277 a year. The newsletter is also available on-line from NewsNet, Inc., at 800-345-1301.

Multichannel News

300 South Jackson, Suite 450   303-393-6397
Denver, CO 80209

   A weekly newspaper, this publication offers information and news on all aspects of the cable television field, and presents the latest developments in the industry. Subscription is $27 a year.

SAT Guide
PO Box 1048   208-788-9531
Hailey, ID 83333

A magazine published for the cable television industry and its alternative distribution systems, *SAT Guide* is devoted exclusively to the domestic satellite industry. It addresses the issues, problems, and changing technologies. Regular feature articles include programming, reviews, hardware, trends, a yellow pages reference, and reviews of regulatory matters.

TITSCH Communications
2500 Curtis Street, Suite 200
PO Box 5727 T.A.
Denver, CO 80217

The publications available from TITSCH include:

- *Cable Vision*—this weekly publication contains information and data on the cable television industry. The subscription price is $64 a year for 50 issues.
- *CableFile* and *CableFile/Update*—an annual two-volume directory with more than 2,000 pages of information. It includes detailed profiles of operating companies and of over 5,700 cable systems, as well as information on suppliers, service companies, and consultants. The *Update* contains information on the cable systems listed in *CableFile* and is published twice a year. Subscription information can be obtained from TITSCH Communications.

Data Bases

CIS Data Base (Cable Information Service)
TITSCH Communications   202-835-0900
1701 K Street, NW, Suite 505
Washington, DC 20006

This data base provides all the information contained in the *CableFile* directory and *CableFile/Update* in data-base form. It contains statistics, facts, and profiles on virtually every operating system in the U.S. The data is delivered on computer printouts, tapes, and by on-line access. All key data are updated quarterly, and the entire data base is updated annually.

Additional Experts and Resources

The Science and Technology Division of the Library of Congress has put together many tracer bulletins that provide guides to the materials on various technological topics. A "Bullet" has been prepared for cable television, and is available free of charge from:

Reference Section
Science and Technology Division
Library of Congress
Washington, DC 20540
202-287-5670

CAD/CAM

Organizations and Government Agencies

Bureau of Industrial Economics

U.S. Department of Commerce   202-377-0314—Mr. Thomas J. Gallogly, Director of Metal Working
Office of Producer Goods   202-377-0315—Mr. John Neuman, Industry Specialist
Washington, DC 20230   202-377-4073—Mr. Franc Manzolillo, Trade Specialist

The specialists listed above develop industry and market analysis, and disseminate information covering a wide range of the CAD/CAM sector, such as machine tools, robotics, weldings, power hand tools, and many others. The Office of Producer Goods and the experts noted above can be contacted for information and referrals on CAD/CAM.

Center for Manufacturing Engineering

U.S. Department of Commerce   301-921-3421
National Bureau of Standards   301-921-3691—Mr. Brad Smith, Computer Aided Design and
Washington, DC 20234   Computer Aided Manufacturing

The Center has several specialists in the various areas of CAD/CAM who develop research work in subjects such as application of advanced control manufacturing systems, and robot manipulators. For referral to these experts on specific technical topics, please write the Center. A brochure titled "Publications of Center for Manufacturing Engineering" lists the papers, books, and articles developed by the Center's members. It can help identify experts to contact. A copy of the brochure can be ordered for $10 by writing to: National Technical Information Service, Springfield, VA 22161.

Computer Aided Manufacturing—International, Inc.

611 Ryan Plaza Drive, Suite 1107   817-860-1654
Arlington, TX 76011

CAM-I is a nonprofit organization dealing with research and development in technology. It has offices in Europe, Japan, Australia, and in the U.S., and is supported by funds provided by its members, which include Kodak, IBM, GM, Westinghouse, and McDonnell Douglas. It is regulated by the federal government, and information about its technological developments is public domain. The staff will provide technical information and referrals on a limited basis. The library can furnish information about CAD/CAM technical literature available to the public. CAM-I has international group meetings every fall and spring, when papers are presented on various areas of CAD/CAM. Workshops are also sponsored on CAD/CAM, in subjects like robotics and machine tools. Copies of the proceedings of the meetings as well as other publications are available, but you must pay for reproduction costs.

Computer Automated Systems Association (CASA)
Society of Manufacturing Engineers   313-271-1500
One SME Drive
PO Box 930
Dearborn, MI 48121

This is an educational and scientific association of professionals in the manufacturing community. It deals with computers and automation in the advancement of manufacturing. The Association holds conferences, workshops, and educational seminars. CASA has a list of consultants who can be contacted at the address listed above. It publishes:

- *CAD/CAM Technology*—a quarterly magazine for professionals dealing with CAD/CAM. It has features on current business events related to manufacturing systems developments. Also included is a current calendar of major educational events in the U.S. and abroad. The subscription price for nonmembers is $24, and for members the cost is included in the annual membership fee of $43.

- *AUTOFACT Proceedings*—bound volumes of technical papers presented at the annual AUTOFACT meetings and events sponsored nationally and internationally by this Association. The prices for these proceedings vary from $40 to $90.

National Machine Tool Builders' Association
7901 Westpark Avenue   703-893-2900
McLean, VA 22102

This Association represents the machine tools industry in the U.S. The Technical Department follows the latest technological innovations in the domestic and international machine tools sector. A summary report is issued every month covering technical products, and other publications are periodically released to the members of the Association. An International Machine Tools Show is sponsored annually by NMTBA. Mr. Anthony M. Bratkovich at the Technical Department is the main resource person, who may also give referrals.

Publications

American Machinist
McGraw-Hill, Inc.   212-512-2000
1221 Avenue of the Americas
New York, NY 10020

This magazine is oriented toward manufacturing technology. Its contents focus heavily on control systems, manufacturing engineering, and metalworking technology involved in CAD/CAM. The magazine is available free to managers of manufacturing corporations, and the subscription for individuals is $38 a year.

American Metal Market/Metalworking News

Fairchild Publications   212-741-4000
7 East 12th Street
New York, NY 10003

Fairchild publishes *American Metal Market*, a daily newspaper covering manu-facturing technology and business news on the metalworking products industry, with a weekly insert entitled *Metalworking News*. This weekly issue is a magazine that reports and analyzes special topics of technology in the area of CAD/CAM. The cost is $310.

A Survey of CAD/CAM Systems

Leading Edge Publishing, Inc.   214-341-9606
324 Forest Central Two
11551 Forest Central Drive
Dallas, TX 75243

This book consists of answers to 175 questions about CAD/CAM products, including hardware, software, displays, input/output devices, and nongraphic pro-cessing. It has some fundamental information for those new to the world of CAD/CAM, besides a chapter on technical considerations, and a final section that deals with the future of CAD/CAM. The cost of the book is $96.

Computer Graphics Glossary

Oryx Press   602-254-6156
2214 North Central at Encanto, Suite 106
Phoenix, AZ 85004

The *Glossary* explains, among other things, the specialized language of CAD/CAM technology. Available for $18.50.

The CAD/CAM Industry Directory

Technical Data Base   409-539-9688
PO Box 720
Conroe, TX 77305

This *Directory*, which is updated annually, lists consultants, experts, systems analysts, and other types of professionals working in the CAD/CAM industry, as well as hardware, software, modems, technical systems, systems components, and other varieties of systems related to this segment. A magazine is sent every other month to those who buy the *Directory*, which costs $35.

Data Bases

ISMEC—(Information Services in Mechanical Engineering)

Cambridge Scientific Abstracts (CSA)—Producer   301-951-1400—Pam Mckeehan
5161 River Road, Bldg. #60
Bethesda, MD 20816

ISMEC is a data base that covers international technical literature in mechanical engineering, production engineering, and engineering management. About 90 percent of the contents is from journal articles, and the remaining 10 percent is from papers and monographs presented at conferences, as well as theses. It maintains many records on specific categories of CAD/CAM, and it is updated monthly. The data base is accessible on DIALOG, but can also be leased through the producer.

NEXIS

Mead Data Central   800-227-4908
PO Box 1830   513-865-6800
Dayton, OH 45401

NEXIS consists of the full text of over 110 newsletters, journals, magazines, newspapers, and wire services. It covers a wide variety of sources on the various areas of CAD/CAM. The data base is updated continuously and covers information since 1975.

Calibration Systems

See: Standards and Measurement

Cancer-Related Technology

See also: Medical Imaging; Medical Technology; Nuclear Medicine; Pharmaceuticals; Toxicology

Organizations and Government Agencies

American Association on Cancer Research

Temple University School of Medicine   215-221-4565
West Building, Room 301
Philadelphia, PA 19140

The Association comprises clinicians, researchers, and physicians involved in the field of cancer research. The organization holds a conference each spring, at which papers are presented and symposia are held on various aspects of cancer research. Topics covered include: clinical research, epidemiology, biochemistry, therapeutics, and pharmacology. The proceedings are published and are available to the public. The Association also publishes:

• *Cancer Research Journal*—a monthly publication featuring the latest cancer-related developments in basic biology, biochemistry, oncology, and immunology. The *Journal* contains important findings of carcinogenesis diagnostic treatments for both human and animal cancers. The annual subscription price is $100 for individuals and $175 for libraries. Order from Waverly Press, Inc., 428 East Preston Street, PO Box 64473, Baltimore, MD 21264, 301-528-4255, 800-638-6423.

American Cancer Society

777 Third Avenue   212-371-2900
New York, NY 10017

The Society provides funding and grants for cancer research. It also promotes public education concerning cancer, and acts as a support group for cancer patients and their families. The Society publishes free information brochures about the various forms of cancer and related treatments, such as radiation and chemotherapy. An information brochure about the organization, its services, and how to apply for cancer research funding are available upon request. The Society is a good information source, and the staff will answer questions and make referrals to other groups covering specific aspects of cancer. The Society also publishes:

• *CA: A Cancer Journal for Clinicians*—this bimonthly Journal contains cancer-related papers and research results covering such topics as colon cancer, urological cancers, tumors, and treatment programs such as hospices. Available free of charge, the *Journal* can be obtained from your local chapter of the American Cancer Society.

Association of Community Cancer Centers

1160 Nebel Street, Suite 201   301-984-9496
Rockville, MD 20852

This membership organization consists of hospitals, institutions, and people active in the field of diagnosing and treating cancer as well as managing community cancer centers. The Association can provide information about clinical research organizations, new technology, cancer data management systems, research results, federal funding, and a broad spectrum of clinical practice issues. In addition to books and special reports, it also publishes:

• *Cancer Program Bulletin*—a quarterly newsletter that provides members with information about federal legislation, regulations, new funding opportunities, and clinical research programs. The publication also includes information on community cancer programs and their components as well as Association news. It is available free of charge to members and is not available to nonmembers.

• *Community Cancer Programs in the United States*—published annually, this 250-page directory provides detailed information about each member institution of the Association, including program components, personnel, and history. The directory is available free of charge.

Cancer Information Service

Johns Hopkins Oncology Center

550 North Broadway, Suite 307

Baltimore, MD 21205

202-636-5700

800-638-6070—Alaska

800-524-1234—Hawaii

800-4 CANCER—Elsewhere

   Funded by the National Cancer Institute, this Service has a toll-free telephone inquiry/referral service established to supply cancer-related information to the public, cancer patients and their families, and health professionals. It consists of 21 regional offices associated with the major cancer centers across the U.S. The staff can answer questions about the latest cancer technologies; refer you to experimental programs, experts, local resources, and literature; provide specific information about particular types of cancer and treatment; and give you facts about the patient referral process. Publications, geared toward both the general public and health professionals, are available from CIS.

Publications

American Journal of Clinical Oncology

Masson Publishing, USA, Inc.

133 East 58th Street

New York, NY 10022

212-838-8510

   This bimonthly publication covers the latest advances in pathologic, surgical and clinical cancer research. The *Journal* provides information on all aspects of cancer diagnosis and treatment, and contains photographs, graphs, and charts for clarity. The publication is indexed in Index/Medicus and Current Contents/Clinical Practice data bases. The 90-page publication is available for $68 a year.

Cancer

J.B. Lippincott Publishers

East Washington Square

Philadelphia, PA 19105

215-238-4200

800-638-3030

   Published twice a month, *Cancer* is a collection of original articles and research papers geared toward individuals involved in clinical research. The journal seeks to bridge the gap between the investigator and the clinician, and it covers all aspects of cancer. The 250-page publication is available for $60 a year, or for $5 an issue.

Cancer Bulletin

Medical Arts Publishing Foundation

1603 Oakdale

Houston, TX 77004

713-492-6014

   Geared toward physicians, this highly technical bimonthly provides the latest information about cancer-related research developments and treatment. Each issue explores a specific topic in depth, and examples of subjects covered include: tumors, critical care of cancer patients, interarterial chemotherapy for cancer, cytogenetics, breast cancer, and new cancer treatment techniques. The 50-page publication is available for $12 a year, or $3 an issue.

Cancer Genetics and Cytogenetics

Elsevier Scientific Publishers   212-867-9040
52 Vanderbilt Avenue
New York, NY 10017

Published ten times a year, this journal consists of original articles about human and animal molecular structures, biomedical genetics, and cytogenetics. The journal provides comprehensive and concise information about the latest research and development in the fields of cancer genetics and cytogenetics, and each issue carries editorial reviews of literature discussing outstanding developments. The 100-page publication is available for $495 a year.

Cancer Letter, Inc.

P.O. Box 2370   703-620-4646
Reston, VA 22090

The following publications are available from this company:
- *Cancer Letter*—is a weekly publication covering government developments affecting cancer research and general policy established by the federal National Cancer Program. It provides information about the work of the committees, the advisory board, and The President's Cancer Panel. A monthly calendar profiles important conferences in the field of cancer research, as well as grants, contracts, and program developments in the field of cancer. The 8-page publication is available for $150 a year.
- *Clinical Cancer Letter*—a monthly publication covering the latest advances in clinical research, protocol, and studies by investigators for practicing oncologists. Conference proceedings relating to advances in clinical cancer research are also published. The 8-page publication is available for $30 a year.

Cancer Research

Williams and Wilkins Company   301-528-4000
428 East Preston Street
Baltimore, MD 21202

Published monthly, this journal is a collection of research papers covering new developments, advances, and applications in cancer research. It contains a section on basic sciences, cellular investigation, and a calendar announcing upcoming conferences related to cancer research. The 400-page publication is available for $100 a year.

Cancer Treatment Reports

NIH NCI Bldg. 81, Room 203   301-496-6975
9030 Old Georgetown Road
Bethesda, MD 20205

Compiled by the National Cancer Institute, this 100-page quarterly follows cancer treatment research and developments. Twenty-five to 30 investigative reports are covered in each issue. Topics have included: animal studies and pharmacokinetics, surgery, radiotherapy, chemotherapy, supportive care, pharmacology, cell biology, kinetics, mechanisms of drug actions, and medicines and their relationships to treatment. The annual subscription is $29; the quarterly can be ordered from the Superintendent of Documents, U.S. Government Printing Office, Washington, DC 20402, 202-783-3238.

Cancer Treatment Review

Academic Press, Inc.   212-614-3000
111 Fifth Avenue
New York, NY 10003

Geared toward physicians, urologists, clinicians, and researchers, the publication provides current information about the advances and applications of cancer treatment. All methods of treating tumors are covered, including surgery, radiotherapy, chemotherapy, immunotherapy, and combined modality approaches. The 100-page journal is available for $69.50 a year.

Current Problems in Cancer

Yearbook Medical Publishers   312-726-9733
35 East Wacker Drive
Chicago, IL 60601

This monthly journal reports on diagnosis, treatment, and medication problems in the field of cancer. It covers a broad range of topics, providing the latest information on radiation, radio-biology, hairy-cell leukemia, surgical treatments, genetics, tumors, radiotherapy, breast cancer, and nuclear medicine. The 60-page publication is available for $47 a year for individuals, and $60 for hospitals.

Hematological Oncology

John Wiley & Sons, Ltd
Bassins Lane, Chichester
Sussex PO19 IUD
England

John Wiley & Sons, Inc.   201-469-4400
1 Wiley Drive
Somerset, NJ 08873

This quarterly journal is a collection of research papers and reports covering antibodies, leukemia, all aspects of cancerous tumors, diseases affecting the blood, and related disorders. Although the journal must be ordered from its publisher in England, information is available from the New Jersey office listed above. The 100-page publication is available for $99 a year.

International Journal of Cancer Research & Treatment: Oncology

Karger Publishing Company   914-948-0138
Albert J. Phiebig, Ordering Agent
PO Box 352
White Plains, NY 10602

> International in scope, this bimonthly journal covers the latest in experimental and clinical cancer research. It provides articles and reports on cancer research results and applications, as well as information about cancer detection and treatment. The 100-page publication is produced by Karger Publishing Company, and is available for $222 a year.

Journal of the National Cancer Institute

National Cancer Institute   301-496-6641
Westwood Bldg., Room 850
5333 Westbard Avenue
Bethesda, MD 20816

> Published monthly, this scientific journal covers treatment and research in the field. The highly technical publication presents investigative studies of both human and nonhuman cancers. The 200-page journal is available for $59 a year from the Superintendent of Documents, U.S. Government Printing Office, Washington, DC 20402, 202-783-3238.

Oncology Times

Herlitz Publications   212-532-9400
404 Park Avenue South
New York, NY 10016

> Published monthly, this newspaper is geared toward physicians involved in the diagnosis and care of cancer patients. It covers all aspects of cancer including the latest developments, hospital care, AIDS, leukemia, tumors, and breast cancer. The 10-page newspaper is available for $10 a year.

Primary Care and Cancer

Dominus Publishing Company   516-621-1080
79 Powerhouse Road
PO Box 67
Roslyn Heights, NY 11577

> This monthly journal provides information to help the primary-care physician detect and treat a wide variety of cancerous diseases. Each issue includes articles on the latest diagnosis and treatment techniques, as well as equipment. The 50-page journal is available for $30 a year.

Seminars in Oncology

Grune and Stratton, Inc.   212-614-3000
111 Fifth Avenue
New York, NY 10003

This quarterly journal consists of papers covering diagnostic and treatment techniques for all types of cancer. It also contains cancer research results. The 100-page publication is available for $49 a year.

Springer-Verlag Publishers
44 Hartz Way   212-460-1500
Secaucus, NJ 07094

Springer-Verlag Publishers puts out the following journals, which deal with specific aspects of cancer research and treatment:

• *Cancer Chemotherapy and Pharmacology*—this bimonthly publication is a collection of original articles and general reviews of recent advances in chemotherapy and pharmacology. It provides coverage of such studies as metabolism of human leukemia cells, pharmacokinetics of children, cancer in adolescents, and colon cancer. The 50-page publication is available for $196 a year.

• *Cancer Immunology and Immunotherapy*—a bimonthly journal covering experimental and clinical research data relating to immunology, immunotherapy, and other biological response modifications. The 50-page publication is available for $196 a year.

• *Journal of Cancer Research and Clinical Oncology*—this bimonthly journal presents experimental and clinical research data collected from developments in cancer diagnosis and treatment. It provides information about recent developments in tumor therapy, general diagnosis techniques, diagnostic and experimental pathology, and epidemiology. The 50-page publication is available for $321 a year.

Data Bases

CANCERLINE
MEDLARS   301-496-6308
National Library of Medicine
8600 Rockville Pike
Bethesda, MD 20209

CANCERLINE is the largest collection of cancer research information currently available. This computer-based system provides rapid retrieval of abstracts of cancer literature, research project descriptions, and summaries of clinical protocols. CANCERLINE is accessed by using typewriter-like terminals linked to a central computer facility located at the National Library of Medicine (NLM) in Bethesda, Maryland. These terminals are available at some 1,500 terminals throughout the U.S. and in 13 other countries. The CANCERLINE system includes three computer data bases:

• *CANCERLIT* (Cancer Literature)—contains more than 300,000 abstracts of published literature dealing with all aspects of cancer, selected from more than 3,000 biomedical journals. It includes abstracts of books, technical reports, theses, and papers presented at meetings. *CANCERLIT* is updated monthly, with about 45,000 abstracts being added each year.

• *CANCERPROJ* (Cancer Projects)—contains approximately 20,000 descriptions of ongoing cancer research projects around the world. It includes over 5,000 cancer-related projects from more than 83 different countries outside the U.S. It includes descriptions of both federally and privately supported grants and contracts. *CANCERPROJ* is updated quarterly.

  • *CLINPROT* (Clinical Protocols)—contains summaries of more than 2,400 experimental clinical cancer therapy protocols supported by the Division of Cancer Treatment (DCT) of the National Cancer Institute or by cancer centers outside the U.S. Protocols can be retrieved by type of cancer, modalities, type of agent(s) used, investigator name, protocol identification numbers, and several other search fields.

Careers and Employment

See also: Labor and Technological Change

Organizations and Government Agencies

American Society of Association Executives (ASAE)
Information Central   202-626-2723
1575 I Street, NW
Washington, DC 20005
      ASAE can direct you to an association that can provide you with information about a career or industry that interests you. This organization also maintains a job bank with listings of openings at its member associations. A fee is charged to users of the job clearinghouse.

Bureau of the Census
Data User Service Division   301-763-5820
U.S. Department of Commerce
Washington, DC 20233
      This Bureau collects and publishes statistics about the growth rate of specific industries. It also has statistics about the growth rate of geographical areas, ranging from states to city blocks.

Bureau of Industrial Economics
U.S. Department of Commerce   202-377-4356
14th Street and Constitution Avenue, NW, Room 4845
Washington, DC 20230
      Staff members in this office can refer you to an industry specialist who can tell you about industry trends and guide you to information about companies.

Bureau of Labor Statistics

U.S. Department of Labor 202-523-1239

441 G Street, NW

Washington, DC 20212

This government agency can supply you with wage, employment, productivity, and labor force data for most occupations and industries.

Congressional Caucus for Science and Technology

House Annex Bldg. #2, H2-226 202-226-7788

2nd and D Streets, SW

Washington, DC 20515

The Caucus, established to serve members of Congress, can share its information about the impact of science and technology on economic development and employment trends, including issues of training and retraining. The Caucus plans to develop "employment roadmaps," which will highlight employment opportunities, as well as training and educational requirements, for current and prospective jobs. Staff members can answer questions about high-tech-related legislation, and they will refer science- and technology-oriented inquiries to the Caucus's research arm, the Research Institute for Space, Science and Technology.

National Center for Education Statistics (NCES)

U.S. Department of Education 202-254-6057

400 Maryland Avenue, SW

Washington, DC 20202

The Center collects and publishes statistical information about all aspects of education, including the number of individuals receiving education and training for various occupations. Its publications and data tapes are available to the public, some free of charge.

Office of Opportunities in Science

American Association for the Advancement of Science (AAAS) 202-467-5438

1776 Massachusetts Avenue, NW

Washington, DC 20036

The primary concern of the Office is to increase the status and participation of women, minorities, and the handicapped in the sciences. However, the staff will help anyone seeking employment in scientific fields. The Office maintains a file of recruitment openings from universities, the private sector, and the federal government. It also keeps a listing of resources for women in science and sponsors several projects, such as support groups for women in mathematics, engineering, science, and health. Upon request you can receive a free listing of its publications, many available at no cost. Materials include:

• *Career Information in the Sciences*—Updated once every year or two, this listing of about 100 entries provides information about a variety of scientific disciplines. It covers engineering, computers, chemistry, physics, aerospace, health sciences, agricultural sciences, and other fields. The 15-page publication is available free of charge.

● *Mathematics, Engineering, Science and Health (MESH)*—Published approximately twice a year, this newsletter for the national MESH support group gives information about recruitment, financial aid, and career development. It also lists new publications and studies, as well as future dates of conferences. Eight to 20 pages long, the publication is available free of charge.

● *Info Memo*—Published once or twice a year, this newsletter for the National Network of Minority Women in Science covers timely areas of interest, such as the crisis in education, computer technology, and computer science. It gives information about recruitment, financial aid, and so on, and it lists new publications and studies. The 10-page publication is available free of charge.

Office of Scientific and Engineering Personnel

National Research Council   202-334-2700
2101 Constitution Avenue, NW
Washington, DC 20418

This office is concerned with more effective development and utilization of scientific and engineering personnel. It provides the National Research Council with expertise about the status of scientific and engineering personnel and methodologies for assessing both current and projected employment demand and supply. Its focus is in four major areas: Statistical and Survey Programs, Fellowship Programs, Associateship Programs, and Studies.

Public Employment Service Offices
Check your telephone book under state and local government listings, or call or write:

Employment and Training Administration (ETA)   202-376-6289
U.S. Employment Service
U.S. Department of Labor
Washington, DC 20213

Government-sponsored public employment service offices exist in communities throughout the U.S. The centers facilitate the matching of people and jobs. The staff can provide information and publications about labor market conditions, jobs available, and training programs. The ETA office listed above coordinates the local, regional, and state employment services. If you cannot locate your local office in the phone book, call this office for information about publications and/or your nearest center.

Scientific and Technical Personnel Studies Section

Division of Science Resource Studies   202-634-4622—Editorial and Inquiries Unit for Publications
National Science Foundation (NSF)
1800 G Street, NW
Washington, DC 20550

This NSF office conducts studies of scientific and technical personnel in the U.S. The resulting reports are available and provide data about the training and deployment of individuals in science and engineering fields. Staff members in the Studies Section are highly specialized, and they will respond to written inquiries. A listing of reports is available by writing the Studies Section or by calling NSF's Editorial and Inquiries Unit. The following publications are available free of charge:

● *Science and Engineering Personnel: A National Overview*— Updated every two years, this 66-page publication provides a comprehensive overview of the status of U.S. scientific and technological efforts as they relate to the employment and other characteristics of science and engineering personnel. It consists of text and appendix tables, and provides a framework for analyzing issues relating to these personnel.

● *Women and Minorities in Science and Engineering*—Updated every two years, this 182-page report looks at current employment in science and engineering. Consisting of text and tables, it includes labor market indicators, comparing salary, experience, and unemployment rates. This publication contains a large section about the education and training of women and minorities, including precollege preparation.

● *Survey of Recent Science and Engineering Graduates*— Published biannually in even-numbered years, this survey reports on graduates from the previous one-to-two years. It looks at demographic data, including type of employer, work experience, age, primary work activity, sex, race, and ethnic groups, as well as field of employment versus field of degree.

● *Survey of Doctoral Scientists and Engineers*—Published biannually in odd-numbered years, this survey looks at demographic data, including type of employer, work experience, age, primary work activity, sex, race, and ethnic groups.

● *Survey of Experienced Scientists and Engineers*—This publication surveys scientists and engineers who were in the labor force when the 1980 Census was taken. It provides the following information: type of employer, work experience, age, primary work activity, sex, race, and ethnic groups.

● *U.S. Scientists and Engineers*—Published biannually in even-numbered years, this in-house model looks at the total population of scientists and engineers, combining information from the surveys described above. It provides the same basic information, including type of employer, work experience, age, primary work activity, sex, race, and ethnic groups.

Publications

Career Opportunities in Science & Technology
Tracer Bullet #ISSN 0090-5232   202-287-5639
Reference Section
Science & Technology Division
Library of Congress
10 First Street, SE
Washington, DC 20540

A brief guide to information sources covering careers in science and technology, this Tracer Bullet includes a bibliographic listing of books, journals, articles, and abstracting and indexing services. It also includes places you can contact for information, as well as the subject headings used by most libraries to locate science and technology career materials in their card, book, and on-line catalogs. The publication is updated as needed and available free of charge. The Library of Congress prepares Tracer Bullets on a variety of subjects, and it may have one on a particular field of interest to you, such as computers.

Careers in the Electrical, Electronics and Computer Engineering Fields

The Institute of Electrical and Electronics Engineers (IEEE)   212-705-7870
IEEE Headquarters
345 East 47th Street
New York, NY 10017

This precollege guidance brochure gives students an idea of job opportunities in the electrical, electronics, and computer fields. It also discusses which courses should be emphasized, particularly in the areas of math and science. Single copies of the 11-page publication are available free of charge.

Computer Careers: Putting the World at Your Fingertips

CBEMA—Careers   202-737-8888
311 First Street, NW, Suite 500
Washington, DC 20001

Updated approximately once a year, this brochure, written for children, covers about 20 different careers. Examples are: programmer, designer, information manager, technical writer, computer operator, and repair technician. For each job it gives a description, educational requirements, and a short interview with someone who holds the position. The 20-page publication is available for $3.

Health Careers and Guidebook

Employment and Training Administration (ETA)   202-376-6289
U.S. Employment Service
U.S. Department of Labor
Washington, DC 20213

This publication offers an overview of the health field, along with a brief picture of what is happening in the field today and basic facts about the industry. It discusses new and changing job opportunities and provides information for career planning. Individual career descriptions for more than 100 occupations and a reference list of 150 health organizations that provide career information are also included. The 220-page publication (#029-000-00343-2) is available for $7.50 from the Superintendent of Documents, U.S. Government Printing Office, Washington, DC 20402, 202-783-3238. It can also be obtained from your local ETA office, which you can locate by checking the "U.S. Government" listing in your telephone book.

Peterson's Guides

PO Box 2123   800-225-0261—Book Order Department
Princeton, NJ 08540   609-924-5338—in New Jersey, Alaska, Hawaii, and outside the U.S.

This company publishes several guides to high-technology careers:

● *Engineering, Science and Computer Jobs 1985*—updated annually, this publication links academic specialties with specific job openings. It contains information from about 1,000 manufacturing, research, consulting, and government organizations that are currently hiring technical graduates. Fields included are the various disciplines of engineering, computer science, and physical and biological sciences. The 685-page publication (no. 2480) is available for $14.95.

● *The Engineering/High Tech Students' Handbook: Preparing for the Careers of the Future*—this publication defines the new careers in engineering and high technology, and gives full information about how to prepare for them. It contains information about more than 40 academic majors that are useful for finding employment in technical areas. The 160-page publication is available for $8.95.

Superintendent of Documents

U.S. Government Printing Office (GPO)   202-783-3238
Washington, DC 20402
(or check your telephone book under "U.S. Government" for a local listing)

The following government-produced publications are available from GPO. Many are also available in libraries, especially at schools and guidance centers. Most are updated on a regular basis, either yearly or quarterly.

● *Dictionary of Occupational Titles*—this comprehensive directory lists, defines, and describes all the occupations currently in existence in the U.S. Each occupational entry also includes the educational requirements and work experience generally needed to obtain a job. The publication was prepared by the Department of Labor's Employment and Training Administration. This publication (# 029-013-0079-9) is available for $23.

● *Exploring Careers*—describes the qualifications an individual must have to be successful in 250 different occupations. The publication (# 029-001-02224-7) is available for $11.

● *Occupational Outlook Quarterly*—published four times a year, this publication covers employment trends, training and educational opportunities, new and emerging jobs, and salary trends. Examples of recent articles are: "Careers in Commodity Futures Trading"; "Fringe Around the Paycheck: Employee Benefits"; "The Job Hunter's Guide to the Library"; and "A Checklist for Going into Business." This publication is available at most libraries and schools. Single copies can be purchased for $4.50 from the Superintendent of Documents, U.S. Goverment Printing Office, Washington, DC 20402. The annual subscription rate is $9.

● *Occupational Projections and Training Data, 1984 Edition*—provides information about job openings in 240 occupations. Supply and demand data are provided for each job category, including projected employment for 1990, the percentage change from 1978 to 1990, the average annual openings for 1978 to 1990, and the number of people being trained each year for these jobs. The publication (# 029-001-02804-1) costs $6.

• *Occupations Outlook Handbook*—updated yearly, this publication describes nearly 300 occupations and 25 industries. For each job, information is provided about the nature of the work, working conditions, training and qualifications required, employment outlook, earnings, related occupations, and sources of additional information. For each industry, the *Handbook* describes the nature and location of the industry, occupations in the industry, working conditions, employment outlook, earnings, and sources of additional information. This publication (# 029-001-02325-1) is available in most libraries. It can be purchased for $8.50, paperback, and $10, hardback.

• *1984 U.S. Industrial Outlook*—describes the outlook for 200 industries in the U.S. Basic information and a financial profile are provided for each industry, along with growth projections through 1990. The cost of this publication (# 003-008-00190-4) is $14.

Additional Experts and Resources

Associations

Many professional associations publish surveys and informational booklets about career opportunities and employment trends in their respective fields. Some organizations also offer employment referral services, and many run help-wanted ads in their association newsletter. Staff members are often available to answer questions about career opportunities, and they can refer you to a member who may be able to give you an insider's view of the field. For further information, consult the major associations listed under topics of interest to you in this book.

Federal Information Center
Check you telephone book under "U.S. Government, Federal Information Center" for a local number, or call or write:

Federal Information Center   202-783-3238
General Service Administration (GSA)
7th and D Streets, SW
Washington, DC 20405

The federal government sponsors a network of information centers throughout the U.S. to help citizens find their way through the federal bureaucracy. By calling, writing, or visiting your local Federal Information Center you can reach staff who will direct you to an expert in the government, with information about a particular career or training you seek.

Libraries, Career Centers and Counselors

The reference librarian at your local library can help you identify written materials about fields of interest to you. He or she may also be able to guide you to career centers and guidance counselors in your area.

Local Sources

Personnel departments are one of the best information sources for determining the outlook of a specific occupation in a given geographical area. Check with companies, the government, or organizations for which you would like to work.

State Occupational Information Coordinating Committees

Each state maintains a committee to provide information about employment opportunities within the state. Call or write your local state government office.

CAT Scanners

See: Cancer-Related Technology; Medical Imaging

Cellular Radio

See: Mobile Phones/Cellular Radio

Ceramics

Organizations and Government Agencies

American Ceramics Society, Inc.
65 Ceramics Drive   614-268-8645
Columbus, OH 43214

This is a professional society of scientists, engineers, plant operators, and others interested in ceramics systems. Its purpose is to disseminate scientific and technical information through its publications and technical meetings. The Society sponsors more than 130 technical meetings annually. Its staff can answer technical inquiries. The Society has a library which contains bound volumes, periodical subscriptions, as well as reports, brochures, booklets, and catalogs. Duplication services of noncopyrighted material are provided for a fee. The Society also has a data base that covers areas such as ceramics, glass, refractories, porcelain enamels, cements, composites, whitewares, nuclear and electronic ceramics, and instrumentation for high-temperature reaction. The staff can provide limited reference on-line literature searching, as well as referral services.

Inorganic Materials Division

National Bureau of Standards (NBS)   301-921-3181—Roger Rensberger, Public Information Office
Department of Commerce   301-921-2318—Norma Redstone, Technical Information and Publications
Washington, DC 20234

Staff members in the Division provide industry, governmental agencies, and scientific organizations with data, measurement methods, standards, reference materials, and fundamental concepts related to processing, structure, chemistry, physics properties, and performance of ceramics, glasses, and other inorganic materials. The Division pursues studies of inorganic materials to foster their safe, efficient, and economic use, and to address the national concern for productivity, critical materials, and environmental quality. Cooperative programs are conducted with the American Ceramics Society at the Phrase Diagrams for Ceramists Data Center. Those seeking in-depth technical advice should contact the Public Information Division Office.

Metals and Ceramics Information Center

Battelle Memorial Institute   614-424-5000—General Information
505 King Avenue   614-424-6376—Technical Inquiries Service
Columbus, OH 43201

The Center is sponsored by the Department of Defense, and its mission is to provide technical assistance and information about materials within the Center's scope, with emphasis on applications to the defense community. The main areas of interest with respect to ceramics at the Center include borides, carbides, carbon/graphite, nitrides, oxides, sulfides, silicides, intermetallics, selected glasses and glass ceramics, and other materials mutually agreed upon by Battelle and the government. Battelle has an on-line data base that contains technical journals, reports on government-sponsored research, patents, trade literature, and technical books. The data base can be accessed by most terminals. For details about this service, contact the general information number listed above.

The Center for Ceramics Research

College of Engineering   201-932-2724—Dr. John B. Wachtman, Center Director
Rutgers University
PO Box 909
Piscataway, NJ 08854

A university-industry cooperative program, the Center is funded through a grant from the National Science Foundation and from industrial membership fees. Research areas of interest include sol-gel technology; powder processing; strength, toughness, and microstructuring; corrosion of ceramics; electronic ceramics; surfaces; nonoxide ceramics; rheology; and thin ceramics films. The Center's main purpose is to perform basic generic research selected to support the interests of its sponsoring members. A packet of information describing the Center's activities, research, and personnel is available. The packet contains a ceramics bibliography developed by the Center's scientists.

Chemotherapy

See: Cancer-Related Technology

Chips and Integrated Circuits

See also: Microelectronics; Semiconductors

Organizations and Government Agencies

Center for Integrated Systems (CIS)
Stanford University   415-497-4212
Stanford, CA 94305
   A teaching and research institution on the Stanford campus, CIS is primarily supported by a group of high-technology companies, as well as the U.S. government and the University. It represents an integration of computer science, information science, and physical science. Research is being conducted on every aspect of semiconductor/computer systems relationships, which will lead to state-of-the-art design and fabrication of VLSI chips. Questions are answered on a limited basis. Technical reports, as well as publications about CIS and Stanford's Electrical Engineering Department, are available upon request.

Computer Science Option
California Institute of Technology (Caltech)   818-356-6704
Computer Science Department
Pasadena, CA 91125
   Caltech's Computer Science Option is involved in the research and study of chips and integrated circuits. Information can be obtained from the library at the address above. A current listing, *Technical Memos and Technical Reports*, is available free upon request. The reports vary in price from $2 to $10.

Data Quest
1290 Ridder Park Drive   408-971-9000
San Jose, CA 95131
   Data Quest is a subsidiary of Dun & Bradstreet, which provides consulting services regarding peripherals, information systems, design and manufacturing automation, and semiconductors. The Semiconductor User Information Service provides

information and analyses regarding the industry's markets, products, technologies, and strategies. The Information Service is useful to those who buy and use semiconductors, and the fee charged depends on the amount of information needed. This service includes a continually updated data base, several newsletters dealing with the latest developments, and inquiry privileges. Conferences are sponsored annually and are open to the public.

Duke University

Department of Electrical Engineering   919-684-3123—Dr. Peter Marinos
Durham, NC 27706

The current areas of research at Duke include computer-aided design, integrated circuit design and fabrication, solid state devices and materials, and VLSI circuit design. All inquiries should be directed to Dr. Marinos, who can refer you to other information sources if necessary.

Electronics Research Laboratory (ERL)

University of California   415-642-2302
341 Corey Hall
Berkeley, CA 94720

Extensive research in computer-aided design of integrated circuits, particularly circuit design and process development, is being conducted at the laboratory at the University of California at Berkeley. Work is also being done on design MOS and bipolar integrated circuits, and in device research and modeling. Faculty and graduate students can answer your questions on a limited basis, or you can order an annually updated listing of the *ERL Memos*. The listing includes abstracts of the technical reports, which are available for a minimal cost.

IEEE Electron Devices Society

Institute of Electrical and Electronics Engineers   212-705-7890—Technical Affairs
345 East 47th Street   212-705-7866—Public Affairs
New York, NY 10017

The Society is an organization of electrical engineers, physicists, chemists, and metallurgists from industry, government, and universities. Its area of interest is electronic and ion devices, which include electron tubes, solid state and quantum devices, energy sources and converters, and other related technical aspects. The Society seeks the advancement of research, development, design, manufacture, materials, technology, and application of electron devices. It sponsors several conferences, including the annual IEEE International Electron Devices Meeting, after which all proceedings are published and made available for sale. The Device Research Conference is also held annually, and the abstracts are then published in *IEEE Transactions on Electron Devices*. The president of the Society is available to answer your questions. The Society publishes the following:

● *IEEE Transactions on Electron Devices*—a monthly publication, which includes information on the theory, design, and performance of electron devices.

Subject areas include electron tubes, solid state devices, integrated electron devices, energy sources, power devices, displays, and device reliability. Issues average 150 pages, and an annual subscription is $85 for nonmembers.

• *Electron Device Letters*—a monthly publication of short communications on new results in the electron devices field. Each issue has approximately 40 pages, and an annual subscription is $72 for nonmembers.

National Research and Resource Facility for Submicron Structures (NRRFSS)

Phillips Hall, Knight Laboratory   607-256-2329
Cornell University
Ithaca, NY 14853

NRRFSS is a national research effort by government, academic, and industrial sponsors. The facility conducts and promotes research to advance submicrometer fabrication and circuit fabrications. One of NRRFSS's research projects is the study of submicron silicon process and device technology, including the feasibility of fabrication, the physical properties of individual devices, and integrated circuits with minimum feature sizes. The facility is available as a resource for individuals in the government and in university and industrial laboratories across the country. The staff can give you general information about the research being conducted and can send you publications, including technical reports.

Semiconductor Industry Association (SIA)

4320 Stevens Creek Boulevard, Suite 275   408-246-1181
San Jose, CA 95129

This trade association of U.S.-based semiconductor manufacturing firms deals with public policy issues, health and safety programs, and international trade practices relevant to the industry. It sponsors several conferences each year and cosponsors the World Semiconductor Trade Statistics program. It publishes numerous books, statistical periodicals, and newsletters, and will send you a free flyer that describes them. Some of these publications are:

• *SIA Yearbook/Directory*—an annual comprehensive overview of the U.S. semiconductor industry. It includes industry statistics, outlook information, and public policy issues. The directory section contains names, addresses, phone numbers, senior executives, and key products of U.S. semiconductor firms. The cost of the 60-page publication is $50.

• *SIA Circuit*—a quarterly newsletter that covers association and industry activities, upcoming events, pending legislation, and general industry news. The 4-page newsletter is available free of charge.

• *World Semiconductor Trade Statistics*—a four-part subscription series that includes: (1) *The Blue Book*—a detailed monthly publication of bookings and billings for major semiconductor product families and key products; (2) *The Flash Report*—a monthly summary of leading indicators; (3) *Complete Forecast*—a semi-annual historical summary, and three-year WSTS Committee projection, detailed by product family; and (4) *Historical Data*—a complete set of year-end summaries of semiconductor trade activity by U.S.- and European-based manufacturers. An annual subscription to this series is $1,200.

Semiconductor Research Corporation

Manager, Information Services   919-549-9333
PO Box 12053
Research Triangle Park, NC 27709

This nonprofit cooperative research organization is supported by companies that manufacture or use integrated circuits. It plans and implements developmental research, which is contracted through local universities to advance the technology of integrated circuits. Research is conducted in three primary areas: microstructure sciences, design sciences, and manufacturing sciences. Specific areas of study include: CAD systems, packaging, semiconductor processes, reliability, development of simulations and modeling, manufacture automation, and materials and phenomena. Some technical literature can be obtained from the universities doing the research, and there is also literature available upon request from the organization.

Semiconductor Technology Program

U.S. Department of Commerce   301-921-3357—Robert Scace, Deputy Director
National Bureau of Standards (NBS)   301-921-3786—Frank Oethinger, Semiconductor
Washington, DC 20234   Materials & Processes Division
            301-921-3541—Kenneth Galloway, Semiconductor Devices & Circuits Division

The Semiconductor Program at the NBS works on measurement-related topics for the semiconductor industry, its suppliers, and its customers in other industries and in the government. The Semiconductor Materials and Processes Division develops measurement methods, data, physical standards, and models and theory characterizing semiconductor materials and processes. This Division develops standardized test methods and standard reference materials, and maintains facilities for material processing and solid state device fabrication. The Semiconductor Devices and Circuits Division develops and evaluates measurement methods, data, reference artifacts, models, and theory for the design, characterization, and performance assurance of electron devices and solid state circuits. It also disseminates and fosters application of results and assists in the development of standardized test methods for electron devices and solid state circuits. Staff members within these two Divisions are available to answer questions. Workshops and seminars are sponsored, and topical papers are available for sale. A complete listing of these papers is available in the free *NBS List of Publications #72: Semiconductor Measurement Technology*.

University of Illinois

155 Electrical Engineering Building   217-333-3097
1406 West Green Street
Urbana, IL 61801

The Engineering Department of the University of Illinois is conducting research in compound semiconductors, VLSI design, and related work on physics of small devices. Two conferences are sponsored each year: the Annual Review of Electronics, which presents research papers by the faculty; and the Physical Affiliates Program, which involves a presentation of research papers by students and industry people. The staff is available to answer your questions on a limited basis.

VLSI Technical Committee

IEEE Computer Society   301-589-8142
1109 Spring Street, Suite 300
Silver Spring, MD 20910

The VLSI Committee addresses the interaction of the semiconductor process and system design in VLSI. Emphasis is on integrating the design, fabrication, application, and business aspects of VLSI from both hardware and software points of view. This Committee sponsors the International Conference on Circuits and Computers, as well as special sessions and workshops for other conferences of the Computer Society.

Publications

Advanced Microprocessors

IEEE Service Center   201-981-0060
445 Hoes Lane
Piscataway, NJ 08854

This book, part of a series called the IEEE Press Books, is a collection of selected reprints that document the history of microprocessors as well as current trends. There are several sections: overview, 16-bit microprocessors, 32-bit microprocessors, performance evaluation, related technologies, and system issues. The 368-page publications costs $41.50.

Electronic Design Magazine

Hayden Publishing Co., Inc.   201-393-6051
Mulholland Drive
Hasbrouck Heights, NY 07604

Electronic Design Magazine is a biweekly trade magazine for electronic design engineers and directors of design projects. It provides the latest information about new and emerging developments in electronic design. Articles cover chips and applications of chips, test instruments, data communications, CAD design, robotics, and other technical areas. Subscription is free of charge to those who qualify, and $45 for others.

Electronics Week

McGraw-Hill, Inc.   212-512-2000
1221 Avenue of the Americas
New York, NY 10020

This weekly magazine focuses on original equipment manufacturing. Articles deal with technology and business, including design, manufacturing, application of chips and integrated circuits, and sophisticated aspects of computer-aided design. The average issue has 140-160 pages, and an annual subscription is $32.

Semiconductors/Integrated Circuits

Communications Publishing Group   617-787-0138
1505 Commonwealth Avenue, Suite 32
Boston, MA 02135

This monthly directory presents newly published U.S. and foreign patents related to semiconductor and integrated circuit technology. It consists of abstracts, bibliographic data, and drawings taken from patent documents. It varies between 16 and 24 pages an issue, and the annual subscription rate is $257.

Solid State Technology

Cowan Publications   516-883-6200
14 Vanderventer Avenue
Port Washington, NY 11050

This publication is directed to the semiconductor industry manufacturers of solid state devices and integrated circuits. Articles deal with fabrication processes and product technologies. Each issue averages 200 pages, and is available free of charge to qualified subscribers.

VLSI Design Magazine

CMP Publications   516-365-4600
111 East Shore Road
Manhasset, NY 11030

This monthly technical industry journal is geared to the electronic systems designer and engineering manager. It provides current information on the development and application of custom and semicustom integrated circuits. It features technical articles as well as product coverage and news on computer-aided engineering and computer-aided design tools, design methods, IC technologies, and training and education in IC design. Each issue averages 130 pages, and is available free of charge to qualified industry subscribers.

Additional Experts and Resources

Robert Burger, Ph.D., Acting Director

Manufacturing Sciences Program   919-549-9333
Semiconductor Research Corporation
PO Box 12053
Research Triangle Park, NC 27709

Dr. Burger is a physicist who administers the Manufacturing Sciences Program at this nonprofit cooperative research corporation. His specialty is integrated processing of integrated circuits and assurance reliability. He is an expert in VLSI packaging, metrology, and materials control, as well as manufacturing processes and equipment.

David L. Carter, Ph.D.
AT&T Bell Labs, Room 3B-319   201-582-6826
600 Mountain Avenue
Murray Hill, NJ

    A physicist by training, Dr. Carter has worked in microelectronics and applied physics. His work at Bell Labs involves following the technology of integrated circuits. He is editor of *IEEE's Transaction of Electron Devices* (see description above). In addition to being a good resource for developments in the field, Dr. Carter can also refer you to other information sources.

Ralph Cavin, Ph.D
Director, Design Sciences Program   919-541-7084
Semiconductor Research Corporation
300 Park Avenue, Suite 215
PO Box 120533
Research Triangle Park, NC 27709

    Dr. Cavin is an electrical engineer specializing in VLSI algorithms and architectures. He is involved in the development of VLSI realization for special computing applications such as artificial intelligence, signal processing, control devices, and communications devices. Dr. Cavin can respond to inquiries on a limited basis. He has written extensively and can refer you to the journals and proceedings in which his work has been published.

Dr. William Holton
Program Director, Microstructure Sciences Program   919-541-7495
Semiconductor Research Corporation
PO Box 12053
Research Triangle Park, NC 27709

    Dr. Holton is a physicist with specialties in solid state physics and design technology for CAD systems for IC design. As Program Director of the Microstructure Sciences Program, he is involved in device physics, materials systems, process technology, and device and circuit modeling. Dr. Holton has written extensively, and can refer you to other sources of information.

CIM

See: Computer-Integrated Manufacturing

Cloning

See: Biotechnology; Reproductive Technology

Coastal Engineering

See: Marine Technology

Codes, Deciphering

See: Cryptology

Commercial Product Development

See: Technology Transfer

Compact Discs

Organizations and Government Agencies

Compact Disc Group
C/o Exposé, Inc.   800-872-5565
919 3rd Avenue   212-355-0011—in New York
New York, NY 10022
    The Compact Disc Group is a nonprofit trade organization comprised of manufacturers of compact disc hardware, record companies, accessory manufacturers, publishers, and trade associations. CDG's primary goal is to further consumer aware-

ness, understanding, and acceptance of the compact disc system. Consumers can contact the Group's information hotline for answers to questions they may have about compact discs and the technology involved.

A brochure listing all current compact disc titles is available from CDG. In order to receive a copy, send your request with a self-addressed, stamped envelope to the above address. The organization also publishes:

• *Compact Disc Group Newsletter*—a quarterly that contains general and retail information about compact discs. Each issue runs six pages and is available free of charge. Ask to be placed on the mailing list and include a self-addressed stamped envelope to receive a copy.

Consumer Electronics Group

Electronic Industries Association   202-457-4919
2001 Eye Street, NW
Washington, DC 20006

This trade association serves the entire spectrum of electronic manufacturers, which represents more than 1,000 participating companies. It sponsors two annual conferences, with workshops, seminars, and an exhibit. Questions can be answered by the technical staff. There are several publications that pertain to the compact disc system, as well as statistical information on production and sales.

Denon America, Inc.

PO Box 1139   201-575-7810
West Caldwell, NJ 07007

This manufacturer of compact discs was the first company to introduce the Compact Disc ROM (Read Only Memory). The staff can provide you with general information about compact discs, as well as explain how to prepare material for compact disc transformation. Some technical literature is available and can be sent to you upon request.

Polygram Records, Inc.

8335 West Sunset Boulevard, 3rd Floor   213-656-3003
Los Angeles, CA 90069

Polygram Records, Inc. was among the first to develop compact disc software. It is both a manufacturer and a wholesale distributor of compact discs, and has the world's largest compact disc plant, located in Germany. The staff in Los Angeles can answer questions or refer you to another information source.

RCA Records

1133 Avenue of the Americas   212-930-4000
New York, NY 10036

RCA is involved with the entire spectrum of digital audio compact disc systems, with the exception of manufacturing. The company does everything from digital recordings to digital mastering. The Compact Disc Club is operated by RCA Direct

Marketing, Inc. To receive an enrollment brochure and membership information write to RCA at PO Box 91412, Indianapolis, IN 46291.

Sony Corporation of America

Sony Drive   201-930-1000—Mark Finer, Products Communications Manager
Park Ridge, NJ 07656

Sony is the co-inventor of the compact digital format and offers an extensive product line of compact disc players, including auto and portable compact disc players. Mark Finer can answer your questions about audio digital technology, as well as send you literature about Sony's products. Sony also sponsors the Sony Digital Audio Club, which serves to educate its 11,000 members about digital audio technology. The membership fee is $15 a year, which includes a copy of *The Sony Book of Digital Audio Technology*, a reference manual about compact discs and digital recordings. This book can be ordered from Sony without membership for $11.95. Another publication that is available through the Club is *The Sony Pulse*, a quarterly newsletter that contains articles about new product developments and emerging digital technology. The newsletter is free to Club members.

Compilers

See: Programming Languages (Computer)

Computer and Technology Law

See also: Technology Transfer

Organizations and Government Agencies

Association for Data Processing Service Organization (ADAPSO)

1300 North 17th Street, Suite 300   703-522-5055
Arlington, VA 22209

A trade association for the computer services industry, ADAPSO focuses on computer software, covering timesharing, systems integration, microcomputer software, and customized programs. The organization often represents the industry before the U.S. Congress, government agencies, and the courts. ADAPSO's three staff attorneys specialize in computer law, and they are available to answer questions and make referrals. The Association publishes:

• *ADAPSO Newsletter*—a biweekly publication providing information about computer law issues. The six-page publication is available free of charge.

Center for Computer Law

1112 Ocean Drive, Suite 201   213-372-0198
Manhattan Beach, CA 90266

International in scope, the Center is an excellent resource for information on computer law and crime. It provides research services to government agencies and corporations on a contractual basis. Mike Scott, an attorney specializing in computer law, is available to provide assistance, answer questions, and make referrals. The Center publishes:

• *Computer/Law Journal*—a quarterly law review that contains articles written by noted authorities in the computer law field. Since it is edited by students at the University of Southern California Law Center, it also contains articles and case studies written by USC law students. The 200-page journal is available for $66 a year.

Computer Law Division

American Bar Association (ABA)   312-988-5000—Susan J. Walkowski,
750 North Lake Shore Drive   Staff Liaison for Section of Science and Technology
Chicago, IL 60611

ABA's Computer Law Division consists of several committees, each dealing with a specific aspect of computer law. Ms. Walkowski can refer you to committee members specializing in the area of computer law you are interested in. The attorneys will answer information inquiries and refer you to an appropriate lawyer in your geographical area. The Computer Law Division committees are: Computer Crime; Contracting for Computers; Data Communication Network; Electronic Funds Transfer; International Data Flow; National Market Systems; Original Equipment Markets; Professionalism of Computer Specialists; Proprietary Rights in Software; Software Licensing Practices; Taxation of Computer Systems and Products; Tort Liability for Use of Computer Systems and Services; and Use of Computer-Produced Data. ABA's Science and Technology Section publishes:

• *Jurimetrics, Journal of Law, Science, and Technology*—a quarterly providing legal analysis of questions relating to high-tech issues. The journal includes articles written primarily by attorneys. ABA members receive a complimentary subscription; others can subscribe for $26 a year.

Computer Law Institute

University of Southern California Law Center   213-743-2582
Room 105, University Park
Los Angeles, CA 20089

The Institute is a two-day conference sponsored each spring by the University of Southern California Law Center. The conference features more than 20 speakers, covering a wide variety of topics on computer law. The Center also sponsors joint conferences with Harcourt Brace Jovanovich, Inc. All conferences are open to the public, and brochures describing them are available from the Center. The Center publishes:

• *Computer Law Institute Annual Program*—this includes papers presented at the annual Computer Law Institute. Examples of topics covered include: proprietary rights, antitrust and international issues, structuring distribution arrangements, procurement, and financing high-technology companies. Available in a three-ring binder, the 500-page volume costs $45 to $55.

Rutgers Law School
15 Washington Street   201-648-5549
Newark, NJ 07102

Focusing on high technology and the law, Rutgers Law School covers legal issues relating to subjects such as computers, technology transfer, biotechnology, communications, and aerospace. The School plans to begin offering annual seminars focusing on specific high-tech topics. Staff members are available to answer questions and make referrals. The School publishes:

• *Rutgers Computer and Technology Law Journal*—published three times a year, the *Journal* covers the full scope of high tech and the law, including computers, biotechnology, communications, aerospace, and much more. Articles have appeared on software, patent systems and copyrights, product liability, electronics, and computer bank transmissions. The 200-page *Journal* is available for $40 a year.

Publications

Computer Negotiations Report
Sunscape International   305-898-1700
1513 East Livingston Street
Orlando, FL 32803

Published monthly, the *Report* covers all aspects of contract negotiating for data processing. Information is provided on everything from consulting to contract provisions, with advice on litigation and arbitration techniques. Focused mainly on computer negotiations, the publication also covers procurement of large equipment. Annual subscription fee is $265.

Computer Law Reporter
1519 Connecticut Avenue, NW, Suite 200   202-462-5755
Washington, DC 20036

This bimonthly publication covers a wide scope of issues pertaining to computer law, including patents; copyrights; trademarks; privacy; and criminal, civil, and contract law. It includes articles about recent cases, decisions, and developments that are of interest to computer law litigators. The publication, l50-200 pages long, is available for $475 a year.

Law and Technology Press
1112 Ocean Drive, Suite 201
Manhattan Beach, CA 90266

Law and Technology Press publishes:

• *Scott Report*—this monthly publication provides legal analysis and overviews of the latest developments in computer and telecommunications law. Industry and legal trends are provided for both fields. Topics have included: software protection, contracting, privacy, international data regulations, and legislative developments. Each issue contains book reviews and a calendar of events. The 16-page publication is available for $176 a year.

• *Software Protection*—a monthly publication covering the legal aspects of protecting computer software. Analytical articles are provided about new developments and cases, with an emphasis on patents, copyrights, trade secrets, trademarks, and contracting. Each issue also contains book reviews and a calendar of events. Annual subscription is $72.

Computer and Technology Law Journal
University of Santa Clara School of Law   408-554-4077
Santa Clara, CA 95053

Published twice a year, the *Journal* provides information about the often changing legal decisions in high-technology and computer law. Plans are in effect to expand the *Journal* to a quarterly.

Data Bases

AMBAR
American Bar Association   800-621-6159
750 North Lake Shore Drive   312-988-5158
Chicago, IL 60611

AMBAR is a very comprehensive data base, covering legal areas of interest to the American Bar Association such as: aerospace law, communications law, life sciences and physical sciences law, technology transfer law, and computer law. AMBAR contains abstracts of publications and coverage of ABA committee activities. There are more than 11,000 documents in the system, which is updated quarterly. Searching is done by key words. AMBAR is hooked up with WESTLAW and LEXIN, and is also available on the communicating law network ABA/NET.

Additional Experts and Resources

John Lautsch
Computer Law Division   714-432-5623
American Bar Association (ABA)
1730 Adams Avenue
Costa Mesa, CA 92626

Chairman of ABA's Computer Law Division, Mr. Lautsch is familiar with experts in the major areas of computer law. Mr. Lautsch's primary interest, as an attorney, is the computer software industry. He frequently lectures about software business law and microcomputer usage.

The following university law schools are involved in aspects of the law and high technology, particularly computer law:

Arizona State University College of Law   **Franklin Pierce Law Center**
Tempe, AZ 85287   2 White Street
602-965-6181   Concord, NH 03301
                                                     603-228-1541

Computer Animation

See also: Holography

Organizations and Government Agencies

Computer Graphics Center
Lawrence Livermore National Laboratory (LLNL)   415-422-1100
PO Box 808
Livermore, CA 94550

Using a system of computers including four Cray-1's, a CDC 7600, and several other supercomputers, the Lab's Computer Graphics Center services thousands of users within the lab compounds. The main use for the lab's computers is in modeling complex physical properties and processes. The Center does research and development in computer architecture, language, and algorithms, with the Lab's computer scientists writing the programs and explanations needed. Information on the software utilized and the applications of graphics in the laboratory environment may be obtained by writing the Lab's Computer Graphics Center with specific requests.

Computer Graphics Lab, Inc.

405 Lexington Avenue
New York, NY 10174

212-557-5130—Sales Department
516-686-7833—Development Office

This commercial sales venture markets the Images II machinery developed by the New York Institute of Technology Computer Graphics Laboratory. Base price of the processor is about $50,000, with extensive optional hardware and software available. Images II is a computer painting system, capable of sophisticated graphics and 3-D animation for both slides and film. Used primarily for drawing advertising layouts and business graphics, Images II is also used in film recording. The sales staff at the New York address can discuss the capabilities of their equipment, while the staff at the development office can give specifications on the equipment itself.

Computer Graphics Laboratory

New York Institute of Technology
PO Box 170
Old Westbury, NY 11568

516-686-7644

This division of the New York Institute of Technology deals with many of the recent developments in the field of computer graphics. The Lab supplies computer assisted animation systems, such as cartoons, and two-dimensional characters as a rapid and economical method of animation. It also deals with computer generation of animation, a three-dimensional modeling of images as created by the Images II equipment from the aspects of light, design, and rendering. The staff at the Lab can refer you to specialists and publications in the computer animation field, as well as to commercial vendors and producers of computer animation. Many of the technical reports have appeared in the journals and presentations of the annual SIGGRAPH proceedings. (See entry below.)

Cranston/Csuri Productions, Inc.

1501 Neil Avenue
Columbus, OH 43201

614-421-2000

Formed by Professor Charles Csuri of Ohio State University's Computer Graphics Center, this commercial production company specializes in three-dimensional computer-generated graphics. Using graduates and students from the adjoining Ohio State Program, Cranston/Csuri does advertising, special films, and other imagery applications of computer animation. Professor Csuri will make referrals to other animation specialists, as well as address specific application requirements.

Digital Effects, Inc.

321 West 44th Street
New York, NY 10036

212-581-7760

This commercial producer of computer graphics is used by many of the major networks and advertising agencies in the production of computer-generated animation. Specializing in live action imagery, Digital Effects relies heavily on Harris and IBM equipment to produce animation. The company also does extensive work in the area of optical deposits. For examples of work done, general information on animation capabilities, or price quotations for specific job requirements, write or call the sales division within Digital Effects.

Lucas Film's Computer Graphics Project

Box 2009   415-457-5282
San Rafael, CA 94912

This special project of the Lucas Film company was started in 1980 as an effort by Lucas to provide competent technicians to aid moviemakers. Directed by Fred Catmull, formerly of the New York Institute of Technology, the project encompasses four separate areas of examination: (1) computer animation; (2) digital audio; (3) video editing; and (4) computer games. In addition to working closely with filmmakers on such blockbusters as *Star Wars*, *Indiana Jones*, and *Star Trek II*, the project has also been successful in developing new technology to meet the demands of animators. A recent project, Pixar, involved the development of a machine for Industrial Light and Magic for computer-generated graphics. Lucas Film contributes heavily to the annual conference of the ACM Special Interest Group on Computer Graphics, SIGGRAPH. In addition, many of its researchers and technicians sponsor lectures and contribute articles to periodicals such as *Scientific American*. A spokesman for the project, Thomas Porter, will make referrals to graphic specialists with whom he has been associated in the field of special effects animation.

The Mathematical Applications Group, Inc.

3 Westchester Plaza   914-592-4646
Elmsford, NY 10523

This commercial vendor of computer services offers many graphic applications for public consumption. In addition to computer animation, the company also specializes in CAD/CAM applications, design graphics, direct marketing through computers, and slide and film presentations. The staff members are well versed in the commercial applications of computer graphics and also the accompanying technology. They will try to answer queries and make referrals to specialists.

Ohio State Computer Graphics Center

Ohio State University   614-422-3416
Cranston Center
1501 Neil Avenue
Columbus, OH 43201

Ohio State offers a graduate program in computer animation through a combination of its computer sciences and art education departments. The Center's director, Professor Charles Csuri, will provide information about specific courses of study, as well as make referrals to other specialists in the computer animation field. He will also provide bibliographies and reading lists on current technology and applications in the science of computer animation.

Pacific Data Images
550 Weddell Drive, Suite 3   408-745-6755
Sunnyvale, CA 94089

    This commercial production company specializes in computer-generated animation for advertising and broadcasting media. In addition to promotional packages designed for specific companies or products, it does the opening graphics for many local network television productions. An in-house research and development division is constantly creating and improving software applications for computer animation. Much of the work has been presented at the Association for Computer Machinery's SIGGRAPH conferences on computer graphics, and is available in the published proceedings of this group. In addition, the sales staff will make referrals to other production companies and specialists in the field of computer animation.

SIGGRAPH (Special Interest Group on Computer Graphics)
Association for Computer Machinery (ACM)   212-869-7440
11 West 42nd Street
New York, NY 10036

    This special-interest group of the ACM serves as a forum for the promotion and dissemination of current computer graphics research, technologies, and applications. State-of-the-art hardware and software subjects encompass animation, business, CAD/CAM, cartography, fine arts, geometric modeling, image processing, and image synthesis and simulation. SIGGRAPH sponsors the largest annual conference within ACM, including technical presentations, tutorials and seminars, a large exhibition of hardware and software products, and extensive film and video presentations. The latest computer graphics achievements are communicated to membership through publications, slide sets, fiche, and video tapes, listings of which are available upon request. Membership in SIGGRAPH is available to non-ACM members for $45, which includes subscription to the newsletter:

    • *Computer Graphics*—a quarterly report of ACM SIGGRAPH. Each issue, about 60 pages, contains contributions from members on current research in computer graphics, including reference citings to other publicized material, on topics ranging from animation to image synthesis. There is also a section on current activities containing minutes from SIGGRAPH meetings, and an announcement section containing news of meetings and upcoming events, reviews, and editorials. The subscription rate is $30 a year for non-SIGGRAPH members.

Publications

The S. Klein Newsletter on Computer Graphics
Technology and Business Communications, Inc.   617-443-4671
730 Boston Post Road
Sudbury, MA 01776

    This newsletter is published twice a month and covers a variety of topics in the computer graphics field. In addition to computer-aided design and computer-aided manufacturing, it emphasizes computer animation. Topics frequently studied include modeling, 3-D environments, image processing, animation rendering, and animation systems. The newsletter also contains information on related events and meetings, conference proceedings, and recent publications. It is about eight pages in length, except when double issues focus on a new development area; the annual subscription is $155.

Computer Architecture

See also: Computers (General); CAD/CAM

Organizations and Government Agencies

Architecture Technology Corporation (ATC)
PO Box 24344   612-935-2035—Dr. Harvey, Vice President
Minneapolis, MN 55424

ATC is involved in the architecture of frontend processors, computers, and equipment involved in networks. ATC sponsors seminars about personal computers, local networks, and new developments in communications. Dr. Freeman is an expert in computer architecture and is available to answer questions and make referrals. ATC publishes:

● *Localnetter*—a monthly publication covering local network news, product developments, standards, and financial news. Annual subscription is $300.

● *Token Perspectives*—a monthly publication providing news about IBM network efforts. Annual subscription is $312.

● *Localnetter Designers Handbook*—an annual publication describing all products relating to local networks. Information is provided about product use, effectiveness, and implementation. The 45-page handbook is available for $77.

● *Localnetter Update Service*—a quarterly publication providing in-depth reports about 80 companies involved in computer architecture and local networking. Each issue is approximately 800 pages, and the annual subscription is $1,500.

Center for Computer Systems Engineering
Systems Components Division   301-921-2705
National Bureau of Standards
Gaithersburg, MD 20899

This government agency is concerned with the measurement and prediction of various computer architectures, especially multiprocessor architectures. The staff in the research and development office can provide you with information about the latest research and development activities in the field.

Computer Architecture Technical Committee (CATC)

IEEE Computer Society   301-589-8142
1109 Spring Street, Suite 300
Silver Spring, MD 20910

A technical committee of IEEE, this group is involved in the research and development of integrated hardware, as well as the software design of general-purpose and special-purpose digital computers. It sponsors workshops on such topics as computer arithmetic and interconnection networks. CATC cosponsors, with ACM's SIGARCH, an annual International Symposium on Computer Architecture. The Committee publishes:

• *Computer Architecture Newsletter*—a quarterly newsletter containing technical papers submitted by Committee members, information about new developments in the field, and announcements of upcoming activities and conferences related to computer architecture. The 64-page publication is available free of charge.

SIGARCH

Association for Computing Machinery (ACM)   212-869-7440
PO Box 12115
Church Street Station
New York, NY 10249

A special-interest group within the computer organization ACM, SIGARCH is involved in the architecture of computer systems. The group is concerned with constituents and arrangements of the physical resources of the computer system; partitioning and organization of a processor; arithmetic; hierarchy of memory; addressing; control and sequencing; parallel and multiple stream organization; and concurrent execution of dependent procedures. SIGARCH publishes:

• *Computer Architecture News*—a quarterly newsletter containing articles and papers written by SIGARCH members, information about technical advances in computer architecture, and a calendar of technical events and conferences. Annual subscription is $20 for members and $26 for nonmembers.

• *International Symposium on Computer Architecture*—published annually, this contains the papers and proceedings of the yearly computer architecture conference sponsored by ACM and IEEE. It is available for $25.50 to members and for $45 to nonmembers.

Publications

Digital Equipment Corporation

Digital Press   617-663-4123
12 A Esquire Road
Billerica, MA 01862

This corporation publishes the following materials related to computer architecture:

• *Computer Programming and Architecture*—this two-part publication contains information about programming and architecture for VAX-11. It is geared toward students and hardware and software professionals. Part one, *User Architecture*, discusses basic concepts of computer programming in architecture. Part two, *Systems Architecture*, covers the more advanced aspects of computer architecture. The publication is available for $28.

- *VAX Architecture Handbook*—an introductory-level publication, this provides programming examples, data types, addressing modes, and native mode instruction sets. The 506-page publication is available for $15.
- *PDP-11 Architecture Handbook*—provides an overview of all PDP systems as well as programming examples. The 1983 edition is 272 pages long and is available for $15.
- *Microvax Handbook*—provides information about the new Microvax Digital System, and outlines its size and Q-VUS features. The 1984 500-page edition is available for $15.

Computer Crime

See also: Computer Security; Smuggling of High Technology

Organizations and Government Agencies

Bureau of Justice Statistics
U.S. Department of Justice
633 Indiana Avenue, NW
Washington, DC 20531

The Bureau of Justice Statistics acts as a clearing house for information on computer crime and security, and puts out five one-time publications dealing with these topics. The Bureau is an excellent source for legal and technical information. The five volumes are:

- *Computer Crime: Expert Witness Manual*—deals with issues relevant to the selection of witnesses to use in computer crime trials.
- *Computer Crime: Legislative Resource Manual* (Pub. #027-000-01135-7)—deals with issues relevant to the selection of witnesses to use in computer crime trials.
- *Computer Crime: Criminal Justice Resource Manual* (Pub. #027-000-00870-4)—describes different types of computer crime strategies that the prosecution and defense must consider. The 150-page publication is available for $10.
- *Computer Crime: Computer Security Techniques* (Pub. #027-000-01169-1)—describes different security techniques used to control computer crime. The 150-page publication is $7.50.
- *Computer Crime: Electronic Funds Transfer Systems and Crime* (Pub. #027-

000-01170-5)—describes EFT crimes and discusses available sources of data. The 150-page publication is $7.

These reports are available from Superintendent of Documents, U.S. Government Printing Office, Washington, DC 20402, 202-783-3238.

National Center for Computer Crime Data (NCCCD)

2700 North Cahuenga Boulevard, Suite 2113   213-850-0509—Jay Bloombecker
Los Angeles, CA 90068

NCCCD is a research organization dedicated solely to the study of computer crime. It serves as a clearing house for information about products, services, employment, training opportunities, books, and periodicals. The Center also maintains an expert index, which enables users to find investigators, prosecutors, and other experts to consult on computer crime problems. The following are published:

• *Computer Crime Law Reporter*—a compilation of texts of all existing and pending computer crime laws. The cost is $45.

• *Computer Crime Digest*—a monthly that covers news and analyses of the latest crimes, laws, security products, seminars, investigative strategies, and books. The cost is $120 a year.

Computer Security Institute (CSI)

43 Boston Post Road   617-845-5050
Northborough, MA 01532

CSI is a membership organization for those interested in computer security. Each year the Institute holds a conference, several seminars, and workshops. A "hotline service" provides assistance and referrals, but is available to members only. CSI publishes:

• *Computer Security: A Newsletter for Computer Professionals*—covers all aspects of computer security, including news, case histories, reading suggestions, and legal developments. It is available free to members.

• *Computer Security Journal*—a semiannual journal devoted solely to computer security, with new ideas and useful information about current products and packages. It covers such topics as risk analysis, disaster recovery, fraud prevention/detection, laws and legislation related to privacy and security, auditing techniques, and physical and environmental controls. It is available for $65 a year.

• *Computer Security Handbook*—features case studies, checklists, product information, samples and extracts from actual security policies and job descriptions, and hard-to-find material from the government and other sources. The *Handbook* is divided into ten sections, which discuss computer security, protecting the data center, software management, communications security, disaster recovery planning, auditing, and information sources. It costs $95.

Publications

Computer Crime

Lexington Books   617-862-6650
125 Spring Street
Lexington, MA 02173

This book, by August Bequai, explores the strengths and weaknesses of the existing legal apparatus for investigating and prosecuting computer crimes, and analyzes the federal and state laws that govern computer crime. The cost is $12.

Computer Crime Digest

Washington Crime News Service   703-941-6600
7620 Little River Turnpike
Annandale, VA 22003

This monthly newsletter delivers up-to-date information on computer crime in the U.S. Topics covered include legislation, products, specific criminal cases, seminars, job listings, and publications. The subscription is $125 a year.

Computer Fraud and Security Bulletin

Elsevier International Bulletins   212-867-9040
Journal Information Center
52 Vanderbilt Avenue
New York, NY 10017

This monthly international newsletter reports on the methods of computer criminals and the best means of defense against them. Each month a major case study of a recent computer crime is presented. Special Computer Fraud Supplements are also provided, which deal in depth with a particular aspect of computer crime. The subscription is $190 a year.

Fighting Computer Crime

Charles Scribner's Sons   212-486-2868
115 Fifth Avenue
New York, NY 10003

A comprehensive analysis by Donn B. Parker, this book concentrates on all facets of computer crime today. Topics covered are computer abuse, crime methods, in-depth looks at criminals, ethical conflicts in computing, computer crime in the law, and future vulnerabilities and risk reduction. It is available for $17.95.

Data Bases

LEGIS

2nd and D Streets, SW   202-225-1772
House Annex 2, Room 696
Washington, DC 20515

So far, eight bills, dealing with computer crime have been proposed by the U.S. Congress. Current information about the status of such legislation is maintained in the Congressional database LEGIS. Upon your request, staff members will search the database for you and either tell you the status over the phone or mail you a printout. Although the service is free, there is a printout charge of 10 cents per page for mailed copies. Once you learn what legislation is being considered by Congress, you can request a free copy by writing to House Document Service, U.S. Capitol, Room H-226, Washington, DC 20515.

Additional Experts and Resources

Dr. Harold Joseph Highland
562 Croydon Road   516-775-1313
Elmont, NY 11003
Dr. Highland, a specialist in computer crime, modeling, and simulation, is very helpful and is willing to answer questions and make referrals to other information sources. His articles have appeared in many computer journals, and he is the author of several books.

Donn B. Parker
SRI International   415-859-2378
Menlo Park, CA 94025
Donn B. Parker, the author of *Fighting Computer Crime*, is one of the country's top experts in the field of computer crime and is a very good source of information on more detailed and technical aspects of computer crime.

Computer Graphics

Organizations and Government Agencies

National Computer Graphics Association (NCGA)
8401 Arlington Boulevard   703-698-9600
Fairfax, VA 22031-4670
This professional organization serves as a national clearinghouse for information about computer graphics hardware and software. It holds an annual conference and publishes the following:
  • *Computer Graphics News*—a bimonthly with information on market trends and investment, legal issues, industry news, university programs, new products, and reviews of recent publications. Free to members.
  • *IEEE Computer Graphics and Applications*—free to members.
  • *Graphics Network News*—information about NCGA headquarters and

local chapter activities, and computer-graphics-related events. Free to members.

● *Biographical Directory of the Computer Graphics Industry*—includes listings for over 5,000 computer graphics professionals. Available for $125.

Other computer-graphics-related books and audiovisual materials are also available. The annual membership fee is $45, and $12 for full-time students.

Special Interest Group on Computer Graphics

Association for Computing Machinery   312-644-6610
111 East Wacker Drive
Chicago, IL 60601

The Group acts as a forum for the promotion and dissemination of current computer graphics research, technologies, and applications. State-of-the-art hardware and software subjects encompass animation, business, CAD/CAM, cartography, fine arts, geometric modeling, human factors, image processing, image synthesis, simulation, and standards. The Group holds annual conferences and seminars. It publishes a quarterly newsletter, *Computer Graphics*, available for $30 a year, as well as *ACM Transactions on Graphics*, a quarterly that presents the significant and original work on all aspects of the development and use of computer graphics, which costs $65 a year. A helpful resource person is Thomas A. Defanti, Chairman of Information Engineering at the University of Illinois, Chicago Circle, Box 4348, Chicago, IL 60680.

World Computer Graphics Association

2033 M Street, NW, Suite 399   202-775-9556
Washington, DC 20036

This nonprofit association is a clearinghouse for the dissemination of information about global computer graphics activities. It holds several annual joint international conferences in different countries, and conducts seminars and expositions.

Publications

The Anderson Report

4505 East Industrial Street, Suite 25   805-581-1184
Simi Valley, Ca 93063

This monthly newsletter covers industry news, new products, software news, orders, and installations, and it has a calendar section. Every issue includes an "Inside Report" on a major computer graphics vendor, or a "Special Report" about a business segment. Annual subscription is $125.

Computers and Graphics

Pergamon Press, Inc.   914-592-7700
Maxwell House
Fairview Park
Elmsford, NY 10523

This quarterly journal disseminates information on the application and use of computer graphics techniques. The emphasis is on interactive computer graphics, using CRT-type consoles and manual input devices such as light-pens, tablets, and function keyboards, as well as graphical models, data structures, attention-handling languages, picture manipulation algorithms, and related software. Available for $170 a year.

Computer Graphics World
1714 Stockton Street   415-398-7151
San Francisco, CA 94133
This monthly magazine is devoted to computer applications. It covers all aspects of this technology and the industry, including new products and technology, the latest in hardware and software, CAD/CAM, image processing, business graphics, videotex, book reviews, and a calendar of events. Available for $30 a year.
● *Computer Graphics Directory*—each annual edition contains over 1,200 listings of hardware, software, consultants, service bureaus, associations, conferences, and educational sources. Other topics covered include recent start-up companies, detailed company profiles, and the latest trend in technology. Available for $80.

Computer Vision, Graphics, and Image Processing
Academic Press   212-741-6800
111 Fifth Avenue
New York, NY 10003
This monthly journal publishes papers on the computer processing of pictorial information. Topics covered include image compression, image enhancement, pictorial pattern recognition, scene analysis, and interactive graphics. Available for $260 a year.

Directory of Computer Graphics
Marquis Who's Who, Inc.   800-428-3898
Department HG   312-298-5483
200 East Ohio Street
Chicago, IL 60611
The directory covers all levels of computer graphics, from users and vendors to programmers and educators. More than 5,000 industry-specific biographies are included. It is indexed by area of interest and application, product expertise, and geographic location. Available for $125.

Oryx Press
2214 North Central at Encanto, Suite 106   602-254-6156
Phoenix, AZ 85004
Oryx Press publishes the following:
● *Computer Graphics Marketplace*—this annual resource directory lists, clas-

sifies, and describes the manufacturers, consultants and services, professional organizations, educational programs, conferences and conventions, and publications related to the computer graphics area. Each edition is available for $25.

• *Computer Graphics Glossary*—explains the specialized language of CAD/CAM technology. The cost is $18.50.

Principles of Interactive Computer Graphics

McGraw-Hill, Inc.   609-426-5000
Hightstown, NJ 08520

The authors, William Newman and Robert Sproull, offer comprehensive coverage of the technology behind interactive computer graphics to enable readers to develop design capability. The book also details solutions to all the problems arising in the construction of graphics software systems. It costs $35.95.

Technology and Business Communications, Inc.

730 Boston Post Road   617-443-4671
Sudbury, MA 01776

Technology and Business Communications, Inc. publishes the following:

• *The S. Klein Newsletter on Computer Graphics*—this twice-monthly publication covers the latest developments in computer graphic hardware, software, applications, systems, services, new companies, new business ventures, industry studies, surveys, and new books and journals. The price is $155 a year, for 24 issues.

• *The S. Klein Directory of Computer Graphics Suppliers*—an annual directory that features detailed listings on more than 500 computer graphics suppliers; a comprehensive cross-referenced index to computer graphics hardware, software, turnkey systems and services; international listings of firms and organizations; and market research and educational resources, industry demographics, and trends. It is available for $60 a volume.

Transactions on Computer Graphics and Applications Magazine

IEEE Headquarters   212-705-7900
345 East 47th Street
New York, NY 10017

This monthly publication covers computer graphics hardware and software, display technology, computational geometry, geometric data structures and data bases, industrial applications, animation methodology, human factors for graphics, interactive graphics languages, graphic arts, graphics support of management information systems, and distributed graphics techniques. It costs $70 a year.

Additional Experts and Resources

Several universities conduct research in the area of computer graphics, and can be excellent sources of information for the most recent developments in the area. Some of these universities are:

Rensselaer Polytechnic Institute
Center for Interactive Computer Graphics
School of Engineering
Troy, New York 12181
518-266-6751

Program of Computer Graphics
Cornell University
120 Rand Hall
Ithaca, New York 14853
607-256-7444

The Berkeley Computer Graphics Laboratory
Computer Sciences Division
Department of EECS
University of California
573 Evans Hall
Berkeley, CA 94720
415-642-0344—Dr. Brian Barsky

Computer Science Department
University of Utah
Salt Lake City, UT 84112
802-581-8224—Richard Riesenfeld

Computer Graphics Research
Computer Science Department
University of North Carolina
102 New West Hall
Chapel Hill, NC 27514
919-962-2211—Henry Fuchs

Lab for Computer Graphics & Spatial Analyses
Harvard University Graduate School of Design
48 Quincy Street
Cambridge, MA 02138
617-495-2526—David Schodek

Computer Graphics Research Group
Ohio State University
Kramston Center
1501 Neil Avenue
Columbus, OH 43201
614-422-3416—Charles Csuri

Sheridan College of Applied Arts
Computer Graphics Department
140 Trafalgar Road
Oakville, Ontario L6H 2L1
Canada
416-845-9430—Robin King

Computer Hardware

See: Computers (General); Peripherals

Computer-Integrated Manufacturing

Begins next page. See also: CAD/CAM, Robotics

Organizations and Government Agencies

American Society of Mechanical Engineers (ASME)

345 East 47th Street   212-705-7722
New York, NY 10017   212-705-7470—Computer Engineering Division

A technical and educational society of 111,000 members, ASME is composed of mechanical engineers and other people working in the field of mechanical engineering. Several technical divisions of ASME publish codes and standards, and the Society conducts one of the largest technical publishing operations. Contact the Computer Engineering Division for conference papers relating to computer-integrated manufacturing. Staff members are available to answer questions and/or make referrals. The Society publishes *Computers in Mechanical Engineering*, a bimonthly publication directed toward engineers, which costs $30 a year.

Computer Aided Manufacturing—International, Inc. (CAM-I)

611 Ryan Plaza Drive, Suite 1107   817-860-1654
Arlington, TX 76011

This nonprofit organization, with offices in Europe, Japan, Australia, and the U.S., is supported by the funds provided by its members. It is regulated by the federal government, and information about its technological developments is in the public domain. Staff members will provide technical information and referrals on a limited basis. The library can furnish information about technical literature available to the public. CAM-I has several relevant publications, including proceedings of meetings and special reports by the Advanced Technical Planning Committee (ATPC).

Computer Automated Systems Association

Society of Manufacturing Engineers (CASA/SME)   313-271-1500
One SME Drive
PO Box 930
Dearborn, MI 48121

CASA/SME is an educational and scientific association covering the field of computers and automation in the advancement of manufacturing. Members are professionals in the manufacturing community. It holds conferences and educational seminars, and has video tapes of the sessions available. CASA/SME has several relevant publications:

• *CAD/CAM Technology*—a quarterly magazine for professionals, with features on current business events related to integrated manufacturing and regular feature articles related to manufacturing system developments, including flexible manufacturing systems. A current calendar of major educational events nationally and internationally is also included. The regular subscription price is $24 a year for nonmembers, and is included in the annual fee of members.

• *AUTOFACT Proceedings*—bound volumes of technical papers given at the annual AUTOFACT meetings and events sponsored nationally and internationally by CASA/SME. Prices vary from $40 to $90.

- *CASA/SME CIM Glossary*—includes 2,000 terms relating to CIM. The cost is $10.
- CASA/SME has lists of consultants who can be contacted.

National Bureau of Standards (NBS)

Washington, DC 20234   301-921-2577—Automated Production Technology Division

301-921-2381—Industrial Systems Division

The U.S. Department of Commerce has two divisions within NBS that are involved in building a small automated factory and developing the principles of computer integrated manufacturing. The Automated Production Technology Division has experts in machine tool metrology, or precision machining and robot metrology. The Industrial Systems Division concentrates on high-level factory control. Members of this Division work on machine vision, or robot vision, robotics, and data bases. Two-day seminars on a full range of topics are held bimonthly. The staff at both divisions can answer technical questions or refer you to other resources.

Publications

IEEE Spectrum

345 East 47th Street   212-705-7900

New York, NY 10017

This monthly magazine covers articles on the state-of-the-art and applications chosen for their utility to a wide range of engineers and scientists in the electrical and electronics field; book reviews and listings; letters; and news of IEEE activities, the industry, and the profession. For ordering and price information call or write IEEE Service Center, Publication Sales Department, 445 Hoes Lane, Piscataway, NJ 08854, 201-981-1393.

Data Bases

INSPEC (Information Services in Physics, Electrotechnology, Computers and Control)

Producer: Institute of Electrical Engineers (IEE)   011-44-4-625-3331

Station House, Nightingale Road

Hitchin, Hertfordshire

SG5 1RJ United Kingdom

Distributor: INSPEC/IEEE Service Center   201-981-0060

445 Hoes Lane

Piscataway, NJ 08854

The coverage of the INSPEC data base is centered on four main subject areas: physics, electrical engineering and electronics, control theory and technology, and computers and computing. One can access journal articles, conference papers, books and monographs, and reports and dissertations. This data base is available through DIALOG Information Systems, SDC Search Service, and BRS.

COMPENDEX

Engineering Information, Inc.	212-705-7600
345 East 47th Street	800-221-1044
New York, NY 10017	800-221-1046

This data base provides the engineering and information communities with abstracted information on engineering developments, from significant engineering and related technological literature. COMPENDEX provides abstracts of articles from journals, monographs, reports, standards, and conference proceedings. The data base is accessible through nine commercial vendors, including DIALOG, SDC Orbit, BRS, and ESA. Engineering Information, Inc. may be contacted directly if a search is needed.

Computer Music

See also: Electronic Keyboards; Music, Instructional Computer; Synthesizers

Organizations and Government Agencies

Acoustical Society of America

335 East 45th Street	212-661-9404
New York, NY 10017	

The Acoustical Society of America has technical committees specializing in all aspects of acoustical applications, including musical acoustics. The Society holds two meetings a year, featuring technical sessions on various aspects of acoustics, including sessions on musical acoustics. The Society publishes:

• *Journal of the Acoustical Society of America*—a monthly publication dealing with the measurement, quality, production, control, transmission, reception, and effects of sound. The *Journal* covers many different aspects of acoustics, including computer music, synthesizers, and measurement of musical instruments, as well as speech, physiological, psychological, architectural, and physical acoustics. The *Journal* is available free of charge to members. The subscription for nonmembers is $240 a year.

American Music Conference (AMC)

150 East Huron Street	312-266-7200
Chicago, IL 60611	

This nonprofit organization represents 200 members of the music industry, including manufacturers, retailers, and wholesalers. It is concerned with promoting the sale and use of musical instruments by the general public and educational systems. It collects statistics and compiles reports on all different aspects of the industry. Staff members are available to answer inquiries and make referrals. The Conference publishes:

● *Music USA*—an annual statistical survey containing information about all aspects of the music industry. Data are provided on sales, imports and exports, the music industry as a whole, specific instruments, household use, and more. Profiles are provided about specific electronic instruments, such as synthesizers and keyboards. The publication also includes a directory, with contact information, of all music associations and companies. The 33-page publication is available free of charge.

Computer Music Association

PO Box 1634   415-441-6275
San Francisco, CA 94109

This small nonprofit association is composed of some 2,000 people who exchange information about computer music. It sponsors the annual International Computer Music Conference and publishes the resulting conference proceedings. The organization also publishes members' papers about computer music, as well as:

● *Computer Music*—a quarterly newsletter providing timely information about computer music and synthesizers, as well as announcements and classified ads of interest to the community. This newsletter is included in the subscription to the MIT Press publication *Computer Music Journal*, described below.

Creative Audio & Music Electronic Association

10 Delmar Avenue   617-887-6459
Framingham, MA 01701

This association of audio and music electronics manufacturers and distributors catering to the creative audio market, sponsors educational seminars and conferences. Subjects have included synthesizers, computer music, and electromagnetic technology. The organization publishes:

● *Creative Dictionary of Audio Terms*—this defines more than 1,000 audio and musical electronic terms. The 100-page publication is available for $4.95.

National Association of Music Merchants

500 North Michigan Avenue   312-527-3200
Chicago, IL 60611

This trade association represents manufacturers of musical instruments and equipment. The staff in the Information Department can answer questions relating to products, producers, distributors, and various aspects of the computer music industry. The organization sponsors trade shows and conferences. It publishes several reports relating to the music retail market, as well as a reference source:

● *Program Directory*—updated frequently, this lists all companies involved in manufacturing high-tech musical instruments and equipment. It includes manufacturers of synthesizers. The 70-page publication is available free of charge.

Publications

Computer Music Journal

The MIT Press   617-253-2889
28 Carleton Street
Cambridge, MA 02142
   This quarterly publication covers all applications of computers in music, including digital audios, synthesizers, and electronic keyboards. The *Journal* contains research articles, features about computer music, book and product reviews, meeting announcements, and news of interest to the field. The 90-page publication is available for $25 a year.

Keyboard Magazine

20085 Stevens Creek Boulevard   408-446-1105
Cupertino, CA 95014
   Published monthly, the *Magazine* profiles all keyboard instruments and provides information about electronic music and synthesizers. It contains consumer guidelines on subjects such as buying a synthesizer and electronic keyboard instruments. The publication is geared toward musicians. The annual subscription to the 90-page magazine is $18.95.

Upbeat Magazine

22 West Adams   312-346-7820
Chicago, IL 60606
   Geared toward the music retailer, this monthly publication focuses on new products and new features of musical equipment. It covers synthesizers, electronic keyboards, and computer music and musical instruments. Each issue focuses on a specific topic or musical instrument and profiles the product's retailers. Business and trade show information is also provided. Each issue is approximately 75 pages, and the annual subscription is $8.50.

Additional Experts and Resources

James Beauchamp

University of Illinois   217-344-3307
Urbana, IL 61801
   Mr. Beauchamp is an associate professor in the University's School of Music and Department of Electrical Engineering. His expertise includes analyzing musical instrument tones and synthesis models.

Judith Corea
Consumer Goods Office   202-377-2132
Durable Goods Division
U.S. Department of Commerce
IPA Room 4312
14th and Constitution Avenue, NW
Washington, DC 20230

   An industry trade specialist with the Department of Commerce, Ms. Corea is responsible for following developments of the musical instrument industry as a whole. She can provide you with import/export and product shipment data, as well as an overview of the industry and its new developments and trends. Ms. Corea can also refer you to other sources.

Dr. David Peters
Chairman of Music Education   217-333-0675
University of Illinois
Urbana, IL 61801

   An expert in computer music, Dr. Peters is active in developing music instruction programs for instrumental music. He is founder of the National Consortium for Computer Based Music and chairman of the Music Education Department at the University of Illinois. Dr. Peters is available to answer questions and make referrals.

John Strawn
Lucas Film   415-499-0239
PO Box 2009
San Rafael, CA 94912

   Mr. Strawn's expertise is computer music and digital audio recordings. A digital audio programmer at Lucas Film, he uses the computer to analyze musical instruments and digital recordings. Mr. Strawn edited a series of books, published in spring 1985, titled *The Computer Music and Digital Audio Series*. He is available to answer questions and make referrals to other experts.

Computer Network Technology

Organizations and Government Agencies

Computer and Business Equipment Manufacturers Association
311 First Street, NW, Suite 500   202-737-8888
Washington, DC 20001

   This association of 43 companies includes manufacturers and assemblers of information processing, business and communications products, supplies, and services. It works with the Federal Communications Commission to establish regulations for network systems. Staff member are available to respond to inquiries regarding FCC regulations.

IEEE Computer Society

Institute of Electrical and Electronics Engineers (IEEE)   301-589-8142
1109 Spring Street, Suite 300
Silver Spring, MD 20910

The Computer Society is the largest professional society in the computer field providing technical interaction for its members. Each year the Society and National Bureau of Standards (NBS) sponsor the Computer Networking Symposium, and the Society sponsors an annual conference called Local Computer Network. The proceedings are published. The Society has a publications list describing all its publications. Publications can be ordered from IEEE Computer Society, Customer Service, 10662 Los Vaqueros Circle, Los Alamitos, CA 90720, 714-821-8380. Several tutorials, listed below, are published and updated regularly.

● *Computer Networks* by Marshall Abrams and Ira Cotton—offers the most recent information on computer networks. It provides a basic understanding of networking and is directed to consumers as well as technical experts. The tutorial deals with utility, application, and future directions of computer networks. The 500-page publication costs $20 for members and $30 for nonmembers.

● *Local Computer Network* by Kenneth Thruber and Harvey Freeman—contains 40 reprints on local computer network products, including distributive processing and system enhancement. This tutorial has 371 pages and costs $15 for members and $20 for nonmembers.

● *Local Network Technology* by William Stallings—describes innovations regarding local networks, including network types and design approaches, and technical and architectural developments. This tutorial also includes the common principles underlying the design and implementation of local networks. The 320-page publication costs $20 for members and $32.50 for nonmembers.

● *Security of Data in Networks* by Donald Davies—this tutorial, containing 30-percent original material, explains the main components required to make a secure data network. It also discusses potential weaknesses in the security of networks. The 250-page publication costs $15 for members and $20 for nonmembers.

IEEE Project 802

c/o IEEE Computer Society   301-589-8142
1109 Spring Street, Suite 306
Silver Spring, MD 20910

Project 802 is a local network standards committee, with members who are users of networks as well as manufacturers of networks and components of networks. In its efforts to develop local network standards, IEEE Project 802 is divided into eight Working Groups, each considering a separate but related topic: 802.1—High Level Interface; 802.2—Logical Link Control; 802.3—CSMA/CD for baseband, broadband and fiber optics; 802.4—Token Bus for baseband and broadband; 802.5 —Token Ring; 802.6—Metropolitan Area Networks; 802.7—Technical Advisory Group for broadband technology; and 802.8—Technical Advisory Group for fiber optics. Many documents have resulted from the work of 802 groups, such as minutes, reports, proposals, and draft standards, which may be obtained by contacting the specific Working Group chairperson via the IEEE Computer Society. The draft computer standards can be obtained from the IEEE Computer Society, West Coast Office, 10662 Los Vaqueros Circle, Los Alamitos, CA 90720, 714-821-8380. The 802 standards, after they are approved by the IEEE Standards Boards and published, are available from The IEEE Standards Office, 345 East 47th Street, New York, NY 10017, 212-705-7966.

Institute of Computer Science and Technology
National Bureau of Standards (NBS)   301-921-2731
Gaithersburg, MD 20899

The Institute is a technical center for computers and network technology, which primarily serves the federal government, and to a limited extent offers assistance to industry and the general public. The publications listing is available upon request; it includes more than 20 selections on the standardization of network protocols. These can be ordered through the Government Printing Office, Washington, DC 20402, or the National Technical Information Service, 5285 Port Royal Road, Springfield, VA 22161.

Network Users Association (NUA)
PO Box 76828   800-638-8510
Washington, DC 20013   703-683-8500

This association of users and manufacturers of network systems promotes the establishment of working protocol standards in order to provide the capability of information exchange among different manufacturers, products, and systems. Association services are primarily for members, although the annual conference and exposition are open to the public. The Association publishes:

• *NUA Update*—a four-page monthly newsletter that deals with current issues and developments in the field of networking. The subscription is free to members.

OMNICON, Inc.
501 Church Street, Suite 206   703-281-1135
Vienna, VA 22180

This company advises the public and industry about standards on open systems interconnection architecture, protocols, and applications. OMNICON's staff can provide information and working documentation for standards development in the area of data communications, distributive information systems, and networking. Prepublication drafts, reports, and other technical papers from ANSI, ISO, ECMA, EIA, CCITT, and other organizations may be obtained from OMNICON. Seminars are conducted for industries and the general public.

Special Interest Group on Data Communications (SIGCOMM)
Association for Computing Machinery (ACM)   212-869-7440
11 West 42nd Street
New York, NY 10036

SIGCOMM is an international technical group with membership from industry, government, and universities. The Group provides information about data communications, including computer networks. ACM staff can refer you to an appropriate expert on computer networking. Two conferences are sponsored: the biannual Data Communication Symposium (cosponsored with the IEEE Communications Society) and the annual SIGCOMM Conference. Both are open to the public, and the proceedings are later published and available through ACM. Relevant publications are:

- *Computer Communications Review*—a 40-page quarterly newsletter, which deals with draft networking standards, conference and workshop information, as well as technical contributions from ACM members. The cost for members is $25 a year and $40 for nonmembers.

- *ACM Transactions and Office Information Systems*—a quarterly journal that focuses on new concepts and ideas in office information, including computer networking with practical applications. The annual subscription to the 85-page journal is $20 for members and $70 for nonmembers.

- *Transactions on Computer Systems*—a 90-page quarterly journal that reports significant and original works on the design, implementation, and use of computer systems, including networks. The subscription rate for members is $20 and $70 for nonmembers.

X353, Data Communications
c/o American National Standards Institute (ANSI)   212-345-3300
1430 Broadway
New York, NY 10018

X353 is a technical committee of volunteers which develops ANSI standards for the interconnection of computers, terminals, and similar devices. The committee comprises seven task groups with specific assignments, including computer networks. Questions should by directed to ANSI in New York, and referrals will be made to members of X353. More than 30 publications dealing with data communications technical standards are available through the Institute.

Publications

Computer Communications
Prentice-Hall, Inc.   201-592-2000
310 Sylvan Avenue
Englewood Cliffs, NJ 07632

Computer Communications, by Dr. Wushow Chou, is a two-volume publication. Volume I addresses the principles of various aspects of computer network systems and deals with extending the use of computer resources. It also instructs the reader how to interconnect independent computers to communicate with each other. Volume II emphasizes the application of computer networks, offering recent research and information on networks, local networks, data-voice-image systems, distributive data base design, digital radio, and switching. Volume I (430 pages) is $35, and Volume II (496 pages) is $37.

Computer Networks
For subscription information:
Elsevier Science Publishing Company, Inc.
PO Box 1663
Grand Central Station
New York, NY 10163

For editorial information:
Dr. Philip A. Enslow   404-894-3187
School of Information and Computer Science
Georgia Institute of Technology
Atlanta, GA 30332

This journal is published ten times a year, and contains material on all aspects of the design, implementation, use, and management of computer networks, the communications subsystem, and all supporting activities. In addition to technical articles, the journal covers nontechnical topics such as economics, legal and regulatory issues, and social impacts. The annual subscription is $81.25 for institutions, with reduced rates for individuals.

Data Communications
McGraw-Hill, Inc.   212-512-3149
1221 Avenue of the Americas
New York, NY 10020

Data Communications is a monthly magazine about distribution, data processing, and information networks. It is geared to data communications managers and their staff. Issues average between 300 and 350 pages, and the annual subscription is $24.

Journal of Telecommunication Networks
Computer Science Press   301-251-9050
11 Taft Court
Rockville, MD 20850

The *Journal* includes articles on technological advancements, research/development results with practical implications, significant system implementation, and important regulatory, social, and policy issues. Each issue averages 96 pages, and the annual subscription rate is $100.

Networks

Journals Department   212-692-6000
John Wiley & Sons, Inc.
605 Third Avenue
New York, NY 10158

Networks is a 150-page quarterly journal devoted to a broad range of topics related to the design and analysis of any large-scale network. Articles cover data, video, vehicle, computer, and power networks, as well as any interconnection of many components. The annual subscription is $98.

Additional Experts and Resources

Harold Folts

Executive Director, OMNICON, Inc.   703-281-1135
501 Church Street, Suite 206
Vienna, VA 22180

Mr. Folts is an internationally recognized authority on standards technology in the field of distributive communications. He and his staff can answer questions about computer networking standards. Mr. Folts is the editor of *McGraw-Hill's Compilation of Data Communications Standards*, which is a reference book containing photo reprints of international standards for distribution information systems, including computer networks. It is 1,933 pages long, is updated every four years, and sells for $295.

John F. Heafner, Ph.D.

Chief, Systems and Network Architecture Division   301-921-3537
National Bureau of Standards
Bldg. 225, Room B218
Gaithersburg, MD 20899

The Systems and Network Architecture Division is responsible for the following activities in computer/communications networking: research, standards development, agency assistance, and technology transfer to government and industry. Dr. Heafner is an expert in these areas of networking, and he and his staff can answer questions and/or make referrals.

Darrell Icenogle

Western Behavioral Science Institute (WBSI)   619-459-3811
1150 Silverado Street
PO Box 2029
La Jolla, CA 92037

Mr. Icenogle is the director of Education/Communication Systems, as well as the vice president of Strategic Information Systems, Inc., a division of WBSI, which develops networking systems. He has extensive experience in the design of computer conferencing systems and is noted for his development of Onion software, which greatly facilitates computer networking. Literature about computer networks is available upon request. Mr. Icenogle and his staff are available to answer questions about networking.

Murray Turoff, Ph.D.
Center for Information Age Technology (CIAT)   201-596-2929
New Jersey Institute of Technology
323 Martin Luther King Boulevard
Newark, NJ 07102

Dr. Turoff, who developed the first computer conferencing system, is currently Director of Research for CIAT. He is noted for his work in information systems, networking systems, and technological forecasting and assessment. He is the designer of EIES (Electronic Information Exchange System), which is an internationally used conferencing system, and he also is co-author of the award-winning book *The Network Nation: Human Communications via Computer*. Dr. Turoff and his staff are available to answer questions about computer networks.

Computer Performance Technology

See also: Computer Simulation

Organizations and Government Agencies

Computer Measurement Group
11242 North 19th Avenue   602-995-0905
Phoenix, AZ 85029

This international group focuses its attention on computer performance evaluation, including computer simulation. It holds an annual conference, and publishes two journals, *Conference Proceedings* and *Transactions*. The annual membership is $20, and the *Conference Proceedings* is $45. A newsletter is published periodically.

SIGMETRICS
Association for Computing Machinery   212-869-7440—Membership and publication information
11 West 42nd Street   512-343-0860—Dr. Herbert D. Schwetman, Chairman
New York, NY 10036

SIGMETRICS is a special-interest group of the Association, geared toward measurement and evaluation. It provides a forum for those interested in problems related to the measurement and evaluation of computer system performance. Both those concerned with the application and adaptation of existing methodologies and researchers interested in developing new methodologies are represented. Relevant areas of interest include the role of simulation; design and interpretation of benchmarks; analysis of outputs from hardware and software monitors; and the development of suitable analytic models. The group sponsors an annual symposium, after which the proceedings are published. It also publishes:

• *Performance Evaluation Review*—a quarterly newsletter that includes papers on measurement, modeling, and evaluation of computer system performance. It presents the proceedings of the annual symposium in a special 200-page issue in the fall. The subscription cost is $9 a year for members and $23 a year for nonmembers.

Publications

Journal of Capacity Management

Institute for Information Management   408-749-0133—Carol Kaplan, Editor
400 Oakmead Parkway
Sunnyvale, CA 94086

This quarterly publication is designed for corporate managers and processing personnel concerned with information systems management. It presents information on the capacity management function and on computer performance evaluation. The *Journal* provides a comprehensive forum for promoting skills in these fields. Articles deal with research theory and practice. Book reviews are also included, addressing the latest thought in the two fields indicated above. The subscription cost is $95 a year for nonmembers, while members receive it for free.

Performance Evaluation

Elsevier Science Publications   212-867-9040
52 Vanderbilt Avenue
New York, NY 10017

This quarterly journal, published in the Netherlands, is oriented toward systems theorists, designers, implementors, and analysts who are concerned with performance aspects of computer systems, computer communication, and distribution systems. It is the only journal devoted to performance evaluation. The subscription cost is $92.70 a year.

Additional Experts and Resources

Jon D. Clark, Ph.D.

North Texas State University   817-565-3103
College of Business
PO Box 13677
Denton, TX 76203

Dr. Clark, an expert in computer performance evaluation, is a teacher of Business and Computer Information Systems. He has written several articles on data base and computer system evaluation, which have appeared in *Data Management*, *Per-*

formance Evaluation Review, and various conference proceedings. He also has co-authored the book *Computer System Selection—An Integrated Approach* (published by Praeger). Dr. Clark can answer technical inquiries and make referrals.

Dr. Herbert D. Schwetman, Ph.D.
MCC   512-343-0860
9430 Research Boulevard
Austin, TX 78759

Dr. Schwetman is a specialist in the area of measurement and evaluation of computer system performance. He can answer inquiries from those seeking technical information, and can make referrals to other resources.

Computer Programming Languages

See: Programming Languages (Computer)

Computer Security

See also: Computer Crime

Organizations and Government Agencies

American Society of Industrial Security
1655 North Meyer Drive, Suite 1200   703-522-5800
Arlington, VA 22209

This 20,000-member professional association consists of people in business and industry who are responsible for protecting the assets of their employers. The Society sponsors an annual four-day seminar, with exhibits, workshops, and other activities dealing with computer security issues. It publishes:

 • *Security Management*—a monthly publication that features articles about computer security written by experts in the field. The annual subscription to the 92-page publication is $27.

Automation Training Center
11250 Roger Bacon Drive, Suite 17   703-471-5751
Reston, VA 22090

The Center sponsors an annual international conference in the spring in conjunction with the Electronic Data Processing (EDP) Auditors Association. The con-

ference consists of a full week of exhibits, seminars, and keynote speakers to disseminate the latest information in the field of computer security. It publishes:

• *EDPAC Newsletter* (EDP Audit Control & Security)—a monthly publication
providing articles, abstracts, book reviews, and news items on audit controls and
security for computers. The annual subscription to the 20-page publication is $72.

Computer Security Institute
43 Boston Post Road   617-845-5050
Northborough, MA 01532

The Institute's members are people and organizations interested in computer
security. Each year it holds a conference, several seminars, and workshops. A "hotline service" provides assistance and referrals, but is available to members only. It
publishes the following:

• *Computer Security: A Newsletter for Computer Professionals*—covers all aspects of computer security, including news, case histories, reading suggestions, and
legal developments. It is available free to members.

• *Computer Security Journal*—a semiannual journal devoted solely to computer security, with new ideas and useful information about current products and
packages. It covers such topics as risk analysis, disaster recovery, fraud preven-
tion/detection, laws and legislation related to privacy and security, auditing techniques, and physical and environmental controls. Available for $65 a year.

• *Computer Security Handbook*—features case studies, checklists, product information, samples and extracts from actual security policies and job descriptions,
and hard-to-find material from the government and other sources. The *Handbook*
is divided into ten sections, which discuss computer security, protecting the data
center, software management, communications security, disaster recovery planning,
auditing, and information sources. The cost is $95.

Department of Defense Computer Security Center
9800 Savage Road   301-859-6883
Fort Mead, MD 20755-6000

The Center is involved with research and development of computer security
techniques and procedures. It publishes:

• *The Trusted Computer System Evaluation Criteria*—a one-time technical
publication dealing with the levels of security in various computer systems; it provides standards of security for both the government and private industry. The 100-
page publication is free.

EDP Auditor Foundation
373 South Schmale Road   312-653-0950
Carol Stream, IL 60188   312-935-6668

The Foundation is a professional association for those involved in EDP auditing,
and data security and management. It holds seminars on various aspects of auditing
and sponsors a series of courses for qualification as CISA-Certified Information Sys-

tems Auditors. Staff members are available to answer questions and make referrals to experts in the field. The Foundation publishes:

- *EDP Auditor Journal*—a quarterly publication that includes articles covering many aspects of auditing and targeting key areas such as computer security and necessary controls. The *Journal* is 60-70 pages long; subscription is $40 a year.
- *Control Objective*—updated every two years, provides a comprehensive list of areas where controls should be installed to prevent theft in computers. It deals with management controls, resource management, planning, policies, procedures, standards, technical services, application controls, and operational controls. The 220-page publication is available for $24 an issue.
- *EDP Auditor Update*—a quarterly publication that covers chapter and association news. The five-page update is available to members only.

IEEE (Institute of Electrical and Electronics Engineers)

345 East 47th Street

New York, NY 10017

212-705-7890

The IEEE maintains several committees and interest groups that are involved with different aspects of computers and computer security. Many of the committee members are experts in the field of computer security and can answer questions or make referrals to other experts. Contact the IEEE Technical Activities person for further information.

Institute for Computer Science and Technology

National Bureau of Standards (NBS)

Technology Building

Gaithersburg, MD 20899

301-921-3861—Stuart Katzke, Ph.D. (Mail Stop B366)

301-921-3427—Dennis Branstad, Ph.D. (Mail Stop A216)

301-921-3427—Miles Smid (Mail Stop A216)

The Institute is involved in the overall area of computer science and technology, and in the establishment of standards and guidelines for the field. Among its computer security experts are: (1) Dr. Katzke, a specialist in internal computer security; (2) Dr. Branstad, an expert in the communications aspect of computer security; and (3) Mr. Smid, an expert in cryptology. This government office has produced more than 45 publications, covering computer security guidelines, standards, and technology. A free listing of their publications is available.

Institute of Internal Auditors

249 Maitland Avenue

PO Box 1119

Altamonte Springs, FL 32701

305-830-7600

This professional organization of 27,000 members is made up of system analysts, auditors, accountants, CPAs, and management consultants. It sponsors five conferences a year, consisting of seminars, workshops, and speakers covering auditing practices and procedures. The Institute also holds courses and seminars on over 30 topics, such as fraud detection and investigation, audit computerized systems, information systems auditing concepts and applications, auditing practices, and data base auditing. A complete seminar directory is available for free from the Institute. It publishes:

> • *Security, Accuracy, and Privacy in Computer Systems*—a one-time publication dealing with security and privacy of computer systems. It details hardware and software controls, programming, physical security requirements, legal and social requirements, and administrative controls. The 626-page publication is available for $39.

> • *Computer Control and Audit*—a one-time publication that analyzes various controls used in computerized and manual information systems. The 489-page publication is available for $35.25.

> • *Computer Fraud and Counter Measures*—a one-time publication outlining various computer fraud techniques and countermeasures that can be taken to prevent computer crime. The 509-page publication is available for $40.95.

International Association of Computer Crime Investigation

PO Box 30726   213-553-0449
Oakland, CA 94604

This nonprofit organization disseminates information about computer crimes to police departments and security agencies. It sponsors two three-day conferences a year, with workshops and seminars on the latest technology advances in computer crime. Staff members can answer questions and make referrals to experts in the field of computer crime investigating and security. The Association publishes:

> • *CompuCrime Newsletter*—a monthly that provides the latest information and developments in computer crime investigation. The 10-page publication is available free to members.

National Center for Computer Crime Data

JFK Library #4053   213-850-0509
CSULA
515 State University Drive
Los Angeles, CA 90032

The Center is a research and development organization that maintains a data base of over 800 case studies on computer crime and computer security. It is an excellent source of information on computer security, and referrals can be made to other experts in the field. Over 25 publications on computer crime are released, and a free publications list is available. Some of the publications are:

> • *Computer Crime Law Reporter*—updated every two years, a collection of computer crime laws and their effects in the U.S. The 100-page publication is available for $35.

> • *Annual Statistical Report*—an annual publication that provides statistical information about computer crime. The 50-page publication is available for $15.

> • *Investigation of Computer Crime*—a one-time publication that provides procedures for investigating computer crime. The 100-page publication is available for $10.

- *Trial of Computer Crime*—a one-time publication that provides information on how to take a computer crime to trial. The 100-page publication is available for $5.
- *International Computer Crime*—a one-time publication that focuses on computer crime in the international arena. The 100-page publication is available for $25.

The Plagman Group

1450 Broadway   212-921-4980
New York, NY 10018

The Plagman Group is an internal management consulting firm that offers over 200 courses on computers and business practices, some covering such computer security topics as auditing security control, audit controls, security software, and data security. A free course directory is available.

SIGSAC

Association for Computing Machinery (ACM)   212-869-7440
11 West 42nd Street
New York, NY 10036

SIGSAC stands for the Special Interest Group of Security Audit and Control. Many of its members are excellent sources of information about computer security. Contact the Association for further information, and for the name and number of the chairman of the Group. The SIGSAC newsletter is published three times a year and provides information about the technology of computer security. The annual subscription to the 52-page newsletter is $18.

Publications

Balance Sheet

Total Assets Protection, Inc.   817-261-2556
500 Brookhollow One
2300 East Lamar Boulevard
Arlington, TX 76011

Balance Sheet is a monthly newsletter that covers many aspects of computer security and design, such as the construction of data centers, data security, advance technology in research and development, computer crime, and computer law issues. The eight-page publication is available for $175 a year.

Computer Protection Systems, Inc.

711 West Ann Arbor Trail   313-459-8787
Plymouth, MI 48170

A consulting firm, the company provides extensive training programs for auditors and EDP auditors in computer security. It also publishes the following, which are available from Territorial Imperative Publishers, Assets Protection, PO Box 5323, Madison, WI 53705, 608-231-3817:

• *Computer Security Digest*—a monthly publication, 8-12 pages long, which provides information on tools and techniques pertaining to computer security, and information on the hardware and software packages available. The *Digest* has abstracts on articles in the field and provides reviews of products. The annual subscription is $75.

• *Computer Crime Investigation Manual*—a one-time technical publication that deals with computer operation, covering procedures and techniques used when documenting a computer crime. The 375-page publication is available for $39.95.

• *Computer Crime Wave of the Future*—a one-time nontechnical publication that covers computer-related crimes. It examines such issues as the privacy aspect of computerized information and computer fraud. The 100-page publication is available for $100.

Computers and Security

Elsevier Publishing   212-867-9040
Journal Information Center
52 Vanderbilt Avenue
New York, NY 10017

This quarterly journal deals with information processing technology, system design, evaluation, and hardware and software developments—and how these relate to computer security. The annual subscription to the 60-page publication is $89.

Department of Defense Computer Institute

National Defense University   202-695-9468
Washington Navy Yard Bldg. 175
Washington, DC 20374

The Computer Institute publishes the following:

• *Industrial Security Manual for Safeguarding Classified Information*—this publication, updated every few years, provides information on handling classified information and industrial security. The March 1984 edition (# 5220.22-M) is available for $10 from the Superintendent of Documents, U.S. Government Printing Office, Washington, DC 20402, 202-783-3238.

• *Security Requirements for Automated Data Processing Systems–Manual and Directive*—the *Directive* sets forth policy for classified information processed by computer systems and provides guidelines to design and establish a comprehensive security plan. The 15-page publication is available free of charge. The *Manual* provides more detailed procedures and amplification of the *Directive*; this 40-page publication is also free. These can be ordered from: Director of U.S. Naval Publisher and Forms Center, 5801 Tabor Avenue, Philadelphia, PA 19120.

Pocket Guide to Computer Crime Investigation

Bruce Goldstein   817-261-2556
9600 Royal Lane, Suite 708
Dallas, TX 75243

This one-time publication is used by investigators of computer crimes, and contains practical information related to the investigation, warranty and documentation of evidence, and the use of expert witnesses. The 19-page publication is available for $5.

Privacy Journal
PO Box 15300   202-547-2865
Washington, DC 20003

This monthly journal covers privacy from three different perspectives: legislative, individual, and business, with an emphasis on computer security. The subscription rate to the eight-page journal is $89 a year. A research service is also available, which includes the monthly journal, and allows access to resource files and reports work performed by the staff on the status of computer security, legislative issues, and new publications. The charge for this service is $300 a year.

Security Systems Digest
Washington Crime News Service   703-941-6600
7620 Little River Turnpike
Annandale, VA 22003

This biweekly newsletter provides information on security applications for large businesses and government in the areas of terrorism, law enforcement, and computer security. The subscription to this ten-page publication is $95 a year.

Additional Experts and Resources

Dr. Harold Joseph Highland
562 Croydon Road   516-775-1313
Elmont, NY 11003

Dr. Highland, a specialist in computer security, modeling, and simulation, is very helpful and is willing to answer questions and make referrals to other information sources. He writes books in his areas of expertise, as well as articles in many computer journals.

Donn B. Parker
Senior Management Consultant   415-859-2378
SRI International
Menlo Park, CA 94025

A specialist in computer security research, Mr. Parker has produced more than 65 computer security reviews for clients. He is the author of several books on the subject and advises many members of Congress on developing computer-security-related legislation.

Representative Ron Wyden

U.S. Congress   202-225-4811
1406 Longworth
Washington, DC 20515

Congressman Wyden of Oregon is the author of the "Small Business Computer Crime Prevention" bill, which was passed by the House and the Senate in June 1984, and was signed by President Reagan on July 16, 1984. The law establishes a Computer Security and Education Advisory Council within the Small Business Administration (SBA), made up of representatives from federal agencies and the private sector. This Council will provide information about the problems of computer security and education to SBA. The legislation also establishes a computer security and education program designed to encourage SBA to share the information that is now available to the business community.

Computer Simulation

See also: Computer Performance Technology

Organizations and Government Agencies

Association for Business Simulation and Experiential Learning

Department of Finance and Real Estate   316-689-3219
Arizona State University
Tempe, AZ 85202

This professional organization is dedicated to the development and use of experiential teaching techniques, both computerized and noncomputerized, in business education. It sponsors intercollegiate Simulation Gaming and Competition. The Association publishes a quarterly journal, *Simulations and Games*, as well as the proceedings from its annual convention.

Computer Measurement Group

11242 North 19th Avenue   602-995-0905
Phoenix, AZ 85029

This international group focuses on computer performance evaluation, especially computer simulation. It holds an annual conference, and puts out two publications: *Conference Proceedings* and *Transactions*. Modeling and simulation are extensively covered in both the conference and the resulting publications. The annual membership is $20 a year, and the *Conference Proceedings* are $45.

CONDUIT

Oakdale Campus   319-353-5789
University of Iowa
Iowa City, IA 52242

CONDUIT is a nonprofit educational software publisher that distributes high-school-level and college-level educational software packages. With outside grants, CONDUIT conducts research in the field of computer simulation in education, and then publishes, markets, and distributes the software. It makes available two catalogs, one for Basic and Fortran, and the other for a variety of small computers.

IEEE Computer Society

1109 Spring Street, Suite 300   301-589-8142
Silver Spring, MD 20910

The Simulation Technical Committee of the IEEE Computer Society promotes all aspects of research, development, and application of analog and digital computer simulation. The Committee organizes technical sessions at Computer Society conferences, and cosponsors the annual Simulation Symposia. It publishes a newsletter, *Modeling*, three times a year.

International Association for Mathematics and Computers in Simulation (IAMCS)

c/o Robert Vichnevetsky   201-932-2081
Department of Computer Science
Hill Center, Busch Campus
Rutgers University
New Brunswick, NJ 08903

IAMCS is a professional organization for people specializing in scientific computation, with a special emphasis on the simulation of systems. This includes mathematical modeling, numerical analysis, approximation theory, consideration about computer hardware and software, programming languages, and compilers. The society facilitates exchange of scientific information among specialists, and organizes national meetings and exhibits. It publishes a monthly journal, *Mathematics and Computers in Simulation* (see separate entry below), as well as *IAMCS News*, which keep readers informed of the activities of the Association and of a large selection of related events. It also publishes issues proceedings, other scientific publications, and a directory. A free listing of books on the topic of computer simulation is also available. The annual membership fee is $26.

International Simulation and Gaming Association

c/o Dr. Cathy Stein Greenblat   201-932-4034
Department of Sociology
Rutgers University
New Brunswick, NJ 08903

The Association is interested in every facet of simulation and gaming. It maintains resource lists, conducts specialized education, presents discussion papers, and sponsors workshops and research activities. It publishes a quarterly journal, as well as *Conference Proceedings* annually.

North American Simulation and Gaming Association
c/o Professor W. Nichols   412-946-8761
Box 100
Westminister College
New Wilmington, PA 16172

The Association of 200 members aims to promote training of specialists in the field of computer simulation and gaming, to facilitate communication between these specialists and policymakers, and to promote the development of better techniques in the field of simulation and gaming. It publishes an annual *Conference Program* and an annual *Membership Directory*. It holds a conference each October.

SIGSIM (Simulation and Modeling)
Association for Computing Machinery (ACM)   212-869-7440
11 West 42nd Street
New York, NY 10036

SIGSIM is one of 32 special-interest groups of the ACM, and is devoted to the advancement of state-of-the-art in simulation and modeling; discrete and continuous applications in computer systems, industry and business, networks, socioeconomics, and other fields; experimental design and validation of simulations; techniques of digital, analog, and hybrid simulation; random number theory and generators; simulation languages and packages; statistical distribution analysis and related computational mathematics.

SIGSIM publishes *Conference Proceedings on Simulation*, which deals with practical papers on simulation in a variety of applications designed for exchanging experience. Other topics include advancement of state-of-the-art simulation and modeling; discrete and continuous applications in computer systems; industry and business networks; socioeconomics and other fields. SIGSIM also publishes a monthly newsletter, *SIMULETTER* (see separate entry under *Publications*).

The Society for Computer Simulation
PO Box 2228   619-459-3888
La Jolla, CA 92038-2228

This technical society is devoted to the advancement of simulation and allied computer arts in all fields. It holds two major conferences each year: the Multiconference and the Summer Computer Simulation Conference. The annual membership is $35. The Society publishes the following:

• *Simulation*—a monthly technical journal that contains papers and articles dealing with all aspects of modeling and simulation, improved mathematical methods, simulation languages, pitfalls to avoid, validation of models and simulations, tutorial papers, simulation as an aid in teaching and designing, simulation hardware, and applications and programming in new and old disciplines. The subscription is $42 a year for nonmembers; the publication is free to members.

• *Transactions*—a quarterly journal that concentrates on the highest technical aspects of simulation. Available for $80 a year to nonmembers and $20 to members.

• *Simulation Series*—a semiannual series of hardcover books on various topics in computer simulation and modeling, with emphasis on newer areas of application. Each book is $30, with a 25-percent discount for members.

• *Prominent Names in Simulation*—a comprehensive 30-page directory of the best-known people in the field of simulation, with addresses and background information.

• *Index to Simulation Literature*—a reference source spanning 1976 through 1981, indexing the *Simulation* journal, the SCS *Simulation Series*, and proceedings of conferences.

Publications

Mathematics and Computers in Simulation

Elsevier Science Publications
PO Box 211
1000 AE Amsterdam
The Netherlands
Attn: Journals Subscription

This bimonthly journal, a publication of the International Association for Mathematics and Computers in Simulation, covers topics including mathematical tools and in-the-foundation system modeling, as well as specific applications to science and engineering, and supporting analysis of numerical methods. Also covered are computer hardware for simulation, and special software and compilers. Articles cover the general philosophy of systems of simulation and its impact on disciplinary and interdisciplinary areas. The journal also includes a bibliography and book reviews. The annual subscription rate is $115.80.

SIMULETTER

Association for Computing Machinery (ACM)   212-869-7440
11 West 42nd Street
New York, NY 10036

The quarterly newsletter of the Simulation Interest Group of the ACM is devoted to modeling and simulation. The annual subscription rate is $10. Call Richard Nance, the chairperson, at 703-961-5605, for further information about the newsletter, or call ACM to find out who is the current chairman.

SIMSNIPS

CACI   619-457-9681—Hall Duncan, Product Manager
3344 North Porrey Pines Court
La Jolla, CA 92037

This quarterly newsletter covers general information on simulation modeling. It includes abstracts of modeling projects, reports of conferences, book reviews, new product reports, and so on. It is available free to those interested in simulation modeling.

Simulation and Games

Sage Publications, Inc.   213-274-8003
PO Box 776
Beverly Hills, CA 90210

> This interdisciplinary journal of design and research in the areas of simulation and gaming is available for $4.20 an issue. *Simulation and Games* is the official publication of the International Simulation and Gaming Association.

Additional Experts and Resources

Dr. Harold Joseph Highland

562 Croydon Road   516-775-1313
Elmont, NY 11003

> Dr. Highland, a specialist in computer security, modeling, and simulation, is very helpful and is willing to answer questions and make referrals to other information sources. He writes books in his area of expertise, as well as articles in many computer journals.

Thomas Nichols

Box 96   412-946-8761
Westminister College
New Wilmington, PA 16172

> Dr. Nichols, the treasurer and co-editor of *Simulation and Games* magazine, is willing to answer general questions about simulation. He also has a large membership list that he can consult for names of experts in the field, and he can refer callers to other sources of information.

Pritsker and Associates

PO Box 2413   317-463-5557
West Lafayette, IN 47906

> This consulting and software publishing firm specializes in computer simulation. It makes available several brochures, catalogs, and price lists that describe its business and products. Alan Pritsker is an expert in the area of computer simulation and is willing to answer questions. He has published six books on the topic, and has developed several languages. His area of specialty within computer simulation is language development and the uses of languages for industrial problem solvers.

Professor Thomas J. Schriber

Graduate School of Business   313-764-1398
University of Michigan
Ann Arbor, MI 48109

> Dr. Schriber is a professor of computer and information systems at the University of Michigan. His principal area of interest is discrete-event simulation, and

includes the development, delivery, and dissemination of teaching materials in this area; the application of simulation modeling in problem-solving situations; simulation-related research; and professional service. He has written articles and a book on modeling, using GPSS-based simulation. He can answer technical inquiries and make referrals on the subjects listed above.

Susan L. Solomon, Ph.D.
Department of Management Information Systems   509-458-6200
Eastern Washington University
Cheney, WA 99004
Dr. Solomon, an expert in computer simulation, can answer technical inquiries and make referrals. She has written several technical articles on simulation, as well as a textbook, *Simulation of Waiting Line Systems*, published by Prentice-Hall in 1983. She is a past president of ACM's SIGSIM, and is currently an associate technical editor for *Simulation* magazine of the Society for Computer Simulation. She is also a professor with the Management Information Systems Department of Eastern Washington University.

Computer-Aided Design and Manufacturing

See: CAD/CAM

Computerized Automation

See: Robotics

Computers for the Handicapped

Organizations and Government Agencies

Artificial Language Laboratory
Michigan State University   517-000-0000
405 Computer Center
East Lansing, MI 48824-1042
The Laboratory builds computerized speech systems for people that are nonvocal or speech-impaired. Dr. John Eulenburg is an authority in the field, and is the

director of the Artificial Language Laboratory. The staff can provide assistance, answer questions, and make referrals to experts in the field. It publishes:

• *Communication Outlook*—a quarterly magazine that focuses on communication aids and techniques for the handicapped worldwide. The annual subscription is $12.

SIGCAPH

Association for Computing Machinery (ACM)   212-869-7440
11 West 42nd Street
New York, NY 10036

The Special Interest Group on Computers for the Physically Handicapped is a subcommittee of the Association. It publishes:

• *SIGCAPH*—a quarterly journal that contains articles on computers and the physically handicapped, as well as a bibliography of international scientific literature on computers and the handicapped. The annual subscription to the 32-page publication is $22 for nonmembers and $10 for members.

Center for Computer Assistance to the Disabled (C-CAD)

2501 Avenue J, #100   817-640-6613
Arlington, TX 76011

The Center is involved with the collection, evaluation, and dissemination of information about computers for the disabled. The staff can answer questions and make referrals to experts in the field of computers for the handicapped. The information is collected in a data base for easy access, and there is no charge for using the information resource center. Computers are available on loan to those disabled individuals who are trained to use the terminals. The following is published:

• *C-CAD Newsletter*—a quarterly publication that highlights special applied applications of computers for the handicapped. It also announces workshops and demonstrations taking place across the country. The four-to-eight page publication is available free of charge.

Closing the Gap

PO Box 68   612-248-3294
Henderson, MN 56044

Focusing on computer technology with special emphasis on the needs of the deaf, this organization organizes seminars and workshops throughout the U.S., ranging from one day to 16 weeks in length. Topics such as beginner awareness, and specific peripheral training on keyport and adaptive frameworks are covered. The staff can provide assistance, answer questions, and make referrals to experts in the field. The organization operates a "help line," accessible 24 hours a day, to provide information about the latest hardware and software technology. It sponsors a major international conference each year in conjunction with Johns Hopkins University and the Technology and Media Division of the Council of Exceptional Children. The following is published:

● *Closing the Gap*—a bimonthly newsletter that provides information about microcomputer technology for special education and vocational rehabilitation for the handicapped. The 24-page publication is available for $15 a year.

Computing and the Handicapped Technical Committee
IEEE   301-589-8142
1109 Spring Street, Suite 300
Silver Spring, MD 20910

The Committee, made up of 250 members, promotes use of modern computer technology in the rehabilitation, education, and employment of handicapped persons. The organization encourages interchange among scientists, engineers, clinicians, and the handicapped community. It sponsors annual workshops cosponsored by ACM's Special Interest Group on Computers for the Physically Handicapped (described above), and the proceedings are published for the members. A newsletter is published irregularly, covering the latest issues.

COPE 2
Committee on Personal Computers and the Handicapped   312-477-1813
2030 Irving Park Road
Chicago, IL 60618

COPE 2 is a consumers' organization concerned with the use of personal computers for the handicapped and disabled. It can provide free technical assistance, and it maintains a resource library and a networking system for the dissemination of information to its members. Computers are available on loan to the members on a limited basis. COPE 2 is also actively involved in advocating legislative measures for the handicapped. The following is published:

● *Link and Go*—a bimonthly publication that provides information on how people with disabilities are using computers. It reports on pertinent conferences and meetings. The publication, 10-24 pages long, is available free of charge to members.

Council for Exceptional Children (CEC)
1920 Association Drive   703-620-3660
Reston, VA 22091

This membership organization consists largely of teachers and administrators of special education. It has a large professional library oriented toward education and the handicapped. The Council maintains on-line data bases that the staff will search for the public. The data bases are: Exceptional Child Resources—a bibliographic data base with abstracts of literature relating to the handicapped; and ERIC Clearinghouse on Handicapped and Gifted—a bibliographic data base of non-commercially published materials. CEC has produced books, bibliographies, guides, manuals, audiovisual materials, and other materials. A free catalog is available from the Council. Publications include:

● *Microcomputer Resources Book for Special Education*—covers the full spectrum of software and adaptive devices. Approximately one third of the book is devoted to a series of appendices, which provide information about more than 200 publishers of software products. The 224-page publication costs $15.95.

• *Micromputing and Special Education: Selection and Decision Making Process*—provides information and guidelines helpful to administrators and others involved in deciding what kind of computer system to set up in school. The 112-page publication costs $7.95.

• *Proceedings of the National Conference on the Use of Microcomputers in Special Education*—includes papers presented at the conference, results of teacher-training sessions, material on computer literacy, and so on. The 250-page publication costs $20.

International Council for Computers in Education

University of Oregon　　　　　　　　　　　　　　　　　503-686-4414
1787 Agate Street
Eugene, OR 97403

This professional organization, which is dedicated to the effective and proper use of computers in education, can supply you with information about computers, learning, and the handicapped. The staff can direct you to resource people and publications. The Council has produced two publications relating to computers and the handicapped:

• *Resource Guide*—an annotated bibliography covering the latest publications about computers and the handicapped. The 190-page publication costs $7.

• *Learning, Disabled Students and Computers: A Teacher's Guidebook*—introduces simple and practical ways to use computers in the education of disabled students. The 50-page publication costs $2.50.

International Society for Augmentative and Alternative Communications

New York State Association for Retarded Children, Inc.　　516-585-0100, ext. 577
2900 Veterans Memorial Highway
Bohemia, NY 11716

The Society focuses attention on the work being done to help people throughout the world with communicative difficulties. A three-day conference is held each year, which is a good source of information about computers and the handicapped. Susan Sansone is the membership chairperson, and is an excellent information source. She can answer questions and make referrals to experts in the field. Computer services are also available for mentally handicapped adults to train them in various skills. The following publications are available from William and Wilkens Publishers, 428 East Preston Street, Baltimore, MD 21202:

• *Augmentative and Alternative Communication*—a quarterly journal that provides information for those involved with nonverbal communication, speech pathology, special education, rehabilitation engineering, and occupational therapy. The publication is available to members for $20, and to nonmembers for $30.

• *Communicative Outlook*—a quarterly publication that provides information about various computer and mechanical aids for the handicapped and disabled. The magazine includes articles by experts in the field, and profiles various groups and agencies involved in this area. The 24-page publication is available for $12 a year.

Minnesota Education Computing Corporation
3490 Lexington Avenue North   612-481-3660
St. Paul, MN 55112

This firm creates, designs, and distributes educational software for elementary and secondary school levels, and also provides training for teachers and administrators on the use of computers and instructional programs in the classroom. Programs are for normal classroom use and for special education for the handicapped. A free directory is available, describing its software programs. Examples of these are:

● *Special Needs Volume I—Spelling*—oriented toward motor-impaired students. Geared for grades two through six, the program costs $44.

● *Special Needs Volume II—Simulation and Drills*—oriented toward motor-impaired students; presents programs with arithmetic practice. Geared toward grades two through six, the program costs $44.

● *MECC Newsletter*—a bimonthly publication that provides information on instructional computing and new products that are available. The six-page publication is free.

Western Center for Microcomputers in Special Education
1259 El Camino Real, Suite 275   415-326-6997
Menlo Park, CA 94025

The Center assists districts and individuals in acquiring specialized equipment and peripherals to be used with computers in special education for the handicapped and disabled. The Center publishes:

● *The Catalyst*—a quarterly publication that covers the applications of microcomputer technology in the fields of special education and rehabilitation. The 18-page publication is available for $15 a year to agencies and organizations, and for $10 a year to individuals.

Additional Experts and Resources

Handicapped Educational Exchange (HEX)
11523 Charlton Drive   301-593-7033
Silver Spring, MD 20902

HEX is a computer bulletin board which serves both as a clearing house for information about the use of computers to educate the handicapped, and an extensive electronic system that is accessible to deaf people through the use of TDD-telecommunication devices. HEX maintains a large data base of resource material and the staff will access it for you. The system is available 24 hours a day, is free of charge, and requires no access code.

Dr. Everett L. Johnson

Wichita State University   316-689-3415

Department of Electrical Engineering

Wichita, KS 67208

Dr. Johnson is associate professor of electrical engineering at Wichita State University and is the director of research programs involved in the modification of keyboards for the handicapped. Dr. Johnson is also the acting chairperson for the IEEE Computing and the Handicapped Technical Committee, and is an expert in the field. He is available to answer questions or make referrals to other experts in the area of computers for the handicapped.

Computers (General)

Organizations and Government Agencies

American Federation of Information Processing Societies (AFIPS)

1899 Preston White Drive   301-620-8900

Reston, VA 22091

The Federation is composed of 11 national organizations engaged in the design and/or application of computers and information-processing systems. Dedicated to nonprofit scientific and educational purposes, the Federation acts in behalf of its constituent societies in carrying out programs designed to advance the information-processing profession. It organizes two major annual conferences in the computer field. The National Computer Conference, geared toward the computer professional, focuses on what has transpired in the information-processing industry over the past year, as well as future trends. The Annual Office Automation Conference encompasses automation, communications, and integration. The staff will answer inquiries and/or make referrals to appropriate members and associations. The Federation has produced more than 50 reports, studies, and books dealing with information processing. These publications are available to the public for a fee. The following newsletters are available free of charge:

• *The Washington Report*—a monthly publication, eight to ten pages long, containing articles about issues of importance to the information-processing profession. The newsletter covers what is happening in the field, with articles about research and development, education, security, telecommunications, and international issues.

• *Legislative Update*—a quarterly newsletter listing all pending legislation that would affect the information-processing profession. The quarterly is automatically sent to Washington subscribers.

Association for Computing Machinery (ACM)

11 West 42nd Street   212-869-7440
New York, NY 10036

      ACM's 65,000 members are mostly computer programmers involved in software. Members also include engineers, physical scientists, business system specialists, computer analysts, and others interested in the computer field. The Association is concerned with the advancement of information-processing, including the study, design, development, construction, and application of modern-technology computing techniques; appropriate languages for general information-processing, storage, retrieval; transmission/communications, and processing of data of all kinds; and the automatic control and simulation of processes. ACM has 32 special-interest groups covering various aspects of computing as well as local chapters throughout the U.S. The Association's staff can answer technical questions and refer you to an appropriate member for additional information. ACM sponsors more than 40 conferences annually, with topics ranging from computer architects to computer education. Their conferences, courses, and workshops are open to the public. The office listed above will send you a free calendar of events. ACM publishes a journal, a newsletter, conference proceedings, and reports. Its special-interest groups publish their own newsletters, reports, and sometimes a journal. ACM's materials vary in technical content, and prices range from $2.50 to $20. A free catalog of publications is available from ACM.

Computer and Automated Systems Association

Society of Manufacturing Engineers   313-271-1500
One SME Drive
PO Box 930
Dearborn, MI 48128

      This educational and scientific association provides comprehensive and integrated coverage of the field of computers and automation in the advancement of manufacturing. Its members are engineers, managers, and other professionals involved with computers and automation systems. The Association functions as a technology transfer agent, accumulating technical information and disseminating it throughout the manufacturing environment via numerous vehicles. These vehicles include educational conferences and expositions of hardware and software, technical papers on specific topics, books, videotapes, magazines, and newsletters. The Society maintains a research library and a manufacturing data base, which the staff will search for the public. The staff will also answer inquiries and/or make referrals.

Computer-Assisted Training Programs

Call or write your local U.S. Small Business Administration field office, listed in your telephone book,
    or U.S. Small Business Administration (SBA)   202-653-6894—Charles Liner
1441 L Street NW, Room 317
Washington, D.C. 20416

      Under an agreement between SBA and Control Data Corporation (CDC), the federal government will assume most of the fees for small businesses taking computer-assisted management courses offered by CDC. Training under the program, called PLATO (Program Logic for Automatic Teaching Operations), involves four series of learning topics: building your own business; selling, the psychological approach; accounts-receivable collection techniques; and contract bidding. Students enrolled in the program work with both printed materials and a computer terminal.

Computer and Business Equipment Manufacturers Association (CBEMA)

311 First Street, NW   202-737-8888
Washington, DC 20001

CBEMA is a trade association representing the interests of 42 member companies, generally large entities like IBM, Sony, Panasonic, Texas Instruments, AT&T, and Honeywell. CBEMA is involved in issues of concern to the computer industry, including legislation, government regulations, telecommunications, taxation, international activities, ergonomics, product service and distribution, and educational programs. The staff will answer inquiries and/or make referrals. CBEMA sponsors the American National Standards Institute's (ANSI) Secretariat for information-processing standards. Standards and information can be obtained from the ANSI Secretariat at CBEMA's address, listed above.

CBEMA publishes the following materials, which are available to the general public:

• *Computer Careers Booklet*—a six-page publication aimed at the junior-high-school-level through college-level student. Available for $3.

• Set of publications about visual displays—CBEMA publishes fact sheets about video display terminals or cathode ray tubes. Included are scientific reference sheets, recommendations to users, recommendations to employers, and a brief flier entitled *The Facts About Visual Display Units*. The entire set is available for $2.

• *Industry Marketing Data Book*—provides statistics about all aspects of the industry, including imports, exports, employment, sales of various types of products, and educational data. The directory also contains much background reference information about the computer and business equipment industry. The 1985 edition (200 pages) is $195.

• *Dictionary for Information Processing*—a 150-page book of terms used in the industry. Available for $27.

The Computer Language Company

140 West 30th Street   212-736-8364
New York, NY 10001

This company focuses on helping a non-user of computers easily understand complex computer information and terminology. It conducts in-house computer literacy seminars to help a client's staff learn about computers, teaching everything from how computers work to how they can be used to manage a business. The company has produced videotapes, coloring books, and other materials to easily guide novices through the computer field. Examples of publications include:

• *The Computer Glossary*—this paperback book reduces complicated computer concepts to words readily understood by everyone. Directed toward computer beginners, it contains illustrations and more than 1,000 definitions, each of which is cross-referenced. The 298-page publication is available for $14.95.

• *The Computer Coloring Book*—this paperback publication is directed toward anyone wanting to understand computer basics. It contains 25 full-page illustrations as well as 50 clear-cut terms and explanations. The 50-page publication is available for $3.95.

The Computer Museum

300 Congress Street   617-426-2800—Registrar
Boston, MA 02210

The first international museum solely dedicated to computers and computer applications, The Computer Museum contains a continually changing and growing collection of exhibits and artifacts. Exhibits currently cover a cross section of activities in the computer industry from 1950 to the present. Future exhibits may date back to the 1940s and earlier. The Museum's archives section, which will become open to the public in late 1985, contains a variety of artifacts, such as the abacus, calculators, and other computing devices, as well as early software, hardware manuals, and so on. The Museum also maintains a collection, which is open to the public. The staff can answer questions.

Data-Based Computers

505 West Busse Highway   312-825-8124
Park Ridge, FL 60068-3191

Data Processing Management Association (DPMA) is the largest association in the field of information management, with 715 chapters mainly in the U.S. and Canada, but also abroad. DPMA's members include managers, supervisors, systems analysts, educators, and others involved in information processing. The staff will answer questions and make referrals. Each year the Association sponsors an international conference and business exhibition, at which there are more than 200 exhibits on computers, software, computer equipment, and accessories. DPMA's publications, which are available to the public, include:

• *Data Management*—a monthly publication for the data-processing professional, which covers management and technical trends in the industry. Examples of topics from recent issues are decision support systems; telecommunications; performance evaluation; and data-processing managers-in-charge. The cost is $16 a year.

• *Comp-U-Fax*—an industry trends newsletter for data-processing executives. The semimonthly covers data-processing budgets, salary surveys, legislative trends, and developments in the industry. The cost is $10 per year.

IEEE Computer Society

1109 Spring Street, Suite 300   301-589-8142
Silver Spring, MD 20910

This is one of several Societies of the Institute of Electrical and Electronics Engineers (IEEE). With a membership of 80,100 computer engineers, scientists, educators, and managers, the Society is the largest association of computer professionals in the world. The organization supports 31 technical committees, each dealing with a different aspect of computers. Committee members are experts in their respective areas, and they will respond to inquiries. Call or write the Society's office listed above for referral to an appropriate member.

Each year the Computer Society sponsors approximately 80 conferences, workshops, and symposia in the U.S. and abroad. These events are open to the public. The organization has published more than 500 books, journals, reports, conference proceedings, and so on. At least 60 new publications are produced each year. A free catalog of publications is available from the Society. Below is a sample of the Society's publications:

● *Computer Magazine*—a monthly that presents survey and tutorial articles covering a broad range of computer hardware, software, and system design and application. Special issues focus on such topics as VLSI design, software engineering, local area networks, computer communications, and computer architecture. Regular departments present new product announcements, book reviews, and a professional calendar. Available for $90 a year, for 12 issues, and free to members.

● *Computer Graphics and Applications Magazine*—a monthly that covers computer graphics hardware and software, display technology, computational geometry, geometric data structures and data bases, industrial applications, animation methodology, human factors for graphics, interactive graphics, languages, graphic arts, graphics, support of MIS, and distributed graphics techniques. Available for $70 a year, for 12 issues, and available at a discount to members.

● *Design and Test of Computers Magazine*—a quarterly that covers methods, practical experience, research ideas, and commercial products that aid in the design and test of chips, assemblies, and systems—for example, design automation, CAD work stations, design software, computer-aided tests, test equipment, self-tests, and designs for testability. Available for $30 a year, for four issues.

● *Transactions on Computer-Aided Design of Integrated Circuits and Systems* —a quarterly that contains information on methods, algorithms, and man-machine interfaces for physical and logical design, including planning, synthesis, partitioning, modeling, simulation, layout, verification, testing, and documentation of integrated circuit and system designs of all complexities. Practical applications of aids, resulting in producible analog, digital, optical, or microwave integrated circuits are emphasized. Available for $68 a year, for four issues.

Institute for Computer Science and Technology

National Bureau of Standards (NBS)   301-921-2731—Shirley Radack
Washington, DC 20234

The Institute for Computer Science and Technology sponsors two free electronic bulletin boards, which provide information about conferences, articles, and other literature dealing with new standards and technology in computers and software. One bulletin board deals with computer performance evaluation, and the other with microprocessors.

National Bureau of Standards (NBS)
U.S. Department of Commerce   301-921-3112
Route 270
Gaithersburg, MD 20234

The National Bureau of Standards (NBS) provides the measurement standards and data necessary to create, make, and sell U.S. products and services at home and abroad. This involves the in-house research and development necessary to understand the physical quantities involved and to develop means of characterizing and measuring them. It also involves extensive interactions with users at every level to assure that the system is meeting their needs.

The NBS has several offices that specialize in computers and software. The experts in these offices can be extremely useful in identifying new trends, technology, or related information sources:

Data Management and Programming Languages Division
301-921-3553—Helen M. Wood

The Division deals with data management technology, programming languages, and end-user applications.

Systems and Software Technology Division
Center for Programming Science & Technology
301-921-2208—Allen Hankinson

The Center follows software development, hardware maintenance, computer security, risk management, and management selection of computers.

Systems Components Division
Center for Computer Systems Engineering
301-921-2705—John Riganati

The Division deals with storage media, data communications, systems interconnections, computer security techniques, speech input/output, and advance architecture.

Systems Network Architecture Division
Center for Computer Systems Engineering
301-921-3537—John Heafner

The Division follows computer-to-computer networks, office automation, local area networks, and satellite communications.

Publications

Computer Careers: Putting the World at Your Fingertips
CBEMA—Careers   202-737-8888
311 First Street, NW, Suite 500
Washington, D.C. 20001

Updated approximately once a year, this brochure, written for children, covers about 20 different careers. Examples are programmer, hardware designer, information manager, technical writer, computer operator, and repair technician. For each job, it gives a description, educational requirements, and a short interview with someone who holds the position. The 20-page publication is available for $3.

Guide to Free Computer Materials

Education Progress Service, Inc.   414-326-3126
214 Center Street
Randolph, WI 53956

This 350-page directory lists and describes 198 free computer materials, including films, videotapes, disks, pamphlets, booklets, magazines, posters, and books. Some materials are available, but others are loaned free and must be returned to the supplier. The directory also includes a listing of user groups in the U.S. and a glossary with definitions of 1,000 terms used in connection with computers. The guide has 12 subject categories, ranging from business to software. It is available for $30.25.

National Bureau of Standards Publications (NBS Publications)

A 209 Administrative Building   301-921-2834
U.S. Department of Commerce
Washington, DC 20234

NBS has produced several publications dealing with various subjects in the computer and software field. One of the publications is a free newsletter, which is available from NBS at the above address. All other publications listed below must be purchased from the Superintendent of Documents, U.S. Government Printing Office, Washington, DC 20402, 202-783-3238.

• *ICST Newsletter*—NBS's Institute for Computer Sciences and Technology publishes a free newsletter that keeps readers up to date on the Institute's current activities. Information is provided about new publications, conferences, seminars, guides, and standards.

• *Selection of Local Area Networks*—describes features of networks and advises how to determine requirements and evaluate proposals for networks. The cost is $6.

• *Care and Handling of Computer Magnetic Storage Media*—this comprehensive reference guide answers many questions about physical care and handling, storage, transportation, and recovery of damaged media and archival factors. The cost is $5.50.

• *Microcomputers: A Review of Federal Agency Experiments*—provides results of a six-month study of the management and technical issues associated with the use of microcomputers. The cost is $5.50.

• *Future Information Processing Technology—1983*—contains forecasts of hardware and software technologies and products, and changes in the information industry and markets. It covers legal, regulatory, and management issues and costs $6.50.

● *Software Validation, Verification and Testing Technique and Tool Reference Guide*—provides detailed information about the requirements and capabilities of a wide variety of techniques and tools. It is available for $6.

Datamation

Circulation and Editorial Department   212-605-9400
875 Third Avenue
New York, NY 10022

Datamation is a general-interest magazine servicing the field of information processing in the U.S. and abroad. The 200-page biweekly focuses primarily on information management and information technology. The magazine covers topics of interest to executives and companies using the services of data processing, word processing, and communications. *Datamation* also has features of interest to designers and vendors of related products and services. Each issue has feature articles about data-processing management, technology, and industry issues, along with a comprehensive review of new products and services. International news and information about forthcoming events are also provided. The annual subscription rate is $50.

Electronic News

PO Box 1013   201-899-5566
Manasquan, NJ 08736

Electronic News is a weekly business newspaper, one quarter of which is devoted to computers. The computer section is divided into three topics: (1) systems; (2) peripherals; and (3) software. Geared toward people in the electronics industry, the paper mainly covers business trends. Articles also feature new products and their technology. *Electronic News* also covers components, communications, materials packaging and production, test and measurement, financial news, consumer electronics, government electronics, distribution of electronic products, and worldwide industry developments. The annual subscription rate is $30.

Computer Decisions

10 Mulholand Drive   201-393-6030
Hasbrouck Heights, NJ 07604

Computer Decisions is a 220-page monthly magazine serving managers of data processing and management information systems, as well as executives in the computer industry, government, and nonprofit organizations. Regular columns appear about data communications, personal computing, software, office automation, careers, and new products. The magazine is available free of charge to qualified subscribers, or for $35 a year to others.

Computer World Newspaper

375 Cochituate Road   617-879-0700
Box 880
Framingham, MA 01701

Computer World Newspaper is a weekly publication geared toward people working as managers of data processing departments in large organizations, chiefly Fortune 500 companies. The newspaper covers data-processing management and computer technology, including software, hardware, telecommunications, microcomputers, and office automation. The weekly follows industry trends, new developments, financial news, events such as conferences and meetings, and news of vendors. The annual subscription rate is $44.

MIS Week

7 East 12th Street   212-741-4000
New York, NY 10003

MIS Week is a newspaper for people in the computer industry, especially those involved in information management. Regular sections in the weekly cover information systems; computer software, data bases, and services; telecommunications; computer-related applications; and computer industry news, such as new products, financial information, and personnel changes. The newspaper is available free of charge to qualified subscribers. The annual subscription rate for others is $40.

Computer Media Directory

COMPUMEDIA
3220 Louisiana Street, Suite 213
Houston, TX 77006

The *Directory* is a quarterly list of computer publications and other publications that report computer news regularly. The major categories are general computer publications; publications concerning computer manufacture, design, and theory; publications concerning computer marketing and distribution; and general business and noncomputer trade publications. Available for an annual subscription rate of $149.95.

Additional Experts and Resources

Since most educational institutions offer programs in computer science, your area college is probably a good source of information. Listed below are the major university computer science departments in the U.S.

Carnegie-Mellon University
Computer Science Department
Pittsburgh, PA 15213
412-578-2565

Stanford University
Computer Science Department
Stanford, CA 94305

University of California, Berkeley
Computer Science Department
438 Evans Hall
Berkeley, CA 94720
415-642-7214

Massachusetts Institute of Technology (MIT)
Lab for Computer Science
545 Technology Square
Cambridge, MA 02139
617-253-5851

MIT's computer science center is involved in research and development concerning all aspects of artificial intelligence and computer applications in the twentieth century and beyond.

University of Illinois
1304 West Springfield Avenue
Urbana, IL 61801
217-333-4428

The staff in the university's information office will answer questions and make referrals to experts in the field. Both the staff and graduate students publish papers and reports throughout the year, and a bibliographic listing is available from the information office. The reports, which range from 20 to 250 pages, are available free of charge.

Data Bases

Computer SPECS
Data Resources, Inc. (DRI)   617-863-5100—Christina Huston
24 Hartwell Avenue
Lexington, MA 02173

Data Resources, Inc. (DRI) and GML Corporation have developed an on-line service called Computer SPECS. This service is designed to provide vendors, buyers, and users with the capability of accessing and analyzing current information on computers of all sizes, microcomputer software, printers, terminals, and various storage devices.

Computer SPECS contains 11 data bases, including detailed specifications for minicomputers, mainframe computers, and microcomputers; microcomputer software; disk drives, magnetic tape units, cassettes; display and teleprinter terminals; and line and serial printers. Each of these data bases contains comprehensive technical data about thousands of products, including company addresses, prices, communication features, and detailed physical characteristics and applications. The information is continually updated by GML Corporation.

Computers in Education

Organizations and Government Agencies

Association for the Development of Computer-Based Instructional Systems (ADCIS)
Miller Hall, Room 409   206-676-2860
Western Washington University
Bellingham, WA 98225

ADCIS is an international association of 2,000 members involved in the instructional uses of computing. Its members are from educational institutions, government, and industry. ADCIS has 13 special-interest groups, each dealing with a different aspect of education computing. Its members will respond to inquiries and make referrals. The Association sponsors an annual conference and publishes a journal and newsletter. The publications are available to members only, and membership cost is $40 for individuals and $75 for institutions. ADCIS publications are:

• *Journal of Computer-Based Instruction*—a scholarly quarterly containing research studies, literature reviews, and other articles of interest to those in the field. Subjects have included education of the handicapped, health education, computer-based training, and music.

• *ADCIS Newsletter*—published six times a year, the newsletter contains up-to-date information relating to educational computing. It includes literature abstracts, new developments in the field, and information about forthcoming events and conferences.

Association for Educational Data Systems (AEDS)

1201 16th Street, NW   202-822-7845
Washington, DC 20009

AEDS is an international association of professionals concerned with computers, as they relate to education, particularly their use in administration and education at all grade levels. Membership consists of 2,500 teachers, professors, directors of data processing, department of education officials, administrators, and others interested in the field. The organization has a small staff, but it will try to meet the public's requests for information and referrals. Most of the Association's information is dispersed through publications, meetings, workshops, and its annual convention. AEDS publications include:

• *AEDS Journal*—a technical publication with an emphasis on research and actual experience. This quarterly contains research papers on the instructional and administrative uses of computers, as well as the latest developments in the application of computer technology to all levels of education. The annual subscription to the 60-page journal is $32.

• *AEDS Monitor*—a bimonthly publication with articles that explore current developments and directions in educational computing. Its editorial divisions are Instructional, Administrative, and Computer Science. Each issue contains reviews of new software and other new products. The annual subscription rate is $28.

• *AEDS Newsletter*—a bimonthly publication with timely news and information of interest to the educational computing audience. The newsletter contains Association news, chapter activities, current news items, and information about forthcoming educational computing conferences, workshops, and meetings. The annual subscription to the six-page newsletter is $9.

International Council for Computers in Education (ICCE)

University of Oregon   503-686-4414
1787 Agate Street
Eugene, OR 97403

ICCE is an international professional organization dedicated to the effective and proper use of computers in education. Its membership includes 23,000 individuals and 44 organizations, ranging from statewide organizations to small user groups in Japan. ICCE's primary goal is to connect people with helpful resources, experts, and organizations. In addition to being a resource and referral center, ICCE also produces publications of interest to professionals in the field and the general public. It has produced publications on computer technology for the handicapped, parents' guide to computers in education, and computers and composition instruction. The publications generally cost about $3 and ICCE will send you a free catalog describing the materials. Of particular note is ICCE's journal *The Computing Teacher*, which is published nine times during the school year. It focuses on kindergarten through the twelfth grade, and is geared toward anyone interested in the instructional use of the computer. The journal emphasizes using computers, teaching about them, the impact of computers on the curriculum, and teacher education. Free to members, the journal is available to others for $21.50 a year (which is the same price as membership).

Publications

Classroom Computer Learning

Pitman Learning, Inc.   415-592-7810
19 Davis Drive
Belmont, CA 94002

This magazine contains articles for teachers concerning computer use in the classroom, interesting research and innovative philosophies in computer teaching and learning, as well as discussions of controversial issues in computer education. Twice a year an extensive software directory is included in the publication. The yearly subscription fee is $22.50, for nine issues and two directories.

Electronic Education

PO Box 20221   904-878-4178
Tallahassee, FL 32316

Electronic Education, a monthly magazine about computers and education, is directed toward secondary-school-level through college-level teachers and administrators. It contains book reviews, a calendar of events, news briefs of general and specific interest, software reviews related to curriculum planning, industry news, new product and new software listings, briefs about administrative uses of computers, and information about pending federal legislation. The annual subscription is $18, for eight issues.

Scholastic, Inc.
Editorial Office   212-505-3000
730 Broadway
New York, NY 10003

Scholastic, Inc. publishes the following:

● *Electronic Learning*—is geared toward decision makers in educational technology at the kindergarten level through the twelfth-grade level. Its audience includes school administrators, teachers, computer coordinators and laboratory people, and others interested in the field. Published monthly during the school year, the magazine provides a blend of buying information and practical help to educators trying to learn how to use technology, especially microcomputers, creatively and effectively. Eight issues are published yearly for a subscription fee of $19.

● *Teaching in Computers*—is published monthly during the school year, and focuses on kindergarten through the eighth grade. It contains teaching units, quick tips, and general-feature articles—all showing teachers how to use computers in the classroom with children. Departments include a question-and-answer column, reviews of software and new products, lessons in Logo, passcards teaching Basic language, and instructional posters and calendars. Eight issues are published yearly, for a subscription cost of $19.

T.H.E. Journal
PO Box 17239   714-261-0366
Irvine, CA 92713

The *Journal* concentrates on computers, interactive video, robotics, and other technologies, either as used in the administration of education, as a means for aiding education, or as a curriculum subject. Feature articles are written by educational administration, and each issue contains information of practical use to readers, such as new products, book reviews, and applications. The 150-200-page magazine is published monthly during the school year, and is available free of charge to school and training administrators in educational institutions and in industrial, vocational, and commercial training programs.

Computers, Used

See: Used Computers

Conferences, Meetings, and Trade Shows on High Tech

Publications

Proceedings in Print

Proceedings in Print, Inc.   617-646-0686
1026 Massachusetts Avenue
Arlington, MA 02174

Proceedings in Print, a bimonthly publication, cites conferences, symposia, institutes, and meetings held in all fields of knowledge, covering events worldwide in all languages. Each issue contains a minimum of 500 entries, and the following information is provided for all entries: title of conference, where and when held, names of sponsors, and a complete bibliographic citation. The bibliographic citation includes title of proceedings, editors, place of publication, publisher, date of publication, and price (when possible). Each issue contains an in-depth index, including all editors, sponsors, and subjects. A *Cumulative Index* is available (on a subscription basis only) shortly after publication of the sixth issue of *Proceedings in Print*. The staff maintains a computerized data base, which it will search for clients, free of charge. *Proceedings* is $425 a year, and the *Index* is $125.

World Convention Dates (WCD)

79 Washington Street   516-483-6881—Kevin Ward
Hempstead, NY 11550

WCD publishes reference dictionaries, which can help you identify upcoming events in high technology:

• *U.S. Geographic Events Guide*—published each spring, this *Guide* is a geographical calendar of conventions, trade shows, meetings, and expositions to the year 2000. The directory includes data on over 28,000 projected events, presented in geographical format by state, city, and date of event. Each entry includes estimates on size of event and indicates if exhibits and banquets are planned. Available for $30.

• *Generic Directory of Conventions, Trade Shows and Expositions*—issued each fall, this directory includes over 28,000 events, classified by 132 industry fields, including high-technology topics such as aerospace, computers, and electronics. Events are covered through the year 2000, and special indexes and cross referencing allow you to locate particular events by name, as well as by related fields of interest.

• *Canadian and International Events Guide*—counterpart to the *U.S. Geographical Events Guide*, this directory profiles international meeting activities by place and time, for Canada and major countries throughout the world. Issued each summer, the *Guide* gives the names of event planners and attendance estimates. Available for $10.

Note: WCD offers a variety of subscription services through which you can purchase their publications and monthly updates at discounts.

Successful Meetings

633 Third Avenue   212-986-4800—Edward Murray, Book Division
New York, NY 10017

Successful Meetings publishes two directories containing information about high-technology conventions and trade shows. They are:
• *Directory of Conventions*—lists over 13,000 North American-based organizations in 82 industries, holding more than 23,000 conventions throughout the world during the upcoming two years. Information provided includes date, place, expected attendance, and contact information. The 382-page book costs $80.
• *Exhibit Schedule*—this annual directory lists over 6,000 associations and exhibition sponsors, holding more than 10,000 trade shows and exhibits in the U.S. and abroad. The 250-page hardcover book costs $85.

National Trade and Professional Associations of the U.S.

Columbia Books, Inc.   202-737-3777
777 14th Street, NW, Suite 236
Washington, DC 20005

This directory, revised each January, contains 6,000 entries covering the whole range of trade and professional associations. Information is provided about each organization's scheduled meetings and future plans. The book is cross-referenced by geographical location, budget, and interests. Available for $40.

World Meetings

866 Third Avenue   212-702-2000, ext. 4296
New York, NY 10022

World Meetings publishes the following guides, each on a quarterly basis:
• *World Meetings: United States and Canada*—offers details on all important scientific, technical, and medical meetings and conferences to be held in the U.S. and Canada during the next two years. Includes contact information, content of each meeting or technical session, exhibits and products to be shown, deadlines for submitting papers, and dates on which conference proceedings will be published. Available for $130 a year.
• *World Meetings: Outside U.S. and Canada*—contains information on all forthcoming major scientific, technical, and medical meetings and conferences to be held outside the U.S. and Canada. It includes contact information, content of each meeting or technical session, deadlines for papers, and dates on which abstracts, papers, reprints, preprints, and conference papers will be published. Available for $150 a year.

● *World Meetings: Social and Behavioral Sciences, Human Services, and Management*—covers worldwide meetings and conferences of all the social and behavioral sciences, including information science, communications, management and public administration, statistics and operations research, and more. Information for each entry includes location, date, agenda, and deadlines for papers.

● *World Meetings: Medicine*—covers the meetings and conferences to be held during the next two years. Details are provided on location, date, contents, and goals of meetings.

Data Bases

World Convention Dates (WCD)
79 Washington Street   516-483-6881—Kevin Ward
Hempstead, NY 11550

This firm maintains a computerized data bank with information on over 32,000 events worldwide. Data are available for conventions, trade shows, meetings, and expositions held within the past three years, and for events projected for the next 20 years. Retrievable information includes place of event, dates, agenda, and contact information. The data base is searchable by geographic zip code, type of event, budget, and more.

Additional Experts and Resources

Most associations hold meetings, conferences, and educational seminars. You can contact directly the individual association listed elsewhere in *Info High Tech*.

Many trade and professional magazines, newsletters, and journals contain announcements about events of interest to their readers. See the appropriate topic in this book for a listing of periodicals to check.

Construction

See: Architectural Engineering

Consumer Electronics

Organizations and Government Agencies

Consumer Electronics Group (CEG)
Electronic Industries Association   202-457-4919
2001 Eye Street, NW
Washington, DC 20006

CEG is a membership group of manufacturers of consumer electronics products. It tries to keep the public informed of new technological developments, and is an excellent source for consumer electronics statistics. The Public Affairs office will answer questions and refer people to appropriate experts and resources. The Group's publications include:

• *Consumer Electronics News*—a free bimonthly newsletter that tracks technological developments, new products, and marketing trends in all consumer electronics product categories.

• *EIA/Consumer Electronics Group Quarterly Review*—provides news of the Association's activities in government affairs, marketing services, engineering, and industry developments programs. It is available for free.

• *Consumer Electronics Annual Review*—provides production and sales data and marketing information for all consumer electronics product categories, covering a 12-year span. It costs 50 cents.

• *Electronic Market Trends*—monthly publication with articles on lay electronics markets, new technologies, international developments, and important policy decisions. It is available for $150 a year.

• *Electronic Market Data Book*—statistical information on the electronics industry, covering production, sales, foreign trade, research and development, and U.S. government markets. Published annually; costs $55.

• *Electronic Foreign Trade*—a monthly publication that identifies import/export trends. It is available for $150 a year.

• *Electronic Industries Information Source*—annual directory of organizations and publications that provide information on electronic industries and markets. The cost is $20.

Consumer Electronics Society (CES)
IEEE   212-705-7900
345 East 47th Street
New York, NY 10017

The Society is concerned with all aspects of consumer electronics, which include electronic equipment for audio and video signal generation reproduction, and processing as well as related devices, normally used by the consumer for educational, informational, and leisure-time entertainment purposes. CES has chapters world-wide, and their members will answer questions from the public. For a referral to an expert in your area, call the IEEE headquarters listed above. The Society publishes the following:

● *Transactions on Consumer Electronics*—covers the design and manufacture of consumer electronics products, components, and related activities, particularly those used for entertainment, leisure, and educational purposes. The cost of this quarterly is $64 a year, and it is also available in microfiche.

● *Proceedings from the International Conferences on Consumer Electronics*—the papers cover such areas as TV signal processing, cable television, video recording and playback systems, electronic photography, electronic tuning systems, integrated circuit technology and components, and advanced audio techniques. It is available from 1982.

Publications

Consumer Electronics
135 West 50th Street   212-957-8800
New York, NY 10020

This monthly magazine reports many aspects of the consumer electronics industry. Topics regularly covered are general news, people in the industry, technology advances, statistical information, industry, buying guides, marketing information, special reports on different products, and so on. The cost is $50 a year.

Television Digest, Inc.
1836 Jefferson Place, NW   202-872-9200
Washington, DC 20036

● *TV Digest with Consumer Electronics*—this weekly report covers the broad-cast cable and consumer electronics fields. Half of the publication is devoted to many consumer electronics topics, with information on companies, conferences, legislation, international news, new products, technology advances, and so on. It is available for $406 a year.

● *Consumer Electronics Video Data Book*—a source of information for market statistics and projections covering color and monochrome television, VCRs, videocas-settes, videodiscs and players, and audio equipment. Published annually, it costs $125 a copy.

Consumer's Resource Handbook
Consumer Information Center (CIC)
Pueblo, CO 81009

The *Consumer's Resource Handbook*, prepared by the U.S. Office of Consumer Affairs, is an excellent resource for identifying whom to contact in the government or industry regarding product complaints. It lists corporate consumer contacts; trade associations; federal, state, county, and city government consumer protection offices; private and voluntary consumer groups; and much more. Single copies are available for free, but you must make your request in writing.

Additional Experts and Resources

International Trade Administration

U.S. Department of Commerce, Room 4044   202-377-0570—E. McDonald Nyhen
Washington, DC 20032

A specialist in the consumer electronic industry is always available to answer questions about the field and suggest other sources of information. The expert prepares the section in the *Industrial Outlook* on consumer electronics.

U.S. Consumer Product Safety Commission (CPSC)

Washington, DC 20207   800-638-CPSC
800-638-8270 (Teletype Number for the Hearing Impaired)

The CPSC Hotline staff can answer questions about product safety and refer you to other sources of information. The staff can also provide information about product recalls.

Contraception

See: Reproductive Technology

Cooperative Programs (Government/Industry/University Partnerships)

See also: Technology Assessment, Technology Transfer

Organizations and Government Agencies

Business—Higher Education Forum

One Dupont Circle, Suite 800   202-833-4716
Washington, DC 20036

The Forum, consisting of 80 corporate and academic chief executive officers, was founded in 1978 in affiliation with the American Council on Education. Its major objective is to provide an opportunity for interchange among its members. The group frequently issues policy statements about the high-technology and business-university cooperation. The following two reports are available to the public:

● *Corporate and Campus Cooperation: An Action Agenda*—details the ways in which American business and institutions of higher education can work together for their mutual benefit. The 40-page report, published May 1984, makes a case for partnerships between the two communities, and it outlines the advantages of joint initiatives in such vital areas as contract research, the sharing of facilities and staff, adult education and training, corporate financial support, corporate recruitment, and more. Available for $9.

● *The New Manufacturing: America's Race to Automate*—analyzes the rapid transformation of the American workplace and assesses the implications of the computer revolution on the U.S. work force. The 52-page report, published June 1984, addresses such questions as: How rapidly are robots and other forms of advanced automation taking over America's manufacturing facilities? Whose jobs will be lost? What jobs will be created? What impact does automation have on the management of companies and on the very structure of the U.S. economy? It is available for $12 a copy.

Industrial Liaison Program

Massachusetts Institute of Technology (MIT)   617-253-2691—Professor James Utterback, Director
Bldg. E38, 4th Floor
292 Main Street
Cambridge, MA 02139

This program is an integral part of MIT's Resource Development operation. Its basic function is to help member companies of the program assess the importance of emerging fields of science, technology, and management. Companies pay a fee for the service, which ranges from $10,000 to $50,000. The program acts as a source of information in many high-tech areas, as well as a liaison with all departments at MIT in which industry might be interested. It is a good overall referral source for much of the high-technology information available at MIT, but the services are available on a fee basis only. Departments of particular interest are electrical engineering and computer science, mechanical engineering, materials science and engineering, chemical engineering, aeronautical and astronautical engineering, civil engineering, ocean engineering, biology, chemistry, physics, nutrition and food science, and earth and planetary sciences.

National Governors Association (NGA)

Hall of States   202-624-7814—Task Force on Technological Innovation
444 North Capitol Street, NW   202-624-5311—Committee on Economic Development
Washington, DC 20009   and Technological Innovation
                        202-624-5395—Office of Research and Development
                                202-624-7880—Publications Office

The Association of governors of the United States and its territories is organized to promote the interests of the states. Charilyn Cowan directs the NGA Task Force on Technological Innovation. The largest Task Force at NGA, this group comprises more than half the governors. Ms. Cowan is a good resource for obtaining information about what each state is doing regarding technological innovations and cooperative programs. Richard Geltman is Staff Director for NGA's Committee on Economic Development and Technological Innovation. He is another good resource person at the Association. Several publications pertaining to cooperative programs include:

● *Technology and Growth: State Initiatives in Technological Innovation*—a 120-page 1983 report, which surveys state initiatives to promote technological innovation. Issues covered are: (1) development of state policies; (2) support for education; (3) support for education, worker training, and employment; (4) technical and management support to small businesses; (5) economic support for new business growth. The publication can be ordered from NGA's Publications Office for $8 a copy.

● *Annotated Bibliography*—Charles Watkins, a fellow in NGA's Office of Research and Development, has prepared a critical annotated bibliography, which is available free of charge to the general public. The bibliography cites publications related to state roles and economic development through the use of high technology. Included are reports of think tanks; universities; industries; and federal, state, and local governments.

National Science Foundation (NSF)

Division of Industrial Science & Technological Innovation   202-357-9805—Lou Tornatzky
Productivity Improvement Research Section   Section Head
1800 G Street, NW, Room 1237
Washington, DC 20550

This NSF office conducts and funds technological innovation studies of university/industry relations. The office frequently publishes reports about cooperative research. Lou Tornatzky and his staff have a great deal of expertise in looking at the organizational aspects of this problem, and they are concerned with how the process of technology innovation is managed. They will answer questions and make referrals.

National Science Foundation (NSF)

Industrial Science and Technological Innovation   202-357-7527
Industry/University Cooperative Programs
1800 G Street, NW, Room 1250
Washington, DC 20550

This office can provide information about NSF-sponsored industry/university cooperative programs, free publications and other information sources. NSF sponsors two major programs:

● *Industry/University Cooperative Research Projects Program*—directed by Fred Betz, the program is one-on-one, in which NSF funds collaborative research projects between a university scientist and an industry scientist.

● *Industry/University Cooperative Research Centers Program*—directed by Alex Schwartzkopf, this program is cofunded by NSF and industry sponsors. The funds are considered seed money and are used to initiate large research programs within a university. Funded programs must address industrially relevant research.

Office of Technology Assessment (OTA)

U.S. Congress   202-224-8996—Publications Office
Washington, DC 20510   202-226-2173—Paul Phelps,
                                        Director for Regional High Technology Development Projects

OTA is a nonpartisan analytical agency that supports the U.S. Congress by providing objective analysis of major public policy issues related to scientific and technological change. Paul Phelps, OTA's expert on cooperative research and development, will answer inquiries and make referrals. OTA has published several reports of interest to people involved in cooperative research and development. Summaries of these reports are available free of charge from OTA's Publication Office. The full text of these reports must be purchased from the Government Printing Office, Washington, DC 20402, 202-783-3238. Publications include:

• *Technology, Innovation and Regional Economic Development*—a report that identifies and describes the efforts of state and local governments, universities, and private-sector groups to promote the creation, expansion, and retention of high-technology firms and industries.

• OTA has published several reports on specific technologies, such as robotics, biotechnology, microelectronics, and computer science. In addition to discussing the particular technology, these reports contain information about institutional arrangements for development of these technologies. OTA also publishes a free catalog describing its reports.

Small Business Development Centers (SBDCs)

U.S. Small Business Administration (SBA)   202-653-6768—Ms. Johnnie Albertson,
1441 L Street, NW, Suite 317   Deputy Associate Administrator for SBDCs
Washington, DC 20416

SBDCs are the result of a three-way cooperative agreement among the federal government, a state governor's office, and educational institutions. Centers exist in approximately 40 states, and part of their role is to provide technical assistance to small businesses through counseling, training, and economic development activities. Small business owners or potential entrepreneurs can use their state's SBDCs service for little or no cost. Call your local SBA office or SBDC listed in the phone book under "U.S. Government, Small Business Adminstration." For further information, contact the Washington, DC office listed above.

State Government "One-Stop-Shops" for Small Businesses

Most state governments have established small business service/help/action centers, often called "one-stop-shops" for small businesses. Many of these offices are involved in creating a research and development environment, and promoting partnerships among business, industry, and universities. The staff can provide information, publications, and referrals. Since the centers often try to entice new businesses to their states, their services are for both in-state and out-of-state residents. These centers are often part of a state's Department of Commerce. To find a particular state's small business help center, contact the appropriate state capitol's information operator or a state representative's office. (Note: The names of these centers vary from state to state.)

Additional Experts and Resources

Carlos Kruytbosch
Science Reports and Special Projects Officer   202-357-7791
Office of Planning and Special Projects
National Science Foundation (NSF)
1800 G Street, NW, Room 1225
Washington, DC 20550

Dr. Kruytbosch is a sociologist studying organizational patterns of science and technology. He has coordinated two major National Science Board (the policy-making board of NSF) reports about industry-university research relationships. Dr. Kruytbosch will answer inquiries and make referrals. Free copies of the following reports can be obtained from his office while the supply lasts:

● *Industry/University Research Relationships: Myths, Realities and Potentials* (Pub. #NBS82-1)—this 35-page report provides an overview of the state of the industry/university research relationships.

● *Industry/University Research Relationships: Selected Studies* (Pub. #NBS82-2)—this publication includes six background studies. They are a large field study of 450 cases of research relationships, a study of the history of the industry/university relationship in chemistry and chemical engineering, a study of the relationship between nondoctoral schools and industry, a study of intellectual property issues, a paper on industry/university/state government consortia in microelectronics research, and an annotated bibliography with an index of the overall report.

Community Partnerships Resource Center (CPRC)
U.S. Department of Housing and Urban Development   202-755-4370—Joe Panaro
Office of Policy Development and Research
451 7th Street, SW
Washington, DC 20410

The Center helps communities and neighborhood groups solve problems through the development of public-private partnerships. NPRC's main goal is to put cities, businesses, neighborhood groups, and leaders in touch with each other to learn about new solutions and to set up programs. The staff members have researched partnerships throughout the U.S., and they can put you in contact with groups working on a problem similar to yours. The Center maintains an on-line data base of examples of local problem-solving groups and partnerships. The NPRC data base is available on LOGIN, a subscriber service, and Partnership Data Net (see next entry), a nonprofit membership network with some free services for the public. If time permits, NPRC will also search their data base for interested parties.

Partnership Data Net (PDN)

1015 18th Street, NW, Suite 300   202-293-8380
Washington, DC 20036   800-223-6004

     This is a nonprofit organization established to encourage collaborative problem-solving efforts among the private, public, and nonprofit sectors. PDN operates a national computer network to link these three sectors. The network includes several data bases, an electronic bulletin board and mail service, and teleconferencing capability. Two major data bases are the Community Partnership Resource Center Data Base (CPRC) which contains examples of local partnerships throughout the U.S., and the Private Sector Initiative Data Base, which contains information about partnerships in education, economic development, jobs, crime prevention, and more. The Washington office can give you information about joining the network (a $200 fee) and data base contents, and can direct you to appropriate PDN members and partnerships. Members can search PDN's data bases and use the electronic services for a minimal fee. The staff at the toll-free number listed above will search PDN's data bases, free of charge, for the public. The organization plans to offer conferences, workshops, and seminars on partnerships.

Copyrights

See: Patents, Trademarks, and Copyrights

Crime

See: Computer Crime

Crop Genetics

See: Agriculture; Biotechnology

Cryogenics Technology

Organizations and Government Agencies

ALCOR Life Extension Foundation
4030 North Palm, Suite 304   714-738-5569
Fullerton, CA 92635

     The Foundation is involved in the low-temperature preservation of humans after death, in the hope that they can be restored to life and health in the future. The Foundation also conducts research in the fundamental aspects of the low-temperature preservation of living systems. Through meetings and participation in annual conferences, the Foundation provides information on cryonics. In addition, an information handbook and brochures are available upon request. Staff members can answer your questions and make referrals. Publications include:

     • *Cryonics*—a monthly journal containing information about the scientific, technical, and philosophical aspects of cryonics. Articles also deal with cryobiology and life-extension systems. Each issue runs 23 to 32 pages. The annual subscription is $15.

Bay Area Cryonics Society
c/o Dr. Paul Segall   415-525-7114
Secretary/Treasurer
1098 Euclid Avenue
Berkeley, CA 94708

     This nonprofit membership organization promotes cryonics and cryonics suspension. The Society can arrange for qualified organizations to provide cryonic suspension and maintenance services. It also supports scientific research in cryonics. Dr. Segall can answer your questions or refer you to an appropriate source. Literature can be provided upon request.

Cryoelectronics Section
Division 724.03   303-497-3988
National Bureau of Standards
Boulder, CO 80003

     The Cryoelectronics Section is involved in the research and development of superconducting devices that operate in a cryogenic environment. Among these Josephsen devices are ultra-sensitive magnetometers, voltage standards, and high-speed digital electronic devices. The staff can answer your questions. Technical reports are available upon request.

The Cryogenic Society of America
1033 South Boulevard   312-383-7053
Oak Park, IL 60302

   This organization of engineers, researchers, manufacturers, and users is involved with cryogenics and cryogenic applications. Areas of interest include liquid gases, electromagnetism using superconductors, medical techniques, and food freezing. The staff can refer you to an appropriate cryogenic expert. The Society sponsors conferences, after which the proceedings are published and made available for sale.

Electricity Division
National Bureau of Standards (NBS)   301-921-2701
Methodology Building
Room B 258
Gaithersburg, MD 20899

   The Electricity Division is involved in research and calibration regarding the maintenance of the unit voltage using the Josephsen effect. The staff can answer your questions and make referrals. Technical reports can be sent to you upon request.

Fermilab
M.S. 208   312-840-3200
Box 500
Batavia, IL 60510

   The primary functions of the laboratory have to do with research and development in high-energy physics. With respect to cryoengineering, Fermilab has a unique system for the production of high-energy particles called the Superconducting Energy Saber, which requires a major helium refrigeration system to cool it. The lab is a leader in the development of superconducting cable, and it also develops superconducting magnets. Research is being done involving low-temperature materials and properties. Inquiries should be directed to Mr. Carrigan. Literature is available regarding cryogenics and the work being done at the lab. Access to the lab is for those Fermilab Industrial Affiliates who pay $1,000 for membership.

Goddard Space Flight Center
National Aeronautics and Space Administration (NASA)   301-344-5405
Code 713
Greenbelt, MD 20771

   The Flight Center is involved with cryogenics related to space activities. Applications include long-term storage of cryogens on space stations; long-term cooling of infrared instruments with large-aperture heat, which involve earth observations, atmospheric science, meteorological satellites; gamma-ray and x-ray astronomy, and long-term storage of liquid helium, which requires thermal shielding and the elimination of aperture heat loads with a mechanical cooler. The staff at Goddard can answer your questions, and technical papers are available upon request.

Jet Propulsion Lab

M.S. 183-901   213-354-6581
California Institute of Technology
4800 Oak Grove Drive
Pasadena, CA 91109

The Jet Propulsion Lab develops cryogenic systems in space craft and cryocoolers for space and ground applications on behalf of NASA. The Lab also conducts experiments using liquid helium in space. The staff can answer questions and refer you to specific publications.

Liquid Carbonic Corporation

135 La Salle Street   312-855-2500
Chicago, IL 60603

The Liquid Carbonic Corporation manufactures industrial and medical gases, as well as related equipment. The company has a complete line of food-freezing equipment, which involves low-temperature processes. Technical literature is available about the gases and equipment from the Advertising and Public Relations Department. The staff can answer your questions and refer you to an appropriate information source.

Oxford Superconducting Technology

600 Milik Street   201-541-1300—Dr. David Andrews, Dr. Philip Sanger
Carteret, NJ 07008

This company produces superconducting magnets for magnetic resonance imaging, which is a recent technique for imaging without using x-rays. The magnets are made of cryogenic fluid. The firm also produces superconducting wire for the magnets. Dr. Andrews and Dr. Sanger can answer questions regarding superconducting magnets. Company literature is available upon request.

Society for Cryobiology

c/o New York Blood Center   212-570-3079
310 East 67th Street
New York, NY 10021

The Society, whose membership consists of scientists from medicine, industry, and universities, is devoted to the research and dissemination of knowledge regarding cryobiology. An annual meeting is sponsored. The Society's staff can answer your questions and make referrals. The following publication presents current information:

● *Cryobiology*—a bimonthly journal that contains original research and review articles dealing with all aspects of cryobiology. This includes studies of freezing, freeze drying, cryoprotective additives and pharmacological actions, medical applications of reduced temperatures, cryosurgery, hypothermia, perfusion of organs, hibernation, and frost hardiness in plants. The annual subscription to the approximately 110-page journal is $47 for members, and $94 for nonmembers.

Society for Low Temperature Biology

MRC Medical Cryobiology Group   011-44-0233-63919—Dr. John Armitage
Douglas House
Trumpington Road
Cambridge, England CB2AH

The Society is a small international organization, which has three or four meetings a year. Interest areas include low-temperature storage of plants, seeds, and germ plasma; cryosurgery; storage of blood and other products for transportation; storage of organisms; embryo and sperm preservation for the improvement of livestock production; and human ova and embryo preservation. Members of the Society can respond to your inquiries.

Temperature Division

National Bureau of Standards (NBS)   301-921-3315
Physics Building, Room B128
Gaithersburg, MD 20899

The Temperature Division is involved in the measurement of cryogenic temperatures. Standard reference devices are developed based upon the superconducting transition of certain materials. These standard reference materials can be purchased from The Office of Standard Reference Materials, National Bureau of Standards, Gaithersburg, MD 20899, 301-921-2045. The staff can answer your questions and make referrals. Technical reports can be sent to you upon request.

Ultra-Low Temperature Systems and Cryogenics Committee

American Society of Heating, Refrigeration & Airconditioning Engineers   404-636-8400
1791 Tullie Circle, NE
Atlanta, GA 30329

The Committee is concerned with all aspects of refrigeration at temperatures below 260°F. Members are primarily mechanical and chemical engineers who work in the cryogenics industry. The staff can refer you to an appropriate expert in cryogenics. The Society publishes:

- *The ASHRAE Handbook* in several volumes, which are updated periodically. These handbooks serve as a guide, with data references that include information about the applications of cryogenics, and the fundamental properties of cryogenic fluids. Each volume costs $80.

Publications

Cryo-Letters

Dr. Felix Franks, Editor
Department of Botany
University of Cambridge
Cambridge, England CBT 3EA

Cryo-Letters is published bimonthly and contains articles dealing with low-temperature biology, as well as low-temperature technology. The annual subscription is $40.

Cryogenics

Business Press International   212-867-2080
205 East 42nd Street
New York, NY 10017

Cryogenics is a monthly journal, published in London, England, which contains articles dealing with varied applications of temperatures below 100°C, including low-temperature plants, delicate engineering of dilution refrigerators, superconducting motors, and cryosurgery. Each issue contains commissioned review articles on the latest developments in cryogenics, as well as original papers in theoretical and technological advances. The annual subscription is $318.

Additional Experts and Resources

Cryogenic Engineering Conference

c/o Ms. A. M. Dawson   617-253-5547
Massachusetts Institute of Technology
Plasma Fusion Center
NW17-203
Cambridge, MA 02139

This is an annual conference, at which papers are presented on superconductivity and all areas of cryogenics. The conference is open to the public, and its proceedings are published in the following:

• *Advances in Cryogenic Engineering*—a series of volumes, which contain many of the papers presented at the conference. Subject areas include resource availability of helium magnets for fusion and physics research, applications of superconductivity, heat transfer in helium I and liquid nitrogen, heat transfer in helium II, refrigeration and liquefaction, cryocoolers, cryogenic applications in space science and technology, storage and transfer of cryogenic fluids, cryogenic instrumentation and data acquisition, and properties of cryogenic fluids. Each volume is approximately 1,000 pages long and costs $95. The books can be ordered from Plenum Press, 233 Spring Street, New York, NY 10013, 212-620-8016.

Dr. Max Glasser

Goddard Space Flight Center   301-344-8378
National Aeronautics & Space Administration (NASA)
Code 713
Greenbelt, MD 20771

Dr. Glasser is a chemical engineer who specializes in cryo coolers, especially for space application. He has edited proceedings and written extensively regarding this field, and is available to answer questions and make referrals to other possible sources.

Eric James, Ph.D.
Department of Ophthalmology   803-792-3206
Medical University of South Carolina
171 Ashley Avenue
Charleston, SC 29407
  Dr. James is an expert in general cryobiology, with a particular specialty in medical parasitology. He conducts research using low-temperature-preservation techniques with parasitic organisms. He is on the faculty of Medical University of South Carolina, and has written technical papers regarding his research. Dr. James can answer your questions about low-temperature biology.

James K. Missig
Assistant Director of Cryogenic Systems Development   312-855-2500
Liquid Carbonic Corporation
Chicago, IL 60629
  Mr. Missig is an expert in low-temperature processes and food freezing. He is involved in equipment development for industrial gases used as cryogenic fluids, and has edited the book *Applications of Cryogenic Technology*. He can answer your questions about food freezing, and can refer you to other sources of information.

Allan Sherman, Ph.D.
Head of Cryogenics Propulsion & Fluid Systems Branch   301-344-7256
Goddard Space Flight Center
National Aeronautics and Space Administration
Code 713
Greenbelt, MD 20771
  Dr. Sherman is an aerospace engineer who is involved in the research and development of space flight. His specialties include cryocoolers, liquid helium systems, and ultra-low-temperature systems. He and his staff can respond to your inquiries.

Romuald J. Szara
The James Franck Institute   312-962-7180
University of Chicago
5640 South Ellis Avenue
Chicago, IL 60637
  A cryogenic engineer, Mr. Szara specializes in liquefaction of helium, and helium transfers and storage. He is active in cryogenic professional associations and conferences, and has published in the field. Mr. Szara will answer questions and make referrals to other information sources.

Cryptography

Organizations and Government Agencies

American Cryptogram Association (ACA)

39 Roslyn Avenue   216-650-1866—Eugene Waltz
Hudson, OH 44236

ACA is an association of hobbyists interested in cryptography and cryptoanalysis. It holds an annual convention and regularly publishes works on cryptography. The bimonthly publication *The Cryptogram* covers many aspects of cryptography, including problems to solve; it includes articles on computers and their history, as well as book reviews. The annual membership fee is $9.50, and includes subscription to the magazine.

IEEE Information Theory Group

Bell Labs   201-582-3000—Andrew Odlynko, Editor on Cryptology
Room 2C-380
Murray Hill, NJ 07974

The Information Theory Group regularly publishes papers on cryptology. The Group's president usually can recommend experts to contact for further information. Call the IEEE office in New York City at 212-705-7900 to get the name and phone number of the current president. A paper titled "Privacy, Authentication and Introduction to Cryptography," published by the IEEE is an excellent reference article, with a complete bibliography.

International Association for Cryptologic Research

c/o Robert Jueneman   703-237-2000
1522 Hardwood Lane
McLean, VA 22101

The Association is open to anyone interested in the field of cryptology. Two conferences are held annually, one in the U.S. (Crypto 84, and so on), and one in Europe (Eurocryt 84, and so on). The proceedings from the conferences are available in an annual book, *Advances in Cryptology—Proceeding From*, published by Plenum Press in New York City. The quarterly newsletter keeps members up to date on developments in the field. The annual membership fee is $15.

National Security Agency (NSA)

Fort Meade, MD 20755   301-688-7838—Mike Levine, Public Relations

NSA is responsible for deciphering the coded communications of other nations and safeguarding those of the U.S. and of academic cryptologists. The Agency does not publish anything on the topic, and is reluctant to answer questions, but it will send you a bibliography that lists books and articles on the topic.

Publications

Aegean Park Press

PO Box 2837   714-586-8811—Wayne G. Barker
Laguna Hills, CA 92654-0837

Aegean Park Press is dedicated solely to books dealing with cryptology. It sells a series of over 60 books on the topic, ranging from historical and classical books to the latest state-of-the-art. Fliers are available listing and describing the books.

Cryptography

Box 641
David, CA 95617

Cryptography is a hobbyist magazine that presents cryptograms for solving and articles on different aspects of cryptography. William Esterbrook is the editor of this bimonthly. The annual subscription rate is $12.

Cryptologia

Rose-Hulman Institute of Technology   812-877-1511
Terre Haute, IN 47803

Cryptologia is the only scholarly journal devoted to all aspects of cryptography. Areas covered include computer security, history, codes and ciphers, mathematics, military science, espionage, cipher devices, and ancient languages. Dr. Brian Winkel, co-editor, teaches a course in cryptology, and may answer limited questions about the subject. Available for $28 a year, for four issues.

Additional Experts and Resources

Wayne S. Barker

Aegean Park Press   714-586-8811
PO Box 2837
Laguna Hills, CA 92654-0837

Mr. Barker, an author of several books on the topic of cryptology, is an excellent source of information. He does consulting work for clients, and is very knowledge-able on many aspects of cryptology.

Robert Jueneman

1522 Hardwood Lane   703-237-2000
McLean, VA 22101

Robert Jueneman also runs a computer bulletin board for people interested in cryptography. He is trying to develop a central registry of who's doing what in the field of commercial cryptography. Conferences and meetings are announced on the board, and abstracts of interesting articles are included. Hours are from 6:30 P.M. to 7:00 A.M., and 24 hours on weekends. The number of the bulletin board to call on your modem is 703-237-4322.

Louis Kruh   516-378-0263
New York Cipher Society
17 Alfred Road West
Merrick, NY 11566
    Mr. Kruh, president of the New York Cipher Society and the American Crypto-gram Association, as well as co-editor of *Cryptologia*, is very helpful and is willing to answer questions. He maintains a large collection of literature on the topic of cryptology.

Miles Smid   301-921-3151
Institute for Computer Science and Technology
National Bureau of Standards (NBS)
Mail Stop A216
Gaithersburg, MD 20899
    A mathematician at NBS, Miles Smid works in the application of the data encryption standard (a standard used for protecting sensitive nonclassified data). He also works with the banking community in establishing standards in the area of encryption and authentication. Mr. Smid can answer inquiries and make referrals.

CT Scans

See: Cancer-Related Technology; Medical Imaging

Cybernetics

See also: Ergonomics

Organizations and Government Agencies

American Society for Cybernetics
c/o Steven Ruth   703-323-2738
Department of Decision Sciences
George Mason University
Fairfax, VA 22030

The Society is an intellectual association of people trying to understand what cybernetics is. It holds an annual convention and two summer meetings. The Society publishes a bimonthly newsletter, which includes information about its activities and projects, and related academic meetings. After each convention, the proceedings are published. Also available are a bibliography and a glossary pertaining to cybernetics.

Systems, Man and Cybernetics Society

Institute of Electrical and Electronics Engineers (IEEE)   212-705-7900
345 East 47th Street   703-323-2939—Dr. Andrew Sage, Editor
New York, NY 10017

The scope of this IEEE society includes integration of the theories of communication, control, cybernetics, stochastics, optimization, and system structure toward the formulation of a general theory of systems. IEEE publishes:

- *Transactions on Systems, Man and Cybernetics*—a bimonthly which includes papers on theoretical and practical considerations of natural and synthetic systems involving men and machines. The annual subscription rate is $113 a year.

Additional Experts and Resources

Dr. Jon Cunningham

Ohio State University   614-422-0304
1775 College Road
Columbus, OH 43210

Dr. Cunningham is the president of the American Society of Cybernetics, and a professor of Information Systems at Ohio State University. He can answer technical inquiries and make referrals. Dr. Cunningham is developing a cybernetics knowledgeable system, expected to be in operation by 1986.

Professor Steven Ruth

The American Society of Cybernetics   703-323-2738
Department of Decision Sciences
George Mason University
Fairfax, VA 22030

Professor Ruth teaches Systems Analysis, Information Research Management, and Computer Systems Design at George Mason University; he is vice-president of the American Society of Cybernetics, a national lecturer for the Association for Computer Machinery, a delegate to the American Association for the Advancement of Sciences, and a consultant to various private companies and government agencies.

Professor Ruth is a specialist in the areas of business cybernetics, decision support systems, information research management, and human factor and computer system design. He has written several articles and books on these subjects, can answer technical inquiries, and can give referrals to other sources of information.

Dr. Andrew Sage
Office of Academic Affairs   703-323-2939
George Mason University
Fairfax, VA 22030

Dr. Sage is a professor of Information Technology at George Mason University, where he is also an acting associate vice-president for Academic Affairs. He is a specialist in cybernetics and related fields, and can answer technical questions and make referrals. Dr. Sage is also the editor of the journal *Transactions on Systems, Man and Cybernetics* (see separate entry above).

Dr. Leo Steg
Brookings Institute   202-797-6017
1775 Massachusetts Avenue, NW
Washington, DC 20036

Dr. Steg is a specialist in cybernetics, and his areas of expertise also include systems analysis and synthesis, stability criteria, sciences research and development, and space technology. He is the chairman of the Gordon Research Conference on Cybernetics, which studies the frontier of cybernetics. The Gordon Conference has two annual meetings, one in England and the other in California. It also sponsors a workshop on cybernetics. Dr. Steg can answer technical inquiries and make referrals.

Dr. Stuart Umpleby
Department of Management Sciences   202-676-7530
George Washington University
Washington, DC 20052

Dr. Umpleby, considered one of the top specialists in cybernetics in the U.S., is an associate professor of Systems and Cybernetics in the Department of Management Sciences at George Washington University, and a former president of the American Society of Cybernetics. His areas of expertise are cybernetics and systems research; communication control and autonomy, and whether these phenomena occur in individuals, organizations, and machines; group process facilitation; long-range strategic planning; design of artificial intelligence programs and software; and modeling and simulation. He has written various articles on these subjects, and he can answer technical inquiries and make referrals.

Data Base Resources, On-Line

See: On-Line Data Base Resources

Data-Processing Technology

See: Computers (General); Microcomputers; Minicomputers; Office Automation; Software Engineering

DBS

See: Direct Broadcast Satellite

Defense Electronics and Software

See: Military Technology

Design, Computer-Aided

See: CAD/CAM

Dialysis

See: Transplantation Technology

Direct Broadcast Satellite

Begins next page. See also: Satellite Communications

Organizations and Government Agencies

Video Service Division

Distribution SVS Branch   202-632-9356
Federal Communications Commission (FCC)
1919 M Street, NW, Room 702
Washington, DC 20554

FCC's Video Service Division has several specialists in direct broadcast satellites. Several resource people are available, including George Fehlner, an engineer who can answer technical inquiries; and Stuart Bidell and Bruce Romano, attorneys who can answer legal questions as well as some technical inquiries.

Publications

DBS News

Phillips Publishing, Inc.   301-986-0666
7315 Wisconsin Avenue, Suite 1200N   800-227-1617, ext. 213
Bethesda, MD 20814

This monthly newsletter covers all the important developments in the direct broadcast satellite field, such as: programming, earth stations, new technology, Ku-band, SMATV, marketing, regulation, high-definition TV, low-power DBS, LPTV, and news from the FCC. It costs $197 a year, and it is also available on-line through NewsNet, Inc. at 800-345-1301 (in Pennsylvania 215-527-8030).

Institute of Electrical and Electronics Engineers (IEEE)

345 East 47th Street   212-705-7900
New York, NY 10017

IEEE publishes several monthly periodicals, which can be ordered from IEEE Service Center, Publication Sales Department, 445 Hoes Lane, Piscataway, NJ 08854:

• *Transactions on Antennas and Propagation*—this monthly publication covers experiential and theoretical advances in electromagnetic theory and in the radiation, propagation, scattering, and diffraction of electromagnetic waves, and the devices, media, and fields of application pertinent to antennas, plasmas, and radio-astronomy systems. The cost is approximately $127 a year.

• *Transactions on Communications*—a monthly publication devoted to the theory and technology of telecommunications, such as data communications by electromagnetic propagation in radio, lasers, conducted and guided signals, in marine, aeronautical, space, and fixed station services; telecommunication error detection and correction; multiplexing and carrier techniques; and communications switching systems. The cost is approximately $113 a year.

• *Transactions on Microwave Theory and Techniques*—this monthly covers microwave theory, techniques, and applications, as they relate to components, devices, circuits, and systems involving the generation, transmission, and detection of microwaves. The cost is approximately $132 a year.

New Electronic Media

Broadcast Marketing Company   415-777-5400
450 Mission Street
San Francisco, CA 94105

This is a major five-volume reference on different aspects of the electronic media. Volume I, *The Competitive Scramble for the Pay Television Market*, deals with the following areas: pay cable; subscription television; multipoint distribution service; satellite master antenna television systems; low-power television; satellite subscription television, and so on. Volume III, *Direct-to-Home Satellite Broadcasting*, covers the developments and implications of DBS. The cost is $50 a volume.

International Transcription Service, Inc. (ITS)

4006 University Drive   703-296-7322
Fairfax, VA 22030

Several DBS reports are available from ITS, including:

• *Summary of the Technical and Planning Output of the Regional Administrative Radio Conference*—this 1983 report deals specifically with direct broadcast satellites for ITV Region 2, which includes the U.S. ITS is the FCC's contractor for duplicating and disseminating FCC files. Call or write to ITS for the exact price and for ordering information.

Technical Aspects Related to Direct Broadcasting Satellite Systems (pub. # PB81-178980)

National Technical Information Service   703-487-4650
5285 Port Royal Road
Springfield, VA 22161

This report is a collection of technical memoranda relating to Direct Broadcasting Satellite systems. The material gives a general description of a system, which includes ground receivers, advances in the technology, some information on satellite planning for the ITV Region 2, and a description of experimental DBS satellite systems.

Additional Experts and Resources

Bruno Pattan

Office of Science and Technology   202-632-7073
Federal Communications Commission (FCC)
2025 M Street, NW
Washington, DC 20554

Bruno Pattan is an engineer who does technical analysis for FCC in the area of direct broadcast satellites. He is the designated federal employee on the Direct Broadcast Satellite Advisory Committee, a group concerned with the advisability and feasibility of DBS standards. Mr. Pattan can answer questions of a technical nature and/or make referrals.

Directories to Software

See: Software Guides, Directories, and Reviews

Disabled, Computers for

See: Computers for the Handicapped

DNA

See: Biotechnology

Drugs

See: Pharmaceuticals; Toxicology

DTS

See: Telephones

EDP

See: Computers (General); Microcomputers; Minicomputers; Office Automation; Software Engineering

Education and Computers

See: Computers in Education

EFT

See: Electronic Funds Transfer

Electromagnetic Technology

See also: Medical Technology; Microwaves; Telecommunications

Organizations and Government Agencies

The American Physical Society (APS)
335 East 45th Street   212-682-7341
New York, NY 10017

APS is a scientific and educational society that advances and disseminates the knowledge of physics. The Society's members are primarily research physicists, including those with a special interest in electromagnetic technology. The Society sponsors an annual conference on magnetism and magnetic materials. The conference proceedings can be ordered through the American Institute of Physics, Inc. at the above address. APS staff can answer questions and refer you to an appropriate expert regarding electromagnetic technology. The Society publishes the *Physical Review A, B, C,* and *D* plus an *Index* which includes the four journals. The subscription rate to all the four journals together is $1,590 a year. The journals are:

• *Physical Review A*—a monthly journal that focuses on topics in "extranuclear" physics. The annual subscription rate for the 540-page report is $310.

• *Physical Review B*—a monthly journal that includes reports of basic research in condensed matter physics. The subscription rate is $620 a year for the 1,125-page journal.

• *Physical Review C*—a monthly journal that contains papers about experimental and theoretical results on nuclear reactions, including electromagnetic moments. The approximately 400-page journal costs $250 a year.

• *Physical Review D*—a semimonthly journal that offers reports on elementary particle physics and field theory, including the theory of particles and fields, general quantum mechanics, and the general scattering theory. It is approximately 250 pages an issue; the annual subscription rate is $410.

Bioelectromagnetics Society (BEMS)

1 Bank Street, Suite 307   301-948-5530
Gaithersburg, MD 20878

This organization's purpose is to promote scientific study of the interaction of electromagnetic energy (at frequencies ranging from 0 hertz through those of visible light) and acoustic energy with biological systems. Many members are actively involved in research on the effects of electromagnetic and acoustic energies on biological systems. BEMS holds one annual scientific conference, at which approximately 200-250 technical papers are presented. Abstracts of these papers are available at $5 a book. BEMS usually holds a second meeting each year—a joint meeting with organizations doing related work. The staff will respond to written inquiries. Publications include:

• *Bioelectromagnetics*—a highly technical journal directed toward researchers in the field of bioelectromagnetics. The approximately 100-page quarterly publication is available to members for $25. Nonmembers can order it from Alan R. Liss, Inc., 150 Fifth Avenue, New York, NY 10011 for $90 a year.

U.S. Department of Defense (DOD)

Public Affairs/Public Correspondence   202-697-5737
R2E777
Washington, DC 20301

This office is the clearinghouse for public information and correspondence for DOD. The staff members can provide general, basic information about electromagnetic technology, including annual reports, newsletters, and information sheets. They can also make referrals regarding more specific inquiries.

Electromagnetics Branch

Division of Physical Sciences   301-443-3840
Center for Devices and Radiological Health   301-443-3532—Technical Literature
U.S. Food and Drug Administration
Rockville, MD 20857

The Electromagnetics Branch is involved in applied research related to sources of electromagnetic energy, product evaluation, the development of measurement techniques, and the development of technological control techniques. Technical reports can be ordered. These reports include information about theoretical studies, electromagnetic source evaluations, measurements, measurement techniques, hyperthermia equipment for cancer treatment, and microwave ovens. The staff in this Branch can respond to written inquiries.

Electromagnetic Energy Policy Alliance

1800 M Street, NW   202-452-1070
Washington, DC 20036

    A newly formed group of users and manufacturers of electromagnetic equipment, the Alliance is a good source for facts about hazards, standards, regulations, and safety phenomena relating to electromagnetics. The Alliance plans to have a newsletter and publications, which will be available to the public.

Electromagnetic Radiation Branch

Division of Life Sciences   301-443-7132
Center for Devices & Radiological Health   301-443-3532—Technical Literature
U.S. Food and Drug Administration
Rockville, MD 20857

    The Electromagnetic Radiation Branch performs and monitors experiments and studies dealing with the effects of radio frequency and microwave radiation on biological systems. This Branch also reviews and analyzes similar studies performed in other labs. The staff in this Branch can respond to your written inquiries. Technical literature is available upon request.

Electromagnetic Spectrum Management Unit

National Science Foundation   202-357-9696—Dr. Vernon Pankonin
Washington, DC 20550

    The objective of this Unit is to ensure the availability of the radio frequency spectrum for scientific research at radio astronomy laboratories. Research involves areas of communications, as well as the tracking of animals. Radio frequency is also used directly to analyze information that is emitted from natural bodies. Dr. Pankonin can answer your questions and refer you to other information sources. A summary statement about the use of radio frequencies at astronomy laboratories can be sent to you free of charge.

Electromagnetic Fields Division

National Bureau of Standards (NBS)   303-497-3131
Boulder, CO 80303   303-497-3535

    This federal office is primarily responsible for developing measurement techniques for radiated electromagnetic fields irrespective of applications, and has all the measurements related to transmission lines. The staff members provide calibration services related to these standards, and make available papers and reports. You can obtain copies of the publications by writing to the Division listed above. If you would like to be kept up to date about the Division's publications, you can subscribe to the following free quarterly available from the Division's parent, the Center for Electronics and Electrical Engineering (CEEE), in Maryland:

    ● *CEEE Technical Progress Bulletin*—provides abstracts and reference information for all publications produced by the Center. The *Bulletin* contains both recently published materials and those released for publication. This free *Bulletin* is mailed only to U.S. addresses. Order from CEEE, National Bureau of Standards, Metrology Building, Room B-358, Gaithersburg, MD 20899, 301-921-3357.

Health Physics Society (HPS)

1340 Old Chain Bridge Road, Suite 300   703-790-1745
McLean, VA 22101

The HPS is a society of scientists from industry, government, universities and medicine. The Society's objective is to develop scientific knowledge so that radiation and atomic energy serve to benefit humanity and the environment. HPS is also involved in radiation protection. The Society publishes:

• *Health Physics*—a monthly journal containing articles about theoretical and applied physics, including the medical applications of electromagnetic technology. Each issue averages 200 pages. The subscription rate for institutions is $360 a year, and individuals pay $55. The Society's staff can answer questions and make referrals.

IEEE Antennas and Propagation Society

Institute of Electrical and Electronics Engineers   212-705-7890—Technical Activities
IEEE Headquarters   212-705-7866—Public Information
345 East 47th Street
New York, NY 10017

The Society is a professional organization of electrical engineers who are involved in the design and analysis of antennas and antenna systems, the propagation of electromagnetic waves, and the scattering of electromagnetic waves. The Society sponsors numerous international and national conferences, including the annual IEEE Antennas and Propagation Society International Symposium. Proceedings from these conferences are contained in symposium digests, which you can order from the address below. The Society publishes:

• *Transactions on Antennas and Propagation*—a monthly journal dealing with the analysis and design of all kinds of antennas and antenna systems, the propagation of electromagnetic waves in all media, and the scattering of electromagnetic waves. The journal averages 125 pages an issue. An annual subscription is $127 and can be ordered from IEEE Service Center, 445 Hoes Lane, Piscataway, NJ, 08854, 201-981-0060.

IEEE Electromagnetic Compatibility Society

Institute of Electrical and Electronics Engineers   212-705-7866—Public Information
345 East 47th Street   212-705-7890—Technical Activities
New York, NY 10017

This Society of engineers has a special interest in the study of the interaction of electronic and biological systems and subsystems that involve electromagnetic elements, and the preclusion of interference among these systems and subsystems. The Society sponsors and cosponsors several symposia and conferences, including the annual International Electromagnetic Compatibility Symposium. Records of all papers presented during the conferences can be ordered. The Society's president is available to answer questions about electromagnetic compatability. The Society publishes:

● *Transactions on Electromagnetic Compatibility*—a quarterly journal containing articles about the origin, control, and measurement of electromagnetic effects on electronic and biological systems. Each issue has approximately 120 pages. An annual subscription is $68.

IEEE Microwave Theory and Techniques Society

Institute of Electrical and Electronics Engineers (IEEE)   212-705-7866—Public Information
345 East 47th Street   212-705-7890—Technical Activities
New York, NY 10017

This Society comprises electronic engineers and physicists with a specialty in microwave band and millimeter band. They are concerned with theory, techniques, and applications, as they relate to components, devices, circuits, and systems involving the generation, transmission, and detection of microwaves. The Society sponsors several symposia and conferences, including the Microwave Symposium, before and after which workshops are conducted. The proceedings of these conferences and symposia are published, and can be ordered from IEEE. The Society also publishes:

● *Transactions on Microwave Theory and Techniques*—contains articles about microwave theories, techniques, and applications, including research, measurement, developments, and biomedical applications. The journal averages 100 pages an issue, and the annual subscription rate is $132.

Institute for Telecommunication Sciences (ITS)

National Telecommunications Information Administration (NTIA)   303-497-3572—Technical
U.S. Department of Commerce   Publications Office
325 Broadway
Boulder, CO 80303

ITS is the chief research arm on telecommunications matters of NTIA and serves as the federal resource to assist other agencies of the government in the planning, design, maintenance, and improvement of their telecommunications activities. ITS is involved in electromagnetic propagation characterizations of telecommunications. The Institute develops models to study the real effects on telecommunications systems at various frequencies. It publishes:

● *Annual Technical Progress Report*—contains an overview and comprehensive summary of its projects. Many technical reports are also available and can be ordered. The staff at ITS can respond to your inquiries.

Interdependent Radio Advisory Committee (IRAC)

National Telecommunications and Information Administration (NTIA)   202-377-0599
U.S. Department of Commerce
Herbert C. Hoover Bldg., Room 1605
14th Street and Constitution Avenue NW
Washington, DC 20230

NTIA is the FCC counterpart for the federal government. IRAC is a Committee that functions on behalf of NTIA for all federal agencies and is responsible for telecommunications protocol. The chairman and his staff are available to answer questions about federal telecommunications. The Committee publishes:

● *The Manual of Regulations and Procedures for Federal Radio Frequency Management*—designed to cover frequency management responsibilities according to delegated authority under Section 305 of the Communications Act of 1934. The *Manual* explains who has what authority over federal telecommunications, and how to get a frequency assignment within the federal government. The *Manual* is published three times a year, and each issue runs approximately 600 pages. A three-year subscription is $118; the publication is available free of charge to government agencies. The most frequently requested information is the chapter on national and international allocation tables, which is available from IRAC in a pocket-size excerpt.

Ionizing Radiation Branch

Division of Life Sciences   301-443-7159
Center for Devices and Radiological Health   301-443-3532—Technical Literature
U.S. Food and Drug Administration
Rockville, MD 20857

The Ionizing Radiation Branch is doing research with low-dose radiation exposures delivered over an extended period of time. Studies and experiments are being conducted concerning genetic behavior changes as a result of single doses of radiation. The staff can answer your questions, and send you technical literature upon request.

Naval Research Lab

Code 6550   202-767-3284
Washington, DC 20375

The Lab is involved with research of low-power and medium-power solid state lasers, as well as the characterization of infrared detectors. Technical reports concerning these areas are available by contacting the branch secretary at the telephone number indicated above. The staff can respond to your questions and refer you to other sources of information.

Non-Ionizing Radiation Branch

U.S. Environmental Protection Agency   702-798-2440
PO Box 18416
Las Vegas, NV 89114

This government Branch is involved in the development of federal regulations for establishing maximum, permissible exposure limits for radio-frequency radiation, including the possible hazards of electromagnetic radiation. Publications, available from the National Technical Information Service, include technical reports on specific areas of investigation, such as field studies to determine environmental levels of radio-frequency exposures, evaluation of instruments for measurement, antenna model and analysis, and radio frequency including microwave radiation.

Optical Radiation Branch
Division of Life Sciences
Center for Devices and Radiological Health
U.S. Food and Drug Administration
Rockville, MD 20857

The Optical Radiation Branch reviews and recommends new products on the market on the basis of their safety and efficacy. This Branch conducts research with visible and ultra light using biological systems. The staff in this Branch can answer your questions and make referrals, and will send some technical literature upon request.

Spectrum Engineering and Analysis Division
National Telecommunications and Information Administration (NTIA) 202-261-8141
179 Adm. Cochrane Drive 301-224-4305—Robert Mayher, Chief
Annapolis, MD 21401

This Division of NTIA conducts studies that provide a quantitative understanding of the potential of band sharing and of electromagnetic problems in various portions of the government's radio spectrum. Mr. Mayher and his staff can respond to your questions and provide you with a listing of technical reports that pertain to electromagnetic technology. These reports, which include policy recommendations, are available.

U.S. National Committee
International Union of Radio Science (IURS) 202-334-3359—Richard Dow, Staff Officer
National Academy of Sciences
2101 Constitution Avenue, NW
Washington, DC 20418

The U.S. National Committee is organized with nine commissions, which deal with various aspects of radio science. These are Electromagnetic Methodology, Fields and Waves, Signals and Systems, Electronic Optical Devices and Applications, Electromagnetic Noise and Interference, Remote Sensing and Wave Propagation, Ionospheric Radio Wave Propagation, Waves and Plasma, and Radio Astronomy. Mr. Dow can refer you to an appropriate expert in the U.S. Abstracts of papers presented at IURS conferences can be ordered at the above address for $5. Other publications include:

● *Review and Radio Science*—presents a review of the most recent advances in radio science. Papers are submitted by the IURS chairman of each country. The *Review* is published every three years and costs $20. It can be ordered from International IURS, Professor J. Van Bladel, Avenue Albert Lancaster 32, B-1180, Brussels, Belgium, tel. 02374-1308.

Publications

Microwave News
PO Box 1799   212-725-5252
Grand Central Station
New York, NY 10163

    Microwave News, a newsletter published ten times a year, describes recent developments and current research about electromagnetic technology. It contains articles about bioeffects, and compatibility and interference of nonionizing radiation. The annual subscription is $200 for the approximately 12-page newsletter.

Additional Experts and Resources

David Cheng, Ph.D.
Electrical and Computer Engineering   315-423-2655
Syracuse University
111 Link Hall
Syracuse, NY 13210

    Dr. Cheng is a retired professor of communications engineering. His research and study have been in electromagnetic theory, antennas, and antenna array. He has articles published in professional journals, and he wrote the textbook *Field Wave Electro Magnetics*.

Robert S. Elliott, Ph.D.
Boelter Hall, Room 7732   213-825-7880
University of California, Los Angeles
Los Angeles, CA 90024

    Dr. Elliott is an electrical engineer who specializes in microwave antennas and applications to radar. In addition to teaching, he is involved in research and works as a consultant. He has written two books and numerous journal articles. Dr. Elliott will answer questions and/or make referrals.

Eugene D. Knowles
16954 Southeast 149th Street   206-271-3396
Renton, WA 98056

    Mr. Knowles is an electrical engineer with the Boeing Company and has a special interest in electromagnetic compatibility. He is involved in design approaches, cable shielding, and systems testing. He has lectured and taught short courses about electromagnetic technology. Mr. Knowles is available to answer questions and make referrals.

John Osepchuk
Raytheon Research Division   617-863-5300
131 Spring Street
Lexington, MA 02173
Mr. Osepchuk is an expert in the heating applications of electromagnetic technology, especially the use of microwave devices, including electron tubes. He is also knowledgeable about the hazards and safety of microelectronics.

Electronic Access to Information

See: Telecommunications

Electronic Banking

See: Electronic Funds Transfer

Electronic Data Processing

See: Computers (General); Microcomputers; Minicomputers; Office Automation; Software Engineering

Electronic Funds Transfer

Organizations and Government Agencies

American Bankers Association (ABA)
1120 Connecticut Avenue, NW   202-467-4101
Washington, DC 20036
This Association strives to exchange information and upgrade banking skills. It maintains a library and an information services department with over 50,000 volumes and 600 periodicals. The library is open to the public by appointment.

Reference staff will answer questions over the phone or by mail. A bibliography on electronic funds transfer is available upon request. The ABA's Catalog of Publications carries a section on EFT that lists publications and audiovisual material available, as well as workshops and conferences. Publications include:

● *The Banker and EFT*—discussion of the history of EFT regulation and legislation, types of systems, marketing of EFT systems, and future prospects. The cost is $25.

● *Developing a More Efficient EFT Service*—describes standard terminology and usage guidelines to provide users with an understanding of message formats. The cost is $30.

● *EFT Interchange: A Directory of Shared ATM Services*—the cost of this publication on automatic teller machines is $50.

Electronic Funds Transfer Association (EFTA)

1029 Vermont Avenue, NW, Suite 800   202-783-3555
Washington, DC 20005

This multi-industry trade association keeps its members informed about new trends and technologies in automated payment services. EFTA holds an annual convention, and has a resume referral service (job network) for individuals seeking employment. It publishes:

● *Washington Report*—a monthly newsletter, which reviews important issues in EFT and keeps track of EFT legislation and the likely impact of EFT regulatory issues on business.

● *Executive Report*—a monthly newsletter offering highlights of general EFT news and reports on Association activities.

Federal Reserve Board (FRB)

20th Street and Constitution Avenue, NW   202-452-3926
Washington, DC 20551

The FRB can be a good source of information on economic aspects of EFT. Call and ask to speak to someone knowledgeable about EFT. The FRB also maintains a specialized research library of economics-related materials, with some documentation available on EFT. The library is open to the public from 9 A.M. to 5 P.M.

Publications

Bank Administration Institute

60 Gould Center   312-228-6200
Rolling Meadows, IL 60008   800-323-5932

The Institute publishes several books and publications on the topic of EFT, such as:

● *Survey of the Electronic Funds Transfer Transaction System*—includes annual surveys of EFT transactions, concentrating on two EFT services: Automatic Clearing House (ACH) and Automatic Teller Machines (ATMs). The cost is $14.

● *Security and Reliability in Electronic Systems for Payments*—offers practical methods of providing security and reliability in electronic systems for payments. The cost is $20.

- *ATM Directory*—annual guide to vendor equipment and services in the auto-mated-teller-machine field.

Bank Network News

117 North Jefferson Street, Suite 204   312-648-0261
Chicago, IL 60606

This biweekly newsletter covers such aspects of EFT as ATMs, cash dispensers, direct-debit point-of-sale (POS) experiments, telephone-bill-paying services, home bank, check truncation. It covers all major developments related to shared network-ing. It is also available electronically via NewsNet. Call 800-345-1301; in Pennsyl-vania, 215-527-8030. The subscription is $295 a year, for 23 issues.

- *Top 100 Network Directory*—this 36-page *Directory* lists the regional shared bank networks that form the foundation of retail electronic banking in the U.S. It is available for $75.

Computer Crime: Electronic Funds Transfer Systems and Crime

U.S. Department of Justice   202-724-7759—Carol Kaplan
Bureau of Justice Statistics
Washington, DC 20531

This report addresses the issue of criminal abuse of EFT systems and tries to determine the nature and magnitude of EFT crimes. An extensive bibliography is included on EFT to identify sources of relevant data, and further sources of informa-tion are identified. Available from the Superintendent of Documents, Government Printing Office, Washington, DC 20402, 202-783-3238.

EFT Report

Phillips Publishing   301-986-0666
7315 Wisconsin Avenue, Suite 1200N
Bethesda, MD 20814

An eight-page biweekly newsletter on EFT, especially about developing issues in such areas as automatic teller machines, debit cards, home banking, point-of-sale, EFT privacy, legislation, interchange programs, and competitors. It is available for $245 a year.

Payment Systems, Inc. (PSI)

Reistad Corporation   813-875-9069
PO Box 30168
Tampa, FL 33630

PSI is involved in researching the area of home information/transaction ser-vices for clients. It publishes the *Reistad Monitor*, a monthly bulletin that interprets events and predicts their impact on the development of the home information and transaction services business.

Office of Technology Assessment (OTA)

U.S. Congress   202-224-8996—Publications
Washington, DC 20510

OTA was established by Congress to provide analytical assistance on difficult issues involving science and technology. It is constantly researching a varied range of topics, including electronic funds transfer. A result is the publication *Selected Electronic Funds Transfer Issues: Privacy, Security, and Equity*. The background paper #PB 82-202 532 costs $9, and the working papers #PB 82 82-238 312 cost $40.50 and are available from National Technical Information Service, U.S. Department of Commerce, Springfield, VA 22161, 703-487-4650. OTA has several free reports available upon request. Contact its Publications Office for a free catalog.

Revolving Credit/Funds Transfer Letter

PO Box 431   203-227-1237
12 Avery Place
Westport, CT 06880

This biweekly eight-page publication reports on credit cards, debit cards, retail payment, and electronic payment networks. It provides current case histories and in-depth analyses of these issues, covering the following: shared EFT networks, automated teller machines, automated clearinghouses, home banking, in-store banking, and more. The cost is $160 a year.

Transition

American Banker   212-943-6700
One State Street Plaza   800-221-1809
New York, NY 10004

Transition focuses on the field of financial services strategy, and carries articles on the long-range implications of the change in services offered and delivered, what is ahead in deregulations, EFT, bank cards, technology, and more. The cost is $295 a year, for ten issues.

Electronic Games

See: Video Games, Home and Computer

Electronic Keyboards

Begins next page. See also: Synthesizers

Organizations and Government Agencies

Computer and Business Equipment Manufacturers Association (CBEMA)
311 First Street, NW, 5th Floor   202-737-8888
Washington, DC 20001

    CBEMA is a trade association representing the interests of the computer and business-equipment industry before the U.S. Congress. In an effort to promote the office automation manufacturers, it frequently publishes and distributes informative publications to educate the public on the benefits and advantages provided by the industry. The staff is very knowledgeable about computer and business equipment, including uses, availability, technology advances, probable useful life, cost forecasts, and general trends for the future of this machinery. It also publishes a newsletter available to the public.

    ● *CBEMA Comment*—a bimonthly newsletter which primarily covers the news of the Association's lobbying activities in influencing legislation affecting science and technology. Also included are product announcements and new technology advances by member and nonmember companies. About eight pages an issue, the newsletter is available free upon request.

National Association of Music Merchants (NAMM)
5140 Avenida Encinas   619-438-8001
Carlsbad, CA 92008

    This trade association represents the manufacturers of musical instruments and equipment, including electronic keyboards. The Association's staff in NAMM's Information Department can answer questions and make referrals. The organization sponsors trade shows and conferences. Its publications include reports about the management and operational aspects of the music retail market and:

    ● *Program Directory*—a listing of all companies involved in manufacturing musical instruments and equipment. It includes manufacturers of such high-tech products as electronic keyboards and synthesizers. The 70-page publication is available for free.

Publications

Evolving Keyboard Markets
International Resource Development, Inc.   203-866-7800
6 Prowitt Street   WU Telex 64 3452
Norwalk, CT 06855

    This report analyzes the major technolgical and market trends in the keyboard business, and presents ten-year market projections. Bottom-up forecasts are developed by looking at each type of keyboard-using equipment and tracing its probable shipment patterns over the next ten years. The publication contains 16 sections, including the following as examples: Current Keyboard Marketplace and Character; The Impact of Japanese and Far Eastern Keyboard Vendors; Microcomputer Keyboards—Home, Personal and Desktop; Keyboards for Terminals, Word Processors and Work Stations; Keyboard Markets for ATM, POS, CVT, Financial Terminals. The 235-page publication is available for $1850. Refer to report #609.

Keyboard Magazine

20085 Stevens Creek Boulevard   408-446-1105
Cupertino, CA 95014

Published monthly, this magazine profiles all musical keyboard instruments. Geared toward musicians, it contains consumer guidelines on such subjects as buying an electronic keyboard or synthesizer. The annual subscription to the 90-page magazine is $18.95.

Upbeat Magazine

22 West Adams   312-346-7820
Chicago, IL 60606

Geared toward the music retailer, this monthly publication focuses on new products and new features of musical equipment. It covers synthesizers, electronic keyboards, computer music, and musical instruments. Each issue focuses on a specific topic or musical instrument and profiles the product's retailers. Business and trade show information is also provided. Each issue is approximately 75 pages, and the annual subscription is $8.50.

Additional Experts and Resources

Judith Corea

Consumer Goods Office   202-377-2132
Durable Goods Division
U.S. Department of Commerce
14th and Constitution Avenue, NW, IPA Room 4312
Washington, DC 20230

An industry trade specialist with the Department of Commerce, Ms. Corea is responsible for following developments of the musical industry as a whole. She can provide you with import/export and product shipment data, as well as an overview of the industry, and its new developments and trends. Ms. Corea can also refer you to other sources.

The companies listed below produce electronic keyboards. The staff may be able to provide you with information and send you literature.

Kaman Music Distributors

PO Box 507
Bloomfield, CT 06002
203-243-1453
800-447-3456

Kaman is the exclusive distributor of SEIKO electronic keyboards in the U.S. The staff will answer questions and refer you to other manufacturers. They will also send you technical briefs, specifications, and product information.

Kimball
l600 Royal Street
Jasper, IN 47546
812-482-1600
    The staff will answer inquiries and send you company information, technical briefs, specifications, and product information.

Lowrey
707 Lake Cook Road
Deerfield, IL 60015
312-480-6600
800-323-5410

    The staff will answer questions and make recommendations about electronic keyboards. They can also send you information, including technical briefs.

Wurlitzer Company
403 East Gurler Road
Dekalb, IL 60115
800-435-2930
815-756-2771
    The staff will answer questions and send you information including technical briefs and specifications.

Electronic Mail

Organizations and Government Agencies

Electronic Mail Association (EMA)

1919 Pennsylvania Avenue, NW, Suite 300
Washington, DC 20006

202-293-7808—Michael F. Cavanagh

    This trade association promotes research and monitors legislation and relevant activities in the field for the industry. Inquiries will be handled, but the staff is very small and cannot take too many calls. Services are mostly for members.

Publications

International Resource Development, Inc.

30 High Street
Norwalk, CT 06851

203-866-6914

    Several electronic mail publications are available:
    ● *EMMS—Electronic Mail & Message Systems*—this newsletter covers the electronic mail and message system field. It reports on all aspects of electronic mail: telex/TWX, teletext, store-and-forward voice, local area networks, communicating terminals, personal computers, and so on. It also assesses the effects of these technologies on more traditional communications players, such as the telephone companies, the U.S. Post Office, and the International Resource Developments. The cost is $235 a year, for 24 issues.
    ● *Video Print*—a biweekly newsletter covering issues in the field of electronic information delivery. It includes information about viewdata, teletext, electronic

mail, electronic publishing, direct broadcast satellites, and videodiscs, as well as their markets and areas of opportunity, and their growing interrelationships. The cost is $180 a year.

● *Paper-Based Electronic Mail*—this in-depth study and analysis of paper-based message systems and services focuses on those electronic mail systems and services in which messages sent or received utilize paper, such as Telex/TWX, Facsimile, Mailgram, and Intelpost; it analyzes the trends expected to influence electronic mail service markets in general. The 206-page report costs $1,285.

● *Computer-Based Electronic Mail*—reports on the uses, the evolution, and the technology of personal messaging systems. It describes market segmentation and the relationship between private and public market segments, as well as the current market size and future market trends. The 141-page report is $985.

Electronic Mail Executives Directory

Knowledge Industry Publications   800-431-1800
701 Westchester Avenue   914-328-9157
White Plains, NY 10604

This *Directory*, compiled by International Resource Development, lists more than 2,800 managers from top U.S. companies responsible for voice and messages services, word processing, and data processing. Each entry gives the company name and address, sales, number of employees, and managers responsible for voice communications, office automation, facsimile transmission, word processing, in-house printing, purchasing, and telecommunications, including satellite. The cost of the 1982 edition is $95.

The Report on Electronic Mail

The Yankee Group   617-542-0100
89 Broad Street, 14th Floor
Boston, MA 02110

This quarterly journal reveals new technologies in electronic mail, evaluates current products and services, and uncovers vendor strategies. Its four targets are electronic mail systems, electronic transactions/paperless buying, data base access, and funds transfer. The cost is $950 a year, for four issues plus *Special Updates*.

Office of Technology Assessment (OTA)

U.S. Congress   202-226-2115—Public Affairs
Washington, DC 20510   202-224-8996—Publications

OTA was established by Congress to provide analytical assistance on issues that involve science and technology. The topics researched by OTA are later published and made available to the public. Two publications have resulted from a study of electronic mail, and free summaries of these reports are available from OTA:

● *Implications of Electronic Mail and Message Systems for the U.S. Postal System*—a 113-page report published in 1982.

● *Summary, Implications of Electronic Mail and Message Systems for the U.S. Postal System*

Several periodicals regularly carry articles on electronic mail, including:

Data Communications
McGraw-Hill, Inc.
1221 Avenue of the Americas
New York, NY 10020
212-512-4852

The Office
($30 a year, for 12 issues)
Office Publications, Inc.
1200 Summer Street
Stamford, CT 06904
203-327-9670

Computer World
($44 a year, for 52 issues)
C.W. Communications
Box 880
Framingham, MA 01701
607-879-0700

Data Channels
($247 a year, for 26 issues)
Phillips Publishing
7315 Wisconsin Avenue, Suite 1200N
Bethesda, MD 20814
301-986-0666

MultiChannel News
($27 a year, for 52 issues)
300 South Jackson, Suite 450
Denver, CO 80209
303-393-6397

Several companies offer electronic mail services. They will send you literature explaining their services and also the technology behind electronic mail. The following are the four largest services in the United States:

GTE Telenet Communications
Telemail
8229 Boone Blvd.
Vienna, VA 22180
800-TELENET

OnTyme
Tymshare, Inc.
20705 Valley Green Drive
Cupertino, CA 95014
800-227-6185
415-876-7508

E-COM (Electronic Computer Originated Mail)
Office of E-COM Operation
U.S. Postal Service Headquarters
Washington, DC 20260-7140

MCI Mail
2000 M Street, NW
Washington, DC 20036
202-293-4255
800-MCI-2255
800-424-6677

Electronic Publishing

Organizations and Government Agencies

Information Industry Association (IIA)

316 Pennsylvania Avenue, SE, Suite 400   202-544-1969
Washington, DC 20003

IIA is a trade association of data base producers, information brokers, publishers, information system designers and suppliers, and videotex and teletext providers. The staff will answer questions relating to electronic publishing and/or make referrals to members. Members are listed in a directory published annually, *Information Sources*. The price of the directory is $59.75 plus $2.50 for shipping and handling.

Learned Information, Inc.

143 Old Marlton Pike   609-654-6266
Medford, NJ 08055

This firm sponsors two annual events: the International Online Information Meeting in London and the National Online Meeting in New York City. A catalog listing the meetings held during the years 1977 through the present, and the proceedings of these meetings, is available. Other publications of Learned Information include:

• *Electronic Publishing Review, the International Journal for the Transfer of Published Information via Videotex and Online Media*—this publication covers all aspects of electronic publishing, including on-line data base publishing, videotex, teletext, and numerous ways of packaging information by electronic media. It is written from the industry's point of view. The cost is $66 a year.

Publications

Electronic Publisher

Paul Kagan Associates   408-624-1536
26386 Carmel Rancho Lane
Carmel, CA 93923

This semimonthly newsletter on publishing via television has a special emphasis on newspaper-cable television joint ventures. It is available for $395 a year.

EPB: Electronic Publishing and Bookselling

Oryx Press   602-254-6156
2214 North Central at Encanto
Phoenix, AZ 85004

This bimonthly newsletter covers computer applications in publishing, bookselling, and related fields. It includes reports on advances in electronic publishing, on new computer systems developed by electronic publishers, machine-readable codes on book covers and how they are used, as well as interviews, major industry figures, feature articles, case studies, book reviews, and more. The cost is $60 a year, for six issues.

Knowledge Industry Publications, Inc.

701 Westchester Avenue   914-328-9157

White Plains, NY 10604   800-841-1880

This firm offers several different kinds of publications, including books, newsletters, an electronic directory, and a complete catalog of its titles. The catalog is free and is published twice a year. Examples of what is available include:

- *The Birth of Electronic Publishing—Legal and Economic Issues in Telephone, Cable and Over-the-Air Teletext and Videotex* by Richard M. Neustadt, 1982, 146 pages, $32.95.
- *Guide to Electronic Publishing—Opportunities in On-line and Viewdata Services* by Frances Spigai and Peter Sommer, 1982, 163 pages, $95.
- *Information and Database Publishing Report*—a bimonthly newsletter directed toward the business community. It is available for $225 a year, or $410 for two years.
- *Software Publishing Report*—a bimonthly newsletter that analyzes the software business. It is available for $288 a year, or $526 for two years.
- *Database Alert*—a service that lists changes occurring in on-line data bases throughout North America. The service costs $48 a year, or $90 for two years. This service is offered as part of the *Database User Service*, which includes a print and on-line directory of more than 1,800 on-line data bases.

VideoPrint

International Resource Development, Inc.   203-866-6914

30 High Street

Norwalk, CT 06851

This newsletter reports on the field of teledelivery and includes a news sheet covering developments in videotex and electronic publishing. Published twice a month, it covers major trials and innovations, market research, new ventures, new and old faces, relationships with older information retrieval schemes, and so on. It also analyzes the development of home terminals, advertising and news media, and the changing roles of print publications and network TV. The cost is $180 a year.

Data Bases

Electronic Publishing Abstracts

Pergamon Info Line   011-44-1-377-4650
12 Vandy Street
London EC2A 2DE
England

   The *Abstracts* cover scientific, technical, and commercial literature in electronic publishing and information technology. Major topics include input methods, transmission of data, storage and retrieval of text and images, output methods, and electronic alternatives both to the publication of printed documents and the methods of printing published documents. Information is updated twice a month, and covers data from 1975 to date. The cost is $70 a connect-hour.

Electronics and Electrical Engineering

Organizations and Government Agencies

Capital Goods and International Construction

Office of General Industrial Machinery   202-377-4382—Richard Whitley, Industry Specialist
International Trade Administration
U.S. Department of Commerce
14th Street and Constitution Avenue
Washington, DC 20230

   The Office of General Industrial Machinery prepares articles for the *U.S. Industrial Outlook*, as well as competitive assessment studies of selected industries. The Office is also involved in international trade development and in the capital goods activities of the Industrial Sector Advisory Committee (ISAC); it is responsible for responding to the Secretary of Commerce regarding matters affecting the industries covered by the Office.

Center for Electronics and Electronical Engineering (CEEE)

National Engineering Laboratory   301-921-3357
National Bureau of Standards (NBS)
Gaithersburg, MD 20899

   A center involved in measurements related to electrical power, radio frequency systems, antennas, and semiconductor technology, the CEEE produces numerous publications, calibration services, and standard reference materials. Catalogs listing the Center's calibration services and standard reference materials are available from the Office of Measurement Services, 301-921-2606. Publications are generally available for a fee from the National Technical Information Service, 5285 Port Royal Road, Springfield, VA 22161, 703-487-4650.

Electronic Industries Association (EIA)

2001 Eye Street, NW   202-457-4900

Washington, DC 20006

A trade association whose primary purpose is to serve its member companies, EIA is involved in legislative affairs, including lobbying on behalf of its members, and regulatory activities. It is also involved in international affairs, with a specific interest in trade and trade barriers. EIA holds trade shows, as well as periodic conferences and seminars, and serves as a manager for the Electronic Components Conferences. EIA also acts as a focal point in the development of standards. It has more than 500 electronics standards, which are described in its *Standards Catalog*. These are available from the Standards Sales Department, 202-457-4966. EIA also has documents, with most providing information on what to look for when making a purchase. EIA bulletins and publications are listed in the *Publications Catalog*. One important publication is:

● *Engineering News*—a bimonthly newsletter directed toward the electronic engineering industry. This somewhat technical periodical covers committee activities, new EIA engineering standards and publications, new engineering technology, international standardization, and so on. The 40-page publication is available for $30 a year, and is free to members.

Engineering Information, Inc.

345 East 47th Street   212-705-7600

New York, NY 10017   800-221-1044

This abstracting and indexing service provides information and engineering developments from the world's significant engineering and related technological literature. The bibliographic references and abstracts are available in print, microfilm, or machine-readable form. The on-line data base, COMPENDEX, includes abstracts of articles from journals, monographs, reports, standards, and conference proceedings. COMPENDEX is accessible through nine commercial vendors including DIALOG, SDC, ORBIT, BRS, and ESA. Alternatively, Engineering Information, Inc., may be contacted directly if a search is needed.

Institute of Electrical and Electronics Engineers (IEEE)

345 East 47th Street   212-705-7890—Technical Activities Information

New York, NY 10017   212-705-7866—Public Information

                                   202-737-4333—Washington, DC office

A membership organization involved in scientific, educational, and professional activities, the IEEE has approximately 250,000 members worldwide and is the largest engineering society in the world. It is run by volunteers and consists of approximately 34 technical societies and councils. Areas of interest include theoretical research, reports of laboratory experiments, and descriptions of applications in commerce, industry and the consumer sector. While the headquarters staff is not a direct source of technical information, they can refer you to appropriate IEEE publications and members. The society publishes free catalogs of IEEE standards, press books, periodicals, conference proceedings, and education courses. IEEE publications, including books, periodicals, conference proceedings, and the IEEE Index, can be ordered from the IEEE Service Center, 445 Hoes Lane, Piscataway, NJ, 08854, 201-981-0060. A major publication is:

• *IEEE Spectrum*—a monthly periodical containing state-of-the-art, review, and applications articles chosen for their utility to a wide range of engineers and scientists in the electrical and electronics field; book reviews and listings; letters; news of IEEE activities, the industry, and the profession. *IEEE Spectrum* is the core magazine of the Institute. The annual subscription price is $88.

National Academy of Engineering (NAE)
2101 Constitution Avenue, NW   202-334-2198
Washington, DC 20418

A nonprofit honorary organization whose members are elected by their peers in the academy, NAE focuses on eight different engineering categories, including Electrical: Communications/Computers/Control. It has a list of names of members and the categories to which they belong. Staff members can answer questions about the Academy, but do not normally make referrals over the phone. The NAE participates in committees of the National Research Council, which advise Congress of matters of national importance, including those pertaining to engineering. The Office of the Foreign Secretary encourages research and development, and promotes cooperation, an exchange of information, and sharing of ideas between the U.S. and foreign countries. The NAE publishes various reports relating to studies done, some of which are free. An important periodical is:

• *The Bridge*—a quarterly publication featuring articles by engineers from the various disciplines in which the NAE is involved. Papers are presented on the basis of general interest and timeliness in connection with issues associated with engineering. The approximately 33-page publication is available free of charge.

National Electrical Manufacturers Association (NEMA)
2101 L Street, NW   202-457-8400
Washington, DC 20037

NEMA is a trade association for U.S. manufacturers of electrical equipment. Its major activities include the development of domestic and international standards; statistical and market programs; and government relations and international trade. Members produce components or end-use equipment for the generation, transmission, distribution, control, and use of electricity. NEMA holds an annual meeting and approximately 1,000 section meetings a year relating to its product scopes. The standards publications department (202-457-8473) has a standards publications catalog. NEMA also has a variety of other publications, including a membership directory, bulletin, annual report, publications concerning the fields of energy, and others concerned with government affairs. Several publications are:

• *NEMA Directory of Member Companies*—includes names, addresses, and sections and divisions with which the companies are affiliated. The 28-page publication is available free of charge.

● *NEMA News*—a monthly bulletin covering news in the following areas: an overview of the electrical industry; energy; government agencies and legislative affairs; NEMA conferences and conferences with other groups. It also discusses international standards and contains an updated list of NEMA standards publications. The four-to-eight-page publication is available to nonmembers for $10 a year, and is free to members.

Science and Electronics Division

International Trade Administration
U.S. Department of Commerce
14th Street and Constitution Avenue
Washington, DC 20230

202-377-4466

This group produces the *U.S. Industrial Outlook* and prepares competitive assessment studies of selected industries. It is responsible for responding to the Secretary of Commerce regarding matters affecting the industries covered by each office. The group also manages the Industrial Sector Advisory Committee (ISAC), which consists of high-level industry advisers who meet to provide comments, recommendations, and other advice on matters that affect the science and electronics industries, including reviewing legislation and trade policies.

The following four offices function within this cluster:

Office of Computers and Business Equipment
John McPhee, Director
202-377-0572

Office of Components and Related Equipment
Jack R. Clifford, Director
202-377-2587

Office of Telecommunications
William Sullivan, Acting Director
202-377-2006

Office of Instrumentation and Medical Sciences
Roger Stechschulte, Director
202-377-0550

Society of Automotive Engineers (SAE)

400 Commonwealth Drive
Warrendale, PA 15096

412-776-4841

An educational society composed of automotive and aerospace engineers, SAE is involved in all types of engineering. It prepares standards, specifications, and test procedures used in the design and manufacture of ground and space vehicles. Staff engineers can answer most technical questions. SAE has numerous national and international conferences, and their continuing educational courses update engineers on new developments in the field. SAE publishes 1,800 to 2,000 technical papers a year, which are presented at the conferences. A technical papers catalog and a publications catalog are available at no charge. Publications include two directories and two books:

- *Directory of Automotive Consultants*—lists approximately 260 consultants and provides information about their education, experience, areas of expertise, geographic areas in which they work, type of clients they serve, and a list of qualifications. Updated annually, the directory costs $15.

- *Aerospace Consultants Directory*—lists about 125 consultants and provides information about their education, experience, areas of expertise, geographic areas in which they work, type of clients they serve, and a list of qualifications. A new publication updated annually in June, the directory costs $15.

- *Electronic Automotive Reliability* (# SP573)—contains six papers on the subject of its title. One paper is concerned with the development of reliability, prediction, models, or electronic components in automotive applications. The approximately 40-page publication is available to nonmembers for $15 and to members for $12.

- *Displays, Electronics and Sensor Technology* (#SP565)—contains 20 papers dealing with electronics and their use in expanding driver control and information. The approximately 90-page publication is available to nonmembers for $43, and to members for $35.

Publications

CMP Publications
111 East Shore Road   516-365-4600
Manhasset, NY 11031

This company offers a number of publications relating to electrical and electronics engineering, including:

- *Electronic Buyers' News (*also known as the *High Technology Purchasing News Weekly)*—is a weekly tabloid written for purchasing executives in the electronics industry. The publication is available free of charge to qualified subscribers who complete a questionnaire.

- *Electronic Engineering Times*—is a weekly newspaper directed to engineers and technical management. It covers a variety of technical topics. The publication is available free to qualified subscribers who complete a questionnaire.

EDN Magazine
Cahners Publishing Company   303-388-4511
270 Saint Paul Street
Denver, CO 80206

A biweekly magazine, *EDN Magazine*, which stands for Electronic Design News, is a technical publication directed to engineers and engineering designers. Examples of areas covered include computers and peripherals, power sources, instruments, literature, professional issues, integrated circuits and semiconductors. The annual subscription price is $48.

Division of Science Resources Studies

National Science Foundation   202-634-4622—Editorial and Inquiries Unit
1800 G Street, NW
Washington, DC 20550

This Division collects data on all sectors of the economy, on research and development funding, and on science and engineering personnel. A number of publications are available:

● *Academic Science/Engineering Research and Development Funds: Fiscal Year 1982* (NSF 84-308)—this is a survey of 563 institutions having doctorate or master's science and engineering academic programs, or which have $50,000 or more in separately budgeted research and development expenditures and 19 federally funded centers. Detailed statistical tables include sources of funds, expenditures by disciplines, capital expenditures by disciplines, and geographic distribution of funds. Many tables also include presentation of data by rankings of the institution as a recipient of funds. The publication is available free of charge.

● *Academic Science/Engineering: Scientists and Engineers, January 1983* (NSF 84-309)—a companion volume to the preceding one, this time with a sample of 2,190 universities and colleges and 19 university-associated federally-funded research and development centers. Detailed statistical tables provide information on basic trends in the employment of scientists and engineers in universities, as well as on employment by field at the top 100 institutions. Numerous other tables round out this overview. The publication is available free of charge.

● *Characteristics of Recent Science/Engineering Graduates: 1982* (NSF 84-318)—the third in the series, these tables provide insight into what all the research and development funds and employed scientists and engineers have been producing. A total of 24,956 graduates from 1980 and 1981 responded to the survey, providing data on fields and disciplines, degrees obtained, sex, race, employment, type of employer, and primary work activity. The publication is available free of charge.

● *Guide to NSF Science/Engineering Resources Data* (NSF 84-301)—this publication contains the instruments NSF used to collect the information covered in the publications discussed above. The instructions for completing the surveys are also provided. Together they prove a useful tool for people interested in collecting related information or using the survey results. The publication is available free of charge.

Electronic News

Fairchild Publications   212-741-4470
7 East 12th Street
New York, NY 10003

This weekly publication covers the breadth of the electronics industry, including computers, components, communications, instruments, and materials. It is directed to upper-tier management and engineers in the electronics industry and the financial community. The 96-to-112-page publication is available for $36 a year.

Electronic Products

Hearst Business Communications, Inc.   516-222-2500
645 Stewart Avenue
Garden City, NY 11530

A semimonthly periodical, *Electronic Products* interprets technology for electronic engineering decisions. It contains technical articles in addition to covering new products. The publication is free to qualified subscribers who complete a questionnaire.

Electrotechnology Newsletter

National Technical Information Center (NTIS)   703-487-4630
5285 Port Royal Road
Springfield, VA 22161

This is a weekly current-awareness newsletter containing summaries and contact information for electrotechnology-related reports released by federal, state, and local agencies, and obtained by NTIS on an elective basis. It carries bibliographic information for the reports and covers the following subcategories: antennas; circuits; electromechanical devices; electron tubes; optoelectronic devices and systems; power and signal transmission devices; resistive, capacitive, and inductive components; and semiconductor devices. The full text of most reports can be obtained from NTIS, or possibly the author. Availability is mentioned in the citation.

McGraw-Hill, Inc.

Circulation Department   609-426-5000
Princeton-Hightstown Road
Hightstown, NJ 08520

The company offers several periodicals relating to electrical and electronics engineering, including:

● *Electronics*— this biweekly publication reports, interprets, and analyzes new technological trends in the electronics industry. It is directed toward electrical engineering designers, particularly solid state, and managers in the field. The annual subscription price is $24.

● *Engineering News Record*— a weekly publication covering the entire process of construction, including design, financing, and economic aspects. Specific high-technology topics include the development of fiber optics systems, as well as computer systems as they relate to the design and construction process. The approximately 100-page publication is available for $35 a year.

● *Electrical Construction and Maintenance*—this monthly magazine is edited to serve the needs of 74,000 electrical contractors, plant electrical personnel, and electrical consultants who specify and buy for the electrical construction market. The editorial pages consist of up-to-date technical information in the areas of design, installation, and maintenance. The annual subscription price is $18.

Data Bases

COMPENDEX

Engineering Information, Inc.   800-221-1044
345 East 47th Street   212-705-7600
New York, NY 10017

The COMPENDEX data base is the machine-readable version of the *Engineering Index Monthly*, which provides the engineering and information communities with abstracted information from the world's significant engineering and technological literature. COMPENDEX includes Electrical and Control Engineering among the subject areas it covers. It provides worldwide coverage of the journal literature, publications of engineering societies and organizations, papers from the proceedings of conferences, and selected government reports and books. COMPENDEX is available from DIALOG Information Services, as well as SDC Information Services, Bibliographic Retrieval Service, Inc., Pergamon Infoline, Ltd., and DataStar.

Ei Engineering Meetings

Engineering Information, Inc.   800-221-1044
345 East 47th Street   212-705-7635
New York, NY 10017

The *Ei Engineering Meetings* data base includes significant published proceedings of engineering and technical conferences, symposia, meetings, and colloquia. The data base covers all disciplines of engineering, including electrical engineering and electronics. Each meeting included is indexed in a main conference record; in addition, all papers from the meeting are individually indexed. Beginning in July 1982 every main conference record included is also referenced in a COMPENDEX record. The data base is available from DIALOG Information Services, as well as SDC Information Services, and Data Star.

INSPEC (Information Services in Physics, Electrotechnology, Computers and Control)

Producer: Institution of Electrical Engineers (IEE)   011-44-462-53331
Station House
Nightingale Road
Hitchin
Hertfordshire SG5 1RJ
England

Distributor: IEEE Service Center   201-981-0060
445 Hoes Lane
Piscataway, NJ 08854

INSPEC covers four main subject areas: physics, electrical engineering and electronics, control theory and technology, computers and computing. You can access journal articles, conference papers, books and monographs, reports and dissertations. Search services include DIALOG Information Services, SDC Search Service, and Bibliographic Retrieval Service.

NTIS Data Base

National Technical Information Service

U.S. Department of Commerce

5285 Port Royal Road

Springfield, VA 22161

703-487-4600 general information

703-487-4642-specific questions concerning data base or subject content

703-487-4650 to order reports

The NTIS data base includes electrotechnology as well as computers, control, and information theory among the subject areas it covers. It represents the reports of three major U.S. federal government agencies U.S. Department of Energy (DOE), U.S. Department of Defense (DOD), National Aeronautics and Space Administration (NASA) plus many other agencies. The data base consists of government-sponsored research, development, and engineering reports, as well as other analyses prepared by government agencies, their contractors, or grantees. Included in this coverage are federally generated machine-readable data files and software, U.S. government inventions available for licensing, federally generated translations, and reports prepared by non-U.S. governments and exchanged with federal agencies. The NTIS data base corresponds to several printed publications, including *Government Reports Announcements & Index* and 26 abstract newsletters, such as *Government Inventions for Licensing*. The NTIS leases the data base to DIALOG Information Services, SDC Information Services, and Bibliographic Retrieval Service.

Additional Experts and Resources

Richard Whitley, Industry Specialist

Deputy Assistant Secretary

Capital Goods and International Construction

Office of General Industrial Machinery

International Trade Administration (ITA)

U.S. Department of Commerce

14th Street and Constitution Avenue, NW

Washington, DC 20230

202-377-4382

Mr. Whitley, the industry specialist at ITA for electronics, is an expert in the areas of transformers, switchgear, motors, industrial controls, wiring devices, and lighting fixtures. He can answer questions on most types of electrical products, with the exception of appliances.

Electro-Optics

See: Optoelectronics

Embryo Transfer

See: Reproductive Technology

Employment and Training

See: Careers and Employment; Labor and Technological Change

Energy, Renewable

See: Renewable Energy Technology

Entrepreneurism

See also: Research and Development; Small Business Incubators; State High-Tech Initiatives; Venture Capital

Organizations and Government Agencies

Academy of Distinguished Entrepreneurs
Center for Entrepreneurial Studies   617-235-1200
Babson College
Wellesley, MA 02157-0901

The Academy was founded to recognize the achievements of distingished entrepreneurs. Each year, several entrepreneurs are invited to Babson to be inducted into the Academy and give speeches. They are selected by a distinguished panel of editors and publishers from *Business Week*, *Wall Street Journal*, *Forbes*, *Harvard Business Review*, and *The Economist*. The Center has compiled extensive files on thousands of entrepreneurs, with a composite biographical summary supported by research of published data. To date, 30 entrepreneurs have been honored for their contribution to the development of free enterprise.

American Entrepreneur Association

2311 Pontius Avenue   213-478-0437
Los Angeles, CA 90064

This Association is geared toward individuals thinking of starting a business and those just starting an enterprise. It has published more than 200 manuals to assist people setting up a business. The Association also sponsors a telephone network of business counselors who will provide members with advice and referrals to literature, organizations, or people. Subscription to the following magazine entitles you to membership privileges:

● *Entrepreneur Magazine*—a monthly publication aimed at those interested in going into business for themselves. Each issue contains a detailed article about setting up a specific business, as well as feature stories on successful entrepreneurs. There is coverage of operating techniques, such as "How to Get the Best Advertising," "Designing a Brochure," and "Accumulating Venture Capital." Each January, the magazine reports its "Franchise 500," a summary analysis of the top 500 franchises available. About 96 pages an issue; the annual subscription rate is $24.50.

Caruth Institute for Owner-Managed Business

Southern Methodist University   214-692-3326
Dallas, TX 75265-9990

The Caruth Institute offers courses in entrepreneurial studies; sponsors seminars and other educational programs, often for visiting businessmen and students from abroad; and participates in an annual venture capital conference. Two major textbooks in the field of entrepreneurism have been written by the staff at the Institute:

● *The Entrepreneurism Master Planning Guide*—written by John Welsh and Jerry White, and published by Prentice-Hall, this is a blueprint for starting a business. It covers all aspects of the business start-up, from the characteristics of businesspeople to the investment of venture capital; it is used widely among the schools offering entrepreneurism as a field of study. About 360 pages, the text sells for $24.95 hardback and $11.95 paperback.

● *Administering the Closely Held Company*—also written by Welsh and White and published by Prentice-Hall. This is a tutorial on cash-flow management. The text covers all phases of cash flow, including accounting methods and tax implications. About 340 pages, it sells for $24.95.

Center for Entrepreneurial Management

83 Spring Street   212-925-7304
New York, NY 10012

This nonprofit educational organization serves presidents of small businesses. Worldwide membership numbers more than 2,500. It offers a variety of seminars (also available on tape), covering such topics as writing a business plan, financing, and raising capital. Its Chief Executive Officers Club meets frequently in major U.S. cities and provides a forum for the exchange of successful strategy and other venture issues. The Center sells more than 120 materials, including audiotapes and video-tapes, all listed in its publications catalog available upon request. Membership includes publication and seminar discounts, special reports and registers, subscriptions to *INC. Magazine* and *Venture Magazine*, as well as the Center's Newsletter.

• *The Entrepreneurial Manager's Newsletter*—a 12-page monthly designed for action and problem solving from the perspective of the entrepreneur. Monthly topics include financing, taxes, venture capital, start-up, and management. A calendar of upcoming events, a question-and-answer column, and a personality section profiling successful entrepreneurs are included in each issue. It is available free to members; nonmembers can subscribe for $71 a year.

Center for Entrepreneurial Studies (CES)
Babson College   617-235-1200
Babson Park
Wellesley, MA 02157-0901

Babson was one of the first American colleges to grant an undergraduate degree in entrepreneurial studies. Its Center maintains a data base of dossiers about thousands of entrepreneurs throughout the world. The repository is updated continually, and the staff will search the data base and make referrals. Researchers maintain an ongoing resource file of local business people willing to share their expertise and, in return, the Center offers specialized consulting services to local entrepreneurs.

A Speakers Bureau, consisting of faculty, administrators, and entrepreneurs, is also coordinated by the Center. The Center sponsors an annual entrepreneurship research conference, by invitation only, for hundreds of leaders in the academic field. Selected papers are published in *Frontiers of Entrepreneur Research*, published each year by the Center.

Control Data Business Centers (CDBC)
Control Data Corporation   612-853-6397
8100 34th Avenue South
Bloomington, MN 55420

Control Data operates more than 40 business centers, each sponsoring a variety of executive forums, for business owners to meet and discuss topics common to all entrepreneurs. The business centers also offer management training courses. Control Data will be making CDBC programs available on disc, and it has plans to offer on-line management training courses in the near future. Examples of material covered include computer-assisted training for small businesses, building your own business, expanding a business, cash flow, financial management, operations and production, and people management. Note: Due to a contract between the Small Business Administration (SBA) and Control Data, the federal government will pay most of the cost for small businesses to take CDBC's computer-assisted management courses. For further details, see Computer-Assisted Training Programs, under *Computers (General)* in this directory.

Entrepreneurial Activities Task Force

Institute of Electrical and Electronics Engineers (IEEE)   202-785-0017—Thomas Suttle
1111 19th Street, NW
Washington, DC 20036

Consisting of volunteer leaders from IEEE, this task force is engaged in looking at entrepreneurial activities from the perspective of electrical engineering. Mr. Suttle is a staff member for the group, and he can tell you about its activities.

Entrepreneur Program

School of Business Administration   213-743-2451
University of Southern California (USC)
Bridge Hall
Los Angeles, CA 90007

USC offers a degree program, at both the graduate and undergraduate level, in entrepreneurism. The program faculty is an interdisciplinary team of professors with extensive teaching experience, headed by Richard H. Buskirk. There is team teaching, and there are many guest speakers, including entrepreneurs, specialists, and program alumni, who supplement the regular faculty. Members of the faculty and staff will make referrals to texts and other publications, entrepreneurs, specialists, and consultants. USC promotes communication between the academic and business communities through its research, information exchanges, seminars, and other public functions.

Hankamer School of Business

Center for Entrepreneurship and Private Enterprise, Suite 215   817-755-3766
Baylor University
Waco, TX 76798

Baylor offers an undergraduate degree in entrepreneurial studies, and works with other schools toward expanding this field of study. Its Hankamer School carries out research in the field, concentrating on the psychological characteristics of entrepreneurs and the planning of small business. Its innovation-evaluation program aids new ventures in determining the commercial feasibility of their invention and the probability of success. Hankamer also offers venture assistance programs to help those planning to start a business. The Center's referral service will provide information about consultants, funding sources, and manufacturers' or distributors' availability.

International Council for Small Business (ICSB)

Headquarters Office: Small Business Development Center   314-534-7232
3642 Lindell Boulevard
St. Louis, MO 63108
U.S. Affiliate Office: ICSB-US
Brooks Hall
University of Georgia
Athens, GA 30602

This international nonprofit organization is devoted to stimulating and improving management education for small independent businessmen and businesswomen. Membership consists of government employees, consultants, and academicians who assist small businesses. The organization sponsors an international conference annually, as well as various local conferences and seminars. The U.S. affiliate office in Athens, Georgia, will mail referrals to experts in the small business field who can supply information, advice, and/or consulting services. Written requests are preferred. The publications listing, available from the St. Louis office, includes quarterly newsletters, special publications about business management, research papers prepared by members, proceeding from meetings, and a bibliography of small business information sources. ICSB also publishes a journal well recognized in the field:

● *Journal of Small Business Management*—a quarterly professional and academic publication dealing with the problems facing small businesses. The *Journal* usually contains about ten articles about research and development, educating the entrepreneur, business practices, and new directions for small business. It also has editorials, book reviews, reader opinions, resource listings, and abstracts from other publications. About 100 pages an issue. The annual subscription is $20 and the publication can be ordered from the Bureau of Business Research, West Virginia University, PO Box 6025, Morgantown, WV 26506-6025, 304-243-5837.

National Institute for Entrepreneur Technology

2000 P Street, NW, Suite 305   202-833-2326
Washington, DC 20036

This membership organization consists of 1,000 representatives from small high-technology businesses, universities, and government agencies involved in promoting innovations. It focuses on science and technology and keeps its members informed about federal research and development funding. The Institute also provides support to people engaged in entrepreneurial activities. Member services include a telephone hotline service, through which callers can obtain information about the federal government's science and technology activities, legislation, regulations, telephone numbers, and so on. The public can obtain free copies of back issues of the Institute's newsletter:

● *Institute Insight*—a monthly newsletter containing feature articles on topics of interest to members, research reports, a monthly column highlighting a company that has provided money or services to further technology, a legislative update section, and a publications section containing abstracts of recent publications.

● *Letter Report*—a monthly report covering specific topics of interest.

National Small Business Training Program

American Association of Community and Junior Colleges (AACJC)   202-243-7050
1 Dupont Circle, Suite 410
Washington, DC 20036

AACJC has established a small business training network of 300 colleges in an effort to help the schools meet their informational needs and standardize curriculum materials. The Association produces several publications, including a free catalog. Materials include the following:

- *Curriculum Guides for Small Business Management Training*—a 13-volume series of reference guides about topics of study within the small business management field. These guides offer institutions a standard by which to gauge their curriculums. The cost for the entire set is $140. Volumes may be purchased individually.

- *Small Business Management Resource Guide*—published in two separate sets, one for noncredit and nontraditional courses, such as those offered by community and junior colleges, and one for credit curriculums, as offered by four-year colleges. These guides give course descriptions and syllabi for a vast array of study modules in the field of small business management. The cost per set is $60.

- *Training Tools Guide*—a comprehensive resource directory, which covers the curriculum, films, video materials, texts, and other equipment resources available for training. Contact the Association for current prices.

- *Older American-Small Business Management for the Senior Citizen*—a four-volume series available for $70.

- *Women's Entrepreneurism Orientation Directory*—this reference guide is available for $25.50.

The Office of Advocacy

U.S. Small Business Administration (SBA)	202-653-7561
1441 L Street, NW	800-368-5855—SBA Answer Desk
Washington, DC 20416	

The SBA's mission is to help people establish businesses and prosper, and to act as a strong advocate for small business. The SBA's Office of Advocacy is specifically mandated to protect, strengthen, and effectively represent small business with the federal government. It informs the small business community of issues that affect it, and assists the entrepreneur with questions and problems regarding federal laws, regulations, and assistance programs. It makes available research in a number of areas, including tax and procurement issues. The Office of Advocacy also serves as a conduit through which small businesses can make suggestions and criticisms for policy consideration.

Publications

Inc. Magazine

38 Commercial Wharf	617-227-4700
Boston, MA 02110	

This monthly business magazine provides practical "how to" information for managers of small companies to mid-sized ones. Each issue, about 200 pages in length, has articles describing individuals who have become entrepreneurs, innovations that have resulted in new or improved business, and news of the small business world. It provides clarifications of new laws and regulations, as well as new sources of information. The annual subscription is $18.

Lord Publishing Co.
46 Glen Street
Dover, DE 02030

This company publishes two guides written by Dr. Ronstadt:

● *Entrepreneurism: Essential Reading and Bibliography*—contains more than 600 citations, 50 readings, and 7 anthologies relating to entrepreneurism. The cost is $24.95.

● *Entrepreneurship: Text, Cases, and Notes*—this volume is used by colleges and universities establishing curriculum requirements in entrepreneurism studies. It covers entrepreneurial perspectives and operations, financing, the business plan, and entry into new ventures. The cost is $29.50.

Venture Magazine
Arthur Lipper Corp.   212-840-5580
35 West 45th Street
New York, NY 10036

This monthly business magazine is devoted to entrepreneurs. Each issue features case histories and newsworthy stories of entrepreneurial successes, as well as timely information for the owners of small enterprise. *Venture* discusses how to find funding capital, manage a business, and handle tax and legal issues. There are notations of new start-ups in specific industries and fields, as well as forecasts for business trends in general. A calendar of upcoming events and a classified section are also included. About 200 pages an issue; the annual subscription rate is $18.

Data Bases

Procurement Automated Source System (PASS)
Contact your local Small Business Administration Office, or: 202-653-6586
U.S. Small Business Administration (SBA)
Office of Procurement and Technical Assistance
1441 L Street, NW
Washington, DC 20416

PASS is a computerized data base containing capability profiles of small businesses interested in becoming a prime contractor for federal agencies or a subcontractor for companies. It contains more than 72,000 firms nationwide, covering such fields as research and development, manufacturing, construction, and services. The system can be searched by geographic location, type of ownership, labor surplus area, zip code, minority type, and more than 3,000 key words. Upon request, the SBA staff will search the data base free of charge. Small businesses wanting to be included in the PASS data base should contact the above office. Firms are added free of charge to the data base.

Additional Experts and Resources

George Solomon, DBA
George Washington University (GW)   202-676-7375
Department of Management Science
Washington, DC 20052

Dr. Solomon is the only person in the U.S. to hold a Ph.D. in Small Business Management Entrepreneurship. He is currently a professor at George Washington University and the senior Research and Development Specialist in the Office of Management Assistance at the Small Business Administration. Dr. Solomon is an expert in research methodologies for measuring the effectiveness of management assistance to entrepreneurs, as well as the behavior and management development of entrepreneurs. He can answer questions and guide you to literature and experts in the field.

Environmental Engineering

Organizations and Government Agencies

American Academy of Environmental Engineering (AAEE)
PO Box 269   301-267-9377
Annapolis, MD 21404

AAEE is an organization of over 2,200 environmental engineers certified as specialists in their discipline. This group holds frequent seminars open to the public on topics ranging from air pollution to waste water and solid wastes, and it allows advertisers to promote their various products and systems at these seminars. A copy of the membership roster with specialties noted is available for $200; mailing labels with the names of engineers grouped by discipline may be purchased for $.15 each.

American Institute of Chemical Engineers (AICHE)
345 East 47th Street   212-705-7338
New York, NY 10017

The membership of this professional society consists of over 60,000 chemical engineers and chemists, including those in the field of environmental engineering. AICHE holds a national meeting for the public, as well as continual symposia on current engineering topics. It also sponsors continuing education for members, and has now computerized many of its learning modules for member access. The publications listing includes four quarterlies, covering chemical engineering, energy, plant operations, and environmental engineering, as well as symposia papers, and technical documents. The monthly magazine, *Chemical Engineering Progress*, contains papers presenting technical topics in the chemical engineering field. The annual subscription for nonmembers is $35.

American Society of Civil Engineers (ASCE)

345 East 47th Street   212-705-7496
New York, NY 10017

A professional society of over 95,000 civil engineer members, ASCE is devoted to the improvement of the technology of the engineering field and the promotion of the technical capabilities of its members. ASCE holds two conventions a year, at which current engineering topics are discussed, and new products and systems are introduced. The Society will make information referrals to its member experts on a wide range of engineering topics, and it publishes an assortment of technical journals covering the various engineering disciplines. A listing of the publications is available upon request, and includes:

- *Civil Engineering Magazine*—a monthly publication of approximately 120 pages, usually focused on the topic of the cover story. It also includes technical papers and classified ads, as well as information and newly available technology. The annual subscription rate is $24.

American Water Works Association Research Foundation (AWWA)

6666 West Quincy Avenue   303-794-7711
Denver, CO 80235

AWWA is a nonprofit corporation that obtains funds for research, commissions the work, coordinates and directs, and then publishes reports of progress and results in the water safety field. Findings of this research and its impact on drinking water supplies are published in several magazines available to members and the public.

- *Water Research Quarterly*—a 20-page newsletter detailing technical research on the quality of the U.S. water supply. The nonmember subscription rate is $30 a year.

- *Research News*—a bimonthly magazine of approximately 20 pages describing ongoing research, funding sources, and results of specific testings in water use. The annual subscription for nonmembers is $20.

- *Municipal Wastewater Reuse News*—a bimonthly newsletter of approximately 30 pages covering the recycling of public water sources. The newsletter contains articles on systems in use, chemical substance levels, and purification standards for water supplies. The annual subscription rate is $50.

Association of Environmental Engineering Professors (AEEP)

No permanent mailing address or headquarters   217-333-4700—V. L. Snoeyink, President
                                                    404-894-2265—F. M. Saunders, Vice President

AEEP was formed to promote education in the field of environmental engineering. It consists of some 300 university and higher-education teachers in the U.S. and Canada. They sponsor various conferences and workshops, as well as apprise those in the field of the funding policies of the various agencies doing research concerning the environment. A membership directory, including specialties, may be obtained through Professor L. Christensen, Department of Civil Engineering, Villanova University, Villanova, PA 19085. AEEP also publishes a register of environmental engineering graduate programs, which describes ongoing programs and the faculties associated with them. The register is available for $30 through Professor D. Lawler, Department of Civil Engineering, University of Texas, Austin, TX 78712. All of the above-mentioned professors are very helpful and knowledgeable sources for locating specialists and agencies in the area you are interested in.

Council on Environmental Quality (CEQ)

722 Jackson Place, NW   202-395-5700
Washington, DC 20006

CEQ is a policy-making organization within the Executive Office of the federal government, and acts as an oversight mechanism in the enforcement of the National Environmental Pollution Act (NEPA). The Act requires the federal agencies to prepare environmental impact statements. CEQ also advises the President on environmental statement. The results of its studies and other technical reports, including the annual report on the state of the environment, are available through EOP Publications Office, 2200 NEOB, 726 Jackson Place, NW, Washington, DC 20503, 202-795-7332.

Environmental Protection Agency (EPA)

Andrew W. Briedenback Environmental Research Center   513-684-7951—Health Effects Lab
216b West St. Clair Street   513-684-7771—Municipal Lab
Cincinnati, OH 45268   513-684-7418—Industrial Research Lab

These three EPA labs are concerned with the water and hazardous wastes aspects of environmental protection. All research in these areas is directed through these labs, and the staff will give specific information about ongoing research, as well as make referrals to publications and experts in various disciplines of water pollution and waste control. The Center maintains an on-line data base with information about current research at the laboratories. Each lab's computer services staff will search its data base for the public. A fee is charged for this service.

Environmental Protection Agency (EPA) Library

Toxic Substances Library   202-382-5922
401 M Street, NW
Washington, DC 20460

These two EPA libraries contain thousands of reference materials, including books, journals, EPA documents, and legal information sources in the field of environmental protection. The EPA reports, containing scientific data based on EPA research, are available through NTIS File 6, an on-line data base accessible through DIALOG. For further information call DIALOG at 800-227-1927.

International Association for Water Pollution Research and Control (IAWPRC)
Alliance House   011-44-1-405-4552
29-30 High Holborn
London, England

Richard Inglebrecht, President   217-333-3832
University of Illinois
3230 Newmark Civil Engineering Lab
208 North Romine Street
Urbana, IL 61801

This international membership organization of about 2,000 engineers, scientists, and corporations was formed to provide a forum for the exchange of research and new information and developments regarding all aspects of air quality. It holds a biannual conference and many specialized and regional conferences, which are chiefly technical sessions on current research in the field. It will make referrals to specialists in any area of water pollution control. Its publications listing includes a newsletter and two international magazines:

● *Water Research*—a monthly publication of about 100 pages, which contains original articles on current developments and technical data on water pollution. The annual subscription for nonmembers is $50.

● *Water Science and Technology*—a monthly magazine of about 200 pages, which chiefly contains the papers of various conference proceedings held by the Association and papers from the technical sessions of its members. The nonmembers' subscription rate is $50.

National Wildlife Federation
8925 Leesburg Pike   703-790-4000
Vienna, VA 22180

The Federation, with over 4.1 million members and supporters, is involved in conservation education and in promoting wise usage of our natural resources and protection of the environment. A free copy of the conservation education catalog, covering education programs, reference and research material, and other various publications, is available upon request. The Federation also publish several magazines:

● *National Wildlife* and *International Wildlife*—bimonthly magazines published alternately and available at $12 each or at $19 for both. These digests of approximately 50-75 pages cover both national and worldwide conservation topics, and contain numerous paintings and photographic essays.

U.S. Geological Survey
U.S. Department of Interior   703-860-7455
425 National Center
Reston, VA 22092
Water Research Scientific Information Center (WRSIC)

Primarily concerned with research and development in areas of national focus, WRSIC does work resulting in more efficient water reuse by American municipalities and industries. Its findings and recommendations are published as water research abstracts available from this Office. WRSIC also provides the information for data base file 117 through DIALOG, which covers a wide range of water resource topics with emphasis on water planning, the water cycle, and water quality.

U.S. Environmental Protection Agency Laboratory (EPA)

Triangle Park Research Laboratory Complex   919-541-4577
Research Triangle Park, NC 27711   919-541-2281—Health Effects Research Lab
                                                        919-541-2191—Atmospheric Sciences Research Lab
                                                        919-541-2821—Air and Energy Engineering Research Lab
                                                        919-541-2106—Environmental Monitoring Systems Lab

Concerned chiefly with air pollutants, these laboratories do biological and clinical research on exposure to pollutants, they research the impact of pollutants on air quality, land, water and weather; and they monitor networks of pollutant samplings in an effort to establish criteria for assessing the health of the American population. Technical reports and project summaries are available to the public and are listed in the ORD (Office of Research and Development) publications announcements. This bulletin is available upon request from ORD-CERI, 26 West St. Clair Street, Cincinnati, OH 45268. There is also a directory of EPA data bases, currently undergoing an extensive update and revision, called the systems information directory, available from EPA Management Information Support, 410 M Street, SW, Washington, DC 20460, which contains listings of the computerized research done at this laboratory center, as well as the other EPA labs.

Water Pollution Control Federation (WPCF)

2626 Pennsylvania Avenue, NW   202-337-2500
Washington, DC 20037

WPCF is a nonprofit educational organization with over 30,000 members worldwide whose main function is the dissemination of information on water pollution control. Its free publications brochure contains hundreds of manuals on such topics as disinfection, odor control, industrial waste water, and other aspects of water pollution control. The technical department will answer questions, give research information, and make referrals to specialists. WPCF's main publications include:

• *Journal WPCF*—a monthly technical journal containing technical papers on water pollution control and assorted advertising supplements for products and services in the water supply industry. About 100 pages, the *Journal* is available to nonmembers for $120 a year.

• *Highlights*—a monthly newsletter of about ten pages with news on the Water Pollution Control Federation undertakings, as well as current governmental concerns, including congressional debate topics. The annual subscription for nonmembers is $16.

• *Operation Forum*—a monthly directed at operators of waste water treatment plants, with innovations and systems analyses. The annual subscription rate is $50.

Publications

Environmental Science and Technology Magazine

1155 16th Street, NW
Washington, DC 20008

202-872-4600

Published by the American Chemical Society, this monthly magazine is a scientific journal of articles on problems in the environment. In addition to the technical articles, it contains product information, lists of available literature, and advertisements for products in environmental control. About 80 pages an issue; the annual cost is $26 for nonmembers.

Pudvan Publishing

1935 Shermer Road
Northbrook, IL 60062

312-498-9840

● *Pollution Engineering*—a monthly magazine available free of charge to qualified professionals in the field of environmental engineering, covers all aspects of pollution, including air, water, noise, hazardous wastes, and toxic substances. About 100 pages an issue, the magazine contains feature articles, case histories, Washington news and legislation, and advertisements on products dealing with environmental engineering. The annual subscription is $18, which includes complimentary copies of the directories listed below:

● *Consultant/Service Telephone Directory*—is published annually in May; contains listings of companies doing environmental engineering and the services they offer. About 150 pages; the price is $10 for nonsubscribers.

● *Environmental Control Telephone Directory*—is published annually in October; it contains listings of products, devices, and ancillary equipment available in the field. The cost to nonsubscribers is $10.

Additional Experts and Resources

Jim Basilico

Environmental Protection Agency
Environmental Engineering and Technology
Office of Research and Development
401 M Street, SW
Washington, DC 20460

202-382-2583

Mr. Basilico can help you identify people, government agencies, research institutions, and professional groups involved in environmental engineering. He is an expert in the field, and director of the Office's Hazardous Waste Research section.

Edward H. Bryan, Ph.D.
Program Director for Environmental and Water Quality   202-357-7737
National Science Foundation (NSF)
1800 G Street, NW, Room 1130
Washington, DC 20550

As the administrator of NSF's program to fund research, development, and demonstration projects in environmental engineering, Dr. Bryan is very knowledgeable about current research in the field. He and his staff can provide information about current and projected funding, ongoing research at institutions across the U.S., and governmental activities. Dr. Bryan himself is an expert in environmental engineering. Prior to working for NSF, he held a professorship at Duke University and was employed by Dow Chemical.

Environmental Impact Assessment

Organizations and Government Agencies

National Association of Environmental Professionals (NAEP)
PO Box 9400   301-229-7171—Ms. Jane Thurber
Washington, DC 20016

NAEP is an association of approximately 1,000 scientists, engineers, technologists, administrators, and policymakers interested in interdisciplinary environmental affairs. Its members are experts in environment impact assessment, and Ms. Thurber can refer you to an appropriate member for information and further referrals. The Association grants fellowships of $800 to $1,000 annually to 10-15 graduate students selected to present a paper at NAEP's annual conference. NAEP publishes the following:

• *The Environmental Professional*—a quarterly journal containing six to eight interdisciplinary scientific studies; policy statements by federal, state, and local government officials, as well as by officials from industry and conservation groups; and a section on what is happening in the field—new technologies, forthcoming activities, and so on. A 12-page newsletter advertising jobs in the field, conferences, and internal news is included. The cost for nonmembers is $50 a year; for members it is $25 to $50, depending on membership status. Direct journal orders and inquiries should be addressed to Professor R. J. Rajagopal, Department of Geography, University of Iowa, 3020 Main Library Bldg., Iowa City, Iowa 52242, 319-353-3334.

Environmental and Ground Water Institute (EGWI)
University of Oklahoma   405-325-5202
200 Felgar Street, EL #127
Norman, OK 73019

The Institute is active in all types of research and training associated with environmental impact assessment (EIA). Its staff, of 10 to 15 doctoral students, follows international developments and does environmental assessment and management studies. EGWI has helped establish an EIA exchange network of individuals worldwide. The Institute staff members can answer inquiries, make referrals, and recommend appropriate literature. The Institute is setting up an environmental management/assessment data base, which may go on-line. EGWI publishes:

● *EIA Worldletter: The International Newsletter for Environmental Assessment*—an eight-page tabloid published six times a year. The newsletter covers issues in detail, featuring EIA activities in countries throughout the world; controversial activities on the horizon; and environmental development activities on an international-to-local basis. The cost is $30 a year for individuals, and $85 a year for institutions.

● *Directory of EIA Contacts*—EGWI is compiling a directory of 2,500 people worldwide involved in EIA, which should be available soon.

The U.S. Man and The Biosphere Program (MAB)
Secretariat   202-632-2786
U.S. Department of State   202-632-2816
OES/ENR/MAB   202-632-7571
Washington, DC 20520

The MAB program is aimed at providing a bridge between fundamental science and technological application. Its participants—distinguished scientists, resource administrators, policymakers, academicians and others in the environmental community—study management problems arising from the interactions between human activities and natural systems. MAB's structure consists of ten committees, with voluntary participants who are experts in their fields. The group produces publications, provides information about resource materials, sponsors consortia, and publishes a free newsletter, *MAB Bulletin*. Published on an as-needed basis, the *Bulletin* provides information about what MAB is doing, new publications released, resources available, and general information about scholarships and university, government, and industry programs or causes.

You can contact the MAB staff in the Washington office for information and referrals, or call one of the Directorates listed below:

Tropical Forests
Dr. Ariel Lugo, Project Leader
Institute of Tropical Forestry, Box AQ
Rio Piedras, Puerto Rico 00928
809-763-3939

Temperate Forests
Dr. Peter F. Folliott
School of Renewable Natural Resources
University of Arizona 85721
602-621-7276/7255

Grazing Lands
Dr. W. James Clawson
Agronomy and Range Science Extension
University of California
Davis, CA 95616
916-752-3455/0700

Arid and Semi-Arid Lands
Professor Henry Caulfield
Department of Political Science
Colorado State University
Fort Collins, CO 80523
303-491-6806 (Messages: ext. 5156)

Fresh Water Resources
Dr. Edward A. Fernald
Institute of Science and Public Affairs
Florida State University
Tallahassee, FL 32306
904-644-2008

Arctic
Dr. Charles N. Slaughter
Principal Watershed Scientist
Institute of Northern Forestry
U.S. Department of Agriculture
Fairbanks, AK 99701
907-474-7443

Island Ecosystems-Caribbean
William S. Beller, Chief
Ocean Programs Branch WH-585
U.S. Environmental Protection Agency
Washington, DC 20460
202-245-3154

Biosphere Reserves
Dr. William P. Gregg (Co-Chairman)
Office of Science and Technology
U.S. Department of the Interior
National Park Service (498)
Washington, DC 20013
202-343-8115

Dr. Stanley Krugman (Co-Chairman)
Director, Timber Management Research
U.S. Department of Agriculture
Forest Service
PO Box 2417, RPE 811
Washington, DC 20013
703-235-8200

Perception of Environmental Quality
Dr. Ervin H. Zube
School of Renewable Natural Resources
325 BioSciences East
University of Arizona
Tucson, AZ 85721
602-621-5462

Pollution
Dr. C. Bruce Wiersma, Manager
Earth and Life Sciences
Idaho National Engineering Laboratory
E.G. & G. Idaho, Inc.
PO Box 1625
Idaho Falls, ID 83415
208-526-1590

International Association for Impact Assessment (IAIA)
Industrial and Systems Engineering   404-894-2330
Georgia Institute of Technology
Atlanta, GA 30332

   IAIA is an excellent resource for environmental impact assessment. The staff will answer inquiries and/or make referrals. IAIA's *Impact Assessment Bulletin* provides good coverage of all aspects of IAIA. For further information on IAIA, see also the entries *Technology Assessment* and *Technology Transfer*.

UNESCO
UNIPUB   212-916-1600
205 East 42nd Street
New York, NY 10017

UNESCO, the United Nation's Educational, Scientific and Cultural Organization, covers worldwide environmental activities and mobilizes the scientific community to tackle world problems. UNESCO sponsors programs related to environmental impact assessment, including: Man and the Biosphere, which involves 1,000 research projects in 80 countries and 200 biosphere reserves around the world; the Intergovernmental Oceanographic Commission; the International Hydrological Program; and the International Geological Co-relation Program. UNESCO has published more than 100 reports, studies, and so on, relating to environmental issues. A free catalog of its publications can be obtained from UNIPUB.

• *Nature and Resources*—a UNESCO quarterly magazine, provides news and information about UNESCO's environmental activities worldwide. It is geared toward researchers and scientists, and contains papers by scientists of international reputation, general news, and information about forthcoming symposia, conferences, and training courses. The subscription fee is $15 a year.

UNESCO's headquarters in Paris, France, is a good resource for information about all aspects of environmental impact assessment. The staff will answer inquiries and refer you to specialists worldwide. Write to Science Sector, UNESCO, 7, Place de Fontenoy, 75700 Paris, France.

Office of Federal Activities
Environmental Protection Agency (EPA)   202-382-5073
401 M Street, NW
Washington, DC 20460

This EPA Office can refer you to the appropriate federal agency for environmental impact statements. In order to obtain a referral, you must have specific information about what you are seeking, such as the federal department responsible for the statement and the exact geographical area affected.

Council on Environmental Quality (CEQ)
Office of the General Counsel   202-395-5754
722 Jackson Place, NW
Washington, DC 20006

CEQ oversees the implementation of the National Environmental Policy Act (NEPA), as well as issues and regulations governing the environmental assessment process. These regulations are binding on all federal agencies. The staff can answer questions about all legal aspects of the EIA process. The following materials are available for free from CEQ:

• *Regulations for Implementing the Procedural Provision of NEPA Act*— a 44-page booklet that includes regulations, statutes, and pertinent executive orders.

• *Scoping Memorandum—1981*—a detailed discussion of scoping and suggestions for improving the EIA scoping process; 18 pages.

• *The 40 Most-Asked Questions Concerning CEQ's National Environmental Policy Act*—a 12-page booklet geared toward the general public.

• *A Guidance Memorandum—1983*—CEQ guidance regarding scoping, tiering, categorical exclusions, alternatives, and conflict of interest provisions for contractors preparing environmental impact statements; 15 pages.

Publications

Environmental Impact Assessment Review

Laboratory of Architecture and Planning, Room 4-209   617-253-1367
Massachusetts Institute of Technology
77 Massachusetts Avenue
Cambridge, MA 02139

The *Environmental Impact Assessment Review* is a 100-page reference quarterly devoted to approaches to impact assessment, environmental decision making, and the resolution of environmental disputes. It has feature articles on problems in the field. Each issue contains brief summaries of analytical techniques and breakthroughs in environmental decision making, organized under eight departments: (1) Generating Alternative Policies, Programs, and Designs; (2) Impacts on the Natural Environment; (3) Social Impact Assessment; (4) Presenting Technical Information; (5) Decision Making; (6) NEPA: Theory and Practice; (7) International Perspectives; (8) Agenda for Environmental Negotiations. Contact the office listed above for information or for submitting articles. To subscribe, write or call Advertising Manager, Plenum Publishing Corporation, 233 Spring Street, New York, NY 10013, 212-620-8000. The subscription cost for individuals is $35 a year; for organizations, $70 a year. Single issues can be purchased for $22. (Note: a free copy is available for examination.)

Note: Many federal agencies are involved with environmental impact assessment. The agencies set their own guidelines, and the federal government's EIA activities are very decentralized. Since there is no central clearinghouse within the government on this topic, you will need to contact the government agency responsible for the area you are involved in.

Epidemiology

See: Toxicology

Ergonomics

See also: Industrial Hygiene; Social Impact of Technology

Organizations and Government Agencies

Aerospace and Human Factors Division

Ames Research Center   415-965-5094
National Aeronautics and Space Administration (NASA)
Moffett Field, CA 94035

The Aerospace and Human Factors Division conducts a variety of human factors research programs for NASA, DOD, FAA, industry, and other government agencies. The Center publishes an annual publications catalog, which is available free of charge. Dr. Chambers will answer questions about the Center's human factors research program or will refer you to other specialists in the field.

American Industrial Hygiene Association (AIHA)

Public Relations Office   216-762-7294—Stephanie Beidler
475 Wolfledges Parkway
Akron, OH 44311

AIHA is a professional association of 6,000 industrial hygienists throughout the U.S. and 40 foreign countries. The Association promotes the study, evaluation, and control of environmental stresses in the workplace. AIHA conducts continuing courses and holds an annual professional conference on occupational health topics. The courses and conference are open to the public. The Public Relations Office will make referrals to specialists, provide information, and send you a copy of the free publications catalog. AIHA publications include:

• *Work Practices Guide for Manual Listing*—this guide summarizes research in basic ergonomics and makes recommendations on both the selection and training of workers who perform manual materials-handling activities. The 180-page guide costs $12.

• *American Industrial Hygiene*—a monthly journal devoted to technical research papers on occupation health topics. The annual subscription rate is $60 for nonmembers.

Center for Ergonomics

University of Michigan   313-763-2243
1205 Beal
Ann Arbor, MI 48109

This research and training center, within the University's College of Engineering, has an interdisciplinary faculty of industrial hygienists, engineers, physicians, and psychologists. The Center conducts periodical training sessions and monthly seminars on occupational ergonomics. The sessions and seminars cost about $100 a day. Mr. Gallay, the Project Manager, will provide additional information, make referrals, and send you copies of the Center's publications. The publications consist of research findings in the areas of workers' strength, testing, biomechanics, cumulative trauma disorders, and human perceptual processes. Publications are available free of charge except for a duplication fee.

Consumer Product Safety Commission
Human Factors Division   301-492-6468
5401 Westbard Avenue
Bethesda, MD 20207

The Human Factors Division comprises multidisciplinary professionals who conduct human factors evaluations of consumer products. Periodic reports are prepared about how man and machine interface to cause accidents. The director will answer questions and provide free copies of the Division's evaluation reports.

Human Factors Safety Division
Nuclear Regulatory Commission (NRC)
Washington, DC 20555

NRC regulates the ergonomics aspect of nuclear power plants. The agency's involvement includes control-room design, man-machine interface, training and qualification of operating staffs, and procedures. The Commission also conducts ergonomics-related research to develop a technical basis for regulation. The following Division and its five branches are involved in ergonomics:

Division of Human Factors Safety

This office oversees the branches described below, and the staff can refer you to appropriate people. Call Hugh L. Thompson, Director, 301-492-9595 for further information.

Licensee Qualifications Branch
Hal Booher, Chief, 301-492-4816.

This office is involved in overseeing the training and qualifications of people who operate nuclear power plants.

Human Factors Engineering Branch
Voss A. Moore, Chief, 301-492-4813.

This office is concerned with man-machine interface.

Procedures and Systems Branch
Denis Ziemann, Chief, 301-492-4720.

Operator Licensing Branch

Don Beckham, Chief, 301-492-4868.

This office administers qualification and requalification tests to nuclear reactor operators.

Human Factors and Safeguards Research Branch

James A. Norberg, Chief, 301-443-7942.

This branch is the research arm of NRC.

Human Factors Society, Inc. (HFS)

PO Box 1369   213-394-1811
Santa Monica, CA 90406

This is an interdisciplinary professional organization of approximately 4,000 engineers, educators, psychologists, and other professionals in the human factors field. The Society promotes the latest research discoveries, technical growth, and career advancement in human factors. HFS conducts an annual five-day conference featuring state-of-the-art research methods, panel discussions, and tutorials. Special-interest groups affiliated with HFS periodically sponsor one-day and weekend conferences. The Society produces several publications, and upon request will send you a free publications catalog. Examples of the publications include:

- *Human Factors*—a bimonthly journal that consists of original papers covering the spectrum of the human factors field. The journal costs $50.
- *Human Factors Society Bulletin*—a monthly newsletter about human factors and ergonomics. It also includes information about jobs available, book reviews, monographs, and other publications. The 16-page *Bulletin* costs $20 a year.
- *Directory and Yearbook*—an annual publication listing HFS members alphabetically and geographically, with information about their specialties included in each entry. The 230-page *Directory and Yearbook* costs $20.

Institute of Industrial Engineers

25 Technology Park Atlanta   404-449-0460
Norcross, GA 30092

An international nonprofit professional organization with over 43,000 members, the Institute provides a variety of programs, seminars, workshops, and conferences throughout the year for its members and productivity professionals. The staff can give you additional information and refer you to other sources. The Institute publishes four periodicals and approximately ten books a year. Its publications catalog is available free of charge by contacting the Publications Division. Publications include:

- *Industrial Engineering*—a monthly journal of current issues concerning modern engineers, productivity improvements, and new products. The annual subscription rate is $35.
- *Industrial Management*—a 32-page bimonthly periodical containing articles

about better management in industry. The annual subscription is $15 for members and $22 for nonmembers.

National Transportation Safety Board
Office of Government and Public Affairs   202-382-6600
800 Independence Avenue, SW
Washington, DC 20594

The Board has two divisions of professionals who deal with human performance and survival factors in relation to transportation equipment. Questions should be directed to the Office of Government and Public Affairs. Individual accident reports that include the human factors aspect may be purchased from the National Technical Information Service, 5285 Port Royal Road, Springfield, VA 20402, 202-783-3238.

Special Interest Group on Computers and Human Interaction (SIGCHI)
Association for Computing Machinery   212-869-7440
11 West 42nd Street
New York, NY 10036

The scope of SIGCHI is the study of the human-computer interaction process, and includes research and development efforts leading to the design and evaluation of user interfaces. The focus of SIGCHI is on user behavior—mainly how people communicate and interact with computer systems. SIGCHI serves as a forum for the exchange of ideas among computer scientists, human factor scientists, psychologists, social scientists, systems designers, and end users. Members can answer questions and make referrals. SIGCHI produces:

● *SIGCHI Bulletin*—a quarterly newsletter providing information about computers and human interaction. The annual subscription rate is $30.

Technical Information Branch
National Institute for Occupational Safety & Health (NIOSH)   513-533-8326
4676 Columbia Parkway
Cincinnati, OH 45226

A government research and funding agency that conducts research in the human factors areas, NIOSH has a clearinghouse for safety and occupational health information. The staff of multidisciplinary professionals will participate in symposia, seminars, and conferences on a selective basis. They also perform health hazards evaluations for ergonomics services. The Division of Technical Services will answer questions, make referrals, and furnish free publications, which are mostly criteria documents and technical reports of general research on ergonomics.

Additional Experts and Resources

University of Michigan
Professors of Ergonomics
Center for Ergonomics
University of Michigan
1205 Beal
Ann Arbor, MI 48109

The University of Michigan has an extensive program in ergonomics, and many of the professors are excellent sources of information in their areas of expertise. Ergonomics specialists include:

Thomas Armstrong, Ph.D.
313-763-2243

Dr. Armstrong is an expert in cumulative trauma disorders. He will answer questions concerning human factors and refer you to other sources of information in this field.

Don Chaffin, Ph.D.
Director, Center for Ergonomics
313-763-2243

Dr. Chaffin is a specialist in biomechanics. He has conducted extensive research in ergonomics and has pioneered some recent discoveries. He can provide additional information about ergonomic issues and refer you to other specialists.

Gary Herrin, Ph.D.
313-763-2243

Dr. Herrin is an expert in manual material handling, biostatistics, and quality control. He will answer questions and refer people to other specialists.

Dr. Kochaar, Ph.D.
313-763-2243

Dr. Kochaar is a specialist in visual display and human perception. He has done extensive research in this area. He is available to provide additional information and act as a referral to other experts in this area.

Gary Langolf, Ph.D.
313-763-2243

Dr. Langolf is an expert in human perception and occupational ergonomics. He is available to answer questions and can refer you to other specialists in the field.

Lorraine Borman
Vogelback Computer Center
2129 Sheridan Road
Evanston, IL 60201

312-492-3682

An ergonomics specialist, Ms. Borman can answer questions and refer you to other experts and information sources. She is chairperson of ACM's Special Interest Group on Computer and Human Interaction (described earlier in this section).

Ethics and Technology

See: Social Impact of Technology

Export Policy, Laws, and Regulations

See: Smuggling of High Technology

Fabrics, Woven and Nonwoven

See: Textiles

Farming

See: Agriculture

Fashion Manufacturing Technology

See: Apparel

Fear of High Technology

See: Social Impact of Technology; Technophobia

Federal Legislation (U.S.), Pending

See: Legislation

Fertility

See: Reproductive Technology

Fiber Optics

See also: Holography; Laser Technology; Optoelectronics

Organizations and Government Agencies

Optical Society of America (OSA)
1816 Jefferson Drive, NW   202-223-8130
Washington, DC 20036

OSA is a publisher and disseminator of research and other optics data. It holds an annual meeting, and conducts a wide variety of special topical conferences, workshops, and other information programs. Publications include the following:

• *Optics News*—a bimonthly with information about meetings and conferences, technical papers of general interest, and news in the field of optics. The subscription rate is $40 a year.

• *Optics Letters*—a monthly newsletter with news and important results in all branches of optics. The subscription rate is $100 a year.

• *Applied Optics*—a biweekly newsletter with papers on the applications of facts, principles, and methods of optics. The subscription rate is $250 a year.

• *Journal of the Optical Society of America*—a monthly with papers and letters that contribute significant new knowledge or understanding of any optical phenomenon, principle, or method. The subscription rate is $165 a year.

• *Soviet Journal of Optical Technology*—translation of a Soviet optical journal. Available for $250 a year.

• *Optics & Spectroscopy*—a monthly cover-to-cover translation of an optical journal from the Soviet Union. Available for $300 a year.

SPIE—International Society for Optical Engineering
PO Box 10   206-676-3290
Bellingham, WA 98227-0010
    SPIE holds over 70 conferences and 75 short courses a year on fiber optics, image processing, optic design, and X-ray and laser surgery. It publishes all conference proceedings in full, as well as the bimonthly journal, *Optical Engineering*, which is available for $60 a year.

Publications

Lasers and Applications
3220 Sepulveda Boulevard, Suite E   213-534-3700
Torrance, CA 90505
    This monthly magazine of the electro-optics field covers new developments in the laser and fiber optics industry, with an emphasis on applications. Also published is an annual *Buyers' Guide* which identifies over 1,000 companies that provide goods and services to the laser and fiber optics industry. The annual subscription rate is $40.

Laser Focus Including *Electro-Optics Magazine*
Penwell Publishing Company   617-244-2939
Box 1111
Littleton, MA 01460
    This monthly magazine covers technology news, business news, and technological features on fiber optic technology, as well as systems and devices. It also publishes an annual buyers' guide issue, identifying suppliers of goods and services to the laser and fiber optic industry. It costs $45 a year.

Photonics Spectra
Optical Publishing Company   413-499-0514
Berkshire Common
PO Box 1146
Pittsburgh, PA 01202
    This monthly magazine provides information on photonics technology, including optics, electro-optics, lasers, fiber optics, vacuum technology, and related fields. Also published is an annual two-volume directory, *Optical Industry and Systems Purchasing Directory*, which identifies over 1,800 manufacturers and suppliers of some 1,000 industry-related products and services. Subscription to the magazine is $36 a year, and the directory is available for $60.

Fiber/Laser News
Phillips Publishing   301-986-0666
7315 Wisconsin Avenue, Suite 1200N
Bethesda, MD 20814

This biweekly newsletter covers marketing, research and development activities, and technology of the fiber optics and laser industries. It also publishes industrial market studies. The annual subscription rate is $275 a year. The full text of the newsletter is also available on-line through NewsNet, Inc., 800-345-1301 (in Pennsylvania, 215-527-8030).

Communications Publishing Group, Inc. (CPG)
PO Box 383   617-566-2373
Dedham, MA 02026

A data-base publisher of high-technology publications, on-line data bases, and research services, based on worldwide patent literature; available in both print and electronic form (NewsNet, Inc., 800-345-1301). Publications dealing with fiber optics are the following:

● *Fiber Optics*—this biweekly newsletter covers components, equipment, and systems and techniques such as cables, fiber, connectors, couplers, detectors, light sources, data links, splicing, transmitter packages, receiver packages, installation, and handling equipment. The subscription rate is $220 a year.

● *Fiber Optic Patents 1980-1983*—a directory of over 1,000 fiber optic patents granted from 1980 through 1983, this publication provides complete coverage of U.S. patents, with partial coverage of Patent Cooperation Treaty and United Kingdom patent applications. It features patent numbers, titles, bibliographic data, brief descriptions, and patent drawings. It is available for $325.

● *Industry Patent Reports*—each report includes patent numbers, descriptive titles, and assignees of all U.S. patents, granted in the field from 1963 through 1983. Selected patents reproduced in their entirety include (1) Fiber Optic Couplers and Connectors ($125); (2) Fiber Optic Fibers and Cables ($125); (3) Fiber Optic Sources and Detectors ($125).

Information Gatekeepers
138 Brighton Avenue, Suite 212   617-787-1776
Boston, MA 02134

Information Gatekeepers publishes:

● *Fiber Optics & Communications Newsletter*—a monthly newsletter that covers domestic and international news on fiber optics, communications, and related fields. It tracks over 300 trade publications, reporting major activities and analyzing key developments. Typical coverage includes market events, the latest technological developments, new companies, major contracts, a calendar of projections, new products, and so on. The cost is $215 a year and includes the *Fiber Optics Handbook and Buyers Guide*.

● *Fiber Optics and Communications Newsletter*—this weekly news service is a fast source of information for worldwide fiber optics news. Annual subscription rate is $128 per year.

The Electro-Optical Communications Dictionary
Hayden Book Company, Inc.   201-843-0550
50 Essex Street
Rochelle Park, NJ 07662

This comprehensive reference is for the fiber optical and lightwave, data-processing, and related manufacturing technologies fields. The *Dictionary* presents terms and definitions consistent with international, federal, industrial, and professional societies' standards. Processes, devices, and components in electro-optical communications are clearly described. It is available for $16.95.

Additional Experts and Resources

Science and Electronics Division
International Trade Administration   202-377-1333—Arthur Pleasants, Fiber Optics Production
U.S. Department of Commerce, Room 4044   202-377-2990—William Sullivan, Fiber Optics Technology
Washington, DC 20230

These experts are responsible for forecasting the *Industrial Outlook* in the fiber optics industry for the U.S. government. They can also be excellent sources themselves or serve as guides for identifying additional information sources.

Reference Section
Division of Science and Technology   202-287-5670
Library of Congress
Washington, DC 20540

The Division of Science and Technology of the Library of Congress has put together several "tracer bullets" on various technological topics, which list different source materials available. One has been done for fiber optics, and is available free from the Reference Section.

Fifth Generation Computing

See: Supercomputers

Films, Documentaries, and Audiovisuals

Organizations and Government Agencies

Films, Inc.
733 Green Bay Road
Wilmette, IL 60091

800-323-4222
312-256-3200 (in Illinois)

Films, Inc., sells and rents approximately 30 films pertaining to different aspects of high technology. The films range from training programs, showing how to use a computer, to management-oriented films about productivity and the Japanese style. There are films on artificial intelligence, computer literacy and anxiety, and the electronic office. The rental fee is $125 a day or $250 for 2-5 days. Films, Inc., gives a 15 percent discount to nonprofit organizations. Examples of their films are:

- *Painting-by-Numbers*—deals with the application of computer graphics.
- *The Billion-Dollar Bubble*—deals with computer security and ethics, and covers the 1969 Equity Funding Corporation of America's computer fraud scandal.
- *Japan vs. USA—The High Technology Shoot-Out*—examines Japan's successful penetration of the semiconductor market and shows how Japan has cornered the computer market, using American information, ideas, and products.
- *Now the Chips Are Down*—focuses on the microprocessor—from word processors and electronic mail to electronic plows—explaining how the microprocessor is made and how it is applied to business.
- *Electronic Office*—explores use of computers in small businesses, large corporations, and laboratories.
- *Technical Studies*—a series of ten 25-minute programs illustrating the practical application of concepts in materials and engineering science.

California Newsreel
630 Natoma Street
San Francisco, CA 94103

415-621-6196

California Newsreel sells and rents several films addressing issues in the high-technology field. Groups with limited financial resources can book films on a "what you can afford" basis, and secondary schools can request a 50 percent discount. The following four films are of particular interest:

- *New Technology—Whose Progress?*—examines the impact of new technology on the workplace. It addresses changing jobs and job displacements; questions whether new technology is an inevitable part of progress; provides examples of new equipment and systems at work; and presents the views of journalists, technologists, trade unionists, and companies.
- *The Detroit Model*—examines the introduction of robotics into the automobile industry.

● *Clockwork*—a film about Frederick Taylor, who 100 years ago invented scientific management and industrial engineering. The film compares the introduction of his ideas with the effect of introducing new technologies today. It addresses the questions: Do new technologies make jobs more creative or more routine? Must the U.S. emulate Japanese business techniques to revive U.S. productivity?

● *We've Always Done It This Long*—documents the innovative efforts of union stewards in a British aerospace plant. The workers saved their jobs and made the plant more productive by intervening with corporate planning and recommending ways new technologies could be applied to make products more efficiently. The workers' plan also listed 250 new products and marketing strategies the company could pursue.

Publications

Educators Progress Service, Inc.
214 Center Street   414-326-3126
Randolph, WI 53956

● *Educators' Guide to Free Science Materials*—this publication, issued every September, is geared toward educators. It provides information on titles, sources, availability, and contents of approximately 2,000 selected free items categorized by type of science covered. Free items include films; filmstrips and slides; tapes, scripts, and transcriptions; and printed materials. Types of science covered include aerospace education, biology, chemistry, environmental education, general science, and physics. Title, subject, and source indexes are provided. Each issue is $24.

● *Guide to Free Computer Materials*—this annual, issued every spring, is oriented toward the general public. It provides information on titles, sources, availability, and contents of 198 selected free items dealing with computers, categorized by 12 areas of interest. Among the free items are five free-loan films, four free-loan videotapes, and one diskette. Title, subject, and sources indexes are included. The single-issue price is $30.25.

Audio-Video Marketplace
R. R. Bowker Company   212-916-1600
1180 Avenue of the Americas
New York, NY 10036

This annual is directed toward all who need the services of suppliers of audiovisual materials and equipment. In its Classified Index to Section 1: Producers, Distributors and Services, it lists companies that distribute audiovisual software (films, videotapes, cassettes, and so on) related to science topics. The single-issue price is $45. It is available from R. R. Bowker Company, Order Department, PO Box 1807, Ann Arbor, MI 48106.

Media Network

208 West 13th Street   212-620-0877
New York, NY 10011

   ● *Survival Audiovisual Guide*, Mobilization for Survival, Publisher—a biannual geared toward the general public, the guide contains a comprehensive list of materials on topics including ecology, technology, nuclear energy, alternative energy and conservation, militarism, multinationals, organizing, and many other social issues. The price is $2.50 for Media Network members, and $3 for nonmembers.

   ● *Reel Change: A Guide to Films on Appropriate Technology*, Friends of the Earth Books, Publisher—this guide, oriented toward general public use, evaluates 79 films on topics including solar power and other small-scale, efficient energy sources. The *Guide* is $3 for Media Network members, and $4 for nonmembers.

Modern Talking Picture Service (MTPS)

5000 Park Street North   813-541-7571
St. Petersburg, FL 33709

   MTPS distributes free-loan 16-mm sound films on a number of science and technology subjects, including data processing, energy, chemical processes, and ecology. Films may be ordered either on an individual title basis or via the Preferred Program Service, through which film users select the subject they want (such as science) and receive films about that subject, on specified dates or according to a specified schedule. Obtain catalogs and order forms from MTPS at the above address.

Additional Experts and Resources

National Library of Medicine (NLM)

Audio-Visual Resources Section   301-496-4244
Reference Service Division
8600 Rockville Pike
Bethesda, MD 20209

   The Audio-Visual Resources Section maintains a collection of 16-mm films and videocassettes on topics related to technology in medicine. Both types of material may be viewed at the Learning Resource Center of the National Library of Medicine. The videocassettes may be borrowed by the general public via the interlibrary loan system. Under this program, you can request that your local library borrow the material you need from an NLM "satellite library" in your region. To find out the location of an NLM library holding the videocassette you need, contact the NLM Reference Service Division listed above.

Educational Film Library Association

45 John Street, Suite 301   212-227-5599
New York, NY 10038

   This is a professional organization composed of schools, libraries, businesses, users, and producers of educational audiovisual materials, including materials related to high technology. It disseminates information on sources for these materials. For references, contact the association at the above address.

Media Network
Information Center   212-620-0877
208 West 13th Street
New York, NY 10011

The Information Center recommends films, videotapes, and slide shows on a wide range of topics, including alternative energy, automation, chemical and biological warfare, and industrial and appropriate technology. It also provides descriptions, running times, rental and sales costs, and distributors' addresses. The Center uses a cross-referenced filing system with more than 200 topics. To use its services, contact the Center at the above address.

The National Audio-Visual Center
Attn: Information Services   301-763-1896
Washington, DC 20409

The Center is the central distribution source for government audiovisual materials, including films, videocassettes, slides, and so on. All of the materials are for sale, and 16-mm films are for rent; none of the materials is loaned out. The materials are indexed in a comprehensive 3,000 item catalog, as well as in 22 subject-oriented catalogs covering such topics as vocational education, library information science, environment energy conservation, agriculture, industrial safety, and medicine. Catalogs are available from the Information Services section.

Financing for Business

See: State High-Tech Initiatives; Venture Capital

Fish and Fisheries

See: Aquaculture; Marine Technology

Food Safety

See: Nutrition

Food Technology

See also: Nutrition

Organizations and Government Agencies

American Association of Cereal Chemists (AACC)

3340 Pilot Knob Road   612-454-7250
St. Paul, MN 55121

AACC concentrates on scientific and technical research with regard to cereal grains, oil seeds, and related materials, and their processing and products; the study of the development and standardization of analytical methods used in cereal and seed chemistry; promotion of scientific cooperation among workers in the fields of cereal and seed chemistry; and maintenance of high professional standards of the Association members. The members are people in the food-processing and food-manufacturing industries, in universities, and in government. The Association sponsors short courses, as well as annual meetings and workshops. Publications include:

● *Cereal Chemistry*—a bimonthly technical journal that reports on laboratory research results. The approximately 112-page publication is available for $88 a year.

● *Cereal Foods World*—a monthly magazine with feature articles of interest to individuals in technical and administrative positions. The approximately 56-page publication is available for $45 a year.

● *Advances in Cereals Science and Technology Series*—a six-volume set intended to supply intermediary information between the AACC journals and the AACC monograph series. The cost is $189 for members and $236 for nonmembers.

● *AACC Publications Catalog of Books and Journals*—available upon request.

American Chemical Society (ACS)

Agricultural and Food Chemistry Division   202-872-4600
1155 16th Street, NW   800-424-6747—to order abstracts
Washington, DC 20036

The Society's staff can refer inquiries to the appropriate offices of its Agricultural and Food Chemistry Division. This Division participates in the two national

ACS meetings held each year. Abstracts of papers presented at each ACS meeting, including Agricultural and Food Chemistry Division abstracts, are published together and are available at $22 to $24 a volume for members and $32 for nonmembers. The Agricultural and Food Chemistry Division occasionally holds its own symposia.

American Oil Chemists' Society (AOCS)
508 South 6th Street   217-359-2344
Champaign, IL 61820

This professional scientific organization is involved with vegetable, animal, and marine fats and oils, and their derivatives. Members work in industry, government laboratories, and universities. The organization disseminates information to both members and nonmembers by means of its journals and meetings. The Society also establishes standardized methods of chemical analysis for fats and oils; conducts a check-sample program in which standard samples are sent to laboratories, where the results are compared; and has a chemist certification program. One annual meeting is held each year, at various locations in the U.S. There is also an annual world conference. Publications include:

• *Journal of the American Oil Chemists' Society*—a monthly publication devoted to fundamental and practical research, production, processing, packaging, and distribution in the field of fats, oils, soaps, detergents, and protective coatings. It contains technical papers, advertising, and news. The approximately 150-page publication is available for $60 a year. The cost of the *Journal* is included in the membership fee.

• *Lipids*—a monthly publication that disseminates significant findings from original research on the physical, chemical, biochemical, biological, pharmaceutical, medicinal, and physiological characteristics of lipids, and/or their related methodology. The approximately 80-page publication is available for $50 a year for nonmembers and $20 a year for members.

American Society of Agricultural Engineers
Food Engineering Division   616-429-0300
2950 Niles Road
St. Joseph, MI 49085

The Food Engineering Division of this society is concerned with the application of engineering to the processing, storage, packaging, and handling of food products. Approximately 10 percent of the Society's members are involved in this Division. Special sessions at the organization's national meeting are devoted to the topic of food engineering. Occasionally there are specialty conferences on the subject. Publications include:

• *Food Engineering Newsletter*—a quarterly nontechnical publication that contains short articles with information on what is going on in the area of food engineering. The cost of this approximately eight-page publication is included in the membership fee. Nonmembers may contact the Society for a sample issue.

● *Conference Proceedings* and technical papers on the subject are also available.

Eastern Regional Research Center (ERRC)

Agricultural Research Service   215-233-6595
U.S. Department of Agriculture
600 East Mermaid Lane
Philadelphia, PA 19118

The Center's six laboratories are involved in applied, basic, and developmental research. They are concerned with the effects of post-harvest processing, storage and handling, and the quality and safety of consumer products. The laboratories are (1) Food Safety, (2) Food Science, (3) Plant Science, (4) Animal Biomaterials, (5) Engineering Science, and (6) Physical Chemistry and Instrumentation. The Center sponsors seminars and a workshop series, and journals publish their research findings.

Food and Nutrition Information Center

U.S. Department of Agriculture   301-344-3719
National Agricultural Library
10301 Baltimore Boulevard
Beltsville, MD 20705

The Center acquires books, journals, and audiovisual materials dealing with human nutrition, food service management, and food science. Although the Center primarily serves professionals, including scientists and food service managers, its materials are available to the general public. The following organizations, agencies, and individuals may borrow materials from the Center free of charge: U.S. Congress, federal and state government agencies, libraries, information centers, universities and colleges, national officers of professional societies, day care personnel, nutrition education and training program staff, cooperative extension services, research institutions, and others. These organizations, agencies, and individuals may also request photocopies of journal articles be sent to them. Other services include:

● The Reference Service staff—two to three full-time nutritionists who answer questions and/or make referrals. Computer searches of major data bases are available.

● A 24-hour telephone monitor—enables callers to leave messages and have their calls returned.

Food, Pharmaceutical and Bioengineering Division

American Institute of Chemical Engineers (AICHE)   212-705-7335—Publications
345 East 47th Street   212-705-7315—Main number
New York, NY 10017   212-705-7663—Subscription Department

The Food, Pharmaceutical and Bioengineering Division of the Institute sponsors various symposia in its five programming areas: food, pharmaceutics, biotechnology, biomedicine, and fundamentals and life sciences. It also has a Research Institute in Food Engineering, which is primarily a conduit for obtaining funds for research in food engineering and transferring them to appropriate academic researchers. The staff at the Institute can refer inquiries to the current chairman of the Food, Pharmaceutical and Bioengineering Division. Publications include:

• *Biotechnology Progress*—a new quarterly magazine carrying news items and papers that have been presented at the various symposia. The cost of the approximately 120-page publication is included in the membership fee. For information on the nonmember subscription price, contact the subscription department.

Institute of Food Technologists

221 North La Salle Street   312-782-8424
Chicago, IL 60601

This professional, scientific society provides an educational and scientific information exchange for its membership. Members are food scientists and food technologists. The staff can answer questions and/or make referrals for the news media only. The annual meeting is open to the public. Publications include:

• *Food Technology*—a monthly scientific and technical publication directed to food scientists and food technologists. The approximately 120-page publication is available for $40 a year for nonmembers, while for members the cost is included in the membership fee.

• *Journal of Food Science*—a scientific publication directed to food scientists and food technologists. The approximately 250-page publication is available for $45 a year for nonmembers. The cost of the *Journal* is included in the membership fee.

• *Scientific Status Summaries*—covers individual topics relating to current food issues. These publications, 4 to 12 pages long, are available for $1 a copy and are issued irregularly.

Publications

Journal of Agricultural and Food Chemistry

American Chemical Society   614-421-3776
Membership and Subscription Services
PO Box 3337
Columbus, OH 43210

The bimonthly chemistry research *Journal* publishes papers in a broad multidisciplinary area. It is concerned with the chemistry of all aspects of agriculture and food. The approximately 250-page publication is available for $99 a year for nonmembers, and $19 a year for members.

Foreign Trade

See: International Competition; Smuggling of High Technology; Technology Transfer

Foundation Resources

See: Research and Development

Four-Dimensional (4-D) OPTICS

See: Holography

Fuels, Alternative

Organizations and Government Agencies

Bureau of Alcohol, Tobacco, and Firearms (ATF)
Department of the Treasury   202-566-7268—James F. Lynch, Public Affairs and Disclosure
1200 Pennsylvania Avenue, NW
Washington, DC 20226
> The office provides ATF news releases and information about federal laws and ordinances that affect trade in and licensing of alcohol, tobacco, firearms, and explosives. The Bureau also distributes pamphlets and fact sheets about the its history and responsibilities, and the production of alcohol fuel and gasahol.

Coal Division
Office of Coal, Nuclear, Electric and Alternate Fuels   202-252-6860—Charles C. Heath, Director
Energy Information Administration, DOE
FORRESTAL, EI-52, Room 2F021
1000 Independence Avenue, SW
Washington, DC 20585
> This Division manages data and information systems for all supply and demand aspects of coal. It also collects and analyzes data. Staff members perform analyses and projections relating to coal supply, including production, prices, and distribution; identify and analyze coal reserves; examine new technologies that derive energy from coal; and study existing and proposed legislation and regulations affecting coal supply and demand.

Electric Power Division

Office of Coal, Nuclear, Electric and Alternate Fuels   202-252-9863—Mary J. Hutzler, Director
Energy Information Administration, DOE
FORRESTAL, EI-54, Room 2G060
1000 Independence Avenue, SW
Washington, DC 20585

This federal Office directs programs to collect and analyze data on electric power supply, including capacity, generation, distribution, fuel use, finances, and rates. Staff members provide the Federal Energy Regulatory Commission (FERC) with the information that it needs to regulate electric power. FERC develops analytical models and prepares projections of capacity, generation, fuel use, costs, rates, financial requirements, and distribution of electric power, including nuclear, hydropower, and new central station technologies. Staff members also analyze the effects of policy and regulatory actions on the electric utility rates, costs, capacity, generation, distribution, finance, and consumption of input fuels.

Heavy-Duty Transport and Fuels Integration Branch

Technology Development and Analysis   202-252-8055—E. Eugene Ecklund, Chief
Vehicle and Engine R&D
Conservation and Renewable Energy, DOE
FORRESTAL, CE-131, Room 5G030
1000 Independence Avenue, SW
Washington, DC 20585

Staff members in this federal agency manage long-term research and development programs for new energy- and petroleum-saving technology in the heavy-duty transport sector. The Branch also manages research and development projects associated with the use of alternative fuels in advanced and conventional engines.

Inputs and Productivity Branch

Inputs Supply, Demand, and Price Section   202-475-3853
Economic Research Service
U.S. Department of Agriculture
408 GHI Bldg.
500 12th Street, SW
Washington, DC 20250

This Branch provides current and historical statistics on energy supply and use in agriculture. Data include farm production and use of alcohol fuels, fuel consumption in farming and food processing, and agricultural fuel prices. The Section also monitors the fuel supply situation for the agricultural sector.

Nuclear and Alternative Fuels Division

Office of Coal, Nuclear, Electric and Alternate Fuels   202-252-6363—R. Gene Clark
Energy Information Administration, DOE
FORRESTAL, EI-53, Room BG057
1000 Independence Avenue, SW
Washington, DC 20585

    The Division manages the data information systems on all supply aspects of nuclear power and alternative fuels. The Office also prepares analyses and projections about the availability, production, costs, processing, transportation, and distribution of nuclear and alternative energy sources. It prepares projections about energy supply and production from alternative energy forms, including solar, wind, waste, wood, and alcohol.

Office of Alcohol Fuels

Conservation and Renewable Energy, DOE   202-252-1277—Daniel E. Beckman, Deputy Director
FORRESTAL, CE-80, Room 5F043
1000 Independence Avenue, SW
Washington, DC 20585

    This federal Office handles research and development activities related to the production and use of alcohol fuels from renewable resources. Staff members direct long-term, high-risk research and development focused on conversion of nonfuel feedstocks (such as cellulose) into fuel-grade alcohol. The Office implements the Energy Security Act's loan guarantee program, which is designed to encourage the construction of alcohol fuel production facilities by the private sector.

Public Disclosure

U.S. Synthetic Fuels Corporation   202-822-6460—Catherine McMillan, Director
2121 K Street, NW, Suite 303
Washington, DC 20586

    This is a quasi-governmental corporation established by the Congress to stimulate the development of a domestic synthetic fuels industry through financial incentives. These incentives include loan guarantees, price guarantees and purchase agreements, direct loans, or joint ventures. The financing is provided for construction and operation of commercial-scale facilities to produce liquid, gaseous, and solid fuels from coal, oil shale, and tar sands, and hydrogen from hydrolysis. The office responds to public inquiries about the U.S. Synthetic Fuels Corporation and provides referrals within the corporation. Copies of nonconfidential corporation records are available in a public reading room maintained by the Public Disclosure Office.

Renewable Fuels Association (RFA)

499 South Capitol Street, SW, Suite 420   202-484-9320
Washington, DC 20003

    RFA is a privately funded organization that represents its members' interests in legislative and regulatory matters. The Association can send you a packet of information containing articles relating to renewable fuels. For further information and referrals, call or write David E. Hallberg, President.

Technology Development and Analysis
Vehicle and Engine R&D   202-252-8053
Conservation and Renewable Energy, DOE
FORRESTAL, CE-131, Room 5G046
1000 Independence Avenue, SW
Washington, DC 20585

This federal organization analyzes and assesses new energy concepts, and it is directly involved in the research and development of high-temperature materials and components. The alternative fuels utilization program is directed toward lowering risks associated with the development of alternative fuel technologies in the transportation sector and toward encouraging the use of such fuels. Staff members conduct research, development, and design on advanced propulsion systems for both the automotive and heavy-duty transportation sectors.

Data Bases

Alternative Fuel Data Bank (AFDB)
National Institute for Petroleum and Energy Research   918-336-2400
PO Box 2128   918-337-4267—Russell Simkins, Manager
Bartelsville, OK 74005

AFDB, designed for direct public access, can be searched by most home computer users having a telephone linkup. This data bank contains information about the utilization of alternative fuels. It has three types of data: bibliographies, summaries of ongoing research activities, and discussions of topics of current interest. AFDB focuses on the use of nonpetroleum sources as well as nonconventional fuels from petroleum sources used in transportation. Examples of fuels covered include syncrudes from shale, coal, alcohols, hydrogen, ethers, and broadcut. Information is collected from periodicals, abstract news service publications, technical society papers, conference proceedings, project progress reports, and final reports. Data retrieval programs are interactive and designed for easy use by the general public. AFDB contains 1,000 bibliographic citations, 36 topic briefs, and 36 entries regarding ongoing research projects. It is updated continually.

Searches and direct-access privileges are available free of charge. Contact the Institute to find out if your computer and linkup equipment are compatible and to obtain a free users' manual and I.D. number. If you don't have the equipment to search AFDB yourself, the center will query the system for you and send a printout.

Fundraising

See: Research and Development

Futures

See also: Environmental Impact Assessment; Social Impact of Technology; Technology Assessment

Organizations and Government Agencies

World Future Society (WFS)
4916 St. Elmo Avenue   301-656-8274
Bethesda, MD 20814

WFS is a nonprofit scientific and educational organization devoted to the study of the future. It has 3,000 members worldwide, is nonideological, and serves as a clearinghouse for alternative ideas on the future. The staff will answer inquiries and make referrals. The Society has published several books relating to high technology, and it maintains a bookstore of 300 books, one-third of which pertain to high-technology topics, ranging from communications to robotics. A free catalog describing the bookstore's holdings is available from WFS. The Society also provides audiocassettes, journals, bulletins, and a film on the future. WGS materials include:

• *World Future Society Bulletin*—a bimonthly journal that is scientifically, technically, and academically oriented. The cost is $25 a year for individuals and $30 a year for organizations.

• *Future Survey*—a monthly journal summarizing current books, articles, and essays dealing with the future. The subscription includes a directory that indexes the year's issues of *Future Survey* by subject, author, and publisher. The cost is $45 a year for individuals and $65 for organizations.

• *The Futurist Magazine*—a bimonthly geared toward the general public. It contains articles dealing with all aspects of the future, including technology, science, law, business, and sports. Free to members, the magazine is available to nonmembers for $20 a year for individuals and $23 for organizations.

• *Toward the Future: A Film That Automates Tomorrow*—a 20-minute film that provides an introduction to studies of the future. It includes studies methodology, examples of futurist planning in government projects, and illustrations of possible technologies and their impact on society. Rental fee is $30.

• *The Future: A Guide to Information Sources*—the 1979 edition lists and describes organizations, individuals, books, films, educational programs, and other resources dealing with the future. The one-volume directory has both a subject and geographical index, and costs $25. WFS plans to publish an updated version in 1985. It will be in three separate volumes: individuals, organizations, and education. The cost is expected to be $15 a volume.

Congressional Institute for the Future

218 D Street, SE   202-544-7994
Washington, DC 20003

The Institute evolved from the Congressional Clearinghouse on Futures, a caucus of the U.S. Congress. Staff members perform original research to assist Congress in thinking about the future when planning legislation. The staff will answer questions and refer you to the appropriate government agency, private organization, or resource person. The Institute frequently issues reports resulting from its research projects. It publishes:

• *What's Next*—a free bimonthly 12-page quick-reference newsletter that reports on social, political, and technical trends that could be important in the future.

Institute for Alternative Futures (IAF)

PO Box 1417-B42   703-960-7855
Alexandria, VA 22313

IAF is a nonprofit Institute founded in 1977 to develop, evaluate, and disseminate information on the ways in which individuals and organizations can more consciously choose their future. Its focus at the present is legislative foresight and the futures of health, health care, and nutrition. The staff will answer inquiries, make referrals, and work with corporations and citizen groups on all aspects of alternative futures and society. IAF publishes books, conference proceedings, and papers it has prepared. The Institute is planning to publish three different newsletters on a periodic basis. Initially they will be available free of charge. One newsletter is expected to serve as a networking tool to help people interested in anticipatory democracy; a second will focus on the Institute's activities, covering congressional activities, papers presented, and seminars held; the third newsletter will outline research and development contracts received by the Institute, as well as articles in the field.

North American Society for Corporate Planning (NASCP)

11 West Monument Avenue   513-223-4948
PO Box 2307
Dayton, OH 45401

NASCP is an international volunteer association for corporate and strategic planning, especially over the long term. Its 4,050 members are vice presidents, directors, managers, and others involved or interested in planning the future of a business enterprise. The Society spots trends for the next ten years and tries to keep abreast of what is happening in industry. NASCP's 24 U.S. chapters function independently, holding their own seminars, study groups, and so on. The Society's headquarters, listed above, serves as a national clearinghouse, and the staff can answer inquiries and refer you to an appropriate NASCP member or chapter. The organization publishes:

● *Planning Review*—a bimonthly journal that covers all aspects of planning from high technology down to basic business planning. The annual subscription fee to the journal is $50.

Issues Management Association (IMA)

1090 Vermont Avenue, NW   202-682-1548
Washington, DC 20005

IMA is a national network of 500 people looking toward the future, predominantly in terms of industry but also in terms of the government and the academic and voluntary sectors. The Association focuses on foresight and emerging issues that can affect these sectors. IMA's members are mostly practitioners from industry groups, researchers, and suppliers who publish newsletters and maintain data bases on future-related issues. IMA will answer inquiries and make referrals.

Conference Board

845 Third Avenue   212-759-0900
New York, NY 10022

The Conference Board is a nonprofit business research company. It is a membership organization dealing mainly with management and economics topics. It has produced 15 reports about planning and development on such subjects as tracking the strategic plan, and planning and forecasting in small business. A free catalog describing the reports is available. Reports generally cost $20-50 for members, and three times that amount for nonmembers.

Publications

New Options, Inc.

PO Box 19324   202-822-0929
Washington, DC 20036

New Options, Inc., publishes a newsletter entitled *New Options* every three weeks. The eight-page publication looks at local, national, and international politics from a global perspective. Topics covered have included intermediate technology, emerging new values, federal deficits, and middle-class views. Reviews of futurist books are regularly included. The annual subscription is $25, with a $6 additional charge for first-class delivery. New Options, Inc. also sells its mailing list, of 1,600 change-oriented periodicals, for $75.

Gallium Arsenide

See also: Chips and Integrated Circuits; Materials Science; Microwaves; Semiconductors

Organizations and Government Agencies

Cray Research, Inc.
608 Second Avenue South
Minneapolis, MN 55402

612-333-5889—Tina Bonetti, Public Relations

Cray Research does research in the field of gallium arsenide, developing gallium arsenide for commercial purposes. The organization does not make information available to the public yet, although it has published one article on gallium arsenide and will send you a free copy. Some brochures have been printed for distribution and may still be available. Cray Research also publishes a quarterly magazine *Cray Channels*, available upon request.

Gigabit Logic
1908 Oak Terrace Lane
Newbury Park, CA 91320

800-422-7427
805-498-9664

This company produces commercial high-speed GaAs chips using gallium arsenide. It has brochures and printed material on 12 products, and is very helpful with questions about the company and the technology.

Harris Microwave Semiconductor
1530 McCarthy Boulevard
Milpitas, CA 95035

408-262-2222

Harris is very involved in research and development in the area of gallium arsenide and has several products on the market that use GaAs. It has published abstracts resulting from talks and papers presented; can answer technical questions and make referrals to other sources.

IEEE Publishing Services Department
345 East 47th Street
New York, NY 10017

212-705-7354

This engineering society holds an annual Gallium Arsenide Integrated Circuits Symposium and publishes the proceedings of the conference. Articles on gallium arsenide are regularly published in IEEE's many journals, especially in *IEEE Transactions on Electron Devices* ($59 a year) and *IEEE Transactions on Microwave Theory and Techniques* ($126 a year). Susan James in the IEEE in-house library will answer questions and refer you to other sources of information within the IEEE.

ITT Gallium Arsenide Technology Center

7635 Plantation Road 703-563-0371
PO Box 7065
Roanoke, VA 24019

This $2-million research center, sponsored by the federal government, conducts research in the field of gallium arsenide. Hollie Hiems in Public Relations will answer limited questions over the phone, but specific requests for information about publications and research must be sent in writing, and cleared first.

Rockwell International Corporation

Communications and Academic Affairs 805-498-4545—Harry Wugalter
Science Center
1049 Camino Dos Rios
PO Box 1085
Thousand Oaks, CA 91360

Rockwell has been conducting research in the field of gallium arsenide for a number of years. In the commercial field it is developing GaAs to replace silicon as a material in chips. It also conducts defense-related research and has a Department of Energy defense electronics contract. Publications are released as new developments occur, and these publications can be ordered from Rockwell. Send a letter requesting information about its research and development and publications. Calls are also handled, but not much information is made available over the phone.

Semiconductor Industry Association (SIA)

4320 Stevens Creek Boulevard, Suite 275 408-246-1181
San Jose, CA 95129

A manufacturers' association, SIA collects statistics on the production of semiconductor devices, including those using gallium arsenide. It provides information on 108 products and their markets, including some in optoelectronics and microwaves, where gallium arsenide has been used the most.

Strategic, Inc.

10121 Miller Avenue 408-446-4500—Brian Lewis
PO Box 2150
Cupertino, CA 95015-2150

This market research firm in 1983 completed a market study on gallium arsenide titled *GA Technology and Markets* (available for $1,500) and is currently preparing a five-part in-depth study of the topic. The two studies list all suppliers, available products, market predictions technology, applications, and so on.

Texas Instruments (TI)

Central Research Lab 214-995-5550—Public Relations
PO Box 225474
Dallas, TX 75265

TI maintains an active research and development program for gallium arsenide on materials, microwaves, digital devices, and processing. Research is basically for military purposes, and TI receives some funding from the federal government. Papers are published as developments occur, and information is available by writing to the Central Research Lab.

Textronix, Inc.
Gallium Arsenide Integrated Technologies   503-627-6464
PO Box 500
Mail Stop 50-224
Beaverton, OR 97077

Textronix conducts research in the field of gallium arsenide and runs a "foundry service." A customer brings in a design for a product to Textronix, and the company will try to fabricate it.

The firm also publishes papers and symposia proceedings, some through the Institute of Electrical and Electronics Engineers (IEEE). For more information about the "foundry service" or publications, write or call Textronix, Inc.

Additional Experts and Resources

Several universities conduct research in the field of gallium arsenide and can be excellent sources of information about the latest developments in GaAs. Many receive grants and contracts from private companies to do research, and can also direct you to the company involved in the production of GaAs. The following are a few of the top universities conducting research:

Cornell University
Electrical Engineering Department
425 Phillips Hall
Ithaca, NY 14853
607-256-4369

University of Illinois at Urbana
Coordinated Science Laboratory
1101 West Springfield Avenue
Urbana, IL 61801
217-333-0722—Dr. Harris Morkoc

North Carolina State University
Department of Electrical
and Computer Engineering
Box 7911
Raleigh, NC 27695-7911
919-737-2336

University of California at Santa Barbara
College of Engineering
Santa Barbara, CA 93106
805-961-3207—James L. Merz

Stanford University
Solid State Lab
McCullough, Room 208
Stanford, CA 94305
415-497-9775—Jim Harris

Semiconductor Materials and Processes Division
National Bureau of Standards (NBS)   301-921-3738—Dr. Michael Bell
Washington, DC 20234   301-921-3541—Charles Wilton, Semiconductor Devices and Circuits Division
    Dr. Bell, an expert in the materials aspect of gallium arsenide, will answer
questions concerning GaAs, including questions about the technology, markets, and
so on. The secretary can refer callers to others in NBS who may be able to answer
questions. Mr. Wilton will answer highly technical questions and/or make referrals.

Genetic Engineering

See: Agriculture; Biotechnology

Geothermal Power and Hydropower

See also: Renewable Energy Technology; Solar Technology

Organizations and Government Agencies

Geothermal Resources Council
PO Box 1350   916-758-2360
Davis, CA 95617
    The Geothermal Resources Council is a nonprofit educational association open
to anyone who is interested in the development of geothermal resources. It covers
all aspects of geothermal power, from geology and geophysics through financing.
Members include major oil companies, federal employees, venture capitalist firms,
contractors, and consultants. The Council holds an annual meeting and offers hardbound volumes of the transactions of the proceedings. Two to five courses, workshops, and seminars a week are offered on various subjects. The Council publishes
special reports periodically, as well as a bulletin.
    • *Geothermal Resources Council Bulletin*—is a technical periodical dealing
with the development of the whole range of resources. It has articles on exploration,
power plant construction, financing, and so on. The cost for 11 issues is $45 a year,
for both members and nonmembers.

Geothermal and Hydropower Technologies
Renewable Technology   202-252-5340—John Mock, Director
Conservation and Renewable Energy
U.S. Department of Energy
FORSTL, CE-324, Room 5F067
1000 Independence Avenue, SW, M.S. 6B024
Washington, DC 20585

This office oversees the implementation of research and development programs to develop the nation's geothermal and small-scale hydropower resources. The office determines the potential of, and develops the technology for, exploitation of the large geopressured and hot dry rock resources. It makes geothermal resource and reservoir assessments and conducts research and development on advanced geothermal technologies, including conversion systems and improved techniques for environmental control. It also manages the Geothermal Loan Guarantee Program.

Government/Industry/University Partnerships

See: Cooperative Programs

Grant Money Resources

See: Research and Development

Guides to Software

See: Software Guides, Directories, and Reviews

Gynecology

See: Reproductive Technology

Hardware, Computer

See: Computers (General); Peripherals

Health Hazards in the Work Environment

See: Industrial Hygiene

Hearing Aids

See: Acoustics

High-Speed Rail Systems

See: Supertrains

High-Tech Initiatives

See: State High-Tech Initiatives

High-Tech Incubators

See: Small Business Incubators

Holography

See also: Fiber Optics; Laser Technology

Organizations and Government Agencies

Massachusetts Institute of Technology (MIT)
Center for Advanced Visual Studies W11   617-253-4415
40 Massachusetts Avenue
Cambridge, MA 02139

The staff and fellows at this Center perform research in holography, lasers, video, photography, environmental art, inflatable sculptures, and medical performances. Harriet Casdin-Silver, a leading expert in the technical aspects of artistic applications of holography, is one of the Center's many specialists in this field.

Museum of Holography
11 Mercer Street   212-925-0581
New York, NY 10013

This is the only U.S. museum devoted to holography, and it is an excellent information source on the subject. The Museum's alternating exhibits cover the artistic, technical, and commercial aspects of holography. Exhibit catalogs and several other publications are available from the Museum. Examples are:

• *Holography Directory*—a "who's who" listing more than 220 experts in the artistic and technical applications of holograpy. Updated quarterly, the *Directory* is available for $30.

• *Holosphere*—a quarterly magazine covering all aspects and applications of holography. The 50-page magazine is available for $30 a year.

• *Holography Works*—an exhibit catalog containing pictures and descriptions of a museum display devoted to the current applications of holography in industry and commerce. The 70-page publication is available for $75.

• *Holography Refined*—this exhibit catalog, containing pictures and descriptions, covers holography as an art. The 40-page publication is available for $15.

Optical Society of America (OSA)
1816 Jefferson Place, NW   202-223-8130
Washington, DC 20036

OSA is a professional society of more than 9,000 scientists, engineers, and technicians. Many of OSA's members are involved in holography. The Society publishes data and research findings relating to optical sciences and engineering, including holography. A free bibliographic listing of OSA's many publications is available from the above office. The group's major publications include:

- *Applied Optics*—published biweekly, this publication is a collection of papers covering modern optics. Topics have included holography, lasers, electro-optics, image processing, space optics, optical engineering, thin films, adaptive optics, and quantum electronics. The annual subscription fee is $32 for members and $310 for nonmembers.

- *Journal of Lightwave Technology*—this bimonthly contains papers dealing with both theoretical and experimental advances in the science, technology, and engineering of optically guided waves. It covers fiber and cable technologies, active and passive guided wave components, integrated optics, and optoelectronics. The annual subscription fee is $10 for members and $124 for nonmembers.

- *Journal of the Optical Society of America*—this monthly two-part publication contains papers that contribute new knowledge or understanding of optical problems, prinicples, and methods. Part A emphasizes image science and basic material. Part B emphasizes laser spectroscopy and modern quantum optics. Members can purchase them separately for $18 a year or as a set for $30 a year. Nonmembers must purchase the *Journal* as a set for $340 a year.

- *Optics Letters*—a monthly publication covering all aspects of optical science, including atmospheric optics, quantum electronics, Fourier optics, integrated optics, and fiber optics. The annual subscription is $140 for nonmembers and $14 for members.

- *Optics News*—this monthly covers the entire field of optics, ranging from theory to instrumentation and systems applications. It also contains a bibliographic listing of recent papers published in OSA journals. The annual subscription is $45 for nonmembers; it is free to members.

Society of Photo-Optical Instrumentation Engineers (SPIE)

PO Box 10   206-676-3290
Bellingham, WA 98227-0010

SPIE is a professional society of 3,000 optical engineers. The organization sponsors ten conferences a year on a variety of topics, including holography, image processing, micro-lithography, intelligent robots/computer vision, integrated optical circuit engineering, and lasers. Conference proceedings are available upon request. SPIE also publishes:

- *Optical Engineering Report*—a monthly newsletter providing information about innovations in the field of optical engineering. The *Report* also contains organization news. It is available free of charge.

Publications

Laser Focus

Pennwell Publishing   617-486-9501
Advanced Technology Group
119 Russell Street
Littleton, MA 01460

This monthly publication provides an overview of holographic techniques, reviews of industrial applications of holography, profiles on new applications, and articles about holographic testing as an aid to designing mechanical parts. The publication's central theme is lasers and their uses. Information is provided about electro-optical systems, fiber optics, and laser applications in the military establishment, manufacturing, entertainment, and so on. The annual subscription to the 200-page publication is $45.

Lasers and Applications

High Tech Publications   213-378-0261
23717 Hawthorne Boulevard, Suite 306
Torrance, CA 90505

This monthly magazine, geared toward engineers and designers, provides technical information about the applications of lasers in defense, manufacturing, and entertainment. The publication covers holography, lasers, electro-optics, and fiber optics. It includes information about new products, literature reviews, and industrial and financial news. Each December issue contains a directory of laser manufacturers and products. The annual subscription to the 130-page magazine is $40.

Photonics Spectra

Optical Publishing Company   413-499-0514
PO Box 1146
Pittsfield, MA 01202

An international journal, this monthly covers holography, optics, electro-optics, lasers, fiber optics, and imaging. It provides information about the latest scientific and technical applications of holography, as well as the economics of the industry. The annual subscription to the 150-page journal is $40.

Additional Experts and Resources

The following companies produce holograms or holographic equipment used for technical and/or artistic purposes. Holograms have a wide variety of applications, ranging from testing aircraft tires to measuring air pollution or producing three-dimensional photographs. The staff at the following companies may be available to answer questions or make referrals. Company brochures can be a good source of information about the various applications of holography and the technology behind it:

American Holographic Company, Inc.

80 Harris Street
Acton, MA 01720
617-263-2538

This company manufactures holographic defraction grading. The staff can answer technical questions relating to the various applications of holography. A free catalog and product information brochures are available upon request.

Apollo Lasers

9201 Independence Avenue
Chatsworth, CA 91311
213-709-1111

Apollo Lasers is a leader in the manufacture of pulsed ruby lasers for holography. Product information brochures are available upon request.

Blue Bell Holographics

PO Box 14466
Long Beach, CA 90803
213-433-6942

This company is involved in the artistic application and development of scanners phase and volume holograms. Product information brochures are available upon request.

DiKrotek International

12277 South 700 West
Draper, UT 84020
801-572-0921

This company is involved in the artistic and technical applications of holography. Product information brochures are available.

Eastman Kodak Company

343 State Street
Rochester, NY 14650
716-724-4634

A leading company in the development of holography and photographic technology dealing with lasers, Kodak is engaged in holography research and development of such applications as dramatic advertising displays, data storage and retrieval systems, and vibration and stress analysis. Kodak has product information brochures available, and the staff can answer questions about specific applications of holography.

Holo Spectra

7742-B Gloria Avenue
Van Nuys, CA 91406
213-994-9577

This company is concerned with both the artistic and technical applications of holography. Product information brochures are available from the company.

Edmund Scientific

101 East Glouster Pike
Barrington, NJ 08007
609-547-3488

Edmund Scientific is a distributor of holographic laser equipment. A catalog of its products and product information brochures are available upon request.

Jodon, Inc.

62 Enterprise Drive
Ann Arbor, MI 48103
313-761-4044

Jodon produces holographic equipment, helium neon lasers, and optical and electro-optical equipment, and is a distributor of holographic plates, films, and chemicals. The staff will answer technical inquiries and refer you to experts in the fields of holography and electro-optics. Product brochures are available upon request.

Laser Technology

1055 West Germantown Pike
Norristown, PA 19403
215-631-5043

This company manufactures laboratory and industrial holographic products, as well as lasers and optics for use in schools and industry. The company publishes product information brochures.

Newport Corporation
18235 Mt. Baldy Circle
Fountain Valley, CA 92728-8020
714-963-98110
    This company manufactures holographic cameras and holographic equipment, such as optics, mirrors, and lasers. A product information catalog is available from the company.

Spectron Development Laboratories, Inc.
3303 Harbor Boulevard, Suite G-3
Costa Mesa, CA 92626
714-549-8477

This company conducts research in the field of holography, specializing in measurement and analysis through the application of lasers and electro-optics. The staff can answer questions from the public and refer people to experts in the field of holography.

Steller Technologies
PO Box 14466
Long Beach, CA 90803
213-433-6942
    This company is involved in the commercial and industrial applications of holographs. Its main product is the real world laser processor, a computerized system operated by light instead of electronics. Product information brochures are available upon request.

Dr. Stephen A. Benton   617-577-3357
Polaroid Corporation
750 Main Street
Cambridge, MA 02139
    Dr. Benton specializes in the physics of holographic optics and conducts research in holographic imaging for information display purposes. He can provide you with information about the scientific and technical aspects of holography, as well as refer you to other information sources.

Dr. Nickolas George   716-275-2417
Institute of Optics
University of Rochester
Rochester, NY 14627
    Dr. George conducts research on the scientific and technical aspects of holography. He oversees the research performed by the University's Optics Laboratory. Dr. George is available to answer questions and make referrals.

Dr. Emmitt Leith   313-764-9545
The University of Michigan
Department of Electrical Engineering
Ann Arbor, MI 48109
    Dr. Leith developed the first holograph, and he is currently involved in studying the scientific aspects of holography. He is based at the University of Michigan, a leading institution in the development of holographs.

Thomas Lettieri
National Bureau of Standards
Bldg. 220, Room A117
Gaithersburg, MD 20899

301-921-2159

Mr. Lettieri frequently lectures to groups, especially in high schools, about the fundamental principles and applications of holography. His special interest is the dimensional meterology of machined parts. He is a good resource person for leads to experts, literature, and companies involved in holography.

Home Economics

See: Agriculture

Human Experimentation

See: Bioethics

Human Factors Engineering

See: Ergonomics

Hydropower

See: Geothermal Power and Hydropower

IC'S

See: Chips and Integrated Circuits

Ideas, Ownership of

See: Patents, Trademarks, and Copyrights

Illegal Flow of Technology

See: Smuggling of High Technology

Image Processing

See also: Optical Character Recognition; Recognition Technology

Organizations and Government Agencies

International Information Management Congress (IMC)
PO Box 34404   301-983-0604
Bethesda, MD 20817

This nonprofit organization dealing with document-based information systems sponsers seminars and an annual meeting. Its membership includes associations, individuals, and corporate members from 32 countries, including the U.S. Don Avedon is a good resource person, with 25 years' experience in the information systems business. He is available to answer questions and/or make referrals. Publications include:

- *IMC Journal*—a quarterly publication focusing on applications in document-based information systems. It is free for members and costs $40 a year for nonmembers.

- *IMC Newsletter*—features announcements, new products, and new publications. Free to members; the subscription rate for nonmembers is $25, for eight issues.

Association for Information and Image Management (AIIM)
1100 Wayne Avenue, Suite 1100   301-587-8202
Silver Spring, MD 20910

AIIM's 8,000 members include individual managers, manufacturers of microfilm equipment, service bureaus, and users. Harry Kidd, a technical coordinator, can answer questions on information systems, including image management and microfilm in the sense of optical disc technologies or transmitting images of documents. The Association holds an annual conference each spring, which covers topics ranging from the introductory to advanced levels. Over 120 exhibitors participate in the conference. AIIM also develops and publishes more than 30 standards and technical reports. Its publications include:

- *The Buying Guide*—an annual publication listing trade members of the organization, including manufacturing and service bureau members, and consultants. It is available free to members and costs $25 for nonmembers.

- *The Journal of Information and Image Management*—a monthly journal directed toward information generalists. It presents a variety of articles on computer, records management, and micrographics applications; standards; technology and products; book reviews of pertinent literature; new product information; editorials; company and personnel news; and letters to the editor. It is available for $55 to nonmembers and is free for members.

- Tutorial information—books and pamphlets relating to the industry.

Society of Photographic Scientists and Engineers

7003 Kilworth Lane   703-642-9090
Springfield, VA 22151

The Society is a nonprofit membership organization dedicated to advancing the application of science and engineering to the imaging technologies, as well as the appreciation of the imaging technologies to engineering and science. The 3,500 members work in the research field of imaging technology, on both the scientific and engineering aspects. The Society sponsors four to seven conferences a year, each one on a different topic. Its publications include:

- *Journal of Imaging Technology*—published every other month, it includes three types of articles: tutorial papers, technical papers, and technical notes. The *Journal* is referred to a scientific and technical editorial review board prior to publication. The cost is $40 for members and $70 for nonmembers.

- *Photographic Science and Engineering*—a technical journal published every other month. It is directed toward individuals working in imaging businesses, such as photographic imaging, digital imaging, and graphic arts imaging. The nonmember annual subscription rate is $70; for members it is $40. Single issues are available for $18.

American Society of Photogrammetry

210 Little Falls Road   703-534-6617
Falls Church, VA 22046

This professional Society consists of 8,000 individual members working in government, education, and private industry, and 140 corporate members, for which a list is available. The Society's focus includes aerial photography and remote sensing.

Its services include conventions that are held twice a year, and several publications, including a free catalog.

● *Photogrammetric Engineering and Remote Sensing*—a monthly publication that is free to members and costs $65 for nonmembers. This technical journal shows how to obtain information about objects in the environment through recording, measuring, and interpreting photographic images and patterns of electromagnetic radiant energy and other phenomena. It also contains the Society's newsletter, which focuses on news, people, and activities. A section called "Engineering Reports" highlights advances in photogrammetry and remote sensing.

● *Manual of Remote Sensing*—this technical publication is a two-volume set available to members for $99, to nonmembers for $125, and to students for $65.

● *Manual of Photogrammetry*—a technical publication. It costs $55 for non-members, $40 for members, and $30 for students.

Impact Assessment

See: Bioethics; Environmental Impact Assessment; Ergonomics; Labor and Technological Change; Social Impact of Technology; Technology Assessment

Import/Export Information

See: International Competition; Smuggling of High Technology

In Vitro Fertilization

See: Reproductive Technology

Incubator Facilities and Projects

See: Small Business Incubators

Industrial Engineering

See: CAD/CAM; Computer Integrated Manufacturing; Ergonomics; Industrial Hygiene; Office Automation

Industrial Hygiene

See also: Ergonomics

Organizations and Government Agencies

American Conference of Governmental Industrial Hygienists

6500 Glenway Avenue, Building D-5   513-661-7881—Executive Secretary
Cincinnati, OH 45211

This professional society's membership of 2,700 is limited to employees of universities and government agencies. The organization researches current information in the industrial hygiene field, evaluates this literature, and then makes recommendations to promote the health of the worker in the occupational environment. The staff can respond to inquiries and also make referrals to specialists in the field. The organization's guidelines are published for public use in several publications:

• *Threshold Limit Value (TLV) Booklet*—contains a listing of more than 600 chemical substances and the acceptable levels in the work environment. This pocket-sized booklet, published annually, costs $4.

• *Air Sampling Instrument Manual*—contains a survey of currently available instrumentation for measuring the quality of the air in the workplace. It is updated every four years. The cost is $60.

• *Industrial Ventilation Manual*—describes and recommends systems for removing dust, vapors, and so on from the workplace. The cost is $15.

American Industrial Health Council (AIHC)

1075 Central Park Avenue   914-725-1492
Scarsdale, NY 10583

This coalition of industrial corporations and trade associations is concerned with identifying, evaluating, and regulating chronic health hazards in the work environment. Risk assessment, with emphasis on cancerous and reproductive effects, is a primary objective. The organization works closely with the many government research agencies, and will make referrals to specialists among its members and within the Occupational Safety and Health Administration or the Environmental Protection Agency. You may request copies, available at no charge, of the annual report, many technical policy documents, and the newsletter.

● *AIHC Newsletter*—published eight times a year, it presents research aimed at eliminating the negative effects of the workplace environment. It is geared chiefly toward professionals in the industrial hygiene field.

American Industrial Hygiene Association (AIHA)

475 Wolf Ledge Parkway   216-762-7294—Public Relations
Akron, OH 44311

This professional organization serves its 6,000 professional and industrial hygienist members by promoting the study, evaluation, and control of workplace environmental stress as it relates to the health and safety of workers. It sponsors continuing education programs and an annual conference with the American Conference of Governmental Industrial Hygienists staff. It maintains an updated list of consultants, to which the staff will make referrals. You may obtain copies of the publications listing, which includes approximately 25 professional and technical publications, including:

● *AIHA Journal*—a monthly publication that promotes the professions of industrial hygiene. The *Journal* presents current research and commentary on the effects of the workplace environment, as well as articles written by member specialists in the promotion of worker health and safety. About 60 pages an issue, it is available to nonmembers for $60 a year.

American Occupational Medical Association

340 South Arlington Heights Road   312-228-6850—Information Director
Arlington Heights, IL 60005

A national medical subspecialty group, the Association consists of approximately 4,000 occupational physicians, who are helping to ensure quality medical care for workers. In addition to answering questions from the public and making referrals to specialists in the field of occupational medicine, the Association sponsors education programs, seminars, scientific sessions, and annual conferences for its members. In addition to its newsletter for members, it also publishes the following:

● *Journal of Occupational Medicine*—a monthly containing scientific articles and advertisements directed toward physicians in the field of occupational medicine. The annual subscription rate to the 140-page *Journal* is $35 for nonmembers.

Industrial Health Foundation

34 Penn Circle West   412-363-6600
Pittsburgh, PA 15206

This service organization for its 120 member companies promotes the improvement of working conditions in U.S. industry. It conducts industrial hygiene surveys, and holds technical meetings and symposia for its member companies. It also maintains a library and a research laboratory for member use. Various technical bulletins, minutes of the symposia, and a monthly journal are published.

• *Industrial Hygiene Digest*—a monthly that contains abstracts from current literature in the field of occupational health and safety. About 25 pages, the *Digest* is available only to member companies.

National Institute for Occupational Safety and Health (NIOSH)

4676 Columbia Parkway   513-533-8311—Division of Standards Development
Cincinnati, OH 45226   513-533-8328—Technical Information Center

NIOSH is the research agent that provides the technical information to OSHA for the promulgation of its regulations on occupational safety and health. Its findings are available to the public as well as to other government agencies, industry, and academicians, and are published by NIOSH in continual technical reports and criteria documents. NIOSH also sponsors training for health professionals and a grant program for educational centers. NIOSH conducted a national occupational environment survey, querying industries and companies as to what types and levels of chemicals workers are exposed to. The published results are available. The technical information center, NIOSHTIC, will make available detailed research on both chemical and physical health and safety hazards, and will search its internal document file upon request. NIOSH publishes an annual toxic substances list available at no charge, and supplies information for the Register of Toxic Effects of Chemical Substances (RTECS), available through MEDLARS. NIOSH also supplies data for a bibliographic data base containing over 106,000 references on occupational environments, soon to be available as File 160 through DIALOG.

Occupational Health Institute

2340 South Arlington Heights Road   312-228-6850
Arlington Heights, IL 60005

This affiliate of the American Occupational Medical Association (described above) supplies professional support to nurses, industrial hygienists, and physicians in the field of occupational health.

Occupational Safety and Health Administration (OSHA)

U.S. Department of Labor   202-523-6441—Office of Management Data Systems
Room N 3637   202-523-8151—General Information
200 Constitution Avenue, NW
Washington, DC 20210

OSHA's goal is to protect U.S. workers in their occupational environment. The major thrust of OSHA's work is setting and enforcing mandatory standards for employees not only in industry but also in construction, maritime work and agriculture. In addition to compliance assistance, OSHA sponsors training and continuing education for hygienists. OSHA maintains a computerized data base containing inspection reports of companies and industries, which the staff will search for the public, with the only charge being computer time. Pamphlets covering topics concerning safety hazards and protective equipment and toxicity data are available. Single copies as well as information about general standards are available at no charge directly from OSHA, whereas multiple copies are available at a nominal fee from the Superintendent of Documents, U.S. Government Printing Office, Washington, DC 20402.

Publications

Industrial Hygiene News

8650 Babcock Boulevard   800-245-3182
Pittsburgh, PA 15237
    Published every two months, this magazine is mailed at no cost to engineers and other qualifying management personnel to promote new products in the occupational safety and hygiene field. In addition to photographs and descriptions of products, there are often articles describing some of the problems encountered with the technology in the field.

Industrial Safety and Hygiene News

1 Chilton Way   215-964-4060
Radnor, PA 19089
    This monthly trade magazine covers new technologies and products in the field of occupational safety and hygiene. Available free of charge to qualified people in the industry. Others can subscribe for $60 a year.

Occupational Hazards

1111 Chester Avenue   216-696-7000
Cleveland, OH 44114
    This controlled circulation monthly magazine is sent at no cost to safety and health management personnel who qualify. *Occupational Hazards* covers various occupational safety and health topics, including government regulation and workers' compensation. The subscription rate for the public is $35 a year.

Occupational Health and Safety Letter

1331 Pennsylvania Avenue, NW, Suite 509   202-347-3868
Washington, DC 20004
    This biweekly newsletter analyzes legislative and regulatory developments in the field of occupational health and safety. The subscription rate is $135 a year.

Industry/Government/University Partnerships

See: Cooperative Programs

Infertility

See: Reproductive Technology

Information and Records Management

See: Office Automation; Recognition Technology

Information Services and Brokers

Organizations and Government Agencies

American Society for Information Science (ASIS)
1010 16th Street, NW   202-659-3644
Washington, DC 20036

A society of information professionals, ASIS brings practitioners and theorists together to exchange ideas and views, and to enlarge perspectives. Members are administrators, managers, coordinators, information technologists and scientists. The organization has a number of special-interest groups, such as Automated Office of the Future (AOF), Library Automation and Networks (LAN), and Storage and Retrieval Technology (SRT). The Society holds annual fall conferences in key U.S. cities, and meetings in the spring at other locations. Conferences and meetings feature workshops and seminars, presentations of papers, and technical exhibits. The staff can answer questions and/or make referrals. Other services include a Data Base User Service and several publications, all described in the ASIS publications catalog.

• *ASIS Bulletin*—a bimonthly magazine containing articles and regular columns about new data bases, on-line use, copyright management, significant new policies, information systems and services in the U.S. government, and so on. Members receive the 40-page publication free, which costs $45 a year for nonmembers.

• *ASIS Journal*—a bimonthly scholarly journal covering information, science, and related fields. It offers theoretical presentations and practical applications. The 80-page publication is available for $85 a year for nonmembers. The publication is available from John Wiley and Sons, Wiley-Interscience Journals, 605 Third Avenue, New York, NY 10158, 212-692-6000.

● *The Librarian's Guide to Microcomputer Technology and Applications*—this 150-page publication offers information about how libraries are using micro-computers to provide public services, technical services, and library management. Based on a survey of more than 400 academic, public, special, and school libraries, this study describes the machines, operating systems, and applications in current use. It reports libraries' experiences with hardware maintenance, opinions of vendor-developed software, and future plans for microcomputer applications. The cost is $34.50 for nonmembers and $27.60 for ASIS members. The publication is available from Knowledge Industry Publications, Inc., 701 Westchester Avenue, White Plains, New York 10604, 800-431-1880 (914-328-9157 in New York).

● *Database Directory*—this annual publication contains more than 1,500 data bases and can be accessed by subject, company, and data base. The 600-page *Directory* is available for $120 from Knowledge Industry Publications, Inc., 701 Westchester Avenue, White Plains, New York 10604, 800-431-1880 (914-328-9157 in New York).

● *ASIS Handbook and Directory*—updated annually, this publication lists ASIS members, including address, place of employment, and phone number. The *Handbook* also lists officers, chapters, and special-interest groups. The cost is $50; available from ASIS.

Association of Data Processing Service Organizations (ADAPSO)

1300 North 17th Street, Suite 300   703-522-5055
Arlington, VA 22209

This trade association for computer services represents more than 700 companies that market software products, professional services, integrated systems, and processing services. ADAPSO monitors legislation for its members, and provides them with research studies and statistics as well as educational, public relations, and financial services. The staff can answer questions relating to business issues rather than technical ones. The Association holds two annual meetings, which are targeted toward computer executives. It offers several publications, including:

● *Membership Directory*—updated annually, this *Directory* lists all members and services. It includes company names, addresses, and phone numbers. The cost is $95.

Information Industry Association (IIA)

316 Pennsylvania Avenue, SW, Suite 400   202-544-1969
Washington, D.C. 20003

IIA's mission is to contribute to the development, growth, and profitability of businesses involved in the creation, storage, management, and distribution of information. It is also concerned with educating the public about the value of information. IIA sponsors an annual conference, as well as other meetings devoted to specific topics. It has a governmental relations department that represents the industry before the U.S. Congress and regulatory agencies. The publications department offers a free backlist of bound books containing lectures and papers presented at IIA meetings. The books vary in price from $25 to $60. The Association also publishes several surveys, including a financial survey available for $950 and a salary survey available for $240. The directory described below provides an overview and description of Association members:

• *Information Sources*—this annual directory of IIA members includes data base producers and vendors, information retrieval services, information brokers and retailers, market researchers, turnkey services, periodical and newsletter publishers, information technology and computer software producers, videotex, teletext, time sharing, and other information sources. The directory provides alphabetical listings, which describe each member company's activities, services, and branches; a names-and-numbers section that identifies executives with their affiliations and phone numbers; a listing of foreign affiliates by country; and an index. The 400-page publication is available to nonmembers for $59.75 and to members for $29.75.

Publications

Encyclopedia of Information Systems and Services

Gale Research Co.   800-223-GALE
Book Tower   313-961-2242 in Michigan, Alaska, Hawaii, and Canada
Detroit, MI 48226

Updated biannually, this two-volume publication covers about 25,000 systems, services, products, and programs in the U.S. and some 70 other countries. It includes more than 400 on-line services and about 3,600 data bases. The *Encyclopedia* covers several kinds of data bases: on-line and off-line; commercial, government, and private; bibliographic and nonbibliographic. It also provides descriptions of organizations offering information systems, services, products, and programs. These include publishers, associations, government agencies, private firms, organizations producing data bases, on-line vendors, consultants, market research firms that can help you make use of electronic information, videotex/teletext systems, telecommunications networks, library systems, demographic and marketing data firms, and document delivery sources. The 669-page international volume is available for $165, and the 1200-page U.S. volume for $190; the entire set costs $325. A supplement, *New Information Systems and Services*, is issued twice between editions of the *Encyclopedia*. The subscription to this supplement is $250.

The Information Brokers

R. R. Bowker Company   313-761-4700
PO Box 1807
Ann Arbor, MI 48106

This publication tells you how to start and operate your own fee-based information service. It concentrates on the professional qualifications needed to succeed and covers the plans, steps, and methods for identifying and building your clientele. The cost is $15.95 for the paperback edition and $24.95 for the hardback edition.

Additional Experts and Resources

Your local library is an excellent information resource; with the aid of an experienced librarian you can be guided to the best possible use of available materials. Most libraries have reference rooms housing standard reference books, periodicals, microfilm, and files with newspaper clippings and informational brochures. Your librarian generally has access to other library systems, and therefore he or she may be able to obtain additional materials for you. Many libraries operate telephone reference systems designed to quickly supply callers with answers to all types of questions. Today's libraries also may have informational films and video players.

FIND/SVP

500 Fifth Avenue   212-354-2424
New York, NY 10110

Also known as The Information Clearing House, FIND/SVP offers access to a variety of information and research services and products. Its Quick Information Service provides direct telephone access to a comprehensive information center for fast answers to a wide variety of business and general questions. Its Research Projects Service is designed to handle more extensive research and information gathering needs, including custom market and industry studies, surveys, and large information gathering tasks. FIND also produces publications, including several directories and data bases. Its on-line data base indexes and describes nearly 10,000 industry and market studies, product studies, surveys and polls, consumer studies, major multi-client studies, industry and company reports issued by securities firms, major store audit reports, and subscription research services. FIND/SVP Reports and Studies Index is available on Lockheed's DIALOG as File 196.

Information on Demand, Inc. (IOD)

PO Box 9550   800-227-0750
Berkeley, CA 94709   415-644-4500

IOD provides both document delivery and research services. It locates, photocopies, and acquires published information in any form from any country, including journal articles, technical report, government documents, catalogs, annual reports, conference papers, and patents. Research services include providing information to the business, technical, and professional communities, as well as to individuals. Specific services are literature searching, including compilation of bibliographies using published indexes; abstracting services, and computerized data bases; comprehensive market research; current awareness service; fact gathering; telephone interviewing; purchase of product samples; and translations. IOD utilizes all data bases from DIALOG, Pergamon, Info Line, ORBIT, BRS, Dow Jones, NEXIS, MEDLARS, CompuServe, The Source, and NASA/DOE RECON.

The Information Store

140 Second Street   415-543-4636

San Francisco, CA 94105

This company provides information-gathering and research services for the business community. Services include competitive intelligence; customized research services and reports available for strategic planning, market research, law, and other business and professional information needs; document retrieval; and special-interest monitoring. Resources include computerized information banks, research centers and agencies, libraries, and other contacts throughout the world. Publications include:

• *Document Retrieval—Sources and Services*—this comprehensive directory lists worldwide public and private document retrieval suppliers. The cost is $60.

• *Effective Factual Research*—this guide to the location and use of information sources is directed toward lawyers and paralegals. The cost is $25.

• *How to Order On-line from the Information Store*—this guide to the use of five electronic mail systems is available free of charge.

• *Professional Document Retrieval*—this quarterly newsletter covers document retrieval technology, document delivery services, and legal and management issues. It is directed toward information center personnel and those employed by companies providing or using such services or creating information technology. The annual subscription price is $15.

Infosource, Inc.

Two Oliver Plaza   412-562-4170

PO Box 456

Pittsburgh, PA 15230

This specialized research firm offers expertise in developing and implementing research strategies, as well as interpreting and analyzing findings. It combines information from computerized data bases, leading business libraries, and in-depth telephone interviews. Clients use the information the company provides to develop market strategies; monitor competitors; assess technology; analyze the potential of existing and emerging markets; sell products; and evaluate candidates for mergers, divestitures, and joint ventures.

Research Counsel of Washington

226 4th Street, NE   202-544-2700

Washington, D.C. 20002

Customized research on a wide range of topics, focusing primarily on market studies and competitive intelligence, is performed by this company. While most research projects are commissioned by individual clients for their private use, the firm does prepare multi-client studies on a range of topics affecting the financial services industry. Recent projects have centered on aspects of the telecommunications, defense, and banking industries.

Warner-Eddison Associates, Inc.

186 Alewife Brook Parkway   617-661-8124

Cambridge, MA 02138   Telex: 710-320-0094

This company provides several types of services. One is information management consulting, including analysis of information usage patterns and needs, assistance with selection of resources, and help in creating an on-line or printed information delivery system or an entire information center. Other services are indexing and thesaurus development, data base design and production, and information management software. The company has developed software to produce indexes, catalogs, labels, and other information management products. The staff will adapt the software to meet your requirements, or you can buy INMAGIC, the company's information storage and retrieval system. Warner-Eddison also provides assistance to organizations that are starting, expanding, or dealing with any aspect of an information center, including space design, staffing, collection development, user access systems, and cataloging.

Washington Researchers

2612 P Street, NW   202-333-3499

Washington, DC 20007

This firm offers several kinds of information services. Through customized research projects, it locates information for all types of organizations and institutions. Through seminars, the staff members share their sources and teach other business professionals their techniques. Although some of the information compiled by Washington Researchers is for the exclusive use of individual clients, the firm has developed a reservoir of unique and varied informational resources that are available to others. Many of its sources are federal government offices, which can supply information and publications that you can often obtain free of charge. The company's seminars and printed materials are designed to expand the use of these and other resources. Publications include books and a monthly newsletter.

Innovation

See: Productivity and Innovation

Instructional Computing Music

See: Music, Instructional Computing

Instrumentation

See: Standards and Measurement

Integrated Circuits

See: Chips and Integrated Circuits

Interactive Video Technology

Organizations and Government Agencies

Interactive Communications Society (ICS)
PO Box 4520   415-961-9060—Vicky Vance
Mountainview, CA 94040   415-965-8159—ICS

ICS deals exclusively with interactive communications, including interactive video. It holds monthly meetings with speakers, and maintains a library of materials on interactive video, as well as an on-line bibliographic data base. Each month it publishes a one-page update of its running bibliography, which lists all the articles on interactive video. The bibliography dates back to 1975. The Society also publishes a monthly journal.

Interactive Video Association (IVA)
PO Box 1491   219-287-8804—William J. Frascella, President
Evanston, IL 60204

IVA provides information on all aspects of its field. It sponsors educational workshops and seminars; holds an annual meeting; provides use of an informational data base, which includes available hardware and software, funding opportunities, and new developments; and publishes a newsletter, which provides reviews of software and hardware developments, information and funding opportunities, reports on works in progress, and an employment register. Membership is $45; for students it is $15.

Society for Applied Learning Technology
50 Culpepper Street   703-347-0055
Warrenton, VA 22186

The Society holds an annual conference on the current state-of-the-art of video-discs and publishes the proceedings of its conferences. It has formed a special-interest group on interactive video learning, which holds training conferences and soon will publish material on the subject. For answers to questions contact Mr. Fox at the Society or Bob Bellinger, chairman of the special-interest group, at 401-841-4763.

Publications

Interactive Video—1984
Applied Video Technology   314-569-9144
5118 Westminster Place
St. Louis, MO 63108

Interactive Video is a comprehensive, single-volume guide to the interactive video industry. The document contains (1) references to over 400 producers, manufacturers, universities, associations, publishers, and users of interactive video; (2) an index to the interactive videodiscs and tapes that have been produced; (3) an index to individuals in the field; (4) a guide to periodical reports and other documents. The spiral-bound report is $29 and the ring-binder report is $43.

Interactive Video Technology
Heartland Communications   216-567-3732
223 Sunrise Drive
Shreve, OH 44676

The newsletter covers new products, legal issues, changes in the industry, projects in interactive video, and a calendar of events. The annual subscription is $45 a year. This monthly newsletter is also available through the NewsNet at 800-345-1301 (in Pennsylvania, 215-527-8030).

Optical Memory News
Rothschild Consultants
PO Box 14817
San Francisco, CA 94114-0817

This newsletter is devoted exclusively to the computer related read-write optical memory and read-only interactive videodisc technology. The annual subscription rate is $296 for six issues.

Video Disc Monitor
PO Box 26   703-241-1799
Falls Church, VA 22046

Published monthly, this report focuses on nonconsumer industrial and educational applications of the videodisc technology. It emphasizes new products, market statistics, new applications, and comparative analyses. It includes short articles, conference coverage, a calendar of related events, editorials, reviews of discs and projects, reference materials, and a regular listing of vital statistics. The subscription rate is $167 a year, for 12 issues.

Meckler Publishing
520 Riverside Avenue   203-226-6967
Westport, CT 06880

Available publications about video technology are:

● *Videodisc/Videotex*—this bimonthly magazine is devoted to the reporting and critical analysis of research and development in videodisc and videotex systems in business, industry, government, education, libraries, and the home. Available for $75 a year.

● *Videodisc Update*—this companion newsletter provides current news, calendar items, job listings, and conference reports. It costs $96 a year, for 12 issues, or $66 a year for *Videodisc/Videotex* subscribers.

Video Disc Book: A Guide and Directory
John Wiley & Sons   212-867-9800
605 Third Avenue
New York, NY 10016

Written by Rod Daynes, this is a yellow-pages approach to the videodisc industry. The first section includes articles on various videodisc topics, from analyses of the job to be done to the design of discs and implementations, as well as the future of the videodisc industry. The second section lists people and organizations involved in the industry, as well as videodisc players and discs in existence.

Handbook of Interactive Video
Knowledge Industry Publications   914-328-9157
701 Westchester Avenue
White Plains, NY 10604

This book, by Steve and Beth Floyd, describes the technology of interactive video and outlines the available hardware. It discusses authoring systems and how to design, write, and produce an interactive program. Differences between interactive and linear video are clearly explained. The appendix includes detailed case studies, as well as useful lists of manufacturers, software suppliers, and other valuable resources. The 168-page volume costs $34.95.

Additional Experts and Resources

Many of the companies that manufacture interactive videodisc equipment are good sources of information. They usually print brochures describing their products and the technology involved. Their public affairs departments will answer questions about their products. Some of the major manufacturers are:

Pioneer Video
200 West Grand Avenue
Montvale, NJ 07645
201-573-1122

Brochures describing individual products are available; a publication that explains the technology behind many of the products is also available.

Sony Communications Products Company
1 Sony Drive
Park Ridge, NJ 07656
201-930-1000

Sony regularly conducts workshops on different topics in interactive video. Brochures on videodisc products and technology are available.

RCA Commercial Communications Systems
Front and Cooper Streets
Camden, NJ 08088
609-338-2839

RCA is constantly conducting research in high-technology fields, and several brochures on videodisc technology are available, including one on interactive video.

3M
1101 15th Street, NW
Washington, DC 20005
202-331-6900

3M publishes a free directory that lists all suppliers who offer services related to videodisc production. Also provided are two brochures: (1) "Interactive Video Served on a Disc," which explains videodisc technology, and (2) "Producing Interactive Videodiscs," which is a complete guide to interactive videodisc program production.

Several universities and research centers do research in interactive videodisc technology, and are probably among the best sources of information on current developments in the field. The following are some of the places that do research:

Nebraska Videodisc Design/Production Group
1800 North 33rd Street
University of Nebraska
Lincoln, NE 68508
402-472-3611

Department of Instructional Technology
UNC 30
Utah State University
Logan, Utah 84322
801-750-2694—Mike DeBloois

Computer Assisted Instruction Lab
Room 229, Lindquist Center
319-353-3170—Joan Sustic

or
Office of Research and Development
Room 224, Lindquist Center
University of Iowa
Iowa City, Iowa 52242
319-353-3326—Rob Molek

Massachusetts Institute of Technology
Architecture Machine Group
105 Massachusetts Avenue
Bldg. 9, Room 516
Cambridge, MA 02139
617-253-5113—Andrew Lippman

International Competition

See also: Japan; Productivity and Innovation; Smuggling of High Technology; Technology Transfer

Organizations and Government Agencies

Commerce Productivity Center (CPC)
U.S. Department of Commerce   202-377-0940
14th Street and Constitution Avenue, NW, Room 7413
Washington, DC 20230

This information clearinghouse provides business and other organizations with information about how to improve competitiveness, productivity, and quality. As part of the Office of Productivity, Technology and Innovation, the Center has access to over 35 professionals having expertise in such diverse areas as economic policy, industrial competitiveness, research and development, technology transfer, and patent policy. Its collection includes information on both domestic and foreign experience, data industries, and technologies. Upon request, the staff can provide publications, articles, reference and referral services, bibliographies, and reading lists. CPC's services are available free of charge.

Division of Developed Country Trade
Office of Trade   202-632-2718
U.S. Department of State, Room 3822
Washington, DC 20520

This government office tracks high-technology activities in each developed country. The staff can provide you with a general overview of a specific country's activities.

European Space Agency (ESA)
Washington Office   202-488-4158—Ian Pryke
955 L'Enfant Plaza, SW, Suite 1404
Washington, DC 20024

Formed in 1975, ESA is a cooperative association of the following nations: Belgium, Denmark, France, Germany, Ireland, Italy, the Netherlands, Spain, Sweden, Switzerland, and the United Kingdom. ESA's Washington Office will answer specific questions when possible, make referrals to publications and experts, and forward extensive printed material on the agency itself, as well as on specific technologies. Technical manuals on equipment are also available free upon request, as well as several quarterly newsletters on specific satellites and equipment.

GATT Agreement on Technical Barriers to Trade

Standards Code and Information Office   301-921-2092
A629 Administration Building   301-921-3200—Hotline
Gaithersburg, MD 20899

The Standards Code and Information Office assists U.S. exporters and manufacturers by keeping them informed of proposed foreign regulations that may affect U.S. trade opportunities with countries that are signatories to the International Agreement on Technical Barriers to Trade (Standards Code). Federal Register notices of proposed U.S. regulations that may affect trade opportunities of foreign countries are made available to the same countries. The staff can answer questions and make referrals. The GATT Hotline provides a recording with the most recent information on proposed foreign regulations. An important publication is:

● *GATT Annual Report*—describes the activities for the year performed by the Standards Code and Information program in support of the GATT Standards Code. The publication is available free of charge.

Institute for Policy Studies

1901 Q Street, NW   202-234-9382
Washington, DC 20009

The Institute has been conducting a research project on multinational corporations. Staff members have been following the impact of the corporations on trends in the global economy as well as their impact on specific industrial sectors, including high-technology industries. In conjunction with the project, staff members maintain a clip file with articles from newspapers and business magazines throughout the world. Staff members will search their files for the public, and can answer questions and refer you to appropriate sources.

International Comparisons Section

Division of Foreign Labor Statistics   202-523-9291
Office of Productivity and Technology
Bureau of Labor Statistics
U.S. Department of Labor
441 G Street, NW
Washington, DC 20212

The Bureau of Labor Statistics has available for public use an inventory of statistics on international comparisons, which it prepares annually. Most inventories include at least ten of the major industrial countries and have reference periods beginning about 1950. Topics included in the statistics are indexes of productivity, hourly wages, unit labor costs, unemployment rates, labor force participation rates, distribution of employment by economic sector, consumer price trends and indexes, capital investment percentages, and indexes of living costs abroad. A listing of the indexes available may be obtained from the above Office, as well as explanations and sources of reference. The actual documents are published through the Superintendent of Documents, Government Printing Office, Washington, DC 20402, 202-783-3238.

Office of Industry Assessment

International Trade Administration   202-377-5145
U.S. Department of Commerce, Room 1009
Washington, DC 20230

     This federal office is reponsible for following industry trade policy issues that are cross-sectoral and for coordinating the Department's Competitive Assessment Program. It also prepares competitive assessments about such cross-sectoral issues as research and development and its impact on industry. The staff can provide information or advice on cross-sectoral international trade policy problems affecting high-technology industries, and can discuss the issues and concerns these industries have when dealing with international competition. The staff can also refer you to other sources and experts. The Office has produced reports on some specific high-tech industries, as well as some cross-sectoral reports. This Office also oversees the preparation of the *U.S. Industrial Outlook*, and the staff can serve as a source for additional information relating to that publication.

Office of Technology Assessment (OTA)

U.S. Congress   202-226-2115
Washington, DC 20510

     This nonpartisan analytical agency serves the U.S. Congress by providing objective analysis of major public policy issues related to scientific and technological change. OTA has produced more than 100 reports assessing the present and future impact of various technologies. Many of these reports discuss international competition, especially reports on electronics, space activities, biotechnology, and technology transfer. OTA's multidisciplinary staff members, who prepare the reports, are available to answer questions and make referrals. While report summaries can be obtained free of charge, the full text of all reports must be purchased from the Government Printing Office. A free publications catalog is available from OTA. Upon request, OTA will add you to their publications announcement mailing list, which is also free.

Office of Trade Development

International Trade Administration (ITA)   202-377-3808
U.S. Department of Commerce
14th and E Streets, NW
Washington, DC 20230

     The main purpose of ITA is to promote the export of U.S. manufactured products. Counseling is offered to exporters requesting information or assistance, and much of the information is given in personal interviews with specific staff members in their area of expertise. There are 47 district offices of the Department of Commerce that can respond to questions. The Trade Development sector is divided into nine divisions, which are:

Science and Electronics
202-377-4466

This office covers the fields of computers and business equipment, components and related equipment, telecommunications, instrumentation, and medical sciences. Publications include *A Competitive Assessment of the U.S. Fiber Optics Industry*, a similar assessment of the U.S. digital switching industry, and numerous other statistical assessments of U.S. science and electronic capabilities.

Capital Goods and International Construction
202-377-5023

This division focuses on general industrial machinery, special industrial machinery, and international major projects. Periodically, technical reports are made available in these fields.

Automotive Affairs and Consumer Goods
202-377-0823

The Automotive Affairs office deals with motor vehicles, parts, and suppliers. It publishes an annual competitive assessment, *Status of the U.S. Auto Industry*, which is free. The Consumer Goods Office deals with both durable and nondurable goods, and can supply statistical information as to the U.S. competitive position in many major markets.

Trade Adjustment Assistance
202-377-0150

This office is in charge of the certification processes for international trade by U.S. industries. This is the division that pursues unfair practice allegations and provides both technical and financial assistance to companies that export American-made products.

Trade Information and Analysis
202-377-1316

This division is subdivided into five major interest areas: (1) trade finance; (2) trade and investment analysis; (3) trade and industry information; (4) program and resource management; and (5) industrial assessment.

Textiles and Apparel
202-377-3737

This office provides analytical support for international textile negotiations. It monitors textile and apparel imports; provides studies and analyses of domestic and foreign fiber, textiles, and apparel industries; develops and coordinates domestic trade and export expansion programs; and performs several other services.

Aerospace
202-377-8228

This office operates strictly on a consultation basis. One recent publication of note is *A Competitive Assessment of the U.S. Civil Aircraft Industry*.

Basic Industries
202-377-0614

This office has four divisions: (1) chemical and allied products, which has published a *Competitive Assessment of the U.S. Advanced Ceramics Industry*; (2) energy; (3) metals, minerals, and commodities, which has published *A Competitive Assessment of the U.S. Petrochemical Industry*; and (4) forest products and domestic construction.

Services
202-377-5261

The following service industry divisions are part of this office: (1) transportation, tourism, and marketing industries; (2) finance and management industries; and (3) information industries. The latter has published *A Competitive Assessment of the U.S. Software Industry* and *A Competitive Assessment of the U.S. Information Services Industry*. Upon request single copies of both are available free of charge.

Organization for Economic Cooperation and Development (OECD)
1750 Pennsylvania Avenue, NW, Suite 1207   202-724-1857
Washington, DC 20006-4582

OECD is a nonprofit research organization consisting of 24 member countries united to promote economic and social welfare. The major part of its work is carried out in numerous specialized committees in the following areas: economic policy and development, trade, consumer policy, energy policy, industry, agriculture, scientific and technological policy, education, and so on. Each of these committees is engaged in promoting cooperation and improved relations, information exchange systems, cooperative research and development, and general integration of policies among the member countries in their specialized fields. Annual reports, technical documents, and research papers are issued by each of the committees and are available —most of them for a fee—to the general public. An extensive listing of publications is available upon request, with much of the material on microfiche and magnetic tape. The U.S. headquarters will put callers in contact with expert groups and national delegate members specializing in specific areas of research. The following are published:

• *The OECD Observer*—a bimonthly magazine that covers the most important problems dealt with by OECD. Issues contain facts and recommended solutions in such areas as economic growth, employment and unemployment, energy, multinational enterprises, financial markets, the environment, science and technology, and trade with the developing world. The annual subscription rate is $11.

• *Economic Surveys*—annual reports of economic developments in each of the

24 member countries, which provide detailed analyses of recent developments in demand, production, employment, prices and wages, conditions in the money and capital markets, and developments in the balance of payments. The annual subscription to the series is $75, with individual booklets available for $5.

Publications

Competing in the High Technology Era

Price Waterhouse   212-371-2000
1251 Avenue of the Americas
New York, NY 10020

This booklet contains the results of a 1982 Price Waterhouse opinion survey of high-technology companies. Information was collected regarding the factors that affect international competitiveness. The survey includes companies' views about public policy, and tax changes that would enhance competition. Results, narrative, and graphic analyses are included in the booklet, which is available free of charge.

International Trade Administration (ITA)

U.S. Department of Commerce   202-377-3808—Public Affairs
14th Street and Constitution Avenue, NW
Washington, DC 20230

ITA puts out several reports on high-technology topics, including robotics, the telecommunications industry, the semiconductor industry, the computer industry, and biotechnology. One major publication is:

• *An Assessment of U.S. Competitiveness in High Technology Industries*—this study demonstrates the U.S. challenge to maintain its broad technological preeminence. It examines selected high-technology sectors with regard to their importance, their trade performance, and the factors influencing their competitiveness with foreign competitors. Extensive statistical tables support the findings of the study. The report (# 003-009-00358-0) is available from the Superintendent of Documents, Government Printing Office, Washington, DC 20402, 202-783-3238.

U.S. Industrial Outlook

U.S. Department of Commerce   202-377-4356
International Trade Administration
Industry Publications, Room 442
14th Street and Constitution Avenue, NW
Washington, DC 20230

Published every January, this book provides an overview and prospectus for more than 300 U.S. industries. High-tech industries covered include telecommunications, computing and office equipment, electronic components, aerospace, chemicals, photographic equipment, information services, medical and dental instruments, and others. For each topic, the book provides statistics, information about new technologies and trends, and import/export data. Also included are the name, phone number, and address of Department of Commerce experts for almost every industry in the U.S. These specialists can tell you about the latest developments in the field and refer you to other experts and literature. The 1,000-page book costs $14 and is available from the Government Printing Office, Washington, DC 20402, 202-783-3238.

Data Bases

Data Resources, Inc.
Data Products Division Headquarters   202-862-3760
1750 K Street, NW, Suite 1060
Washington, DC 20006

This firm maintains the following three data bases with information relating to international competition:

● *U.S. Central*—this data base contains financial, economic, and demographic indicators, which provide comprehensive measures of economic activity for the U.S. as an aggregate. It contains over 33,000 series, and covers national income and product accounts, retail and wholesale trade, shipments, orders and inventories of manufacturers, labor force, employment, hours and earnings, housing starts and completions, financial and consumer credit data, and the "Quarterly Financial Report" and other industry data. It is available on an annual, semiannual, quarterly, monthly, and weekly basis.

● *DRI*—this current economic-indicators data base provides timely coverage of major financial and economic indicators for Europe, Japan, Canada, the U.S., and selected developing countries. With over 1,700 series, indicators include balance of payments, consumer and wholesale price indices, foreign exchange rates, industrial production indices, money supply and interest rates, national income accounts, retail sales, unemployment, and wages. It is available on an annual, quarterly, monthly, and weekly basis.

● *OECD Main Economic Indicators*—this data base is a compilation of economic, demographic, and financial indicators for the 24 OECD member nations. With over 20,000 series, coverage includes national product and production, construction, domestic and foreign trade, labor force and employment, prices and wages, home and foreign finance, balance of payments, and measures of industrial activity. It is available on an annual, quarterly, and monthly basis.

I. P. Sharp Associates, Ltd.
Suite 1900, Exchange Tower   416-364-5361
2 First Canadian Place
Toronto, Ontario
Canada M5X 1E3

SITC-United Nations Commodity Trade Statistics is a data base that allows the user to monitor the transfer of technology for 28 selected reporting countries by looking at the imports and exports of over 3,000 commodities. Examples of commodities include heavy machinery and office automation products.

Additional Experts and Resources

Industry Specialists
U.S. Department of Commerce   202-377-4356
International Trade Administration, Room 442
14th and Constitution Avenue, NW
Washington, DC 20230

The publications staff can refer you to an industry specialist for nearly every industry in the U.S. These experts can tell you about the latest technological developments, provide you with statistics, give you data on imports and exports, tell you how the U.S. industry compares with industry abroad, and refer you to appropriate experts and literature.

Marjory Searing, Ph.D., Director
Office of Industry Assessment   202-377-5145
U.S. Department of Commerce
International Trade Administration (ITA)
14th Street and Constitution Avenue, NW
Washington, DC 20230

An international economist with 16 years of government experience, Dr. Searing is an expert on high-tech international problems and high-tech competitiveness issues.

Trade reference rooms of libraries located in International Trade Administration (ITA) District Offices across the United States have U.S. import/export data by product, country, or both. Foreign data is available only at the ITA trade reference room in Washington, DC. These libraries have all kinds of trade statistics, but you must obtain the information by visiting the library. To locate your ITA District Offices, look in the U.S. Government section of your telephone book.

Literature and organizations covering innovation and productivity often have information about industrial competitiveness at both the U.S. and international level. For information about these sources, see the section entitled *Productivity and Innovation*.

Japan

Organizations and Government Agencies

Commerce Productivity Center
U.S. Department of Commerce   202-377-0940
14th Street and Constitution Avenue, NW, Room 7413
Washington, DC 20230
  This information clearinghouse provides businesses and other organizations with information about how to improve competitiveness, productivity, and quality. It has information about Japan, and Japanese productivity and management techniques. Upon request, the staff can provide publications, articles, reference/referral services, articles, and reading lists. These government services are available free of charge.

Embassy of Japan
917 19th Street, NW   202-234-2266
Washington, DC 20006
  The Japanese Embassy is an excellent source of information on major matters of concern with regard to U.S.-Japanese relations. The staff can make available the names and addresses of Japanese government agencies, universities and institutes in Japan, Japanese commercial businesses, and other sources of information on high-tech industries. Its extensive library, much of it available in English, contains texts and periodicals covering most major high-tech subjects and is open to the public. In addition, special representatives of the various ministries of the Japanese government maintain offices at this location and will provide information on major issues affecting the relationship between the U.S. and Japan.

Japan Economic Institute of America (JEI)
1000 Connecticut Avenue, NW, Suite 211   202-296-5633
Washington, DC 20036
  This nonprofit research organization is supported by Japan's Ministry of Foreign Affairs and staffed by bilingual experts. The Institute is a reliable source of current information on the Japanese economy and U.S.-Japan relations. The Institute sponsors frequent seminars and informal talks on U.S.-Japanese economic relations. It also publishes detailed reports and analyses in an effort to improve bilateral communications; these include:
  • *JEI Reports A & B*—two weekly publications that are offered as a single subscription. *JEI Report A* focuses on one relevant aspect of U.S.-Japan economic and political relations in each weekly issue. *JEI Report B*, also published weekly, is a

newsletter providing news analyses with names, dates, and sources for each news item. The annual subscription for both reports is $40.

The Japan Productivity Center (JPC)

1-1,3-Chome, Shibuya, Shibuya-Ku   011-81-3-409-1111
Tokyo, Japan
U.S. Office:
1901 North Fort Meyer Drive, Suite 703   703-243-5522
Arlington, VA 22209

    The Center was formed as an integral part of the Japanese productivity movement effort to achieve a self-sustaining economy. Its basic objective is the development and guidance of management in all levels of the Japanese economy. This is accomplished through seminars, courses, in-house company training, and other educational approaches. The international operations center promotes the exchange of information on productivity through overseas study teams, seminars, and training programs on productivity, and through worldwide conferences.

    The Center also has an Electronic Data Processing (EDP) Center, which, in addition to performing routine services for government agencies and businesses and organizations, has initiated its own research and development projects on economic problems. The Productivity Research Institute conducts studies on productivity and the economy, and issues monthly economic index statistics for productivity and trade. The Institute also houses the JPC library, with over 50,000 publications and some 400 periodicals on productivity and management. JPC also distributes a wide variety of printed materials on productivity and technology, much of it translated into English and other languages, and the publications department will forward a publications listing upon request. JPC also publishes:

    • *Japan Productivity News*—a weekly publication that is JPC's principal vehicle for keeping business, labor, and the public informed on productivity matters. Subscription rates vary according to the mailing address and the language of the translations.

Japan Technical Information Center, Inc. (JTIC)

719 8th Street, SE   202-543-3040
Washington, DC 20003

    A private research organization, JTIC specializes in high-tech work areas. Its purpose is to provide information for the U.S. and Japan. Contracted research is done and reports are published on many areas of the high-tech industry. The Center also serves as an interpretation/translation service for Japanese and American businesspeople who require information on competitive productivity. It also escorts study groups visiting the prospective countries. Currently the emphasis is on providing the Japanese constituency with information on U.S. high technology, but work is under way to provide U.S. clients with the same type of data on Japanese business.

Japan Trade Facilitation Committee

International Trade Administration   202-377-5722

U.S. Department of Commerce

14th and E Streets, NW

Washington, DC 20230

     The Committee is the result of a joint effort of the U.S. Department of Commerce and Japan's Ministry of International Trade and Industry, and is intended to aid exporters. The Committee currently deals primarily with problems stemming from industry sectors as a whole, with a focus on such structural issues as energy costs and raw materials availability. The staff can suggest areas of anticipated difficulties in major industries and offer possible solutions, as well as give statistical information on specific industry forecasting.

Japan Work Group on High Technology Industries

International Trade Administration   202-377-5251

U.S. Department of Commerce

14th and E Streets, NW

Washington, DC 20230

     This group was established at the suggestion of the U.S. Foreign Minister to Japan as a result of the 64K Ram semiconductor controversy. The Group acts as a forum for the discussion of relevant issues in the U.S./Japan trade relations, and as an educational vehicle in disseminating information on high-tech industries in both countries. It has attempted to establish a viable relationship between the U.S and Japan to allow possible cooperation in the high-tech industries and has established both short-term and long-term work program agreements to facilitate this relationship. The staff is knowledgeable on matters concerning high-tech industries and will make available nonconfidential agreements and findings of the joint discussions. The staff will also make referrals to other government agency staff.

The Japan-United States Friendship Commission

1875 Connecticut Avenue, NW, Suite 910   202-673-5295

Washington, DC 20009

     The Commission's purpose is to improve the relationship between the U.S. and Japan. There are four main areas for program activities: (1) Japanese studies in American education, (2) American studies in Japanese education, (3) the arts, and (4) research and programs for public education. The Commission supports nationwide fellowship programs for Americans in higher levels of study on Japan in the fields of journalism, law, business, and economics. It also offers funds for research on Japanese subjects, library support of Japanese language materials, and language training. The Commission Center in Washington is an excellent source for obtaining the names and addresses of both American and Japanese universities involved in research. The staff will make referrals to educational specialists and to library collections covering respective areas of study in the U.S. and Japan.

Office of Japan
International Trade Administration (ITA)   202-377-4527
U.S. Department of Commerce
14th and E Streets, NW
Washington, DC 20230

ITA handles trade investment policy with Japan and aids those exporting U.S.-manufactured products to Japan. In addition to consultation work with exporters, this Office is a great resource for general market trends, trade regulations, and other trade data from various sectors of the Japanese industry. The staff will answer questions when possible and make references to experts in other government agencies and publications. Several publications that originate from this Office are available upon request and include:

● *Overseas Business Report on Marketing in Japan*—a study of approximately 100 pages published every two years. It examines marketing trends in general and anticipated areas of growth and problematic areas. It also contains detailed information on Japanese import trade regulations to help the U.S. exporter.

● *Foreign Economic Trends: Japan*—a ten-page publication issued every six months by ITA in cooperation with the U.S. Embassy in Japan. More current than the *Business Report*, it addresses current issues of merit in the Japanese economy as a whole, as well as present areas of flux in the Japanese industry.

● *Data Tables: Japan*—a series of statistics published periodically by ITA covering major aspects of U.S.-Japan trade. In addition to import/export figures, also included are investment data, banking information, input/output statistics, price data, and other tables of trade analysis information.

U.S. Special Trade Representative—Director for Japan
White House Office
600 17th Street, NW
Washington, DC 20506

This Office of the executive branch is responsible for bilateral affairs between the U.S. and Japan in regard to trade policy development and coordination. It oversees negotiations between the two countries in areas of industry, energy, investment, commodities, and agriculture, and it is responsible for interagency coordination in these trade areas. Most of the published material originating from this Office is for interagency use, but the staff will refer you to the proper agency or public document addressing most trade issues with Japan. There is also a private sector advisory group, which serves in an advisory capacity through various government agencies in an effort to promote cooperation between the industry sectors of the two countries.

U.S. Japanese Joint Institute for Fusion Theory
International Fusion Community Computer Exchange Protocol   415-422-1100
National Magnetic Fusion Energy Computer Center
Lawrence Livermore National Laboratory (LLNL)
PO Box 808
Livermore, CA 94550

The agreement to link the Japanese Institute for Plasma Physics at Nagoya University to LLNL's Magnetic Fusion Energy Computer Center supports joint research efforts on magnetic fusion. LLNL staff will provide information sources in the form of referral of a specialist or providing of documentation.

Publications

Comparative Productivity Dynamics: Japan and the United States
Elliot S. Grossman, Ph.D   713-681-4020
George E. Sadler, Senior Economist
The American Productivity Center
123 North Post Oak Lane
Houston, TX 77024

A multiple input productivity comparison, by Elliot S. Grossman and George E. Sadler, this report provides insight into current productivity dynamics and the factors influencing them. The report examines several manufacturing industries of particular importance—primary metals, chemicals, machinery, and transportation equipment. The study quantifies several factors that have contributed to Japan's rapid productivity growth since World War II, including research and development and rapid dissemination and use of new technology, a growth stimulating program by the Japanese government, and so on. Extensive supporting materials, including numerous tables and charts, provide an excellent reference on the Japan-U.S. comparison. The report is available free of charge upon request from the Center.

Industrial Policy: Japan's Flexible Approach
Industrial Policy: Case Studies in the Japanese Experience
U.S. General Accounting Office (GAO)   202-275-5518
411 G Street, NW
Washington, DC 20548

These two reports by GAO are a result of an extensive study of economic policies adopted since World War II by Japan to foster rapid productivity and economic growth. Data are from various government agencies, industry representatives, trade and industry associations, labor groups, and universities in both the U.S. and Japan. The first report discusses monetary and fiscal policies, and also industrial policies implemented by Japan. The second report is a detailed examination of five key industries, including the high-technology industries of computers, aircraft and robotics, and the industry specific policies that have contributed to their rapid growth. The reports may be ordered from the GAO Document Handling and Information Services Facility, PO Box 6015, Gaithersburg, MD 20760, 202-275-6241. Single copies are available free of charge.

Data Bases

Data Resources, Inc. (DRI)
1750 K Street, NW, 9th Floor   202-862-3700
Washington, DC 20006

24 Hartwell Avenue   617-863-5100
Lexington, MA 02173

DRI maintains a massive and expanding collection of economic, financial, and demographic information organized into the largest commercially available data base in the world. This information is offered to government and private clients through the DRI time-shared computer and telecommunications system. It has also developed services allowing clients to download DRI data for display and analysis using a variety of personal computers. A professional staff of consultants, industry and economics experts, and software and data specialists help clients access a vast amount of information and customize it to meet their needs. The data banks dealing with Japan are:

● *Japan*—contains economic, demographic, and financial indicators, which provide a comprehensive economic profile of the Japanese economy. Topics covered include production, labor, prices, finance, trade, household income and expenditures, balance of payments, and national income accounts. Available on an annual, semiannual, quarterly, or monthly basis, the contact person is Nihon Keizai Shimbunsha (Nikkei).

● *Japan Prices*—contains the detailed price indices collected by the Bank of Japan. With over 2,000 series, the coverage includes wholesale price indices by industry and commodity, export price indices by commodity, and import price indices by commodity. Available on a monthly basis, the contact person is Nihon Keizai Shimbunsha (Nikkei).

● There are also data banks on Japanese energy, consisting of 3,300 series covering energy demand, supply and prices, and Japan's finances, presenting the financial position of some 1,700 Japanese companies.

Additional Experts and Resources

Patricia Haigh
Office of Developed Country Trade   202-632-2735
Bureau for Economic and Business Affairs
Room 3822
U.S. Department of State
Washington, DC 20520

A foreign service officer who served three years in Tokyo, Ms. Haigh currently follows U.S.-Japan trade issues, including those in the high-tech sector for the Department of State. Ms. Haigh is available to answer questions and make referrals to other experts following U.S.-Japan trade issues.

Knitting

See: Textiles

Labor and Technological Change

See also: Careers and Employment; Productivity and Innovation; Social Impact of Technology; Technology Assessment

Organizations and Government Agencies

AFL-CIO
Economic Research Department   202-637-5160
815 16th Street, NW
Washington, DC 20006

This large labor union federation approaches the effect of high technology on the workplace from the workers' point of view. This office has data about worker displacement and unemployment resulting from new technologies. The staff can supply you with information about retraining programs, relocation allowances, labor agreements that address the impact of high tech, and union positions on the subject.

Congressional Caucus for Science and Technology
House Annex Building #2, H2-226   202-226-7788
2nd and D Streets, SW
Washington, DC 20515

The Caucus, established to serve Congress members' need for timely information, can provide news about the impact of science and technology on economic development and employment trends, including issues of training and retraining. The Caucus plans to develop "employment roadmaps" that will highlight employment opportunities, as well as training and educational requirements, for current and prospective jobs. The staff can answer questions about legislation related to high tech and will refer science and technology inquiries to the Caucus research arm, the Research Institute for Space, Science and Technology.

Division of Cooperative Labor-Management Programs

U.S. Department of Labor   202-523-6098
200 Constitution Avenue, NW
Washington, DC 20210

Established to assist employees and unions in joint efforts to enhance the quality of work life, this Division can supply information about research concerning the affects of high technology. The Division works closely with trade associations; international and national unions; area labor-management committees; and national, state, and regional productivity or quality-of-work-life centers. It regularly compiles and disseminates information about labor-management cooperation topics, including retraining and relocation programs. A publications listing is available upon request.

Employment and Unemployment Analysis Division

Bureau of Labor Statistics (BLS)   202-523-1944
U.S. Department of Labor
441 G Street, NW
Washington, DC 20212

A division of BLS, the office compiles and analyzes U.S. employment and unemployment statistics. It investigates trends in employment fluctuation based on such factors as age, sex, race, marital status, and education, as well as regional and industrial variances. Given its sizable impact on employment, staff members currently are researching the effect of high-technology infusion in various industries. An inventory of available statistics may be obtained from the above address.

Federal Job Training Centers

Employment and Training Administration (ETA)   202-376-6093
U.S. Department of Labor (DOL)
601 D Street, NW
Washington, DC 20213
or contact your state, local, or regional Government Employment
Service, listed in your telephone book

There are approximately 595 state, local, and regional job-training offices throughout the U.S., and most can provide information about labor market conditions in their geographical area. The Centers are designed to assist the economically disadvantaged, dislocated workers, the unemployed, the underemployed, and those facing special barriers to employment. Since each Center offers different services, it is best to call your local office to find out about eligibility and training programs offered in your area. The ETA office listed above oversees the federal employment training program. Its staff can provide you with statistical information about the training programs they have analyzed and can refer you to an office in your area.

International Labor Organization (ILO)
1750 New York Avenue, NW   202-376-2315
Washington, DC 20006

ILO brings together government, employers, and labor in an attempt to improve working conditions throughout the world. The Organization's activities are carried out by experts in various fields, who make recommendations and work toward establishing international standards of acceptable labor practice. The findings of these various committees are published in a series of special bulletins and other regular publications by the ILO, and by its affiliate, the Institute for Labor Studies. ILO's extensive library of labor-related texts, periodicals, reports, and video materials is open to the public. Information about the collection and a listing of ILO's publications may be obtained by writing to the address above. Major periodicals include:

• *International Labor Review*—a bimonthly publication that is available in English and several other languages. The magazine contains articles about current labor issues, including legislation, management concerns, safety tips, and other worker-environment topics. It also contains notices of current publications in many labor issue areas. About 100 pages an issue, this publication can be obtained at an annual subscription rate of $28.50.

• *Social and Labor Bulletin*—a monthly publication that focuses on current events affecting the labor force. Topics covered include the impact of technology on the work force, legislative happenings, improvement of the work environment through health/safety measures and through socioeconomic factors, and other areas of labor study. Many of the articles are scholarly and have been written by sociologists, economists, and other professionals studying the labor force. The annual subscription rate is $25.75.

Labor Research Association
8 East 11th Street, Room 634   212-473-1042
New York, NY 10003

This nonprofit research organization focuses on the socioeconomic positions of labor and the impact of various social and economic trends on the labor force. Since much of its research is prepared for unions and other labor organizations, its viewpoint is usually pro-labor. Topics covered range from the reduction of labor market demands, due to the increasing use of high technology, to the impact of the increased number of women in the labor force. The Association's research results are contained in its magazine:

• *Economic Notes*—a monthly publication providing in-depth analyses of labor problems. A wide range of topics is covered, including women in the labor force, health and safety issues, economic implications of new labor policies, social trends in the labor movement, and other issues of import to both the worker and management. A recent issue focused on the impact of high technology on the labor force in America. Subscription rates are $10 a year for individuals and $15 for institutions.

National Council for Labor Reform (NCLR)

406 South Dearborn   312-427-0206
Chicago, Illinois 60605

NCLR is a nonprofit organization of more than 2,500 members, primarily employers, that is mainly concerned with labor law problems. The Council keeps members abreast of legislative action and judicial decisions affecting employee-employer relations. It also supports lobbying efforts for fair labor laws and practice. Staff members are knowledgeable about labor law and policies affecting both workers and management. They are also knowledgeable about government retraining programs designed to help workers displaced by new technologies. The staff will answer general questions, recommend literature, and refer you to other labor specialists.

Office of Productivity and Technology Studies

Bureau of Labor Statistics   202-523-9294
U.S. Department of Labor
441 G Street, NW
Washington, DC 20212

Trends in technology and their impact on employment and productivity are investigated by this office. Staff members will send you an inventory of available statistics, and they can tell you about industries affected by and occupations displaced by technological innovations. The Office frequently issues reports and updates about the impact of technological changes and the manpower trends of various industries. A publications list is available upon request. Some of these reports, all available from the Superintendent of Documents, U.S. Government Printing Office, Washington, DC 20402, include:

● *Technology and Labor in Four Industries: Meat Products/Foundries/Metalworking Machinery/Electrical and Electronic Equipment*—Bulletin 2104

● *Technology and Labor in Five Industries: Bakery Products/Concrete/Air Transportation/Telephone Communication/Insurance*—Bulletin 2033

● *Technological Change and Its Labor Impact in Five Energy Industries: Coal Mining/Oil and Gas Extraction/Petroleum Refining/Petroleum Pipeline Transportation/Electricity and Utilities*—Bulletin 2005

● *Technological Change and Its Labor Impact in Five Industries: Apparel/Footwear/Motor Vehicles/Railroads/Retail Trade*—Bulletin 1961

● *Technological Change and Manpower Trends in Five Industries: Pulp and Paper/Hydraulic Cement/Steel/Aircraft and Missiles/Wholesale Trade*—Bulletin 1856

● *Technical Change and Manpower Trends in Six Industries: Textile Mill Products/Lumber and Wood Products/Tires and Tubes/Aluminum/Banking/Health Services*—Bulletin 1817

Women's Bureau

U.S. Department of Labor (DOL)   202-523-6611
200 Constitution Avenue, NW
Washington, DC 20210

    The Bureau follows the impact women have on the U.S. work force, as well as the conditions and special problems encountered by working women. The Bureau has been studying the impact of high technology on labor, and its findings were released in early 1985. The office has studied a variety of issues affecting women, including equal opportunity, women in nontraditional careers, employment goals, and women in management. The staff tracks legislative proposals affecting women, such as sex discrimination laws, job training partnerships, and state labor laws. A listing of publications and bibliographies produced by the Bureau, as well as program models, are available upon request.

Labor Law

See: Labor and Technological Change

Laser Surgery and Medicine

Organizations and Government Agencies

American Society for Laser Medicine and Surgery

425 Pine Ridge Boulevard, Suite 203   715-845-9282
Wausau, WI 54401

    More than 500 physicians, scientists, nurses, allied health professionals, and representatives of commercial companies belong to this organization. Each spring the Society sponsors a conference, which is open to the public. Audiocassette tapes and abstracts of the scientific presentations delivered at the conferences are available. Following a conference, the Society offers a one-day course that is an introduction to laser biophysics. Staff members will answer questions from the public; drawing from their membership list, they can refer you to an appropriate expert in your area of interest. The Society produces the following publications:

    • *Lasers in Surgery and Medicine*—a quarterly journal covering the latest developments in laser surgery and medicine. The publication is geared toward physicians, generally contains ten abstracts, and runs about 80 pages. The subscription price for nonmembers is $85 a year. The journal is available from Alan R. Liss Corporation, 150 Fifth Avenue, New York, NY 10011, 212-741-2515.

● *Laser Medicine and Surgery News*—describes research and developments and the state-of-the-art in laser technology, biology, medicine, and surgery. New clinical work and studies are reported, as well as existing laser safety data, laser laws, and laser instrumentation. The publication focuses on applications in the areas in which the greatest advances are expected, such as treatment of cataract diseases, detection and treatment of cancer, and the rapidly advancing field of cardiovascular surgery. Twelve surgical disciplines are covered. This bimonthly newsletter costs $75 a year and can be obtained from Mary Ann Liebert, Inc., 157 East 86th Street, New York, NY 10028, 212-289-2300.

Laser Institute of America (LIA)
5151 Monroe Street, Suite 118 W   419-882-8706
Toledo, OH 43623

This Institute is an organization devoted to the advancement and promotion of laser technology. Staff members disseminate technical information to those interested in lasers. One of the Institute's areas of research is laser surgery and medicine. It holds an annual technical conference, and publishes the following:

● *LIA Newsletter*—a bimonthly publication.
● *Laser Safety Information Bulletin*—a free brochure giving the basics about laser safety concerns.
● *Laser Safety Guide*—information on laser safety standards, available for $4.
● *Lasers in Materials Processing Information Bulletin*—a free booklet on a wide variety of manufacturing operations that can be accomplished with lasers.
● *Fundamentals of Lasers*—costs $30 for nonmembers and $20 for LIA members.

Midwest Bio-Laser Institute
4550 North Winchester at Wilson   312-878-4300
Chicago, IL 60640

This Institute sponsors and conducts educational and research programs related to the application of lasers to medicine. It conducts professional meetings, conferences, and congresses. It publishes *Bio-Laser News*.

American National Standards Institute (ANSI)
1430 Broadway   212-354-3300
New York, NY 10018

ANSI is a private, nonprofit federation of standards-developing organizations and standards users. Its major functions include coordinating the voluntary development and approval of American standards. The Institute can refer people to technical experts outside the organization, who either write the standards or are otherwise knowledgeable in the field. Several laser standards can be obtained from the Institute.

Publications

Clinical Laser Monthly
67 Peachtree Drive, NE   404-351-4523
Atlanta, GA 30309

This monthly newsletter focuses exclusively on the practical concerns of health professionals involved in laser usage. It keeps readers up to date on the latest clinical advances, and shows new ways to market laser capabilities. It is available for $118 a year.

Education Design
PO Box 31975   303-692-9758
Aurora, CO 80041

• *Laser Network*—published semimonthly, this four-page newsletter provides information about lasers in the operating room, including industry news, new technology, publications, books, journals, and meetings. The cost is $12 a year.

• *A Guide to Lasers in the OR*—this comprehensive 118-page manual provides information about equipment, laser physics, safety, administration, surgical applications, nursing care of the laser patient, and advice on how to buy a laser. The manual costs $32.50 a year.

The Surgeon's Newest Scalpel Is a Laser (606-L)
Consumer Information Center   303-948-3334
Pueblo, CO 81009

The U.S. Food and Drug Administration puts out this four-page free booklet, which is a catalog of basic information for patients and the laypublic.

Data Bases

National Health Information Clearinghouse (NHIC)
PO Box 1133   703-522-2590
Washington, DC 20013-1133   800-336-4797

A variety of government institutes, agencies, and private groups are involved in laser surgery research and development. NHIC maintains a data base with information about health-related organizations, including federal and state agencies, information centers, and private groups. The staff will search the data base to provide you with experts and organizations involved in the particular aspect of laser surgery you are interested in. The federal government has several experts at its various health institutes; these officials can tell you about the latest developments in the field and refer you to other information sources.

Additional Experts and Resources

Reference Section

Reference Services Division   301-496-6095

National Library of Medicine (NLM)

8600 Rockville Pike

Bethesda, MD 20209

   NLM is the largest medical library in the world, and the Reference Section can provide the following quick reference services: verify a title or journal in the NLM collection; verify bibliographic citations; give names and addresses of doctors, scientists, and associations; give doctors' specialties; give definitions of medical terms; give drug ingredients and possible adverse effects.

Dr. Ronald Carey

Center for Devices and Radiological Health   301-427-8228

U.S. Food and Drug Administration (FDA)

Office of Compliance, HSZ-312

8757 Georgia Avenue

Silver Spring, MD 20910

   Dr. Carey is a physicist at FDA who evaluates medical laser products for compliance with the laser performance standards. He can advise manufacturers about engineering performance procedures and other requirements for the safe and effective use of lasers.

Laser Technology

Organizations and Government Agencies

Laser Institute of America (LIA)

5151 Monroe Street, Suite 118 W   419-882-8706

Toledo, OH 43623

   LIA is devoted to the advancement and promotion of laser technology, and disseminates technical information to those interested in lasers. It holds an annual technical conference and publishes the following:

   • *LIA Newsletter*—a bimonthly tabloid.

   • *Laser Safety Information Bulletin*—a free brochure, which provides introduction to laser safety concerns.

   • *Laser Safety Guide*—information on laser safety standards. Available for $4.

   • *Lasers in Materials Processing Information Bulletin*—a free pamphlet discussing a wide variety of manufacturing operations that can be accomplished with lasers.

● *Fundamentals of Lasers*—available for $20 to LIA members, and $30 to nonmembers.

Optical Society of America

1816 Jefferson Place, NW   202-223-8130
Washington, DC 20036

Research and other optics data is disseminated by this society. It holds an annual meeting that covers many optics topics, and it conducts a wide variety of special topical conferences, workshops, and other information programs. It supports 15 technical groups, and 23 local sections; one of the technical groups is dedicated entirely to the topic of lasers. Publications include the following:

● *Journal of the Optical Society of America*—a monthly periodical containing papers and letters that contribute significant new knowledge or understanding of any optical phenomenon, principle, or method. The cost is $165 a year.

● *Applied Optics*—contains papers on the applications of facts, principles, and methods of optics. The cost is $250 a year for six issues.

● *Optics Letters*—a monthly publication with news and important results. It costs $100 a year.

● *Optics News*—a bimonthly publication containing information about meetings and conferences, technical papers of general interest, and news in the field of optics. It is available for $40 a year.

Quantum Electronics and Applications Society

IEEE Headquarters   212-705-7900
345 East 47th Street
New York, NY 10017

The Society is dedicated to researching and disseminating information in the field of quantum electronics. It publishes the following:

● *Journal of Lightwave Technology*—a bimonthly publication that contains articles on current research, applications, and methods used in lightwave technology and fiber optics. Topics covered include optically guided wave technologies, fiber and cable technologies, active and passive components, integrated optics, and opto-electronics, as well as systems and subsystems covering the full range of this expanding field. It costs $96 a year, for six issues.

● *Journal of Quantum Electronics*—a monthly journal of quantum electronics and applications, including opto-electronics theory and techniques, lasers and fiber optics; and design, development, and manufacture of systems and subsystems pertaining to these topics. Subscription is $150 a year.

American National Standards Institute

1430 Broadway   212-354-3300
New York, NY 10018

ANSI is a private, nonprofit institute composed of standards-developing organizations and standards users. Its major functions include coordinating the voluntary development and approval of American standards. The staff can refer people to technical experts outside the organization who either write the standards or are otherwise knowledgeable in the field. Several laser standards can be obtained from the Institute.

Publications

Lasers and Applications

High Tech Publications   213-378-0261
23717 Hawthorne Boulevard, Suite 306
Torrance, CA 90505

Published monthly, this magazine of the electro-optics field covers new developments in the laser and fiber optics industry, with an emphasis on applications. Also available is an annual *Buyers' Guide*, which identifies over 1,000 companies that provide goods and services to the laser and fiber optics industry. The annual subscription price is $45.

Fiber/Laser News

Phillips Publishing   301-986-0666
7315 Wisconsin Avenue, Suite 1200N
Bethesda, MD 20814

This bimonthly newsletter covers marketing, research and development activities, and technology of the laser and fiber optics industries. It costs $275 a year. The newsletter is also available electronically on NewsNet at 800-345-1301 (215-527-8030 in Pennsylvania).

Optical Publishing Co., Inc.

Berkshire Common   413-499-0514
PO Box 1146
Pittsfield, MA 01202

Two laser technology publications are available:

• *Photonics Spectra*—a monthly, the magazine provides information on photonics technology, including optics, electro-optics, lasers, and fiber optics. The emphasis is on techniques, procedures, instrumentation, events, and news coverage in this industry. It costs $36 a year.

• *Optical Industry & Systems Purchasing Directory*—is a reference guide to optics, electro-optics, lasers, fiber optics, and vacuum technology. Book 1 contains a buyers' guide, new products, addresses of companies, and a geographical section. Book 2 contains an encyclopedia, a handbook, a dictionary of photonics technology, and a listing of abbreviations. The cost is $60.

Laser Focus Including *Electro-Optics Magazine*

Penwell Publishing Company   617-244-2939

Box 1111

Littleton, MA 01460

This monthly magazine covers technology news, business news, and technological features on fiber optic technology, as well as systems and devices. Also available is an annual Buyers' Guide issue, identifying suppliers of goods and services to the laser and fiber optic industry. The subscription is $45 a year.

Communications Publishing Group, Inc. (CPG)

PO Box 383   617-566-2373

Dedham, MA 02026

CPG publishes high-technology publications, on-line data bases, and research services based on worldwide patent literature. Its resources are available both in print and electronic form from NewsNet at 800-345-1301 (215-527-8030 in Pennsylvania). Its publication dealing with lasers is:

• *Lasers/Electro-Optics*—this newsletter covers equipment and techniques, such as lenses, beam deflectors, interferometry, spectrometry, light detection and measurement, cooling, and holography. It includes all types of lasers (dye, gas, solid state, and so on), as well as laser systems (medical, industrial, research and development, aerospace, military, energy, and so on). The subscription price is $220 a year, for 24 issues.

The Electro-Optical Communications Dictionary

Hayden Book Company, Inc.   201-843-0550

50 Essex Street

Rochelle Park, NJ 07662

A comprehensive reference for the fiber optical and lightwave, data-processing, and related manufacturing technologies fields. The *Dictionary* presents terms and definitions consistent with international, federal, industrial, and professional societies' standards. Processes, devices, and components in electro-optical communications are clearly described. The book is available for $16.95.

Additional Experts and Resources

Science and Electronics Division

International Trade Administration (ITA)   202-377-1333—Arthur Pleasants

Room 4044   202-377-2990—William Sullivan

14th Street and Constitution Avenue, NW

Washington, DC 20230

These experts are responsible for monitoring and predicting the laser and fiber optics industry. They are excellent sources of information, and can recommend other resources.

Laser Weapons in Space

See: Star Wars: Antimissile Systems in Space

Lawmaking

See: Legislation

Laws Affecting Computers and Technology

See: Computer and Technology Law

Legislation

Organizations and Government Agencies

Bill Status Office
U.S. Congress   202-225-1772
House Annex #2, Room 2650
Washington, DC 20515

A free service, this will tell you the status of any past or current legislation. Congressional researchers will access the data bases, which are updated every evening with the developments of the day. You can find the answer to such questions as:

- How many bills have been introduced on a certain topic
- What is the status of a given bill
- What committees have held hearings on a particular bill
- What similar bills are pending

Congressional Cloakrooms

U.S. Senate   202-224-8601—Republican
                                                     202-224-8541—Democrat

U.S. House of Representatives   202-225-7430—Republican
                                                     202-225-7400—Democrat

These recorded messages tell you hour by hour what is happening on the floors of the House and Senate chambers. They can be used for monitoring votes and other floor activities of legislation identified by the Bill Status Office.

U.S. Senate Committee on Commerce, Science and Transportation

5202 Dirksen Office Building   202-224-5115
Washington, D.C. 20510

The Senate Committee will know about most legislation and hearings that have been conducted on high-technology subjects. The Committee can be contacted for copies of bills, transcripts of hearings, and status reports on the future of specific legislation.

U.S. House Committee on Science and Technology

2321 Rayburn House Office Building   202-225-6371
Washington, DC 20515

This House Committee has primary jurisdiction on most high-technology issues.

Publications

Index to Publications of the United States Congress

Congressional Information Service (CIS)   301-654-1550
4520 East-West Highway
Bethesda, MD 20014

This weekly service indexes all publications (including hearing transcripts) of the U.S. Congress. The *Index* is arranged by subject and will lead you to the committee that investigated or held hearings on specific legislation. It is also available on-line on DIALOG, 800-227-1960 (800-982-5838 in California).

Additional Experts and Resources

Most associations and large corporations in the high-technology field maintain offices in Washington that monitor all federal legislation that may have an impact on the industry. For identifying sources see the listing under specific topics elsewhere in this book.

Libraries

Organizations and Government Agencies

American Library Association (ALA)

50 East Huron Street

Chicago, IL 60611

312-944-6780, ext. 301—Library and Information Technology Association (LITA)

312-944-6780, ext. 319—Resources and Technical Services Division (RTSD)

Approximately 40,000 libraries, librarians, library trustees, and other interested persons belong to this professional association. It has 2 divisions particularly concerned with libraries and high technology. One is its Library and Information Technology Association (LITA), which is chiefly concerned with automation, education technology, and video and cable communications. LITA has about 1,750 members and it sponsors annual conferences, seminars, and publications. The other division, called Resources and Technical Services Division (RTSD), has more than 6,000 members, and it is primarily concerned with the acquisition, identification, cataloging, classification, and preservation of library materials as well as the development of library resources throughout the country. RTSD regularly holds continuing education conferences throughout the United States, and it produces several publications covering new technology.

The ALA staff, especially people in its LITA and RTSD sections, can answer questions and make referrals to experts. Relevant ALA publications include:

- *Library Technology Reports*—a bimonthly subscription service directed to library administrators and others who make purchasing decisions. It contains authoritative information on library systems equipment and supplies. The approximately 100-page publication is available for $145 a year.
- *Library Systems Newsletter*—a monthly publication that focuses on turnkey library automation systems. The eight-page newsletter is available for $35 a year.
- *Information Technologies and Libraries*—a quarterly periodical covering what is new and what is changing in library automation, educational technology, and video and cable communications. The approximately 100-page publication is available for $25 a year.
- *Video and Cable Guidelines Manual*—available for $9.75.
- *Video and Cable Information Kit*—available for $7.50.

American Society for Information Science

1010 16th Street, NW

Washington, DC 20036

202-659-3644

The Society of information professionals brings practitioners and theorists together to exchange ideas and views. Members are administrators, managers, coordinators, information technologists, and scientists. The Society has a number of special-interest groups, including Automated Office of the Future (AOF), Library Automation and Networks (LAN), and Storage and Retrieval Technology (SRT). The Society holds annual fall conferences in key U.S. cities, and meetings in the spring in other locations. Conferences and meetings feature workshops and seminars, and the presentation of papers and technical exhibits. The staff can answer questions and/or make referrals, and provide information about the Data Base User Service. Also available is a publications catalog. Some of the publications are:

• *ASIS Bulletin*—a bimonthly magazine containing articles and regular columns on new data bases, on-line use, copyright management, significant new policies, information systems and services in the U.S. government, and so on. The 40-page publication is available for $45 a year, and is free to members.

• *ASIS Journal*—a bimonthly scholarly journal in information science and related fields. It offers theoretical presentations and practical applications. The approximately 80-page publication is available for $85 a year, and is free to members. The publication is available from John Wiley and Sons, Wiley-Interscience Journals, 605 Third Avenue, New York, NY 10158, 212-692-6000.

• *The Librarian's Guide to Microcomputer Technology and Applications*—an approximately 150-page report offering information on how librarians actually use microcomputers for applications in public services, technical services, and library management. Based on a survey of more than 400 libraries, this study describes the machines, operating systems, and applications in current use. It reports librarians' experiences with hardware maintenance, opinions of vendor-developed software, willingness to share locally produced software, and future plans for microcomputer applications. The cost is $34.50 for nonmembers and $27 for ASIS members. Available from Knowledge Industry Publications, Inc., 701 Westchester Avenue, White Plains, NY 10604, 800-431-1880 or 914-328-9157.

• *Database Directory*—an annual publication that contains more than 1,500 data bases and can be accessed by subject, company, and data base. The 600-page directory is available for $20 from Knowledge Industry Publications, Inc., 701 Westchester Avenue, White Plains, NY 10604, 800-431-1880 or 914-328-9157.

Medical Library Association (MLA)

919 North Michigan Avenue, Suite 3208   312-266-2456
Chicago, IL 60611

A professional society, MLA sets standards of professional practice, promotes the provision of quality service, and provides leadership in solving problems of health information delivery. It works with other professionals and associations to promote health through access to information. Membership includes librarians and others actively engaged in library or bibliographical work in medical or allied scientific fields. The Association activities include an annual conference with exhibitions, continuing education programs, and monitoring legislative issues. Contact the publications department for more information about the following publications:

• *Information Series*—short monographs, each less than 100 pages, on topics of current interest to health-sciences librarians and health-care professionals.

• *Bulletin of the Medical Library Association*—a scholarly quarterly journal containing reports on research in the field of librarianship in the health sciences. The annual subscription rate is $65 a year for nonmembers; it is included in the membership fee for members.

• *MLA News*—a monthly newsletter containing reports, regular feature columns, and news items of interest to members. The annual subscription price is $25 a year for nonmembers; it is included in the membership fees for members.

Online, Inc.
11 Tannery Lane   203-227-8466
Weston, CT 06683

This organization produces practical how-to journals, books, and data bases to help with searches and information management. It sponsors an annual conference each fall. The staff can answer questions and/or make referrals. Publications and data bases include:

• *Online*—a bimonthly journal that keeps abreast of what is new in the field and gives practical advice on using current products and services. Topics covered include microcomputer hardware and software, new data bases, on-line searching tips, and new trends and technologies. The 110-page publication is available for $78 a year.

• *Database*—a quarterly publication that complements *Online* with practical articles on data bases, including those aspects of microcomputing that relate to data bases. The 110-page publication is available for $56 a year.

• *Online Terminal/Microcomputer Guide and Directory*—this extensive guide to terminals, microcomputers, printers, and modems is useful to the information professional. The three main sections are (1) tips on choosing a terminal or microcomputer, (2) a guide containing 16 lists, including equipment, specifications, manufacturers, brokers, abbreviations, a glossary, and a bibliography, and (3) a directory of sales and service offices in the U.S. and abroad. It is available for $40 an issue, plus $30 for two supplements.

• *File SOFT*—this on-line version of the *Online Micro-Software Guide and Directory* is a directory to over 3,200 software programs of interest to the information professional. Updated monthly, it is available on BRS for $40 per connect-hour.

Special Libraries Association (SLA)
235 Park Avenue South   212-477-9250
New York, NY 10003

SLA is one of the largest library information organizations in the world. The Association holds annual conferences that feature sessions of general professional interest, programs devoted to special subjects, discussions of new technology and equipment, division programs and business meetings, exhibits, and tours to outstanding special libraries and information centers. Some of its services include:

Continuing Education Courses

These focus on a wide range of current issues and new technological developments. The Association sponsors an education program at its annual conference and regional education programs throughout the year. It also cooperates in joint education programs with the Medical Library Association and the American Society for Information Science. A five-year continuing education plan offers members at all stages of their careers special direction in their continuing education.

Consultation Service

Chapter consultation officers are provided to assist business and other organizations with their special information needs. Members consult confidentially on a one-day basis as a professional courtesy, working with management and information staff to define problems and find realistic solutions.

Information Services

This is part of a growing clearinghouse for information on special libraries, offering telephone reference assistance and referrals to other sources; a circulating collection of books on library and information science; recent periodicals in the field; and materials from the Association's Management Documents Collection, including job descriptions, corporate organization charts, user survey questionnaires, collection development policies, user guides, promotional pieces, floor plans, budgets, and staff manuals. Publications include:

• *Managing the Electronic Library: Library Management*—a collection of essays that information professionals can find practical to tailor the new technologies to their services. In addition to articles about management, there are essays on marketing and consulting and other topics of interest to library managers who want to reduce the costs and enhance the value of information in their organizations. The 120-page book, published in 1983, costs $15.50.

• *Readings in Technology*—describes trends and analyzes the impact of technology on society. Nancy Viggiano tells how to manage technology and discusses the changes it is bringing. Other noted specialists discuss such topics as success factors, telelibraries, microcomputers, costs and benefits, data base development and user evaluations, nonbibliographic data bases, and utilities. The 224-page book, published in 1984, costs $16.50.

• *Special Libraries*—the official quarterly journal of the Association, which includes scholarly reports of research in librarianship, documentation, education, information science and technology, and so on. The subscription for this quarterly, along with the Association's monthly newsletter, The Specialist, *is $36.*

• *The Specialist*—a six-to-eight page monthly newsletter covering upcoming events, reports on Chapter and Division activities, news of current developments in library automation and technology, and other topical items of professional and Association interest. The annual subscription price for both *The Specialist* and *Special Libraries* is $36.

● *Bulletin*—published by SLA Chapters and Divisions. Also serials, directories, and monographs of professional interest are available.

Publications

The Electronic Library

Learned Information   609-654-6266
143 Old Marlton Pike
Medford, NJ 08055

Published quarterly, this journal is designed especially for librarians and library managers. It delivers information in areas such as the use and impact of microcomputers, minicomputers, software, library automation systems, and networks. It is available for $59 a year.

Hot Off the Computer

Westchester Library System   914-592-8214
8 Westchester Plaza
Cross Westchester Executive Park
Elmsford, NY 10523

This nontechnical newsletter covers library applications of microcomputers. It is published monthly, except in July and August, and is directed primarily toward school and public librarians. The 20-page publication is available for $20 a year.

Library HiTech

Pierian Press   313-434-5530
PO Box 1808
Ann Arbor, MI 48106

This quarterly journal is devoted to the application of high technology in libraries and information centers. Articles include information on robotics, microcomputers, electronic mail systems, future directions in the automation of acquisitions and collection development, videodisc technology, and xerography and preservation. The approximately 120-page publication is available for $19.50 a year.

Meckler Publishing Co.

520 Riverside Avenue   203-226-6967
Westport, CT 06880

Meckler publishes the following items dealing with high-technology applications for libraries:

● *Library Software Review*—a bimonthly publication designed to provide the information necessary to make software evaluation decisions. The approximately 100-page publication is available for $58 a year.

● *M300 and PC Report*—contains anything that relates to using IBM personal computers in libraries, including software and peripherals. It has articles on applications, and software reviews and reports. It is available for $29 a year to institutions, and for $20 a year to individuals.

Microcomputers for Information Management

Ablex Publishing Corporation   201-767-8450
355 Chestnut Street
Norwood, NJ 07648

This quarterly journal is devoted exclusively to the innovative application of microcomputers for information management and processing in all types of libraries and information centers. It highlights the major developments in applications of microcomputers in information processing, organization and dissemination, and decision making. Microcomputer hardware and software relevant to information management are also included. The annual subscription price is $59.50 for institutions and $28.50 for the public.

Online Database Search Services Directory

Gale Research Company   800-521-0707
Book Tower   313-961-2241—in Michigan, Alaska, Hawaii, and Canada
Detroit, MI 48226

The 1,400-page *Directory* provides detailed descriptions of the on-line information retrieval services offered to outside users by public, academic, and special libraries, private information firms, and other organizations located in the U.S. and Canada. The work serves as a link between users in need of on-line information and the libraries or organizations that are able to search the particular data bases that meet those needs.

Falling within the scope of *Online Database Search Services Directory* are libraries, organizations, and private businesses that access one or more publicly available on-line systems; conduct information searches in response to patron requests; and make their search services available outside their own organization, either to the general public or to specific segments such as medical professionals or the local academic community.

Entries furnish up to 17 points of information about each organization, including full name, address, and telephone number; year service was established; key contact person; number of staff members conducting searches; on-line systems accessed; subject areas searched; most frequently searched data bases; associated services; service availability; fee policy; and names of search personnel. Individuals named in the directory can be good sources of information. The cost of the directory is $100.

Data Bases

Micro Use

Graduate School of Library and Information Science   617-738-2224—Dr. Ching-chih Chen
Simmons College
300 The Fenway
Boston, MA 02115

The data base provides current information on microcomputer applications in libraries and information centers. It includes information on hardware and software. Contact Dr. Chen for additional information and cost.

Additional Experts and Resources

Murray Turoff, Ph.D.
Department of Computer and Information Science   201-596-3399
New Jersey Institute of Technology
323 Martin Luther King Boulevard
Newark, NJ 07102

> Dr. Turoff is a professor of computer and information science. His areas of expertise include the design of information systems, networking systems, and technological forecasting and assessment. He is co-author of the award-winning book *The Network Nation*.

Martha Williams
Coordinated Science Laboratory   217-333-1074
1101 West Springfield
University of Illinois
Urbana, IL 61801

> Martha Williams is an expert in information science. Her specialties are on-line systems and data bases.

Licensing of Technology

See: Technology Transfer

Low-Power Television

Organizations and Government Agencies

Low Power TV Branch
U.S. Federal Communications Commission (FCC)   202-632-3894
1919 M Street, NW
Washington, DC 20554

> The Low-Power Branch of FCC is responsible for following all aspects of LPTV, including authority to construct licenses, make interference policy, and grant licenses and renewals. This office will answer questions from the public and make referrals to other sources. The Low-Power Branch chief, Barbara Kreisman, is an expert on the legal aspects of LPTV. Paul Marrangoni and Keita Larson, also with

this office, are experts on the engineering aspects of LPTV. The Low-Power Branch maintains several data bases listing information on all applications and licenses for LPTV. The data bases are available on microfiche or tape from International Transcription Service, Inc., 40006 University Drive, Fairfax, VA 22030, 703-352-2400.

International Transcription Service (ITS)
40006 University Drive   703-352-2400
Fairfax, VA 22030

ITS, the Federal Communication Commission's contractor for duplicating FCC files and disseminating information, provides the following services relating to LPTV:

● *Low-Power Television Information Guide*—discusses opportunities in LPTV and how to take advantage of them. The package includes information about how to prepare an FCC application and apply for a license, how much it will cost, a listing of information sources on FCC, a listing of competent law and engineering firms, a listing of manufacturers that supply appropriate equipment, and the variety of steps to be taken to define a proposed station. It costs $100.

● *Subscription Service FCC Daily Digest with an LPTV Profile*—a daily tracking service covering FCC material released within the past 24 hours. The Service includes FCC's formal releases, which provide information on changes in rules and regulations, administrative law changes, FCC orders to the public, and Notice of Inquiries on proposed regulations. This Service helps subscribers keep track of competition and learn of new LPTV openings throughout the country. The cost is $25 a month, plus 5 cents a page for attached FCC materials.

● *LPTV Times*—this weekly newsletter provides market insight, analysis, and editorial commentary about LPTV. It informs readers about new equipment and how to interpret FCC policy and direction. The newsletter includes information about awards made, rules and regulation changes, and FCC's opening and closing opportunities. The annual subscription is $120 a year.

● *Copying FCC's LPTV Data Bases*—for a fee, ITS will copy FCC data bases and other materials for the public. The cost is $12 an hour for research and information retrieval, plus copying charges of 4.9 cents a page for paper copy; $5 for a fiche, for fiche-to-fiche copy; and 25 cents a page for fiche to paper.

National Institute for Low-Power Television
454 Broome Street   212-925-7751
New York, NY 10013

This membership organization is devoted exclusively to low-power television. It holds two expositions each year, which include seminars on many topics and an exhibition floor with numerous booths representing equipment manufacturers and programmers. It publishes *LPTV Currents*, which gives a bimonthly coverage of the LPTV industry in the U.S.

National Translator/LPTV Association
c/o Michael Couzens   415-621-5999
385 8th Street, 2nd Floor
San Francisco, CA 94103

A trade association, the group can provide information about translators, low-power television, and telecommunications issues. It promotes research and development in the use and regulation of translators and LPTV, and provides legal services and assistance.

Publications

Broadcast Engineering
Intertech Publishing Corp.   913-888-4664
PO Box 12901
Overland Park, KS 66212

The magazine covers the equipment, engineering, and technicalities of LPTV, as well as most larger LPTV conventions. Each March issue includes a review of the LPTV industry. This monthly publication is free to personnel at radio and television stations, and broadcast industry manufacturers. The subscription price is $25 a year.

Broadcast Week
PO Box 5727-TA   303-295-0900
Denver, CO 80217

This weekly publication covers the LPTV industry from time to time, including news, conventions, technical programming, and financial aspects. The annual subscription is $36.

LPTV Magazine
Globecom Publishing Limited   913-642-6611
PO Box 12268
Overland Park, KS 66212

Published bimonthly, the journal is for those interested in community television broadcast and the low-power TV market. It offers timely, practical information on equipment, programming, legal considerations, FCC rulings, station design, sales, and so on. The special fall issue, *The Annual LPTV Directory of Programming, Equipment and Services*, is a complete up-to-date source of equipment manufacturers, dealers, and distributors selling to the low-power television industry.

New Electronic Media
Broadcast Marketing Company   415-957-1400
450 Mission Street
San Francisco, CA 94105

A major reference guide on different aspects of the new electronic media this publication deals with pay cable, subscription television, multipoint distribution service, satellite master antenna television systems, low-power television, satellite subscription television, and so on. It covers all aspects of the industry, including technology and developments in the field. It is available for $295.

New TV Reporter—Low-Power Television and New Broadcast Technology
PO Box 421956   415-626-LPTV
San Francisco, CA 94142
This monthly publication deals with aspects of new broadcast technologies: multipoint distribution service, subscription TV, instructional television fixed services, and so on. The annual subscription rate is $90.

VideoPro Magazine
VidPro Publishing, Inc.   212-734-4440
350 East 18th Street
New York, NY 10028
Published monthly, the magazine is free to qualified professionals in the video industry. Each month a column on LPTV is included. The magazine is directed toward people who operate stations. It covers the technical and financial aspects of LPTV and the state of the industry. Write to Jim Kaminsky at the above address for a sample issue and a subscription card.

Additional Experts and Resources

Professor Robert G. Allen
American University
School of Communications
Nebraska and Massachusetts Avenues, NW
Washington, DC 20016
Professor Allen, a communications attorney, is an excellent source of information on low-power television. Written requests for information about different aspects of low-power television may be sent to him at the above address.

Doug Maupin
618 Ninth Street   405-256-2254
PO Box 881
Woodward, OK 73801
Mr. Maupin, who is very involved in low-power television work in Oklahoma, is a very helpful and knowledgeable person in the field. He can answer questions and refer you to other reliable sources of information in the area of low-power television.

Magnetic Ink Character Recognition

See also: Recognition Technology

Organizations and Government Agencies

American National Standards Institute (ANSI)
1430 Broadway   212-354-3300
New York, NY 10018

The American National Standards Institute is a private, nonprofit federation of standards-developing organizations and standards users. Its major functions include coordinating the voluntary development and approval of American standards. ANSI can refer people to technical experts outside the organization who either write the standards or are otherwise knowledgeable in the field. The following MICR standards can be obtained from ANSI:

• *Print Specifications for Magnetic Ink Character Recognition, ANSI X3.2-1970*—specifies the shape, dimensions, and tolerances for the ten digits and four special symbols printed in magnetic ink and used for the purpose of character recognition. These characters were developed initially for use in banks to permit automatic document handling for bank data processing, but they have application in other automatic handling systems as well. Available for $7.

• *Bank Check Specifications for Magnetic Ink Character Recognition, ANSI X3.3–1970*—complements ANSI X3.2-1970, and covers those design considerations that apply primarily to the placement and location of magnetic ink printing on checks intended for use in bank automation. Available for $8.

X9 Financial Services Committee
American Bankers Association (ABA)
1120 Connecticut Avenue, NW
Washington, DC 20036

ABA is secretariat for the X9 Committee, a standards-developing committee operating under ANSI procedures. The following MICR standard has been developed by the Committee: *X9-13 Specifications for Placement and Location of MICR Printing, 1983*. It is available from the X9 Committee at ABA for $15.

Magnetic Resonance Imaging

Organizations and Government Agencies

American College of Radiology

Commission on NMR   301-654-6900—Bob Bradden

6900 Wisconsin Avenue

Chevy Chase, MD 20815

The Commission on NMR (nuclear magnetic resonance) is responsible for keeping up to date on all developments relating to magnetic resonance imaging (MRI). It has published several reports on MRI, including a listing of all companies in the U.S. and Canada working in the area. The Commission has compiled a *Glossary of NMR Terms*, available for $4.

Radiological Society of America (RSA)

Oak Brook Regency Towers   312-565-1200—Lou Joseph, Public Relations Officer

1415 West 22nd Street, Suite 1150   312-920-2670—Main Office

Oak Brook, IL 60521

RSA is involved in following MRI developments, and the field is generally covered at the Society's annual meeting. The papers presented are published in RSA's journals. Lou Joseph, RSA's public relations official, will answer questions you may have about MRI and the Society. He can also refer you to other sources of information.

Society of Magnetic Resonance in Medicine

15 Shattack Square, Suite 204   415-841-1871

Berkeley, CA 94704

This Society is devoted to the use of nuclear magnetic resonance in medicine. It holds an annual scientific meeting and publishes the proceedings in its publication entitled *Scientific Program*. A newsletter, published irregularly, advises members on the Society's activities and changes. The organization also publishes:

• *Magnetic Resonance in Medicine*—a quarterly scientific journal covering many different aspects of MRI. The journal, which costs $72 a year, is available from Academic Press, Journal Subscription Department, 11 Fifth Avenue, New York, NY 10003, 212-614-3000.

Publications

Journal of Magnetic Resonance

Academic Press   212-614-3000
Journal Subscription Department
111 Fifth Avenue
New York, NY 10003

The *Journal* is a source of up-to-date authoritative and technical information on the theory, technique, and methods of special analysis and interpretation of spectral correlations and results of magnetic resonance. Published 15 times a year, the *Journal* is available for $490 a year.

Medical Resonance Imaging

Pergamon Press   914-529-7700
Maxwell House
Fairview Park
Elmsford, NY 10523

This quarterly journal is multidisciplinary. It encompases physical, life, and clinical sciences investigation relating to the development and use of magnetic resonance imaging. The annual subscription rate is $120.

Mack Publishing Company

RSNA Dues and Subscription Office
20th and North Hampton
Easton, PA 18042

This company publishes the following relating to MRI:

• *Radiology*— a monthly medical journal covering clinical radiology and allied sciences. Each month a section is devoted to magnetic resonance imaging, with several articles about different aspects of MRI. The annual subscription cost is $75 a year.

• *Radiographics*— is a pictorial bimonthly review journal which occasionally devotes an entire issue to magnetic resonance. The annual subscription rate is $35 a year.

Seminars in Ultrasound, CT and MR

Mary Lane
Subscription Department
Grune and Stratton
111 Fifth Avenue
New York, NY 10003

The quarterly journal is directed to all physicists involved in the performance and interpretation of ultrasound, computer tomography, and magnetic resonance imaging. It is a timely source about publications, new concepts, and research findings directly applicable to day-to-day clinical practice. Each issue covers one particular area of importance. Articles generally cover the performance of various procedures. The annual subscription rate is $39.

Additional Experts and Resources

General Electric Company
PO Box 414   414-521-6513
Milwaukee, WI 53201
         GE publishes several brochures that describe, in lay terms, exactly what nuclear magnetic resonance imaging is and how it is used in different fields. The brochures are available free of charge.

Several companies produce equipment for use in the field of magnetic resonance imaging. Many can provide brochures describing the product and its technology, and often the public affairs personnel can answer questions. Examples of companies are:

Diasonics
1656 McCarthy Boulevard
Milpitas, CA 95035
408-946-9001

Picker International
585 Miner Road
Highland Heights, OH 44143
216-473-7700

Johnson & Johnson's Technicare
PO Box 5130
Cleveland, OH 44101
216-248-1800

General Electric Medical Systems Facility
Electric Avenue
Milwaukee, WI 53219
414-647-4000—Switchboard Recording
414-647-4226—General Information

Many university medical centers across the country conduct research in nuclear magnetic resonance. Following is a partial list of some of the major schools:

University Hospitals of Cleveland
Department of Radiology
Cleveland, OH 44106
216-844-1000—Dr. Ralph Alfidi,
   Chairman

New York Hospital
Nuclear Magnetic Resonance
525 East 68th Street
New York, NY 10021
212-472-4594

Hershey Medical Center
Department of Radiology
PO Box 850
Hershey, PA
717-534-8044—Dr. William Weidner

University of Pennsylvania
Pendergrass Lab
3 Medical Education Building
36th and Hamilton Walk
Philadelphia, PA 19104
215-662-6630—Dr. Harold Kundel

Massachusetts General Hospital
Radiology Department/NMR
32 Fruit Street
Boston, MA 02114
617-726-8313

University Hospital X64
Attn: Magnetic Resonance Imaging
926 West Michigan Street
Indianapolis, IN 46223
317-264-2566

University of Chicago
Department of Radiology
5841 South Maryland
Chicago, IL 60637
312-962-6514

University of Iowa
Department of Radiology/MR
Iowa City, IA 52242
319-356-2236—Dr. James Ehrhardt

Duke University Medical Center
Box 3808
Durham, NC 27710
919-684-3403—Dr. Charles Putnam

Department of Radiology
University Hospital
Box 13
Ann Arbor, MI 48109
313-763-0972—Dr. Alex M. Aisen, Clinical
Coordinator, MRI

University of Wisconsin
Department of Radiology
600 Highland Avenue
Madison, WI 53792
608-263-8300

University of California—San Francisco
Department of Radiology
Radiologic Imaging Laboratory
San Francisco, CA 95080
415-952-1366

Mainframe Computers

See also: Computers (General)

Organizations and Government Agencies

Amdahl Corporation
1250 East Arques Avenue   408-746-8918
Sunnyvale, CA 94086

Amdahl is a manufacturer of both dual processor and uniprocessor mainframe computers. The company also produces communication products, and it is a systems supplier to the computer industry. All types of printed material are available to the public, including a company brochure, technical briefs, and product information.

Burroughs Corporation

1 Burroughs Place   313-972-7000
Industry System Headquarters
Detroit, MI 48232

A leading manufacturer of mainframe computers, Burroughs also produces over 100 computer products, including minicomputers and microcomputers. Printed materials about the company's activities and products are available, including company brochures, technical briefs and specifications, and product information.

Control Data Corporation

8100 34th Avenue, South   612-853-8100
Minneapolis, MN 55440

Control Data is a manufacturer of mainframes and super-computers designed especially for use by the scientific and engineering communities. The marketing department will send you company brochures, technical briefs, and product information. Control Data also offers complete engineering and consulting services.

Honeywell, Inc.

Honeywell Plaza at 27th Street   612-870-5200
Minneapolis, MN 55408

A leading manufacturer of mainframe computers, Honeywell is also involved in the manufacture of peripheral equipment and software for its computer lines. Product information and specifications are available, and the following is published:

• *Honeywell Information Systems News*—a monthly newsletter that provides profiles of new products and product specifications. The eight-page newsletter is available free of charge.

• *Honeywell Source*—a monthly publication that provides interviews with industry experts, and product profiles. It also includes a software directory describing applications for Honeywell products. In addition to mainframes, it covers minicomputers and microcomputers. The 50-page publication is available free of charge.

IBM

1130 Westchester Avenue   914-765-1900
White Plains, NY 10604   800-492-5578

IBM is the leading manufacturer of mainframe computers. A variety of printed materials about IBM's work is available, including a company brochure, technical briefs and specifications, and product information. You can call IBM's 800 phone number for information about specific products, or contact an IBM office in your area.

National Advanced Systems (NAS)

800 East Middlefield Road   800-221-6876
Mountainview, CA 94042   415-962-6000

NAS markets Hitachi mainframe computers and peripherals. The staff can give you information about its products.

NCR Corporation

1700 South Patterson Boulevard   513-445-5000
Dayton, OH 45479

A major manufacturer of mainframe computers, and of a full line of data-processing systems, NCR also manufactures specialized terminals for retail and financial applications. Sales and product brochures are available, as well as the following:

• *Alpa*—a quarterly publication covering mainframes, personal computers, retail computers, and specialized terminals. The 40-page publication is available free of charge to customers of NCR.

Publications

Standard EDP Reports

Auerbach Publishers   609-662-2070
6560 North Park Drive   800-257-8162
Pennsauken, NJ 08109

The reports are an eight-volume series, with four volumes dealing with mainframe systems and peripherals, three volumes dealing with minicomputer systems, and one volume dealing with system software. This publication, updated monthly, includes a telephone update service. It is available for $2,995 a year.

Computer World

Computer World Communications   617-879-0700
PO Box 880
Framingham, MA 01701

This weekly newspaper covers the technical computer industry, focusing on management information systems and mainframes. It supplies information on conferences, new products, software, communications, hardware, peripherals, computer architecture, and computer languages. A subscription to the 200-page newspaper is $44 a year. A supplement entitled *Computer World Extra* is published twice a year. The publication focuses on such topics as hardware, data communications, software, and peripherals.

Datamation

Technical Publications   212-605-9400
875 Third Avenue
New York, NY 10022

This semimonthly publication is a general-interest magazine for the data-processing industry. It focuses on mainframes and covers communications, hardware, software, industry storage, trend analysis, and budget and financial surveys. The publication, 150-200 pages long, is available free of charge to qualified professionals.

Man-Machine Systems

See: Cybernetics; Ergonomics

Manufacturing, Computer-Aided

See: CAD/CAM

Marine Technology

Organizations and Government Agencies

American Society of Naval Engineers

1452 Duke Street   703-836-6727
Alexandria, VA 22314

This technical society of marine engineers, naval architects, engineers, and U.S. Navy Department officials promotes the exchange of ideas and information in the area of naval engineering. It sponsors local section meetings, technical symposia and exhibits, a national convention, technical sessions conducted by leaders in the naval engineering field, and a national meeting devoted to a variety of subjects. It also sponsors a scholarship program for students in naval engineering. The staff can answer technical inquiries and give referrals. The Society publishes:

- *Naval Engineers Journal*—a bimonthly publication that includes articles about technology used in naval engineering, proceedings of conferences, and book reviews. It is available free of charge to members and costs $36 a year for nonmembers.

Marine Research Board (MRB)

National Academy of Sciences   202-334-3119
2101 Constitution Avenue, NW
Washington, DC 20418

MRB is a private scientific society that advises the government on matters related to science and technology. Its role is to examine developments in maritime transportation and ocean and coastal engineering. Specialists at the Board can answer technical inquiries and give referrals. They participate in study groups, meetings, and national conferences sponsored by both the Board and the Academy. The annual report, available free of charge, lists ongoing Board activities and meetings.

Marine Technology Society

2000 Florida Avenue, NW, Suite 500   202-462-7557
Washington, DC 20009

This nonprofit organization is dedicated to the dissemination of information about marine science and engineering. It is divided into 14 geographical sections in the U.S. and Canada, and each section holds monthly meetings, at which brief technical presentations are given. The Society has 31 professional committees that sponsor technical conferences, workshops, and short courses that are open to members and nonmembers. The staff can provide technical information and give referrals. The Society publishes:

• *Journal of the Marine Technology Society*—a quarterly publication that presents technical activities of the Society, papers, conferences summaries, book reviews, and so on. It is available free of charge to members and costs $5 an issue for nonmembers.

National Maritime Research Center

U.S. Maritime Administration   516-482-8200
Department of Transportation
Kings Point, NY 11024

A field office of the U.S. Maritime Administration, this Center conducts research and testing on the following: ship control areas, navigation aids, certification and training, instrumentation, ship maneuvering and instrumentation, general management of ships, satellite communications with ships, and preliminary investigation and application of satellite communications in maritime transportation. Its Study Center plans to have an on-line data base set up and in use shortly.

If you need further information about any of the above subjects, contact Dr. Walter MacLean, Director of the Center, or the Senior Information Specialist in an appropriate division. They can answer technical inquiries and give referrals. The Center issues technical reports on the topics indicated above. These publications can be obtained from the Center or from:

Defense Technical Information Center

U.S. Department of Defense
Cameron Station
B Street, Building 5
Alexandria, VA 22314
202-274-6871

National Technical Information Service

U.S. Department of Commerce
5885 Port Royal Road
Springfield, VA 22161
703-487-4807

Ocean Engineering Department

Massachusetts Institute of Technology   617-253-1994
77 Massachusetts Avenue, Building 5225
Cambridge, MA 02139

The Department has several specialists who conduct research work in such areas as ship design and construction, energy, marine transportation safety and port engineering, hydrodynamics, ocean acoustics, seismic exploration, and related subjects. Those seeking technical information on ocean technology can contact Ms. Patti Le Blanc-Gedney. Several technical publications are available from the Department.

Oceanic Engineering Society

Institute of Electrical and Electronics Engineers (IEEE)   201-981-0060—General Information
445 Hoes Lane   201-981-1393—Publication Information
Piscataway, NJ 08854

Members of this IEEE special-interest group are involved in electronics and electrical engineering as applied to the ocean environment. Each year the Society sponsors a series of conferences, at different locations in the U.S., on high-technology applications in ocean engineering. The conferences are open to the public. The Society publishes:

● *Journal of Oceanic Engineering*—a quarterly reference journal providing in-depth technical articles about electrical and electronics engineering applied to the oceanic environment. It presents experimental papers, proceedings of conferences, book reviews, and so on. Each year a supplementary issue is published on high technology, covering such topics as oceanic acoustic remote sensing, beam forming, instrumentation development, and high-level nuclear disposal beneath the ocean floor. The cost of the five issues is $5 a year for members, and $55 a year for nonmembers.

Office of Naval Research

Environmental Sciences Directorate   202-696-4358—Chief
Technology and Support Division, Code 421   202-696-4532—Ocean Sciences Division
800 North Quincy Street   202-696-4120—Geophysical Division
Arlington, VA 22217

The Office of Naval Research supports basic research and technology, including research projects at universities around the U.S. Research reports and publications are available from the Chief's Office. The staff in both the ocean sciences and geophysical divisions can provide you with in-depth technical information.

Society of Naval Architects and Marine Engineers

One World Trade Center, Suite 1369   212-432-0310
New York, NY 10048

This Society is dedicated to the advancement and practice of naval architecture, marine engineering, ship building, naval research and development, and related subjects. It has technical committees in such areas as ship production, marine systems, and hydrodynamics. The Society sponsors an annual technical convention and exposition every November in New York City. A free listing of its technical reports and research studies is available. The Society's chief publications are:

- *Journal of Ship Research*—a quarterly publication with technical articles and papers about naval architecture and marine engineering. The annual subscription is $17 for members, and $30 for nonmembers.

- *Marine Technology*—a quarterly publication presenting technical articles on the technology applied to marine engineering. This publication is not as technical as the above-mentioned *Journal of Ship Research*. The annual subscription is $30; the publication is free to members.

U.S. Coast Guard Research and Development Center

Department of Transportation   203-445-8501—Mr. D'Angelo
Avery Point
Groton, CT 06340

The Center's research and development activities focus on ice technology, marine navigation technology, marine fire and safety technology, marine pollution technology, search and rescue technology, and naval engineering and marine systems technology.

Mr. D'Angelo can supply you with technical information and refer you to experts on the above subjects. The Center issues several technical reports and studies that can be obtained from NTIS, U.S. Department of Commerce, 5885 Port Royal Road, Springfield, VA 22161, 703-487-4807.

Publications

Simmons/Boardman

345 Hudson Street   212-620-7220—Editorial Department
New York, NY 10014

- *Marine Engineering/Log*—a monthly magazine of business and technical articles geared toward vessel building companies, ship operators, and designers. Individuals in the marine sector qualify for free subscriptions. Others can subscribe for $30 a year.

- *Marine Engineering/Log Directory*—this annual publication lists shipyards and ship-operating companies worldwide. The cost is $97.50.

- *Marine Engineering/Log Catalog and Buyers Guide*—an annual listing of manufacturers in the shipping sector. It is available free of charge to subscribers of the *Marine Engineering/Log*.

- *Outlook Letter*—a bimonthly that focuses on sales leads for companies in the shipping sector, including transactions with the federal government. The cost is $150 a year.

Maritime Activity Reports
107 East 31st Street   212-689-3266
New York, NY 10016
     ● *Maritime Reporter and Engineering News*—a bimonthly magazine that covers the latest and most important technical information in the marine engineering field. Major areas of interest include shipbuilding, naval architecture, marine engineering, ship operators, and related areas. Articles cover ships, workboats, naval vessels, and so on. It is available free of charge to those in the maritime industry.
     ● *Maritime Equipment Catalog*—is an annual listing of marine and naval equipment. The cost is $65.

Ocean Science News
Nautilus Press   202-347-6643—John R. Botzman, Editor
1201 National Press Building
Washington, DC 20045
     Published weekly, the newsletter provides worldwide coverage of developments in all areas of ocean science and technology. It presents discoveries in such subjects as ocean development, protection, mining, and fisheries. Its readers include federal, state, and foreign government agencies; private companies; and academicians. The annual subscription cost is $265.

Additional Experts and Resources

Dr. Stanley G. Chamberlain
Raytheon Company   401-847-8000
1847 West Main Road
Box 360
Portsmouth, RI 02870
     Dr. Chamberlain can provide information about ocean engineering issues and technology. He is president of the Oceanic Engineering Society, described earlier in this section.

Hydronautics, Inc.
7210 Pindell School Road   301-776-7454
Laurel, MD 20707
     This private company involved in research and development has a library that contains 5,000-10,000 technical reports on subjects such as off-shore drilling, ship design, fluid design, and hydrodynamics. The library also holds approximately 200 periodicals on marine technology. While the holdings are for in-house use by appointment only, the library will loan materials to libraries and engineers.

University of Texas at Austin
Continuing Engineering Studies   512-471-3506
Cockrell Hall 2102
Austin, TX 78712

    The University offers continuing education courses in subjects such as design of fixed offshore platforms, construction project risk management, and related areas.

Materials Science

See also: Ceramics; Metals; Plastics

Organizations and Government Agencies

Battelle Memorial Institute
505 King Avenue   614-424-5000—General Information
Columbus, OH 43201   614-424-6376—Technical Inquiry

    Battelle is an independent, international organization that performs scientific work for a broad range of research, development, and demonstration programs. In addition to inventing and developing technology, the Institute is involved in educational programs.

    The majority of Battelle's activities are carried out through government- and industry-sponsored research. The Institute's main areas of interest in materials include metals, ceramics, coatings, mechanical and physical properties, materials applications, test methods, design characteristics, processes used in basic materials production, and quality control and inspection.

    The staff will answer technical inquiries free of charge if the questions do not involve extensive work. Battelle maintains an on-line data base with substantial technical data and literature on materials. Users of the Department of Defense DTIC system can directly access Battelle's data base. If you do not have access to DTIC, Battelle's staff will search the data base for you. A fee is charged for this service.

Materials Processing Center
Massachusetts Institute of Technology   617-253-3217
77 Massachusetts Avenue
Cambridge, MA 02139

    This research center is an interdisciplinary laboratory within MIT's school of engineering that carries out basic research in the processing of polymers, semiconductors, ceramics, and metals. The staff can refer inquiries to appropriate faculty members who are associated with the Center.

National Materials Research Laboratory Program (NMRL)
Division of Materials Research   202-357-9791
National Science Foundation (NSF)
1800 G Street, NW
Washington, DC 20550

Under the NMRL program, the NSF funds 14 interdisciplinary materials research laboratories at universities across the U.S. Staffed by physicists, chemists,
metallurgists, ceramists, polymer scientists, engineers, and materials scientists, the
laboratories are an excellent resource for information about all aspects of materials
science. In addition to answering technical inquiries, the scientist often can supply
you with technical papers and refer you to other resources. Listed below is contact
information for each of the laboratories.

Dr. Jan Taud, Director
Materials Research Laboratory
Brown University, Box M
Providence, RI 02912
401-863-2318

Dr. William W. Mullins, Director
The Center for the Joining of Materials
Carnegie-Mellon University
Schenley Park
Pittsburgh, PA 15213
412-578-2541

Dr. Alfred R. Cooper, Director
Materials Research Laboratory
Case Western Reserve University
Cleveland, OH 44106
216-368-4225
216-368-4120

Dr. Ole J. Kleppa, Director
Materials Research Laboratory
James Franck Institute
University of Chicago
5640 South Ellis Avenue
Chicago, IL 60637
312-962-7918

Dr. Robert H. Silsbee, Director
Materials Science Center
Clark Hall
Cornell University
Ithaca, NY 14850
607-256-4272
607-256-2323

Dr. Herb Ehrenreich, Director
Materials Research Laboratory
Division of Applied Sciences
205A Pierce Hall
Harvard University
Cambridge, MA 02138
617-495-3213

Dr. Myron B. Salamon, Director
Materials Research Laboratory
University of Illinois
104 South Goodwin
Urbana, IL 61801
217-333-6186

Dr. J. David Litster, Director
Center for Materials Science and Engineering
Room 13-2090
Massachusetts Institute of Technology
Cambridge, MA 02139
617-253-6801

Dr. Roger S. Porter, Co-Director
Dr. Frank E. Krasz, Co-Director
Materials Research Laboratory
University of Massachusetts
Amherst, MA 02001
413-545-0433

Dr. Stephen H. Carr, Director
Materials Research Center
The Technological Institute
Northwestern University
Evanston, IL 60201
312-492-3606

Dr. James C. Garland, Director
Materials Research Laboratory
Ohio State University
174 West 18th Avenue
Columbus, OH 43210
614-422-5109

Dr. David White, Director
Laboratory for Research on the Structure
of Matter/K1
3231 Walnut Street
University of Pennsylvania
Philadelphia, PA 19104
215-898-8571

Dr. Jacek K. Furdyna, Chairman
Materials Science Council
Physics Building
Purdue University
West Lafayette, IN 47907
317-494-5567

Dr. Theodore H. Geballe, Director
Center for Materials Research
Stanford University
Stanford, CA 94305
415-497-4118
415-497-0215

Materials Science Center
Cornell University   607-256-4272 or 2323—Dr. Robert H. Silsbee, Director
627 Clark Hall
Ithaca, NY 14850

The Center's materials science research activities include surface science, mechanical properties of materials, optical phenomena, phase transitions, amorphous materials, crystal growth, liquid helium, infrared studies, thin film studies, and Rutherford backscattering spectroscopy.

The staff can answer technical inquiries, provide information about research and development in progress, distribute publications, and make referrals to other sources of information. All these services are available free of charge to the public. The Center's publications include technical reports and the following periodicals:

● *Annual Technical Report*—a summary of the technical work performed during the fiscal year. It is available free of charge.

● *Quarterly Report*—abstracts of technical reports submitted to journals each quarter. Available free of charge.

Materials Research Laboratory
University of Illinois   217-333-6186—Dr. Myron B. Salamon, Director
104 South Goodwin
Urbana, IL 61801

The Laboratory's material science activities include solid state physics, solid state chemistry, metallurgy, and ceramics. The staff can answer technical inquiries and make referrals to other sources of information on materials science research. The Laboratory publishes an annual report describing the technical work performed by its major scientists.

Center for Materials Science and Engineering

Massachusetts Institute of Technology   617-253-6801—Dr. J. David Litster, Director
77 Massachusetts Avenue, Room 13-2090
Cambridge, MA 02139

This Center conducts interdisciplinary materials research under the sponsorship of the National Science Foundation's Materials Research Laboratory Section. The Center's current research focuses on high-temperature alloys, semiconductors, phase transition, and polymers. The staff can answer questions and/or make referrals.

Materials Research Center

Northwestern University   312-492-3606—Dr. Stephen H. Carr, Director
The Technological Institute
2145 Sheridan Road
Evanston, IL 60201

Scientific activities on materials science research at the Center are related to solid state physics, solid state chemistry, metallurgy, high polymers, metals and alloys, chemical, physical, and mechanical properties of materials, and so on. The Center's specialists can answer technical inquiries and provide referrals to other information sources. Research reports on individual projects are available free of charge from the Center.

Materials Research Society (MRS)

9800 McKnight Road, Suite 327   412-367-3003
Pittsburgh, PA 15237

MRS is an organization of technical professionals from a wide variety of scientific and engineering disciplines. The Society sponsors several annual symposia on topics of materials research. Through these symposia the Society promotes interaction among physicists, chemists, metallurgists, engineers, and others interested in the various aspects of materials science.

Membership in the Society is open to individuals with an interest in materials development, processing, and characterization. A free brochure listing the various technical symposia, short courses, and meetings sponsored by the Society is available from its office. The staff will provide specific technical information and referrals on a limited basis.

Publications

Scientific American

Scientific American, Inc.   212-754-0550
415 Madison Avenue
New York, NY 10017

Scientific American is a monthly magazine for engineers, scientists, and professionals involved in scientific development. It presents accounts of the work done by scientists in all fields of science. The articles presented in each issue represent the major fields of physics, engineering, life and social sciences, and medicine. The magazine has several departments, such as the amateur scientist, science and citizens, and computer recreation. The annual cost is $24.

Journal of Solid State Chemistry

Academic Press, Inc.   212-614-3147—Circulation
111 Fifth Avenue   317-494-5279—Dr. J. M. Honig, Editor-in-Chief
New York, NY 10003

This *Journal* is one of the most important publications in the broad field of solid state chemistry. Published 15 times a year, it includes articles in such diverse areas as single crystal work, solid reactions, solid solutions, and alloys. The annual subscription rate is $475.

High Tech Materials Alert

Technical Insights, Inc.   201-944-6204
158 Linwood Plaza
PO Box 1304
Fort Lee, NJ 07024

Alert is a series of monthly reports on developments in advanced materials. The reports focus on significant technical information about high-tech materials, and research developments in universities, private companies, and government laboratories. Each report is 8-10 pages and carries 20-40 developments with a variety of detail. Available for $237 a year.

Additional Experts and Resources

Dr. J. M. Honig

Purdue University   317-494-5279
Department of Chemistry
Chemistry Building
West Lafayette, IN 47907

An expert in solid state chemistry, Dr. Honig is editor-in-chief of the *Journal of Solid State Chemistry*, and a professor at Purdue University. He can provide technical advice in this aspect of materials science.

MDS

See: Multipoint Distribution Service

Measurement

See: Standards and Measurement

Medical Imaging

See also: Magnetic Resonance Imaging; Medical Technology; Ultrasound in Medicine

Organizations and Government Agencies

American Association of Physicists in Medicine (AAPM)
335 East 45th Street   212-661-9404
New York, NY 10017

The AAPM is a membership organization that deals with all aspects of physics in medicine. It publishes the following:

• *Journal of Physics in Medicine and Biology*—a monthly 1,000-page publication that deals with medical imaging and the latest results of research in that area. This is available for $21 an issue or $265 a year.

• *Medical Physics*—a bimonthly publication that deals with radiation therapy and the latest results in research pertaining to physics in medicine. Each issue is approximately 450 pages long. This journal is available for $25 an issue or $120 a year.

American College of Medical Imaging (ACMI)
PO Box 27188   213-275-1393
Los Angeles. CA 90027

ACMI is a professional membership association of radiologists involved in the use of medical imaging technology for diagnostic purposes in medicine. The staff can answer questions and make referrals in the field. ACMI publishes:

• *ACMI News*—a quarterly newsletter that covers current advances in the field of medical imaging. It also keeps members up to date on association news. The newsletter is available for $35 a year.

American College of Nuclear Physicians (ACNP)

1101 Connecticut Avenue, NW, Suite 700   202-857-1135
Washington, DC 20036

ACNP is a membership organization whose goal is to disseminate information on the latest advances in nuclear medicine and medical imaging. An annual three-day conference consisting of seminars and meetings is held on varying topics in nuclear medicine. The proceedings are available on audiotape. Also published is:

• *ACNP Scanner*—a monthly newsletter for members that provides information on advances in research, diagnosis, treatments, and legislative issues pertaining to nuclear medicine. The six-to-seven-page newsletter is available free to members.

American College of Radiology

6900 Wisconsin Avenue   301-654-6900—Bob Bradden
Chevy Chase, MD 20815

The College offers educational programs for its members dealing with the latest innovations in medical imaging and other areas of radiology. Clinical application reports are published, as well as the following information brochures: "Computed Axial Tomography (CAT) Scanning," "Ultrasound," "Radiation Oncology," and "Nuclear Medicine for You." These publications are approximately four to six pages long; single copies are available free of charge.

American Institute of Ultrasound in Medicine (AIUM)

4405 East-West Highway, Suite 504   301-656-6117
Bethesda, MD 20814

AIUM, a membership organization of 5,500, has as its purpose the advancement of art and science of ultrasound in medicine. The AIUM has committees handling each of the following areas: archives, educational programs, public relations, and publications. Each year AIUM holds a national meeting, which includes educational and scientific sessions, as well as commercial and scientific exhibits. The proceedings are published and available to the public. The Institute will send you a free listing of its publications, which include:

• *Sonic Exchange*—an annual buyer's guide to equipment in the area of medical imaging. The 30-page publication is available for $10 to nonmembers and $6 to members.

• *Proceedings of the Annual Meeting*—a publication consisting of papers submitted for the annual conference.

• *Safety Considerations for Diagnostic Ultrasound*—an annual publication that discusses the biological effects and considerations when using ultrasound. The 20-page publication is available for $9 to nonmembers and for $5 to members.

• *Medical Ultrasound Bibliography*—an annual publication that consists of a list of books published on ultrasound. The 50-page publication is available to non-members for $9 and to members for $5.

• *Evaluation of Research Reports*—an annual publication that is a collection of reviews of literature on the biological effects of ultrasound. The 36-page publication is available to nonmembers for $9 and to members for $5.

International Trade Administration (ITA)

U.S. Department of Commerce   202-377-5466
Office of Instrumentation and Medical Sciences
14th Street and Constitution Avenue, NW, Room 4424
Washington, DC 20230

ITA oversees trade and production in the U.S., and monitors developments that affect industry, such as legislative issues and trade policies. The staff can refer you to medical imaging specialists or provide you with information from the annual *Industrial Outlook*, which includes chapters on instrumentation and medical sciences, the latest technological equipment, and trade and production figures. The *Outlook* is available from the Government Printing Office, Washington, DC 20402, 202-783-3238.

National Institute for Neurological Communicative Disorders and Stroke (NINCDS)

National Institutes of Health   301-496-5751
Building 31, Room 8A 16
9000 Rockville Pike
Bethesda, MD 20205

The NINCDS Information Office is available to answer questions about various diseases and the diagnostic and treatment techniques used. The Office will send you free informative brochures on the different techniques used to treat diseases. A 20-page brochure entitled *Positron Emission Tomography* (PET) is an example of what is available.

Radiological Society of North America (RSNA)

Oak Brook Regency Towers   312-565-1200—Lou Joseph, Public Relations Officer
1415 West 22nd Street, Suite 1150   312-920-2690—Main Office
Oak Brook, IL 60521

The RSNA holds an annual meeting covering areas such as ultrasound, imaging techniques, radiation, and acoustics. Papers presented are published in either of RSNA's two journals. Lou Joseph, the organization's public relations official, will answer questions about the Society. He can also refer you to other resources. The Society publishes the following journals, which are available from Mack Printing Company, 20th and Northampton, Easton, PA 18042:

• *Radiology*—a monthly medical journal devoted to clinical radiology and allied sciences. It covers various medical imaging techniques, such as ultrasound and scanners. The 355-page scientific text is available to nonmembers for $75 and is free to members.

• *Radiographics*—a bimonthly pictorial review journal that covers the techniques used in radiographics. It provides current information on medical imaging. The publication, 60-70 pages long, is available to nonmembers for $35 and is free to members.

Society of Magnetic Resonance in Medicine

15 Shattack Square, Suite 204   415-841-1871
Berkeley, CA 94704

This Society is devoted to the use of nuclear magnetic resonance (NMR) in medicine. It holds an annual scientific meeting and publishes the proceedings in its publication entitled *Scientific Program*. A newsletter, published irregularly, advises members on the Society's activities and changes. The organization also publishes:

• *Magnetic Resonance in Medicine*—a quarterly scientific journal that covers many different aspects of MRI. The journal, which costs $72 a year, is available from Academic Press, Journal Subscription Department, 11 Fifth Avenue, New York, NY 10003, 212-614-3000.

Society of Nuclear Medicine (SNM)

475 Park Avenue South   212-889-0717
New York, NY 10016

A nonprofit scientific organization, SNM is dedicated to the dissemination of information pertaining to the use of radionuclides (nuclear medicine). Drawing from its membership of physicians, scientists, and technologists in the field, the staff can refer you to experts in specific areas of nuclear medicine. The Society has six councils covering specific areas: universities, computers, instrumentation, correlative imaging, radioassay, and radiopharmaceutical science. SNM sponsors conferences, seminars, audiovisuals, and publishes pamphlets for patients. The Society's major publications are:

• *Journal of Nuclear Medicine*—a monthly journal detailing the trends in both the clinical and basic sciences of nuclear medicine. The publication also contains a newsline section featuring the latest governmental and socioeconomic issues of concern to the field. Each issue is approximately 200 pages, and an annual subscription costs $90.

• *Journal of Nuclear Medicine Technology*—a quarterly journal geared toward technologists, institutions, and technical affiliates involved in nuclear medicine. The publication deals with the practical aspects and day-to-day operation of a nuclear medicine department. Each issue is about 150 pages, and the annual subscription cost is $45.

• *SNM Newsline*—published five times a year, this newsletter is geared toward the layperson interested in nuclear medicine. It covers governmental and socioeconomic issues of concern to the field.

Publications

American Journal of Neuroradiology

American Society of Neuroradiology   312-920-2676
1415 West 22nd Street, Suite 1150
Oak Brook, IL 60521

   This bimonthly 175-page publication covers all forms of neuroradiology, profiling the clinical observations and research in the field. The *Journal* covers the applications of medical imaging in neuroradiology, and is available for $85 a year. This publication can be ordered from Williams and Wilkins Company, 428 East Preston Street, Baltimore, MD 21202, 301-528-4255.

Journal of Computer Assisted Tomography

Raven Press
1140 Avenue of the Americas
New York, NY 10036

   This bimonthly publication provides the latest information on various applications of computer-assisted diagnostic and treatment techniques. The 194-page publication is available to institutions for $159 and to individuals for $98.

Journal of Magnetic Resonance

Academic Press   212-614-3000
Journal Subscription Department
111 Fifth Avenue
New York, NY 10003

   The *Journal* is a source of up-to-date authoritative and technical information on the theory, technique, and methods of special analysis and interpretation of spectral correlations and results of magnetic resonance. Published 15 times a year, the 150-page journal is available for $490 a year.

Mack Publishing Company

RSNA Dues and Subscription Office
20th and North Hampton
Easton, PA 18042

   This company publishes the following relating to NMR:

   • *Radiology*—a monthly medical journal covering clinical radiology and allied sciences. Each month a section is devoted to magnetic resonance imaging, with several articles about different aspects of magnetic resonance. The annual subscription cost is $75 a year.

   • *Radiographics*—this pictorial bimonthly review journal occasionally devotes an entire issue to magnetic resonance. The annual subscription rate is $35 a year.

Additional Experts and Resources

Below are the major companies involved in producing medical imaging products such as PET and CAT Scanners:

General Electric Company
Medical Systems Division
PO Box 414
Milwaukee, WI 53201
414-544-3011

Picker International, Inc.
12 Clintonville Road
Northford, CT 06472
203-484-2711

Siemens Gamma Sonics
2000 Nuclear Drive
Des Plaines, IL 60018
312-635-3100

Technicare Corporation
29100-T Aurora Road
Solon, OH 44139
216-248-1800

Phillips Medical Systems, Inc.
710 Bridgeport Avenue
Shelton, CT 06484
203-926-7674

CTI Company
215 Center Park Drive
Suite 1500
Knoxville, TN 37922
615-966-7539

Dr. Matti Al-Aish, Program Director
National Cancer Institute   301-496-9531
Diagnostic Imaging Research Program
Landow Building, Room 8C-19
Bethesda, MD 20205

Dr. Al-Aish is in charge of the Diagnostic Imaging Research Grant program and oversees the research and development in the program. He is available to provide current information and referrals in the field, and to give further information about the grants available in medical imaging.

Dr. Marvin Reivich, Director
Supervascular Research Center   215-662-2632
Room 429, Johnson Pavilion
School of Medicine
University of Pennsylvania
Philadelphia, PA 19109

Dr. Reivich directs the Supervascular Research Center, where studies are conducted with PET and NMR. Dr. Reivich is available to answer questions and make referrals in the field of medical imaging techniques.

Medical Instruments, Devices, Procedures, and Products

See: Medical Technology

Medical Technology

Organizations and Government Agencies

American Hospital Association

840 North Lake Shore Drive

Chicago, IL 60611

312-280-6000

The primary mission of this management-oriented association is to establish policies for high-quality care in hospitals. While it is not a source of technical information, it is a good resource concerning the management of technology. The Association holds numerous conferences, primarily educational. A complete catalog of their publications and audiovisual materials is available. An example of their publications is:

- *Technology Evaluations Acquisition Methods for Hospitals*—a loose-leaf book providing hospital management with a system for planning and acquiring medical equipment or clinical services. The 212-page publication is available to members for $160 and to nonmembers for $200.

Association for the Advancement of Medical Instrumentation (AAMI)

1901 Fort Meyer Drive, Suite 602

Arlington, VA 22209

703-525-4890

AAMI's purpose is to advance the development and safe use of medical instrumentation and related technologies. It holds numerous conferences and workshops throughout the year, including two conferences in conjunction with exhibit programs. These exhibits give people the opportunity to display or view high-tech medical instrumentation. AAMI's staff members can sometimes answer questions, but they usually make referrals to experts or members. The Association has numerous publications including standards, which they develop, technical assessment reports, and periodicals. Some of these are:

- *AAMI News*—a bimonthly newsletter that reports on regulatory and other government initiatives affecting the health-care profession. It also covers Association news, which includes a section on medical device performance. The 12-page publication is available to nonmembers for $70 a year, and is free to members.

• *Medical Instrumentation*—a bimonthly technical publication that integrates medicine with engineering. It publishes scientific papers on research and development, current technology, and management of instruments and systems. The 64-page publication is available to nonmembers for $60 a year, and is free to members.

Center for Devices and Radiological Health

Food and Drug Administration (FDA)   301-443-4690
5600 Fishers Lane
Rockville, MD 20857

This Center approves new medical devices and radiation-emitting products for use in the U.S. marketplace, and it is a good information source for the safety and effectiveness of new medical devices. Publications include two periodical bulletins that are available from the Center, and some technical papers that can be purchased from the Government Printing Office. The bulletins are:

• *Medical Devices Bulletin* and *Radiological Health Bulletin*—both cover new developments in the regulation of medical and radiological products. They are designed specifically for health professionals, health agencies, and the industry. Both are available free of charge to these individuals and organizations.

The Clinical Efficacy Assessment Project

Division of Scientific Affairs   215-243-1200—Linda White
The American College of Physicians
4200 Pine Street
Philadelphia, PA 19104

The Project evaluates test procedures and therapies used by internists and internal medicine subspecialists to determine the safety, effectiveness, and cost of the technologies. Among other technologies, it evaluates devices, drugs, and surgical procedures. Questions should be directed to Linda White, Manager of the Division of Scientific Affairs.

Diagnostic and Therapeutic Technology Assessment (DATTA)

American Medical Association (AMA)   312-645-4532
535 North Dearborn
Chicago, IL 60610

DATTA is the basic project of the Medical Technology Assessment program of the AMA. The DATTA program evaluates technology with regard to safety, effectiveness, and indications for use. Questions about the relative merits of new and existing technologies in medical practice, medical education, and medical research are reviewed and researched by the staff, and appropriate questions are referred to a panel of physicians or to an outside source of information.

Health Industry Manufacturers Association (HIMA)
1030 15th Street, NW   202-452-8240
Washington, DC 20005-1598

A major trade association, HIMA represents the U.S. medical device and diagnostic product manufacturing industry. Its approximately 320 members, most of whom are involved in research and development, represent about 90 percent of the U.S. market. HIMA has produced numerous publications to keep members abreast of regulatory, technical, and scientific issues that may affect them. The free publications catalog includes abstracts of selected HIMA publications and training programs, as well as a complete index of all available reports and documents. An important publication is:

• *Health Industry Association Annual Report*—contains a list of its members as well as descriptions of HIMA's national affairs programs and its involvement in regulatory and scientific areas. It also features articles on health care. The 18-page publication is available free of charge.

The Health Program
Office of Technology Assessment (OTA)   202-226-2070
U.S. Congress   202-224-8996—Publications Requests
Washington, DC 20510

As an analytical support agency of the Congress, OTA studies science and technology issues. It has nine programs, including a health program, which generate reports that focus primarily on policy issues. These are available from the Superintendent of Documents, U.S. Government Printing Office, Washington, DC 20402, 202-783-3238. Summaries of these reports are available free of charge from OTA. The Office also does analyses of specific technologies, which are printed in the *Health Technology Case Study Series*. Each of these reports is published as a separate monograph. A free booklet containing abstracts of all the case studies in the series is available. An important report is:

• *Federal Policies in the Medical Devices Industry*—a comprehensive report that analyzes the major federal policies affecting the medical devices industry. It presents descriptive information about the companies in the industry and examines the following areas of federal policy: payment, research and development, Food and Drug Administration regulations, policing of providers, and the Veterans Administration.

Foreign Trade Division
Bureau of the Census   202-763-5140
U.S. Department of Commerce
Washington, DC 20233

This office has import and export data for many products. A significant report is:

• *Highlights of U.S. Export and Import Trade*—this publication contains information on the dollar value of products in the broad category of professional, scientific, and controlling instruments and apparatus. Updated monthly, the publication is available for $41 a year from the Superintendent of Documents, U.S. Government Printing Office, Washington, DC 20402, 202-783-3238.

Industry Division
Bureau of the Census   202-763-5850
U.S. Department of Commerce
Washington, DC 20233

This office has data on U.S. manufacturers and their output, including the value and sometimes quantity of products shipped from plants in this country. Reports, including those on medical and dental equipment, are available from the Superintendent of Documents, U.S. Government Printing Office, Washington, DC 20402, 202-783-3238. All publications are also available from federal depository libraries throughout the country.

National Health Information Clearinghouse (NHIC)
PO Box 1133   800-336-4797
Washington, DC 20013-1133   703-522-2590

This clearinghouse, a central source of health information as well as a referral organization, is sponsored by the Office of Disease Prevention and Health Promotion of the U.S. Public Health Service. It can often direct the public, physicians, researchers, and other health professionals to additional sources of information. It maintains a computerized data base of over 2,000 government agencies, advocacy groups, professional societies, and other organizations around the country that can answer questions and provide information on specific health topics.

National Institutes of Health (NIH)
Bethesda, MD 20205

NIH consist of 12 separate Institutes, each covering a specific area of medicine. The staff in the following offices can refer you to experts in their Institute who are knowledgeable about specific medical technologies. Most of these offices also have scientific reports and patient-oriented materials they can send you, often free of charge.

National Institute on Aging
Building 31, Room 5C35
Bethesda, MD 20205
301-496-1752

National Institute of Allergy and Infectious Diseases
Office of Research Reporting and Public Response
Building 31, Room 7A32
Bethesda, MD 20205
301-497-5717

National Institute on Arthritis, Diabetes, and Digestive and Kidney Diseases
Information Office
Building 31, Room 9A04
9000 Rockville Pike
Bethesda, MD 20205
301-496-3585

National Institute of Cancer
Cancer Information Service
Johns Hopkins Oncology Center
550 North Broadway, Suite 307
Baltimore, MD 21205
800-633-6070—Alaska
800-524-1234—Hawaii
800-4 CANCER—elsewhere

National Institute of Child Health and Human Development
9000 Rockville Pike
Bethesda, MD 20205
301-496-1848

National Institute of Dental Research
Public Inquiries and Report Section
Building 31, Room 2C35
Bethesda, MD 20205
301-496-4261

National Institute of Environmental Health Sciences
Public Affairs Office
PO Box 12233
Research Triangle Park, NC 27709
919-541-3345

National Eye Institute
Office of Scientific Reporting
Building 31, Room 6A32
Bethesda, MD 20205
301-496-5248

National Institute of General Medical Sciences
Office of Research Reports
Building 31, Room 4A52
9000 Rockville Pike
Bethesda, MD 20205

National Heart, Lung, and Blood Institute
Information Office
Building 31, Room 4A21
Bethesda, MD 20205
301-496-4236

National Institute of Mental Health
Public Communications Branch
Division of Communications and Education
Parklawn Building, Room 15-102
5600 Fishers Lane
Rockville, MD 20857
301-443-4536

National Institute of Neurological and Communicative Disorders and Stroke
Office of Scientific and Health Reports
Building 31, Room 8A06
Bethesda, MD 20205
301-496-5924

Office of Health Technology
U.S. Department of Health and Human Services   301-443-4990
Park Building, Room 310
5600 Fishers Lane
Rockville, MD 20857

This office reviews and analyzes new medical technologies for the Health Care Financing Administration. The staff members produce technical reports on specific technologies, gathering their information by talking with experts in government agencies and professional societies, reading periodical literature, and studying clinical evidence. Staff members can answer questions or refer you to experts and information sources for topics they have not covered. The office's reports are geared to the medical professional and are available free of charge.

Office of Instrumentation and Medical Sciences

Science and Electronics Cluster   202-377-0550
International Trade Administration
U.S. Department of Commerce
Washington, DC 20230

Information about how the medical equipment industry is doing in this country and in foreign markets is available from this office. The kind of information it has includes the volume of shipments of domestic manufacturers; level of employment; and international trade data, including exports and imports.

Office of Medical Applications Research (OMAR)

National Institutes of Health (NIH)   301-496-1143
Building 1, Room 216
Bethesda, MD 20205

OMAR is the focal point, at NIH, for medical technology assessment. The office coordinates the participation of NIH Institutes and various federal agencies in Consensus Development Conferences at which new drugs, medical procedures, and devices are evaluated for safety and effectiveness. OMAR staff members have evaluated more than 44 technologies, and they can provide technical information and reports on the technologies. The staff can also direct you to experts and organizations covering specific medical technologies.

Research Triangle Institute

PO Box 12194   919-541-6000
Research Triangle Park, NC 27709

A nonprofit contract research organization, the Research Triangle Institute offers a variety of applied research and consulting services. The Institute is involved in such medical technology areas as (1) electronics—mechanical and optical applications in biomedical engineering, (2) technology transfer for treatment—diagnostic and prosthetic devices, (3) preclinical and clinical testing of biomedical devices and drugs, (4) development of sustained release drug delivery devices, (5) economic and social research on the cost effectiveness of medical technologies and programs, and (6) development of applications of polymers in biomedical technology. Free brochures on health care research are available, as well as:

● *Hypotenuse*—a bimonthly publication with articles on various projects, including those in the medical technology areas, as well as brief descriptions of new projects. The 24-30-page periodical is directed to potential sponsors of research.

Scientific Apparatus Makers Association

1101 16th Street, NW   202-223-1360
Washington, DC 20036

This membership organization works to increase marketing opportunities for member companies by means of its legislative activities, educational programs, and market information. About a dozen educational conferences, seminars, and workshops are held each year. The staff can answer questions about the industry, issues, and the Association's member companies. Two publications are:

- *Annual Report*—describes the industry, legislative issues, and activities of the Association. It includes a list of member companies, as well as pictures of some of their products. The 20-page publication is available free of charge.

- *Edit*—a bimonthly newsletter that covers the issues and activities in which the Association is involved. It features news articles, legislative updates, and a calendar of events. The eight-page publication is available to nonmembers for $100 a year and is free to members.

Publications

Annals of Internal Medicine

The American College of Physicians   215-243-1200, ext. 1655
4200 Pine Street
Philadelphia, PA 19104

A monthly journal published in two volumes each year, *Annals of Internal Medicine* consists of papers on clinical, laboratory, socioeconomic, cultural, and historical topics pertinent to internal medicine and related fields. The material includes original articles, case reports, topical reviews, editorials, book reviews, and medical notices for internal medicine and related fields. The annual subscription price is $22.50 a year for medical students and $48 for others.

Medical Devices, Diagnostics and Instrumentation Reports

5550 Friendship Boulevard, Suite 1   301-657-9830—Wallace Werble, Jr.
Chevy Chase, MD 20815

This weekly newsletter covers regulatory agencies and congressional activities, as well as new developments in the medical device field and financial news. The journal is not very technical and is directed toward companies dealing in medical devices. The annual subscription price is $300.

Data Bases

MEDLARS

National Library of Medicine   301-496-6193
8600 Rockville Pike
Bethesda, MD 20209

Access to professional medical technology literature is available through the computerized system MEDLARS. The literature contains information on trends, standards, instrumentation, and so on. The primary data base within the system for medical technology literature is MEDLINE. The system can be searched by numerous terms, including specific medical technologies. Individuals can have searches done on MEDLARS for a fee; contact your nearest Regional Medical Library for further information.

Greater Northeastern Regional Medical Library Network

The New York Academy of Medicine
2 East 103rd Street
New York, NY 10029

Region 1 serves Connecticut, Delaware, Maine, Massachusetts, New Hampshire, New Jersey, New York, Pennsylvania, Rhode Island, Vermont, and Puerto Rico.

Southeastern/Atlantic Regional Medical Library Network

University of Maryland
Health Sciences Library
111 South Greene Street
Baltimore, MD 21201

Region 2 serves Alabama, Florida, Georgia, Maryland, Mississippi, North Carolina, South Carolina, Tennessee, Virginia, West Virginia, and the District of Columbia.

Greater Midwest Regional Medical Library Network

University of Illinois at Chicago
Library of Health Sciences
Health Sciences Center
PO Box 7509
Chicago, IL 60680

Region 3 serves Iowa, Illinois, Indiana, Kentucky, Michigan, Minnesota, North Dakota, Ohio, South Dakota, and Wisconsin.

Midcontinental Regional Medical Library Network

University of Nebraska
Medical Center Library
42nd and Dewey Avenue
Omaha, NE 68105

Region 4 serves Colorado, Kansas, Missouri, Nebraska, Utah, and Wyoming.

South Central Regional Medical Library Network

University of Texas
Health Sciences Center at Dallas
5323 Harry Hines Boulevard
Dallas, TX 75235

Region 5 serves Arkansas, Louisiana, New Mexico, Oklahoma, and Texas.

Pacific Northwest Regional Health Sciences Library Network

Health Sciences Library
University of Washington
Seattle, WA 98195

Region 6 serves Alaska, Idaho, Montana, Oregon, and Washington.

Pacific Southwest Regional Medical Library Network

UCLA Biomedical Library
Center for Health Sciences
Los Angeles, CA 90024

Region 7 serves Arizona, California, Hawaii, and Nevada.

Additional Experts and Resources

Dr. Doris Rouse
PO Box 12194   919-541-6980
Research Triangle Institute
Research Triangle Park, NC 27709

Dr. Rouse has a Ph.D. in physiology and pharmacology, and formerly worked in the area of mass spectrometry for pharmaceuticals. More recently, she has worked with NASA on aspects of technology transfer, and can answer questions on that subject. She is also an expert in her specialty areas, rehabilitation and technology for the elderly.

Medical colleges are often staffed by experts in a variety of specialties who can provide you with information about the latest technology in their field.

Medical journals are an excellent resource for learning about the latest medical products and their producers. Most medical specialties have their own journals, and these publications usually contain advertisements providing pictures, descriptions, and basic information about new products. Check with your area's medical school library for copies of journals.

Meetings on High Technology

See: Conferences, Meetings, and Trade Shows on High Technology

Membrane Separation Technology

Organizations and Government Agencies

Bend Research
64550 Research Road   503-382-4100—Mark Henry
Bend, OR 97701
    Bend is a manufacturer of several membrane separation products, especially in two areas: (1) separation processes; and (2) controlled releases (pharmaceuticals). The company makes available all sorts of free printed material about its work, including a company brochure, technical briefs, and brochures about products.

Business Communications Company
9 Viaduct Road   203-325-2208
Box 2070 C
Stamford, CT 06906
    BCC is a publishing and research firm specializing in several high-technology fields. It publishes technical and economic reports and newsletters, and sponsors conferences in its areas of specialty. In the area of membrane separation technology, it holds an annual conference and publishes the proceedings *(1983 Proceedings—* $125). It also publishes the following:
    • *Membrane Separation Technology News*—see separate entry below under *Publications*.

● *Membrane Separation Technology Directory*—a directory of people in the business, which is updated annually. It is available for $125.

● *Membrane Technology Patent Source Book*—a listing of U.S. and world patents relating to all aspects of membrane separation, except biology, listed by company. The 1984 edition is $200.

● *Membrane As Separation and As Support*—a report on the U.S. and international market of membrane technology. Updated on a three-year basis. The cost is $1,750.

Desalinization Systems

1238 Simpson Way   619-746-8141
Escondido, CA 92025

The company manufactures spiral wound, reverse osmosis, and ultrafiltration products. It maintains a research and development facility and provides technical support to original equipment manufacturers. The company makes available product brochures that describe the products and the technology needed and used.

FilmTec

7200 Ohms Lane   612-835-5475
Minneapolis, MN 55435

FilmTec is a manufacturer of reverse osmosis membranes and ultrafiltration membranes, which are used to purify water for many industrial applications. The personnel are extremely helpful and will answer questions about the company and its research. They also send out information packets, which include an annual report, a quarterly report, technical fact sheets, and brochures describing their work and the technology.

Pall Corporation

77 Crescent Beach Road   516-759-1900
Glen Cove, NY 11542

Pall is a multinational corporation that has been working in the area of membrane separation technology for over 30 years, ranging from sterile filtration to fluid clarification of all membrane types. In addition to being a manufacturer, it maintains research and development facilities. It makes available literature explaining the types of materials it has, as well as the work done and what it will be used for.

Publications

Business Opportunities in Membrane Technology

Battelle Columbus Laboratories   614-424-6549
505 King Avenue
Columbus, OH 43201

This book provides a thorough review of membrane separation processes, and it gives information about the technical and economic merits of different systems and applications. The chapters include coverage of reverse osmosis, electrical filtration, medical and biological applications, and so on. The 560-page book includes 125 tables, and it costs $7,500. The research specialists who worked on the book are available to answer inquiries.

Journal of Membrane Science

Elsevier Scientific Publications   212-867-9040 (New York City office)
1000 AE Amsterdam
PO Box 211
The Netherlands

The *Journal* is a vehicle for the dissemination of information dealing with the science and technology of membrane processes and phenomena. The emphasis is on the structure and function of nonbiological and biological membranes. Papers written by experts in the field form the major part of the *Journal*. The subscription rate is $463.50, for 15 issues.

Membrane Separation Technology News

Chemical Technology Consultants   713-774-3942—Anna W. Crull, Managing Editor
6300 Hillcroft, Suite 603
Houston, TX 77081

A monthly newsletter that explores the science and technology of microfiltration, reverse osmosis, ultrafiltration, gas separations, and so on. Available for $275 a year.

Anna Crull, a consultant specializing in membrane separation technology, is a good source of information in her field. She is also the editor of the newsletter and can answer questions about this publication, and other questions in the field.

Separation Science and Technology

Marcel Dekker Journals   212-696-9000
270 Madison Avenue
New York, NY 10016

A monthly scientific and medical journal that covers medical comatography, ionics charges, fat solids, organics, and related subjects. The subscription cost for 15 issues is $495 a year.

Additional Experts and Resources

See *Business Opportunitites in Membrane Technology* and *Membrane Separation Technology News*, described above, for leads to experts.

Metals

Organizations and Government Agencies

American Society for Metals (ASM)
Metals Park, OH 44073   216-338-5151

A nonprofit educational society, the ASM is dedicated to the collection and dissemination of information about metals and engineered materials, including ceramics, plastics, and any other material used in engineering. The public has access to all the information the association has. Educational seminars are held, which include home study programs and intensive seminars, as well as conferences and expositions. The ASM also offers videotape courses. The Society has a few data bases on metals and other materials as a part of their metals information system. There are 125 technical reference books available for purchase, as well as several engineering journals and other monthly publications. The staff can answer questions about its programs and publications. Described below are two examples of ASM's publications:

● *Metals Handbook*—a multivolume handbook that is a comprehensive source on metals and metalworking. A desk edition of the *Handbook* is available in a single volume, which complements the larger book, offering cross references to it. The desk edition contains the most vital information and can be used alone. It covers many topics in metals and metal production processes needed by engineers on a daily basis. The desk edition is available to nonmembers for $96 and to members for $76. Each book in the multivolume set is sold separately, and is available to nonmembers for $79 and to members for $47.40.

● *Metals Progress*—a monthly magazine covering the essentials of the technology and new developments in materials, processing, and fabrication. The publication is free to members, and nonmembers may request a free sample copy.

Association of Iron and Steel Engineers (AISE)
Three Gateway Center, Suite 2350   412-281-6323
Pittsburgh, PA 15222

AISE is a professional engineering association for engineers and operators in the basic steel industry. It is devoted to the technical advancement of the engineering phases of the production and processing of iron and steel. The Association conducts monthly district meetings on technology issues, as well as two annual national meetings of a technical nature. A group of experts in the Association can answer questions and give referrals to those seeking technical information. The Association publishes:

● *Iron and Steel Engineers*—a monthly magazine devoted to technical topics related to iron and steel processing. It is available free of charge to members, and costs $28 a year for nonmembers.

● Between 12 and 15 technical reports are also issued by the Association each year.

Bureau of Mines

U.S. Department of the Interior   202-634-1004
2401 E Street, NW
Washington, DC 20244

The Bureau's activities with respect to minerals include ensuring that the nation has an adequate supply of minerals for the production of metals; conducting applied and basic research to develop the technology for extraction processing, use, and recycling of the U.S. mineral resources; gathering data, developing research, and preparing reports about the technology of extraction and processing of minerals; and improving the conditions related to the health and safety of personnel employed in the mining and processing of minerals.

There are several specialists on metals and minerals at the Bureau of Mines. The above office can refer you to an appropriate expert at the Bureau. The Bureau publishes the following:

● *Mineral Commodity Summaries*—an annual publication that covers over 90 nonfuel minerals. Each of these minerals is analyzed in 11 categories, including domestic use, statistics, technological events, and trends. The publication is issued every January and is available free of charge.

● *Mineral Industries Surveys*—short publications on various minerals and metals, presenting the most recent production data. The frequency of release of these surveys varies, depending on the product discussed. They are available free of charge.

● *Minerals Yearbook*—an annual publication that presents technical and economic trends and statistics, in three cloth-bound volumes. Volume I is *Minerals Commodities*, and costs $16 a year. Volume II is *Minerals Industry Studies*, and costs $14 a year. Volume III is *Minerals International*, and costs $20 a year.

● *Minerals: Facts and Products*—an encyclopedia type of publication issued every five years (next edition in 1986). It covers the technology used in the extraction and processing of minerals. The cost is $23.

● Several other technical publications, such as studies and reports, are released periodically. People seeking specific information about these publications should call the phone number listed above.

Iron and Steel Society (ISS)

PO Box 411   412-776-1535
410 Commonwealth Drive
Warrendale, PA 15086

ISS is one of the four constituent societies of the American Institute of Mining, Metallurgical and Petroleum Engineers (AIME). Its members are professionals in the field of iron and steel processing and technology. The Society seeks to provide communication and cooperation among those interested in ferrous metallurgy, and materials science and technology.

ISS has five technical divisions, each covering a production phase, from mining to finishing of metal products. Each of these divisions has an annual technical meeting, and the proceedings are published by the Society. ISS releases the following technical publications:

• *Iron and Steel Maker*—a monthly magazine that contains technical information on major topics in the field of metallurgy. It is available free of charge to members, and costs $35 a year for nonmembers.

• *Transactions of the Iron and Steel Society*—a quarterly magazine that presents more in-depth topical papers than the *Iron and Steel Maker*. Available for $45 a year to members, and for $90 a year to nonmembers.

Metallurgy Division

Center for Materials Science   301-921-2811
National Bureau of Standards
Gaithersburg, MD 20899

The Division studies metals and alloys in order to foster their safe, efficient, and economical use to meet U.S. long-term needs within the larger framework of the materials cycle. It develops and maintains competence in four general areas: (1) chemical metallurgy, with special emphasis on phase diagrams and phase stability; (2) process metallurgy, with emphasis on the measurement and science base for innovative processing technologies, such as coating by electrodeposition; (3) quantitative microstructural characterization, including nondestructive evaluation (NDE); and (4) properties and performance, including friction, wear, corrosion, and the application of metals as synthetic orthopedic implants.

The Division serves as a national resource of metallurgical information through its Diffusion Data Center and Alloy Data Center, and provides consultation and assistance to other government agencies and standards organizations in the development of necessary test methods and standards. The major activities in the Division are carried out in discipline-oriented or interest-oriented groups. Division members can answer questions in their areas of expertise or make referrals. The *Annual Report* is available free of charge, as are papers, reports, proceedings, and other publications.

Office of Basic Industries

U.S. Department of Commerce   202-377-0608—Ralph Thompson—ferrous metals
Bureau of Industrial Economics   202-377-0575—Robert C. Reiley—nonferrous metals
Washington, DC 20230

Specialists at the Office of Basic Industries advise policymakers in the U.S. Department of Commerce on issues related to ferrous and nonferrous metals. They also make recommendations about how American industries can become more competitive, prepare analyses and comments for the upper management of the Department, and participate in interagency groups and committees on steels.

The Office of Basic Industries also serves as a liaison between industry and government. The specialists listed above can answer inquiries and give referrals to other government and industry sources.

Publications

Chilton Co.

One Chilton Way   215-964-4000

Radnor, PA 19089

The following magazines are available:

• *Iron Age Manufacturing Management*—released twice a month, it is oriented toward steel producers, and covers the technology involved in the production of steel. The cost is $42 a year.

• *Iron Age Metals Producer*—also released twice a month, this is a magazine for the users of steel materials. The articles cover the technology and management techniques used in the U.S. for processing of finished steel into final materials. The costs is $49 a year.

• *Iron Age Producer International*—a monthly international version of *Iron Age Metals Producer.* The cost is $49 a year.

• *Iron Age International*—a magazine released every two months, covering international articles on the production of steel and its processing for final use. The cost is $60 a year.

Fairchild Publications

7 East 12th Street   212-741-4000

New York, NY 10003

• *American Metal Market*—a daily newspaper that covers news on manufacturing technology, technical aspects of ferrous and nonferrous metals, and business news on the metalworking products industry, with a weekly insert (on Mondays) entitled *Metalworking News*, described below. The annual subscription is $340.

• *Metalworking News*—this weekly reports and analyzes special topics of technology used on ferrous and nonferrous metals. The annual subscription is $32.

33 Metal Producing Magazine

McGraw Hill, Inc.   212-512-2000

1221 Avenue of the Americas

New York, NY 10020

This is a monthly magazine with worldwide circulation among corporations and operating management, and engineering, maintenance, and metallurgy personnel in the basic steel industry. It presents technical articles about the major steel mills in the U.S. The name *33* is derived from the U.S. Department of Commerce's standard industry classification.

The magazine is available free of charge to upper management people who deal with production of ferrous and nonferrous metals. Others can subscribe for $35 a year.

Steel Industry Review
Data Resources, Inc. (DRI)   617-863-5100
29 Hartwell Avenue
Lexington, MA 02173
  This quarterly magazine forecasts the U.S. economy as it relates to the market for steel products. It also covers steel demand, trade, shipments, raw steel production by type of process, new materials use and cost, and steel prices. The magazine is oriented toward producers, users, and consumers of steel, as well as suppliers and government agencies. The subscription cost a year is $1,500, but single copies can also be purchased from DRI.

Data Bases

Data Resources, Inc. (DRI)
29 Hartwell Avenue   617-863-5100—Mr. Carter
Lexington, MA 02173
  DRI has an on-line data base containing reports from the American Iron and Steel Institute (AISI), the International Iron and Steel Institute, and trade associations, as well as U.S. government data. This data base can be accessed by microcomputer users subscribing to DRI.

MICR

See: Magnetic Ink Character Recognition

Microcomputers

Organizations and Government Agencies

Association of Computer Users (ACU)
PO Box 9003   303-443-3600
Boulder, CO 80301
  This independent publishing and consulting organization's membership is made up of users of small computers for business applications. The Association disseminates information about computing products and services, and provides independent consulting services, specializing in small computer product comparisons and competitive evaluations. Membership fees are $75 annually, although a three-

month trial membership is available. Included with membership are two ACU official consumer guides on the selection and management of small computers, as well as the newsletter:

• *Executive Computing Newsletter*—a monthly publication written chiefly for executives, business owners, and office managers. This newsletter provides evaluations of the latest hardware and software, and insights that aid the effective management of the small computer. The *newsletter* is available only to members.

DataPro Research, Inc.

McGraw-Hill Information Systems   609-764-0100
1805 Underwood Boulevard
Delran, NJ 08075

DataPro Research, Inc., offers reports and directories on products and technologies available from the current data-processing and office automation industries. Offered in subscription form, each information service includes loose-leaf volumes of reports, monthly reference supplements with the most current information, and a monthly newsletter reviewing the telecommunications industry. Each subscription allows the subscriber to use DataPro's telephone consulting service for direct access, personalized information, and supplementary information not contained in the comprehensive reports. DataPro also publishes:

• *Directory of Small Computers*—a two-volume report, updated monthly, with descriptions of over 200 computer systems, including specifications and prices. The listings also include notations of related peripherals, software, and services. The annual subscription is $525.

• *Directory of Microcomputer Software*—a two-volume report, updated monthly, with descriptions of more than 3,000 microcomputer software products and more than 1,000 companies involved in the development and marketing of this software. The information is cross-referenced by four indexes: application, product name, vendor, and system supported. The annual subscription is $525.

• *Reports on Microcomputers*—a two-volume set, updated monthly, that is designed to allow management and other professionals to plan, evaluate, buy, distribute, and integrate small computer systems. It includes detailed characteristics, advantages and restrictions, side-by-side comparisons, user reactions and ratings, and other pertinent facts on entire computer systems, including peripherals, software, and related communications. The annual subscription rate is $600.

• *Management of Small Computer Systems*—a two-volume set, updated monthly, of information for the new computer user or manager. It details basic computer system hardware, software and communications concepts, and basic data-processing concepts, including applications. It also includes equipment and software ratings, and vendor listings. The annual subscription is $390.

Institute for Computer Science and Technology (ICST)
National Bureau of Standards (NBS)   301-921-3151
B154 Technology Building
Gaithersburg, MD 20899

      ICST is a small government agency concerned with the federal standards and guidelines by which the microcomputer industy is regulated. Information about the research being done in this field is available through documents and special publications, listings of which are accessible on ICST's electronic bulletin board and publications listing. Also available is the Institute's newsletter:

      ● *ICST Newsletter*—a free quarterly publication containing information about new publications and technical activities, and short articles on federal user perspectives. The newsletter also announces upcoming seminars and workshops for information exchange.

Institute of Electrical and Electronics Engineers (IEEE)
Technical Committee on Microprocessing and Microcomputers   301-589-8142
Computer Society
1109 Spring Street
Silver Spring, MD 20910

      The IEEE Computer Society's Technical Committee is an excellent source of information. For more information see the IEEE entry under *Microprocessing*.

Publications

Business Computing
Advanced Technology Group   617-486-9501
119 Russell Street
Littleton, MA 01460

      Published monthly, this business magazine is for people using an IBM PC or compatible microcomputer in a business or professional setting. It covers the issues involved in using the PC to solve business problems, and offers a mix of tutorials, reports from users, and overviews of the PC's role in various businesses. The magazine is also available on-line through the Source at 703-734-7500 or 800-336-3366, and on CompuServe at 614-457-8600.

Byte
McGraw-Hill, Inc.   603-924-9281
70 Main Street
Peterborough, NH 03458

      This monthly magazine is a technical publication geared toward owners and users of personal computers. Approximately 500 pages an issue, *Byte* addresses the building and purchasing of private computers as well as their use. Each issue contains detailed reviews of software and hardware, tutorials on practical applications, news of clubs and events, and extensive advertising of new technologies. The annual subscription is $21.

Compute!

Compute! Publications
PO Box 5406
Greensboro, NC 27403

A diversified monthly magazine dealing with topics of interest to those in the personal computer field, *Compute!* offers a wide variety of information features. Each month there are regular articles on computers and children; telecom networks; the future of personal computers; and information about programs, games, machine language, and software. *Compute!* emphasizes the current technology and computer market and usually offers helpful hints on purchasing and on programming. The annual subscription rate is $24.

Computer Classified Bluebook

Computer Classified   702-322-8811
PO Box 3395
Reno, NV 89505

The *Bluebook* is a quarterly reference publication with over 200 new and used microcomputer models listed in each edition. The fall and spring editions give completely updated specifications and pricing profiles on models included for the first time; the winter and summer editions update previously published editions, including prices. The annual subscription is $85.

Creative Computing

PO Box 5214   303-449-2979
Boulder, CO 80322

This monthly magazine focuses on computer applications and software at the elementary and advanced levels. Originally a computer education publication, *Creative Computing* still includes articles on designing and evaluating educational software, concepts and terminology in computer education, and computer simulations for the classroom. The magazine discusses in nontechnical language such topics as text editing, animation, graphics, data base and file systems, word processing, and office applications. There are also equipment profiles to aid in the selection and purchase of new products, and articles on computer game software. About 300 pages an issue; the annual subscription cost is $19.97.

80 Micro

Wayne Green, Inc.   603-924-9471
80 Pine Street
Peterborough, NH 03458

A monthly magazine geared toward those with access to the TRS-80 computer system, *80 Micro* covers all levels of experience for users at home, in business, and in education. Regular features include evaluations of available accessories, instruction on the use of the system, and software information. There is also a section aimed at aiding the non-data-processing businessman in the uses of the TRS-80 in the business environment. Additionally, it includes new product analyses, reviews of new books and literature, and a calendar of upcoming events of interest to the microcomputer user. About 200 pages an issue; the annual subscription rate is $35.97.

InfoWorld

Popular Computing Co., Inc.   800-343-6474
375 Cochituate Road   617-879-0706
Framingham, MA 01701

The only news weekly published about microcomputers. Each issue of *InfoWorld* focuses on one theme. Each week includes reviews of hardware and software, with a summary report card issued semiannually. There are also announcements about new and upcoming products, and a calendar section about events and shows of interest to personal computer users. It guides the consumer through buying a microcomputer, setting it up for home or business use, and getting the most from the hardware and software selected. About 100 pages an issue; the annual subscription rate is $22.95, for 51 issues.

Interface Age

McPheters, Wolfe and Jones   213-926-9544
17000 Marquardt Avenue
Cerritos, CA 90701

Directed toward small-business users of microcomputers, *Interface Age* aids the professional in the selection and operation of a small computer system. Regular features include a section on corporate users, consumer focus, industry trends, and a calendar of related happenings. Articles regularly explain programs applicable for business use and peripheral equipment for total office automation. The publication also has frequent reviews of new technology and comparisons of compatible product systems. About 150 pages an issue, this monthly costs $21 a year.

LC Science Tracer Bullets

Science Reference Section   202-287-5670
Science and Technology Division
Library of Congress
10 First Street, SE
Washington, DC 20540

Personal Computing/Home Computers (TB 81-12) and *Microcomputers* (TB 77-3) are free informational guides to the materials available on these subjects at the Library of Congress. The bulletins are excellent information resources: they contain reference listings of basic texts, handbooks, manuals, dictionaries, abstracting and indexing services, periodicals, selected journal articles, and technical reports. The staff at the Library's Referral Center will also recommend names of organizations and other sources of information for each subject heading.

Micro/6502-6809 Journal

Micro, Inc.   617-256-3649
PO Box 6502
Chelmsford, MA 01824

This monthly *Journal* is directed toward the serious microcomputer user, and its technical language defies all but the advanced consumer. Covering only the 6502/6809-based computers, such as Apple, Commodore, VIC, and Atari, *Micro* discusses use at home, in business, and in education. There are regular reviews of new products and of new publications in both the software and hardware fields. About 100 pages an issue; the subscription rate to *Micro* is $24.

Microcomputer Index

Data Base Services   415-948-8304
PO Box 50545
Palo Alto, CA 94303

Microcomputer Index is a subject index derived from over 50 periodical resources. The index gives complete bibliographic information on over 3,000 articles, columns, book reviews, and hardware and software reviews in each issue. It also contains an abstract section with brief summaries of the content of current reviews and articles. The six issues of the 1984 edition are available for $45, and they may also be purchased cloth-bound at the end of the year for $50. The subscription price is $60.

Microcomputing

Wayne Green Publications   603-924-9471—Editorial Information
80 Pine Street   800-258-5473—Subscription Sales
Peterborough, NH 03458

A monthly magazine geared toward the advanced or intermediate user of such microcomputers as the Apple, Atari, Commodore, Heath, IBM, and Sinclair, *Microcomputing* covers computer technology advances, hardware projects, operating systems, computer languages, and more. Tips on programming, literature reviews, publication announcements, and a calendar of events are also included. The annual subscription rate is $24.97.

Ziff-Davis Publishing

1156 15th Street, NW   202-293-3400
Washington, DC 20036

Two magazines of interest to IBM personal computer users are published by this company:

• *PC Magazine*—is the independent guide to IBM personal computers, covering software, hardware, applications, and other topics of interest. Included are evaluations and insights from experts, a special user-to-user section, and news about personal computer clubs, events, and publications. The annual subscription rate is $21.97.

● *PC Disk Magazine*—is a monthly magazine that uses a magnetic disk instead of ink and paper. It gives a library of software programs for the IBM personal computer. Each month's disk contains up to 12 varied ready-to-use programs and files on a floppy disk, along with a complete user manual. The subscription rate is $179 a year.

PC World

PC World Communications, Inc.   415-861-3861
555 De Haro Street
San Francisco, CA 94107

A monthly magazine dealing only with the IBM personal computer and compatible computers, *PC World* covers topics of general interest to the computer community. Sections include information for the novice, those thinking of buying, and test results of new products. There are also reviews of new products, articles on the latest trends in the field and the state-of-the-art, community news with practical applications, and a supplementary *PC Junior World*. About 350-400 pages an issue; the annual subscription rate is $19.

Personal Computing

Hayden Publishing   201-393-6000
10 Mulholland Drive
Hasbrouck Heights, NJ 07604

Published monthly, the magazine is geared chiefly toward the business use of small computers, with regular features on productivity, corporate stories, and educational and leisure uses. Each month, a Buyer's Guide details a particular topic, with articles and summaries of the major technology available under that heading, in addition to the regular reviews of new products. The annual subscription is $11.97.

Popular Computing

McGraw-Hill, Inc.   603-924-7123
70 Main Street
Peterborough, NH 03458

This monthly magazine explains and puts into perspective the technology, capabilities, and applications of small computers for home or small-business use. About 200 pages an issue, *Popular Computing* evaluates new products, in both hardware and software, and their uses, and presents articles of interest on news in the computer field. The magazine often stresses the impact of small computers on education. The annual subscription is $12.97.

Data Bases

Microcomputer Electronic Information Exchange (MEIE)

National Bureau of Standards (NBS)   301-921-3485—Ted Landberg
Institute of Computer Sciences and Technology (ICST)   301-948-5718—ICST
Technology Building B266
Washington, DC 20234

An electronic bulletin board, MEIE enables users to read and exchange information on topics related to microcomputers. The file contains the standard publications on microcomputers from NBS, information on and availability of new products and technology, accessibility of telecomputing services and special-interest groups, publications, and other bulletin boards. There is current information on conferences, seminars, workshops, and other upcoming events and local meetings related to microcomputing. In addition, an electronic message center allows users to read and leave messages, as well as update the information already entered. To determine the computer compatibility required, call Mr. Landberg before dialing the access number.

Microcomputer Index

Data Base Services   415-948-8304
PO Box 50545
Palo Alto, CA 94303

The electronic version of the bimonthly publication *Microcomputer Index* (described earlier), this data base can be accessed through DIALOG as File 233 (800-227-1927) or through Knowledge Index as Comp3 (800-982-5838).

Additional Experts and Resources

PC Telemart

PC National Software Library   703-352-0721
11781 Lee Jackson Highway
Fairfax, VA 22033

This is the first of a proposed network of software resource centers. The Center has about 3,000 software titles and 30 microcomputers, to allow members to test and evaluate actual software and hardware products. Telemart also has an on-line data base with information describing over 6,000 software products and a demonstration downloading data base with which members can read about software products and then actually use the software in a demonstration mode. The company also offers consultation services.

Microelectronics

Organizations and Government Agencies

Electronic Design Center (EDC)

Case Western Reserve University   216-368-2934
University Circle
Cleveland, OH 44106

   EDC is an interdisciplinary research center staffed by engineers, scientists, technical academicians, and graduate students. Its areas of interest include solid state microelectronic transducers, including semiconductor chemical sensors, microelectronic process development for integrated circuits, integrated design and technology, transducer material and micromachining of electronic devices, biomedical transducers, medical instrumentation, biomedical implantable electronics and telemetry, and functional stimulation. EDC's staff members are experts in microelectronics and regularly publish papers about their research in scholarly journals. The staff will answer inquiries, make referrals, and provide consulting services. The Center maintains mailing lists by area of interest.

Electronic Industries Association (EIA)

2001 Eye Street, NW   202-457-4900
Washington, DC 20006

   EIA is a trade association of electronic manufacturers that sets standards for the industry. Currently EIA has six standards relating to microelectronics, and these can be purchased from EIA for approximately $.05 each. The Association also publishes the *JEDEC Bulletin* periodically. The *Bulletin*, produced by the Joint Electronic Device Engineering Council, is on specific topics and ranges from a users' guide to giving standard measurements. The price varies from $5 to $14.

International Society for Hybrid Microelectronics (ISHM)

3305 Atlanta Highway   205-272-3191—Technical Services Office
PO Box 3255
Montgomery, AL 36109

PO Box 670   301-933-8777
Silver Spring, MD 20901

   ISHM is an international technical service organization whose members include 6,000 scientists and engineers, and 300 universities, corporate offices, and service organizations. The Society's primary objective is the dissemination of technological information to its members and the microelectronics community. This is accomplished through publications, audiovisuals, and meetings at the local, regional, and international levels. ISHM also tries to serve the needs of colleges and universities by developing curricula, symposia, manuals, and textbooks. Society materials, including audiovisuals, guides, and text/reference books, are available for sale to the public. ISHM's publications include:

● Technical monographs—each year ISHM produces six to eight new monographs, which provide textbook-like coverage of a specific topic. They generally are 60-300 pages long; past topics have included thermal management, interconnection technology and ceramics for electronics. The cost is $8 for members, and $13 for nonmembers.

● *The International Journal for Hybrid Microelectronics*—a highly technical quarterly journal. Contributors are engineers and scientists throughout the world. The *Journal* generally contains many articles on a specific, subject with an occasional *Journal* devoted to only one topic. The *Journal* is free to members, and costs $60 to others.

● *Technical Proceedings*—each year ISHM publishes the entire proceedings of its annual U.S. Symposium. In May/June of each year the proceedings of the symposium held in Europe or Japan are also published. Volumes are available dating back to 1967; individual volumes cost $30, with discounts given for multiple volume purchases.

Reliability Analysis Center (RAC)

Rome Air Development Center   315-330-4151
Griffiss AFB, NY 13441

RAC is a Department of Defense–contracted information analysis center. The Center maintains several data bases with bibliographic and technical information on the reliability aspects of microelectronics. Its data are collected from manufacturers throughout the U.S. The staff will search the data bases for the public, charging a fee for complicated requests. RAC has published six books (approximately $60 each) relating to microelectronics. A free catalog describing the publications is available from the Center. The staff will answer questions and make referrals.

Publications

Electronic News

PO Box 1013   201-741-4452—Circulation
Manasquan, NJ 08736   201-741-4226—Editorial

Electronic News is a weekly business newspaper that follows the electronic industry and covers many aspects of the microelectronics industry, including semiconductors, computers, consumer electronics, communications, government contracts, test and measurement equipment, and materials equipment. Information includes happenings inside the board room, mergers, acquisitions, personal information, stock developments, and an events section listing upcoming meetings, trade shows, and conferences. The tabloid usually is 80-120 pages long; the annual subscription cost is $38.

Lake Publishing Company
17730 West Peterson Road   312-362-8711
Box 159
Libertyville, IL 60048

The following publications may be useful to those in the microelectronics industry:

● *Microelectronics Manufacturing and Testing* (MMT)—is a monthly magazine of approximately 100-200 pages, which presents in-depth and technical articles, with a strong emphasis on introducing new equipment and materials. The magazine is involved in all aspects of microelectronics, from growing crystals to final packaging and testing requirements. It includes a calendar of events, trade shows, and seminars. Available free of charge to scientists, engineers, technicians, and managers involved in manufacturing and testing microelectronics products. A paid subscription, which includes a copy of the *MMT Desk Manual*, is $50 a year.

● *MMT Desk Manual*—updated yearly, this manual is an encyclopedic listing of various terms, products, and technologies in microelectronics. A technical description is provided for each entry. Suppliers are also listed. Available free of charge to paid subscribers of the MMT magazine; the cost to others is $25.

Solid State Technology
Technical Publishing Company   516-883-6200
14 Vanderventer Avenue
Port Washington, NY 11050

Published by Technical Publishing Company, a division of Dun and Bradstreet, this magazine addresses all aspects of processing, fabrication, assembly, and testing involved in the manufacture of semiconductor devices and circuits. Topics such as bench engineering and corporate level management are covered, as well as areas which interface the manufacture and production of solid state products. Those include testing and evaluation, cost analysis, design, computer control of design, processing, and testing. Engineers and scientists working in the solid state field qualify for a free subscription. For others the yearly subscription fee is $30.

Data Bases

Government Industry Data Exchange Program (GIDEP)
GIDEP Operations Center   714-736-4677
Corona, CA 91720

GIDEP is a cooperative activity between government and industry that provides participants with a means of exchanging information about certain types of technical data, especially the design, development, and operation of the life cycle of systems and equipment. The program is managed by the U.S. government, and participation is open to individuals and organizations able to exchange technical information and willing to abide by GIDEP program guidelines. GIDEP's services, data bases, and publications are available free of charge to participants, and there is no membership fee. The GIDEP staff will perform simple data base searches for nonparticipants, but on a one-time basis. Currently, GIDEP members include industry representatives and U.S. government officials, primarily from NASA and the U.S. Departments of Defense, Energy, Transportation, and Labor. GIDEP services include a bimonthly newsletter covering new developments, and an on-line data base. Participants can obtain free direct on-line access to the data base's main computers, located in Washington, DC, and Corona, CA. Members can also request searches and obtain the entire data base on microfilm and microfiche. GIDEP data base files are:

- *The Engineering Data Interchange*—contains engineering evaluation and qualification test reports, nonstandard parts justification data, parts and materials specifications, manufacturing processes, and related engineering data on parts, components, materials, and processes. This data interchange also includes a section of reports on specific engineering methodology and techniques, air and water pollution reports, alternative energy sources, and other subjects.

- *The Reliability and Maintainability Data Interchange*—contains failure rate/mode and replacement rate data on parts, components, and materials based on field performance information and/or reliability demonstration tests of equipment, subsystems, and systems. This data interchange also contains reports on theories, methods, techniques, and procedures related to reliability and maintainability practices.

- *The Metrology Data Interchange*—contains metrology-related engineering data on test systems, calibration systems, and measurement technology and test equipment calibration procedures, and has been designated as a data repository for the National Bureau of Standards (NBS) metrology-related data. This data interchange also provides a Metrology Information Service (MIS) for its participants.

- *Manufacturer Test Data and Reports*—includes certified test reports from manufacturers detailing test results and inspection conducted on devices of their manufacture. Test data pertain to commercial as well as military and high-reliability devices. The availability of this test data in the GIDEP provides participants with the opportunity to apply the data in every phase of system design, development, production, and support process.

Micrographics

Organizations and Government Agencies

International Information Management Congress (IMC)
PO Box 34404   301-983-0604
Bethesda, MD 20817

IMC is an international micrographic organization whose goal is to improve micrographic understanding and practices around the world. It holds an annual international congress and several regional conferences, and it provides an international clearinghouse for information and advancement in systems and technology. The library is open to the public, offering books, booklets, microfiche, and audiovisual slide programs. The membership fee is $75 a year. Publications include:

- *IMC Journal*—a quarterly publication presenting information on total systems, Computer Output Microfilm, and integrated data-processing technologies. The publication is free to members; it costs $40 a year for nonmembers.
- *IMC Newsletter*—published eight times a year, this *Newsletter* gives a global view of micrographic news, with information about new products, the latest developments in micrographics, international news, association profiles, a calendar of events, and so on. The publication is free to members, and $25 a year for nonmembers.
- *Micrographics*—a text on micrographics. Available for $26.
- *Introduction to Micrographics*—an illustrated primer on fundamentals of micrographics. Available for $5.

Association for Information and Image Management (AIIM)
1100 Wayne Avenue, Suite 1100   301-587-8202
Silver Spring, MD 20910

AIIM covers the field of information and image management. It maintains a Resource Center with 7,000 books, articles, periodicals, research studies, and pamphlets. The Center maintains an on-line data base that will search free of charge for members and at a nominal fee for others. A consulting service is also provided by the Center.

AIIM offers many publications covering all aspects of micrographics. Upon request, it will send you a free catalog describing what is available. Examples include:

- *Journal of Information and Image Management*—a monthly *Journal* with a variety of articles on computer, records management, and micrographic applications; technology and products; book reviews of pertinent literature; new product information; editorials; standards information; and company and personnel news. Available for $55 a year.
- *Reference Series*—special publications dealing with in-depth aspects of information management, including such textbooks as *Micrographics Systems* ($21.75); a workbook, *Micrographic Film Technology* ($12.50); *Comprehensive Registry of Micrographic Equipment, Supplies and Services* ($25); and a *Glossary of Terms*.
- *Special Interest Packages*—called SIPs, these are available for a variety of topics. Examples include *Computers in Micrographics* and *Micrographics in Banking*. Prices range from $18 to $23.
- Consumer Oriented Booklets—AIIM produces several materials geared toward the general public. Examples include *An Introduction to Microform Indexing and Retrieval Systems* ($5.50) and *An Introduction to Micrographics* ($4.50).

Publications

Information Management

PTN Publishing Corporation 516-496-8000
101 Crossways Park West
Woodbury, NY 11797

This monthly publication covers the systems, technologies, software, and media for the file, storage, retrieval, processing, and distribution of information systems, resources, and media. It includes micrographics, computer assisted retrieval, computer output microfilm, the optical disk, and image management. It is directed toward upper and middle-level managers responsible for the management of their organization's information, and is distributed free to that group. It is available for $10 a year.

International Journal of Micrographics and Video Technology

Pergamon Press, Inc. 914-592-7700
Maxwell House
Fairview Park
Elmsford, NY 10523

A quarterly journal reporting on the interrelated advances in electronic information transfer in micropublishing, electronic journals, and the photographic and video fields, including exploitation of data bases by these means. Available for $55 a year.

Meckler Publishing

520 Riverside Avenue 203-226-6967
Westport, CT 06880

Meckler publishes the following publications on the topic of micrographics:

● *Micrographics Equipment Review*—provides critical reviews of micrographic equipment manufactured in the U.S. and around the world. Reviews are of equipment most likely used in library and business applications, ranging from hand-held viewers to the most elaborate reader-printers. The subscription rate is $185 a year, for 20 reviews.

● *Index to Micrographics Equipment Evaluations*—this index includes references and in-depth evaluations of micrographics equipment, printed in several publications. Review journals covered are: *Library Technology Review* and NRCD's *Reprographics Quarterly* and *Technical Evaluation Reports*. The index is cumulative, reissued annually, and available for $35.

● *International File of Micrographics Equipment and Accessories*—provides catalogs of every vendor of micrographics equipment and accessories worldwide. The file is in microfiche, along with a printed index. Available for $250.

● *Microform Reader Maintenance Manual*—addresses maintenance and repair procedures for over 30 of the most commonly found microfilm and microfiche readers and reader-printers. The *Manual* is $49.95.

Microfilm Publishing, Inc.
PO Box 313   914-235-5246
Wykagyl Station
New Rochelle, NY 10804

The following is published by this company:

• *Micrographics Newsletter*—this biweekly newsletter covers new products, developments, companies, computer output microfilm, people, and events, and includes classified ads. The annual subcription is $95.

• *International Micrographics Source Book*—which includes a worldwide directory of job manufacturers, dealers, services, service companies, products, consultants, associations, periodicals, and a key-word index. Available for $59.50.

Microprocessing

Organizations and Government Agencies

Microprocessors and Microcomputers Technical Committee
IEEE Computer Society   301-589-8142
1109 Spring Street, Suite 300
Silver Spring, MD 20910

The Microprocessors and Microcomputers Technical Committee of the IEEE Computer Society is involved with the architecture, design, and application of microprocessors. The Committee organizes and sponsors meetings and workshops in its technical area, actively pursues standardization, and publishes a semiannual newsletter, which covers the activities of the Technical Committee, especially standardization and standards development.

The proceedings from the conferences are published and are available to the public. They concentrate on the design, applications, and utilization of microprocessors and microcomputers. The main emphasis is on the application of small computers to problems such as data bases, graphics, education, the architecture of microprocessor systems, and the investigation of operation system and programming language tools for small computer environments.

Some of the larger manufacturers of microprocessors can be excellent sources of information. Many publish newsletters or make available brochures and pamphlets that describe their product and the technology. Also, several experts are available within each company to answer questions and refer you to other sources of information within the industry.

Texas Instruments
Microprocessor Division
9901 South Wilcrest Drive
Houston, TX 77099
713-879-2000

Intel Corporation
3065 Bowers Avenue
Santa Clara, CA 95051
800-538-1876
408-496-8935

National Semiconductor
Microprocessors Department
Mail Stop 3668, Bldg. D
Santa Clara, CA 95051
408-721-5088

Zilog Corporation
1315 Bell Avenue
Campbell, CA 95008
408-370-8000, ext. 4497

Motorola
Semiconductors M.O.S. Manufacturers
East Ed Blustein Boulevard
Austin, TX 78721
512-928-6000

Publications

Micro Magazine
IEEE Computer Society   301-589-8142
1109 Spring Street, Suite 300
Silver Spring, MD 20910

A bimonthly publication covering microprocessor technology: applications; fabrication and computer-aided design; system support software; interfacing techniques; chip design and fabrication; personal computing; draft standards for hardware, software, and interconnections; and control hierarchies and architectures. Available at $12 a year for members of IEEE and $60 a year for others.

Microprocessing and Microprogramming
Elsevier Scientific Publications   212-867-9040—New York City office
1000 AE Amsterdam
PO Box 211
The Netherlands

A scientific journal that covers the international flow of information about microprocessing and microprogramming, in addition to reports on technical research and progress, computer survey, state-of-the-art reports, short communications, and announcements pertaining to these areas. Topics covered include theory, simulation, emulation, teaching aids, evaluation aid/diagnostics methods, personal computing, and social and economic aspects—all relating to microprocessing, microprogramming, networks, computer structure, modular systems, integrated software/hardware design, and so on. Available for $162.32, for ten issues.

Several personal computing magazines regularly feature articles about microprocessors and their technology. Some of these are:

BYTE
70 Main Street
Peterborough, NH 03458
603-924-9281
  The monthly is $19 a year.

PC Magazine
One Park Avenue
New York, NY 10016
212-725-3500
  The biweekly is $26.97 a year.

Personal Computing
50 Essex Street
Rochelle Park, NJ 07662
201-393-6000
  The monthly is $18 a year.

Dr. Dobb's Journal
PO Box E
Menlo Park, CA 94025
415-323-3111
  The monthly is $25 a year.

PC World
555 De Haro Street
San Francisco, CA 94107
415-861-3861
  The monthly is $24 a year.

Data Base

Microcomputer Electronic Information Exchange
National Bureau of Standards   301-948-5718—Ted Landberg
Institute for Computer Sciences and Technology
Gaithersburg, MD 20899
  This bulletin board, operated by the Systems, Selection and Evaluation Group, is a message board that allows users to communicate about where software can be found, whom to contact, the latest technology, and so on. It is used as a forum for questions and answers, discussions, news, and so on, all related to microcomputers and microprocessors.

Additional Experts and Resources

Dennis Allison
169 Spruce Avenue   415-325-2962
Menlo Park, CA 94027
  Mr. Allison, a consultant, is also a lecturer at Stanford University, and an expert in the area of computer architecture. He has designed languages and operating systems, and is a contributing editor of *BYTE* magazine. He is also a cofounder of People's Computer Company in Menlo Park, CA. He is willing to answer general questions about microprocessing, and can refer you to other sources of information.

Michael Smolin
Chairman, TCMICRO (IEEE)   415-969-5130
98751 Independence Avenue
Mountainview, CA 94043

Mr. Smolin, chairperson of the IEEE Technical Committee on microprocessing, is also an expert in the area of standards in microprocessors, as well as hard-copy computer graphics. He is willing to answer questions, and can refer you to other sources of information.

Microprogramming

Organizations and Government Agencies

IEEE Computer Society
Technical Committee on Microprogramming (TCMICRO)   301-589-8142
1109 Spring Street, Suite 300
Silver Spring, MD 20910

The microprogramming technical committee of the Computer Society addresses all aspects of microprogramming and its support tools, giving particular attention to microprogramming languages, architectures of microprogramming machines, emulation, simulators, control storage technologies, microprogram design tools, and applications of microprograms. The Technical Committee sponsors an annual Microprogramming Workshop, tutorials, IEEE publications, and conferences. The *IEEE Transactions on Computing* occasionally dedicates an issue to microprogramming.

SIGMICRO
Association for Computing Machinery (ACM)   212-869-7440
11 West 42nd Street
New York, NY 10036

SIGMICRO, one of 32 special-interest groups of the ACM, is a forum for practitioners of microprogramming. The group, implementing control logic through the ordered storage of information, cuts across specialties of logic design, operating systems design, and systems architecture. Gerold R. Johnson is currently chair of SIGMICRO, and he can answer questions about the special interest group and its activities. He can be reached at Colorado State University, 202 Engineering, Fort Collins, CO 80523, 303-491-5543. The group publishes:

● *SIGMICRO*—a quarterly journal that includes papers on current microprogramming research and activities. The journal is a sounding board for new technologies for the practitioners in the field. It is available from ACM at the New York address listed above.

Publications

Advances in Microprogramming

Artech House, Inc.   617-326-8220
610 Washington Street
Dedham, MA 02026

The book is a collection of original historical articles on the field of microprogramming, including a description of the origins. Also included is a reference list, and a list of readings. Available for $30.

Microprocessing and Microprogramming

Elsevier Scientific Publications   212-867-9040—New York City office
1000 AE Amsterdam
PO Box 211
The Netherlands

A scientific journal that covers the international flow of information about microprocessing and microprogramming, in addition to reports on technical research and progress, computer survey, state-of-the-art reports, short communications, and announcements pertaining to these areas. Topics covered include theory, simulation, emulation, teaching aids, evaluation aid/diagnostics methods, personal computing, social and economic aspects—all relating to microprocessing, microprogramming, networks, computer structure, modular systems, integrated software/ hardware design, and so on. The annual subscription is $162.32 for ten issues.

Additional Experts and Resources

Scott Davidson

Engineering Research Center   609-639-2221
PO Box 900
Princeton, NJ 08547

Mr. Davidson is an excellent source of information and is willing to answer questions and make referrals to other sources. Mr. Davidson is planning to publish a *Microprogramming Handbook* and can be contacted for further information about the availability of the book, which will include articles on different areas of microprogramming by experts in the field.

Gerold R. Johnson

Chairperson of SIGMICRO   303-491-7708
College of Engineering   303-491-5543
Colorado State University
Fort Collins, CO 80523

Mr. Johnson, the chairman of ACM SIGMICRO, is willing to answer questions about microprogramming. He can also refer you to other sources of information.

William J. Traez
Department of Electrical Engineering
Stanford University
Stanford, CA 94305

Mr. Will Traez, editor of the quarterly journal *SIGMICRO*, is an expert in the field of microprogramming and is very helpful. He will answer questions and make referrals to other sources of information.

Microwaves

Organizations and Government Agencies

International Microwave Power Institute (IMPI)
Tower Suite 520   703-281-1515
301 Maple Avenue West
Vienna, VA 22180

The Institute is concerned about commercial, scientific, and technical applications of microwave power. IMPI holds an annual symposium and various educational short courses. IMPI publications discuss the latest developments in microwave technology and products. They are:

- *Journal of Microwave Power*—a quarterly that includes scientific and technical papers on the newest microwave applications research and engineering. The subscription is $80 a year.
- *Microwave World*—a bimonthly magazine featuring new product developments in cooking equipment. The subscription is $50 a year.
- *IMPI Directory*—lists names, addresses, and phone numbers of all IMPI members. The *Directory* is $10.
- *Digests of Microwave Power Symposia*—the cost of the 1984 edition, which is 240 pages, is $20 for members and $25 for nonmenbers.

IEEE Microwave Theory and Techniques Society
IEEE Headquarters   212-705-7900
345 East 47th Street
New York, NY 10017

The Society's field of interest is microwave theory, techniques, and applications, as they relate to components, devices, circuits, and systems involving the generation, transmission, and detection of microwaves. The Society organizes scientific, technical, and industrial activities. Microwave theory and techniques relate to electromagnetic waves usually in the frequency region between 1-100 GHz; other spectral regions and wave types are included within the scope of the Society whenever basic microwave theory and techniques can yield useful results. The Society publishes the following:

• *Microwave Theory and Techniques*—a monthly publication covering microwave theory, techniques, and applications, as they pertain to components, devices, circuits, and systems involving the generation, transmission, and detection of microwaves. Available for $132 a year.

• Symposia papers—proceedings from the annual International Microwave Symposium. Papers cover such areas as microwave generation, amplification and measurement, filters and components, microwave technology in communication, medicine, transportation and navigation, integrated circuits and computer-aided microwave practices. Available from 1978 to the present.

Publications

Microwave Journal

Horizon House   617-326-8220
610 Washington Street
Dedham, MA 02026

The *Microwave Journal* focuses on the application of devices, subsystems, systems and/or techniques in the electro-magnetic spectrum, from VHF through light frequencies, as well as research, development, and design. Free to qualified recipients, $36 a year for others.

Hayden Publishing Co., Inc.

10 Mulholland Drive   201-288-7520
Hasbrouch Heights, NJ 07604

• *Microwaves and RF*—this monthly magazine provides practical design ideas, timely news, and comprehensive product information to those interested in the high-frequency industry. Application articles that describe the design, specification, and uses of components, subsystems, and systems are emphasized. Free to qualified individuals; $30 a year for others. Subscription includes 11 issues and a directory.

• *Microwaves and RF Product Data Directory*—a worldwide guide on the high-frequency industry, describing each company's product, profile, and offices; published annually. The *Directory* is free to subscribers of the magazine.

Microwave News

PO Box 1799   212-725-5252
Grand Central Station
New York, NY 10163

This monthly report on nonionizing radiation covers the whole radiation spectrum, specializing in the biological effects of the radiation as well as compatibility and interference issues. Available for $200 a year.

Microwave Oven Radiation (539 N)
Consumer Information Center
P.O. Box 100
Pueblo, CO 81002
Information about microwave radiation and safe oven operation is provided in this free seven-page publication.

EW Communications, Inc.
1170 East Meadow Drive   415-494-2800
Palo Alto, CA 94303-4275
● *MSN Microwave Systems News*—this monthly magazine reports vital microwave industry information. It keeps abreast of advances in instrumentation; covers developments; and reviews the presentations, exhibits, and developments of microwave industry shows. Available free to qualified persons.
● *Microwave Systems Designers Handbook (MSDH)*—is written by industry experts to provide a basic reference to modern microwave design. Special articles describe how to use microwave devices effectively, components, and subsystems in design.

Additional Experts and Resources

Several publications regularly carry articles on microwave power and transmission. Some of these are:

Communications News
Harcourt, Brace Jovanovich
124 South First Street
Geneva, IL 60134
312-232-1401
Each October issue presents special, extensive coverage of the field of microwave power; $25 a year, for 12 issues.

Electronic News
Fairchild Publications
7 East 12th Street
New York, N.Y. 10003
212-741-4000
The cost is $28 a year, for 52 issues.

Signal
Armed Forces Communications and
   Electronics Association
5641 Burke Centre Parkway
Burke, VA 22015
703-425-8500
The cost is $23.25 a year, for 12 issues.

International Trade Administration
U.S. Department of Commerce   202-377-1333—Arthur L. Pleasants
14th Street and Constitution Avenue, NW, Room 1015B
Washington, DC 20230

Mr. Pleasants is the industry specialist for radio and television communications equipment. He monitors the industry (SIC code 3662), and is a good contact for information on microwave power and transmissions.

Military Technology

See also: Aeronautics; Star Wars: Antimissile Systems in Space

Organizations and Government Agencies

American Electronic Association

2670 Hanover Street   415-857-9300
Palo Alto, CA 94303

This trade association represents electrical manufacturing companies. Its 2,600 members include computer and defense manufacturing companies, as well as defense software firms. The Association has national committees which monitor defense-related issues, such as procurement, contracting, warrantees, and legislation. These committees frequently produce reports, which are available to the public. The Association also publishes salary surveys, industry productivity reports, and the following:

• *Annual Directory*—this yearly publication provides profiles and product information about each of the Association's 2,600 members. The *Directory* can be used to identify companies involved in defense electronics and applications. The publication is available free of charge to members and costs $90 for non-members.

Association of Old Crows (AOC)

2300 9th Street South, Suite 300   703-920-1600
Arlington, VA 22204

Also known as the National Defense Association, AOC is a professional organization of scientists, engineers, military personnel, and others involved in advancing the science of electronic warfare. AOC has the endorsement of the Department of Defense and it frequently conducts studies for the Pentagon. The Association traces new developments in electronic warfare technology and it maintains a Speakers Bureau to educate others on the subject. It sponsors conferences and publishes a monthly newsletter for members. The Association is compiling a multivolume book titled *History of Electronic Warfare*; the first volume should be available in 1985. AOC also publishes:

• *Journal of Electronic Defense*—a monthly publication providing articles about electronic defense and news of technological advances. The 50-page publication can be found in most libraries and defense agencies. It is available free of charge to members. The annual subscription for nonmembers is $36.

Census/ASM Durable Goods Branch

Bureau of the Census   301-763-7304
Washington, DC 20233

This office prepares the weapons-related sections of the *Census of Manufacturers' Report*. Taken once every five years, in years ending in seven and two, this census covers 450 U.S. industries, four of which are defense-related. Major tables in this *Report* include a table on employment, payroll, hours worked, value of industry shipments, cost of materials consumed, and new expenditures; a table showing detailed types of products; a table with industry breakdown by geographical areas; and a table detailing the material consumed (metals, and so on) in the making of the product. Questions about the contents of the *Report* should be directed to the above office. Sections of the *Report* relating to weapons technology can be purchased by contacting the Superintendent of Documents, U.S. Government Printing Office, Washington, DC 20402, 202-783-3238. Booklets of interest are:

• *Ordnance and Accessories*—this report covers SIC codes 3482, 3483, 3484, and 3489. The 21-page publication (# 34E) costs approximately $5.

• *Ships and Boat Building and Transportation Equipment*— this report provides data on tanks, ships, and other weapons transportation equipment. The 21-page booklet (# 37C) costs approximately $5.

• *Aerospace Equipment—Including Parts*—this publication (# 37B) costs approximately $5.

• *Communication Equipment*—this booklet (# 36D) includes data on search, detection, and navigation instruments. It costs approximately $5.

Center for Defense Information

Suite 303 West   202-484-9490
600 Maryland Avenue, SW
Washington, DC 20024

This nonprofit organization consists of retired military personnel involved in research and analysis of military defense and nuclear weapons. Members deal with defense-related matters, including MX missiles, B-1 bombers, and the overall aspects of defense and nuclear warfare. The Center can provide you with information in these areas. It publishes:

• *Defense Monitor*—issued ten times yearly, this covers the latest technology and defense strategies. A contribution of $25 a year is requested.

Department of Defense

The staff in the Public Affairs Office of each branch of the military will answer questions from the public and make referrals. You can write or call the following offices:

U.S. Department of Defense
Assistant Secretary of Defense for
Public Affairs
Public Correspondence Branch
Washington, DC 20301-1100
202-697-5737

U.S. Department of Defense
Defense Logistic Agency
Kameron Station
Alexandria, VA 22314
703-274-6135

U.S. Department of the Army
Office of Chief of Public Affairs
Washington, DC 20310-1501
202-695-5135

U.S. Department of the Navy
Public Affairs
Office of the Secretary
Washington, DC 20350
202-697-7491

U.S. Department of the Air Force
Director of Public Affairs
Office of the Secretary of the Air Force
Washington, DC 20380
202-694-8010

Marine Corps Headquarters
Division of Public Affairs
Washington, DC 20380
202-694-8010

Electronics Industries Association (EIA)
2001 Eye Street, NW   202-457-4900
Washington, DC 20006

EIA is a trade association that represents all aspects of the electronics industry, including weapons technology and defense electronics. It sponsors conferences and publishes the proceedings. EIA publishes several materials pertaining to military technology, such as:

• *The DOD Electronics Market: A Forecast of Opportunities*—an annual publication containing the proceedings of EIA's fall conference on the subject. It includes an overview of major defense issues; the ten-year outlook for defense electronics; and specific market analysis of U.S. military aircraft and avionics, ballistic missile defense, and space electronics. The 350-page publication costs $30 for members and $60 for nonmembers.

• *The DOD Research and Development Program*—an annual publication containing EIA conference proceedings on the subject. It provides an overview of the DOD research and development budget, as well as presentations by each of the services regarding current and projected research and development budgets. The publication costs $60 for nonmembers and $30 for members.

• *Market Data Book*—published yearly, this covers all aspects of the electronics industry, including defense applications. Articles, data, and graphs are provided along with information about sales, trade, and research and development. The 180-page publication is available for $60.

NASA Scientific and Technical Information Facility

PO Box 8757   301-859-5300
BWI Airport, MD 21240

This Facility has information about reports and other publications concerned with aeronautics, space, and supporting disciplines. The staff can answer questions about a series of continuing bibliographies covering literature in aeronautical engineering, earth resources, management, and so on. An important publication of this office is:

• *Scientific and Technical Aerospace Reports (STAR)*—a biweekly publication, this is NASA's announcement journal. Consisting of a collection of reports about various research and development efforts, it provides comprehensive coverage of aeronautics, space and supporting disciplines, including defense applications. More than 1,000 abstracts appear in each issue. The annual subscription price is $90. Order this and other NASA publications from the Superintendent of Documents, U.S. Government Printing Office, Washington, DC 20402, 202-783-3238.

New England Research Applications Center (NERAC)

Mansfield Professional Park   203-486-4533
Storrs, CT 06268

NERAC is the largest Industrial Applications Center for the National Aeronautics and Space Administration (NASA). As part of NASA's Technology Utilization Program (see description under *Technology Transfer* in this directory), NERAC is dedicated to helping business and industry use aerospace technology. It provides access to PREDICASTS' Defense Markets and Technology (DM&T) Data Base. This data base provides a single source of defense literature on a full range of defense subjects, including land, sea, and air weapon systems, technologies, companies, products, and shipments. Contact the above office for information about services, publications, and data base searches.

Reliability Analysis Center (RAC), IIT Research

Rome Air Development Center/RAC   315-330-4151
Griffiss AFB, NY 13441

RAC collects, analyzes, formats, and disseminates reliability information about microcircuit, discrete semiconductor, and certain electrical/electromechanical components, as well as the equipments/systems in which these components are used. RAC maintains an on-line data base of reliability and experience data, which it will search upon request. The Center's analysis and evaluations are available in the form of compilations, handbooks, and special publications to upgrade and support defense systems reliability. A listing of publications pertaining to defense electronics is available from the Center.

Triservice Industry Information Centers

The Triservice Industry Information Centers provide information about Air Force, Army, and Navy acquisitions, research and development requirements, plans, and future needs. The Centers were established to encourage industry and university participation in solving DOD problems. All Centers have reading rooms where documents can be reviewed. The Centers' services are available to current or potential DOD contractors who are registered for access to DOD information services.

Triservice Industry Information Center
5001 Eisenhower Avenue
Alexandria, VA 22333-0001
Attn: (Specific Branch)
202-274-9305—Air Force Office (AFIFIO)
202-284-9305—AUTOVON
202-274-8948—Army Office
202-274-9315—Navy Office (NARDIC)

Triservice Industry Information Center
1030 East Green Street
Pasadena, CA 91106
Attn: (Specific Branch)
818-792-3192—Air Force Office
818-360-2456—AUTOVON
818-792-7146—Army Office
818-792-5182—Navy Office

Triservice Industry Information Center
Air Force Wright Aeronautical Laboratories
(AFWAL/TST)
Wright-Patterson AFB, Ohio 45433
Attn: (Specific Branch)
513-258-4259—Air Force Office
513-785-5572—AUTOVON
513-258-4260—Army Office
513-258-4261—Navy Office

Publications

Aerospace Daily

Ziff-Davis Publishing Company   202-822-4600
1156 15th Street, NW
Washington, DC 20005

Published Monday through Friday, *Aerospace Daily* provides news about all aspects of the aerospace industry, including defense-related developments. It covers activities of the Department of Defense and NASA, research and development trends, procurement opportunities, and legislative issues. The annual subscription to this eight-page newsletter is $820.

Annual Survey of Manufacturers-Value of Product Shipment

Census/ASM Durable Goods Branch   301-763-7863—Milbren Thomas, Industry Analyst
Bureau of the Census
Washington, DC 20233

This publication provides five-year trend data for all types of manufacturing. Defense-related figures are given for Ordinance and Accessories (SIC 3482, 3483, and 3489), ships, tanks, and communications and aerospace equipment. Mr. Thomas, an industry analyst covering weapons, will provide the data you need, or you can purchase the entire report from the Superintendent of Documents, U.S. Government Printing Office, Washington, DC 20402, 202-783-3238. The cost is $2.75.

Armed Forces Journal International
1414 22nd Street, NW, Suite 104
Washington, DC 20037

International in scope, this monthly provides information about defense and weapons technology, weapons research, military policy, all branches of the U.S. Armed Forces, Pentagon services, and congressional activities. Each February issue is dedicated to defense electronics. The annual subscription fee for the 100-page *Journal* is $19.

Aviation Week and Space Technology
1777 North Kent Street, Suite 710   202-463-1770
Arlington, VA 22209

Published every Monday, this weekly covers aerospace developments, aeronautics, military and weapons technology, avionics and defense electronics, space technology, satellite technology, satellite communication, and missile engineering. The 100-page publication is available for $45 a year.

Commerce Business Daily
Superintendent of Documents   202-783-3238
U.S. Government Printing Office
Washington, DC 20402

This publication contains notices of opportunities and awards for federally sponsored research and development in all areas, including defense. Published Monday through Friday (except on Federal legal holidays) by the U.S. Department of Commerce, in cooperation with the federal agencies, it provides a daily list of U.S. Government RFPs (Requests for Proposals), invitations for bid, contract awards for possible subcontracting leads, sales of surplus property, and foreign business opportunities. Generally, this list is accompanied by a brief description of the proposed procurement action and information about how to get a copy of the formal solicitation or announcement. Copies of the publication are available for reference purposes at Department of Commerce field offices and in most public libraries. The publication may be obtained on a subscription basis for $100 a year via second-class mail or for $175 via first-class mail, from the Superintendent of Documents (address listed above). A purchase order must be accompanied by payment.

An electronic, on-line edition of the *Commerce Business Daily* is available from any of several Department of Commerce contractors. Interested parties may contact them for full details. The contractors are:

DIALOG Information Systems, Inc.
3460 Hillview Avenue
Palo Alto, CA 94304
800-227-1927

DMS/ONLINE
100 Northfield Street
Greenwich, CT 06830
203-661-7800

United Communications Group
8701 Georgia Avenue
Silver Spring, MD 20910
800-638-7728—Joanne Gionnola
301-589-8875—Call Collect

Current Industrial Report-Selected Electronic and Associated Products Including Telephone and Telegraph Apparatus

Data User Services Division   301-763-5353
Customer Services (Publications)
Bureau of the Census
Department of Commerce
Washington, DC 20233

Published each fall, this report contains product shipment data for all known manufacturers of electronic equipment. The survey is done by the Bureau of Census and includes most electronic equipment purchased by the armed forces. The Bureau of Census obtained this information by contacting all known manufacturers in SIC (Standard Industrial Code) 3662. For information about the figures contact the above office. The report (MA 36N) must be ordered in writing, and it costs $2.65.

Defense 84

American Forces Information Service
1735 North Lynn Street, Room 210
Arlington, VA 22209

This monthly publication, produced by the Department of Defense, covers all aspects of defense technology and applications. The publication is distributed free of charge to DOD agencies and contractors. Others can subscribe for $23 a year from the Superintendent of Documents, U.S. Government Printing Office, Washington, DC 20402, 202-783-3238.

Defense Electronics

1170 East Meadow Drive   415-494-2800
Palo Alto, CA 94303

This monthly publication covers the fields of electronic warfare and military electronics, including such disciplines as reconnaissance, surveillance and jamming, distance measuring equipment, computer/microprocessors, navigation, communications, weapons guidance, electro-optics, telemetry, space, and satellites. It provides news, technical reports, and information about new products. The annual subscription is $28.

Defense Week

915 15th Street, NW, Suite 400   202-638-7430
Washington, DC 20005

Published each Monday, this weekly provides coverage of defense technology, policies, strategies, and electronics. It covers news in the U.S., Japan, Great Britain, France, Australia, and Germany. The 20-page publication is available for $595 a year.

Electronic News

7 East 12th Street   212-741-4470
New York, NY 10003

A weekly publication covering the electronics industry, *Electronic News* provides information about defense electronics and the latest developments in weapons technology. It reports on defense contracts and procurement programs, as well as defense appropriations and legislation. The 100-page publication is available for $30.

Electronics Week

1221 Avenue of the Americas   212-512-2484
New York, NY 10020

Geared toward electronic engineers, this weekly publication provides articles and news about the field of electronics. It profiles military applications of electronics and weaponry. The annual subscription fee for the 100-page publication is $32.

Electrotechnology Newsletter

National Technical Information Center (NTIC)   703-487-4630
5285 Port Royal Road
Springfield, VA 22161

This weekly publication contains bibliographic information and summaries of all federal electrotechnology-related reports released during the week. Since a majority of the reports are cited by the Department of Defense or NASA, the *Newsletter* is a good resource for keeping up to date about research and development efforts in defense electronics. The full text of most reports can be obtained from NTIC or the author. The annual subscription rate is $65.

Jane's Publishing Company

135 West 50th Street   212-586-7745
New York, NY 10020   212-247-3087

Jane's publishes several materials pertaining to military technology. In addition to those described below, it also issues annual publications covering fighting ships, sea skimmers, electronics, avionics, and many other areas.

• *Jane's Defense Weekly*—international in scope, this publication covers the wide field of sophisticated modern weaponry. It also contains news about political developments and procurement decisions. The annual subscription fee is $60.

• *Jane's Weapons Systems*—updated yearly, this reference manual provides nomenclature descriptions of weapons systems. The 40-page publication costs $156.50.

Selling to the Military

Superintendent of Documents   202-783-3238
U.S. Government Printing Office
Washington, DC 20402

Updated annually, this publication describes information sources for defense procurements, as well as civilian agency procurements. It contains a complete listing of the major buying offices for the Army, Navy, and Air Force. The publication also describes the basic steps to be taken to find sales opportunities within the military. The 40-page publication costs $6. When ordering, refer to GPO stock number 008-000-00392-1.

Survey of Research and Development

Bureau of the Census   301-763-5616—Bureau of the Census
Industry Division   202-634-4648—National Science Foundation
Washington, DC 20233
1800 G Street, NW, Room L-602
Washington, DC 20550

This annual survey is produced by the National Science Foundation and the Bureau of Census. In terms of military technology, it covers such topics as aircraft and guided missiles, electrical components, communications and telecommunications equipment, ammunitions and ordinances, and electrical equipment. Information is provided about sales, as well as research and development expenditures of government, private industry, and foreign countries. A detailed 84-page report is issued every other year. In alternating years a summary report, six to eight pages long, with tables, is published. Both are available at no cost from the above offices.

100 Companies Receiving the Largest Dollar Volume of Prime Contract Awards

Director for Information Operations and Reports   202-694-5298
Washington Headquarters Services
The Pentagon, Room 1C535
Washington, DC 20301

This report is published annually by the Office of the Secretary of Defense. It presents summary data about the Department of Defense's top 100 prime contractors for the previous fiscal year. A variety of data is provided about these companies and their subsidiaries, such as dollar amounts received, ranking of companies, and a breakdown of dollars awarded to subsidiaries. The 21-page publication (# P01) costs $12.

Data Bases

Aerospace Daily On-Line

Ziff-Davis Publishing Company   202-822-4600
1156 15th Street, NW
Washington, DC 20005

An on-line network, this service provides current, historical, and forecast information about the entire aerospace industry, including programs, technology, research and development, engineering, manufacturing, and industry trends and operations. It offers access to nine data bases: Aerospace Daily Abstracts; Aerospace Daily Historical; Defense Budgetscan; NASA Budgetscan; Defense Electronics Magazine; International Countermeasures Handbook; Commerce Business Daily; Forecast Associates; and Foreign Military Sales Profiles by Country. The data bases provide current information about weapons technology and defense electronics.

Defense Technical Information Center Collection (DTIC On-Line, DROLS)

U.S. government agencies, contractors, subcontractors,
grantees, and universities (DOD contractors may also
contact their contract monitor):

Defense Technical Information Center (DTIC)   202-274-6867
U.S. Department of Defense
Building 5, Cameron Station
Alexandria, VA 22314

Others: 703-487-4600
National Technical Information Service (NTIS)
5285 Port Royal Road
Springfield, VA 22161

DTIC, the Department of Defense's information clearinghouse, maintains data banks with information about planned, ongoing, and completed DOD-related research and develement activities. The collection is multidisciplinary in scope and spans all fields of science and technology, covering such topics as military sciences, aeronautics, missile technology, ordinance, navigation, space and technology, nuclear science, biology, chemistry, environmental science, oceanography, computer science, and human factors engineering. The four principal data banks are the *Technical Reports (TR) Program*, which contains bibliographic information on classified/unclassified reports generated by federally sponsored research; the *Research and Technology Work Unit Information System (WUIS)*, containing information about current research being performed; the *Research and Development Planning (R&DP)* data base, consisting of information about proposed projects; and the *Independent Research and Development (IR&D)* data bank, with information that contractors have supplied to DOD regarding their independent research efforts.

Universities, U.S. government agencies, and associated contractors, subcontractors, and grantees are eligible for most DTIC services. In addition, research and development organizations without current contracts may become eligible for DTIC services by obtaining a military service authorization under the defense potential contractors programs. If you are eligible for DTIC services, you can obtain direct on-line access to the DTIC system, have searches and printouts done free of charge, and purchase DTIC documents for a minimal fee—generally $1.50 to $3. A large percentage of data stored on DTIC is available from NTIS for public access. Therefore, if you are not eligible for DTIC, you can contact NTIS to obtain a limited search of DTIC. NTIS charges for its services. Portions of the DTIC data base are also available on DIALOG, BRS, and SDC.

Data Resources, Inc. (DRI)
24 Hartwell Avenue   617-863-5100
Lexington, MA 02173

DRI offers the following data bases:

• *Defense Data Bank*—provides time series data useful for analyzing defense spending and costs, as well as how these affect defense-related industries. Concept coverage includes: DOD outlays, obligations, personnel; contract awards by program and state; defense purchase price deflators; select acquisition report; and world military expenditures and arms transfers. Coverage is provided at the U.S. national and state levels on an annual, quarterly, and monthly basis. Data sources are U.S. Arms Control and Disarmament Agency, U.S. Department of Commerce, U.S. Department of Defense, and U.S. Department of Treasury.

• *DMS/ONLINE*—offers a comprehensive and detailed collection of textual and statistical information on the defense and aerospace industries. Specific contract awards, defense programs, market analysis, and special studies are included. The three major components of the data base are contract awards, market intelligence, and special services.

Additional Experts and Resources

Arthur Pleasants
Industry Specialist, Electronics   202-377-2872
U.S. Department of Commerce
14th Street and Constitution Avenue, NW, Room 1015B
Washington, DC 20230

Mr. Pleasants is an industry specialist covering section 3662 of the Standard Industrial Code, which includes most defense electronic equipment. He can provide an overview of the industry, provide shipment figures, and tell you where to find lists of defense electronics manufacturers.

Listed below are the leading prime contractors for the U.S. Department of Defense. The staff in each company's Public Affairs Office may be able to answer questions and send you literature.

Boeing-Military Airplane Company
PO Box 7730
Wichita, KS 67277-7730
316-526-2121
Major Products:
C-135 STRATOLIFTER, B-52 STRATO FORTRESS, and E-3A aircraft
Miscellaneous electronics for B-1 bomber

ZAGM-86 missile system
AWACS
Research, development, test, and evaluation for electronics and communication equipment, aircraft, missile and space systems, and space transportation systems.

General Dynamics Corporation

Pierre Laclede Center
St. Louis, MO 63105
314-889-8200—Main number
314-889-8566—Public Affairs
Major Products:
F-16 and F-111 fighter aircraft
Nuclear submarines
TOMAHAWK and STRINGER missile
   systems
MK-15 close-in weapon system
M-1 tank

General Electric

Ordinance Systems Division
100 Plastics Avenue
Pittsfield, MA 01201
413-494-1110
Major Products:
Nuclear reactors for submarines and
   aircraft carriers
J-79, J-85, TF-34, and F-101 turbofan and
   turbojet engines
Space vehicle components
Guided missile subsystems
Transmission components for M-2 infantry
   fighting vehicles
Armament training devices

Grumman Corporation

Bethpage, NY 11714
516-575-0574
Major Products:
F-14 TOMCAT, A-6 INTRUDER, E-2
   HAWKEYE, C-2 GREYHOUND, and
   EA-6B PROWLER aircraft
Electronics for F-16 FIGHTING FALCON
Research, development, test, and
   evaluation for aircraft, electronics and
   communications equipment, and missile
   and space systems

Litton Industries, Inc.

360 North Crescent Drive
Beverly Hills, CA 90210
213-859-5000
800-421-0768
Major Products:
Guided missile cruisers, battleships, and
   destroyers
Various electronics and communications
   equipment

Lockheed-Missiles and Space Company, Inc.

PO Box 504
Sunnyvale, CA 94086
408-742-4321
Major Products:
C-5 GALAXY, C-130 HERCULES, and P-3
   ORION aircraft
TRIDENT and POLARIS missile systems
Amphibious assault ships
Research, development, test, and
   evaluation for missile and space systems,
   aircraft and electronics, and
   communications equipment

Martin Marietta Corporation

6801 Rockledge Drive
Bethesda, MD 20817
301-897-6000
Major Products:
PERSHING and TITAN missile systems
AH-64 APACHE aircraft
Guided missile cruisers
Operation of government ammunition
   facility
Research, development, test, and
   evaluation for missile and space systems,
   and electronics and communications
   equipment

McDonnell Douglas Corporation
PO Box 516
St. Louis, MO 63166
314-232-0232—main number
314-232-5911—Public Affairs
Major Products:
F-18 HORNET, F-15 EAGLE, AV-8
   HARRIER, and KC-10 aircraft
HARPOON missile system

Raytheon Company
141 Spring Street
Lexington, MA 02173
617-862-6600
Major Products:
PATRIOT, HAWK, SPARROW, NATO SEA
   SPARROW, and SIDEWINDER missile
   systems
Various electronics and communications
   equipment

Rockwell International Corporation
600 Grant Street
Pittsburgh, PA 15219
412-565-2000
Major Products:
B-1 bomber
Research, development, test, and
   evaluation for aircraft, and missile and
   space systems
Space vehicles
Various electronics and communications
   equipment

Tenneco, Inc.
1010 Milam Street
PO Box 2511
Houston, TX 77001
713-757-2131
Major Products:
Aircraft carriers and nuclear submarines

United Technologies Corporation
One Financial Plaza
Hartford, CT 06101
203-728-7000
Major Products:
F-100, TF-30, TF-33, and J-52 turbofan and
   turbojet aircraft engines
UH-60 UTTAS, CH-53 SEA STALLION,
   and SH-60 SEAHAWK helicopters

Minicomputers

Begins next page. See also: Computers (General)

Organizations and Government Agencies

American Federation of Information Processing Societies (AFIPS)
AFIPS Press   703-620-8900
1899 Preston White Drive
Reston, VA 22091

AFIPS disseminates information on all aspects of the computer industry, including hardware, software, and minicomputers. It works closely with government agencies on such areas as standardization and legislation. The Foundation sponsors two conferences a year covering many aspects of the computer field, including software and its applications for minicomputers, microcomputers, and mainframes, with a special emphasis on computer architecture. The proceedings, a collection of the technical papers presented at the conference, are published. The 70-page annual publication is available for $80.

Digital Equipment Corporation
146 Main Street   617-264-1751—Information Services
Maynard, MA 01754

Digital is a manufacturer of minicomputers, and is involved in the research and development of these computers, as well as microcomputers and peripherals. All sorts of printed materials about the work of Digital are available, including a company brochure, technical briefs and specifications, and product information. Also published is:

• *Digital Reference Service*—a six-volume publication describing Digital Corporation and its product history, sales, services, software and hardware products, word processors, customer support services, and so on. The publication is updated quarterly, and is available for $300.

General Data Communication, Inc.
Middlebury, CT 06762-1299   203-574-1118

General Data is a leading manufacturer of minicomputer systems. It makes available all sorts of printed materials about the work being done, including a company brochure, technical briefs and specifications, and product information.

Hewlett Packard
3000 Hanover Street   415-857-1501
Palo Alto, CA 94304   800-334-8083

Hewlett Packard manufactures a wide range of computer products, including microcomputers and minicomputers. A wide variety of sales and product information is available upon request. A monthly newsletter, which provides news about the latest developments in industry, is published, as well as:

• *Computer Advance*—a bimonthly publication providing profiles about new products and software available. It also includes a calendar of local trade shows. The 16-page publication is free to customers.

IBM

1130 Westchester Avenue
White Plains, NY 10604

914-765-1900
800-492-5578

IBM is the leading manufacturer of minicomputers and mainframes. It provides all sorts of printed material about its work, including a company brochure, technical briefs and specifications, and product information. There are many toll-free numbers within IBM for those who have questions about specific products. Questions should be directed to the regional IBM offices located throughout the country.

Personal Computing (SIGPC)

Association for Computing Machinery (ACM)
11 West 42nd Street
New York, NY 10036

212-869-7440

One of 32 special-interest groups within ACM, SIGPC focuses on the design and applications of computer systems for personal use, including personal computer systems for the home, clerical work, small business, management, and recreation. Additional areas of interest include the technology of such systems in software and hardware, and emphasis on techniques appropriate to the integration of such tools as graphics, speech, data management, and music systems. The group publishes a newsletter.

SIGSMALL

Association for Computing Machinery (ACM)
PO Box 12115
Church Street Station
New York, NY 10249

212-869-7440

SIGSMALL is a special-interest group dealing with all aspects of small computers. A conference is held each year, as well as a three-day symposium, which covers hardware, software, peripherals, and computer languages. The proceedings are published annually, as well as:

• *SIGSMALL*—a quarterly newsletter that is a collection of papers covering all aspects of minicomputers and microcomputers, with a special emphasis on software. The publication, 32-48 pages long, is available for $9 to members, and for $40 to nonmembers.

Sperry Univac

Division Headquarters
PO Box 500
Blue Bell, PA 19424

215-542-4011

Sperry is a leading manufacturer of minicomputer systems. It will send you all sorts of printed material about the research and work being done, including a company brochure, technical briefs, specifications, and product information.

Publications

Auerbach Publishers

6560 North Park Drive
Pennsauken, NJ 08109

609-662-2070
800-257-8162

Auerbach publishes the following:

• *Data World*—a four-volume publication that provides comprehensive EDP information. Volume I covers computers and peripherals; Volume II covers minicomputers; Volume III covers software; and Volume IV covers data communications. The publication includes charts, product reports, price data, and a comprehensive dictionary of manufacturers. It is updated monthly, and a telephone update service is also available.

• *Software Reports*—is a two-volume report on software for mainframes and minicomputers. Volume I deals with business management and data base use for accounts payable, accounts receivable, and general ledger applications. Volume II deals with operating systems and enhancements with graphics. This service is updated monthly, and a telephone update service is also available. The cost is $725 for both volumes, or $425 a volume.

• *Minicomputer Reports*—this three-volume publication provides evaluation and selection charts, turnkey charts, and a monthly newsletter, *Data Gram*, which focuses on both business and general minicomputers, giving the latest industry news. The service is updated monthly, and costs $745 a year.

• *Complete Technology Library*—is in 14 parts and includes software application and systems, electronic offices and office automation, data communications, peripherals, mainframes, minicomputers, and microcomputers. This service is updated monthly, and is available for $4,610 a year.

Minicomputers

Datapro Research Corporation
1805 Underwood Boulevard
Delran, NJ 08075

609-764-0100
800-257-9406

This three-volume publication, which is updated monthly, reports on minicomputers, microcomputers, and the microprocessor industry. It provides a directory of vendors, product comparisons, product profiles, and user ratings. A monthly *Mininews Newsletter* is also provided, all for $754.

Mobile Phones/Cellular Radio

Organizations and Government Agencies

Mobile Division
Federal Communications Commission (FCC), Room 650   202-632-6450—Steve Markendorff
1919 M Street, NW
Washington, DC 20554

The staff in the Mobile Division of the FCC will answer inquiries about the regulatory aspects of cellular radio and other mobile communications systems, and they will make referrals to experts.

Telocator Network of America (TNA)
Suite 230   202-467-4770
2000 M Street, NW
Washington, DC 20036

TNA is a national trade association representing FCC-licensed radio common carriers in the U.S. It provides general information to the public on all aspects of cellular communications. The library is open to members and students. Others can use it for a small fee. A bibliography of the Radio Common Carrier (RCC) industry is available. TNA publishes:

 • *Telocator*—a monthly magazine that serves the paging and mobile telephone industry. It covers new products, conventions, cellular services, FCC rulings, and important policy issues in the industry. The subscription is $30 a year.

 • *Bulletin*—a weekly newsletter that keeps track of the latest developments in the industry. It covers the same issues as the above magazine but also includes classified ads and a special report each week. The subscription rate is $235 a year.

Publications

TITSCH Communications
PO Box 5727-TA   303-860-0111
Denver, CO 80217

 • *Mobile Radio Handbook*—a standard reference guide for the land-mobile radio industry, this contains complete information on the mobile communications industry, such as relevant government and associations, frequency licensing directory, callbook of industry personnel, suppliers of mobile equipment, manufacturers' representatives, cellular radio data, Specialized Mobile Radio System and Radio Common Carrier data, and a glossary of mobile communications terms. Available for $39.95 a copy.

- *Two-Way Radio Dealer*—a monthly business and technical journal of the mobile communications industry, especially cellular radio. The subscription rate is $24 a year, for 12 issues.
- *Mobile Times*—a monthly news/feature magazine of the mobile radio industry. The subscription rate is $20 a year, for 12 issues.
- *RCR—Radio Communications Report*—a biweekly report covering all aspects of radio, including land-mobile radio, RCCs, cellular radios, new technologies, and marketing. The cost is $19 a year.

FutureComm Publications

4041 University Drive, Suite 304   703-352-1200—Stuart Crump
PO Box 380
Fairfax, VA 22030

This company publishes:
- *Cellular Radio News*—a monthly newsletter devoted exclusively to the mobile and portable radio telephone. It presents a complete and timely coverage of regulatory equipment, RCCs, telco, and other the cellular radio industry. Available for $237 per year.
- *SMR News*—a monthly newsletter covering the specialized mobile radio (SMR) industry. It presents news and features relevant to the SMR industry. The subscription rate is $197 a year.

Telecourier

Shoreline Publishing   305-286-7850
PO Box 308
Palm City, FL 33490

The monthly magazine concentrates on many aspects of the radio common carrier industry. Each issues also has information on the telecommunications field, including the telephone answering service industry. Subjects covered include technical problems, reports on research, regulatory policy, financial analyses, and engineering reports, RCC profiles, marketing and sales analyses, manufacturer and supplier profiles. Available free of charge to RCCs in the U.S. and Canada. Others can subscribe for $25 a year.

Mobile Phone News

Phillips Publishing, Inc.   301-986-0666
7315 Wisconsin Avenue, Suite 1200N
Bethesda, MD 20814

An eight-page biweekly, the newsletter is devoted to new-venture mobilecommunications, with cellular radio being the major topic. It covers the latest market and regulatory developments, new technology products and services, the paging market, updates on cellular battles and full settlements at FCC, and profit opportunities. The subscription rate is $287 a year.

Industrial Communications

A weekly newsletter for the mobile radio industry. It provides information on new technologies, reliable solutions to interface problems, FCC's latest actions and upcoming activities, and so on. The cost is $195 a year.

Articles on different aspects of cellular radio are published regularly in the following magazines:

Communications
PO Box 1077
Skokie, IL 60077

Communications News
124 South First Street
Geneva, IL 60134
312-232-1400

Telecommunication Reports
Business Research Publications, Inc.
1293 National Press Building
Washington, DC 20045
202-347-2654

Additional Resources and Experts

International Trade Administration
14th Street and Constitution Avenue, NW, Room 1201   202-377-4466
Washington, DC 20230

ITA has specialists for the industrial and trade aspects of the mobile communications industry. Linda Bawer is a specialist in cellular communications, and Arthur Pleasants is a specialist in mobile, radio, and television communications. They will answer questions and tell you about other good sources of information.

Modems

Organizations and Government Agencies

Electronic Industries Association (EIA)
Standards Sales Office   202-457-4966
2001 Eye Street, NW
Washington, DC 20006

A trade association consisting of electronics manufacturers, EIA develops internationally recognized engineering standards. Standards are written by committees consisting of representatives of companies belonging to EIA. The following modem standards are available from the Association:

• *RS-232-C*—deals with the 25 pin interface. The cost is $13. Application notes cost $5.

• *RS-449*—deals with the 9 pin and 37 pin interfaces. The cost is $19. Application notes cost $11.

DATAQUEST

1290 Ridder Park Drive   408-971-9000—Anne Whitehurst, Marketing
San Jose, CA 95131

DATAQUEST is a market research firm dealing with modems within its telecommunications group, a subdivision of its Information Systems area. DATAQUEST is also involved in peripherals, semiconductors, and design and manufacturing automation. It can answer questions and supply publications, including reports on modems, to its clients. DATAQUEST sponsors conferences, which are open to the public.

Publications

Cahners Publishing Co.

270 St. Paul Street   303-388-4511
PO Box 17452
Denver, CO 60206

Cahners has several publications covering modems, including:

• *Mini-micro Systems*—modems are likely to be covered in the March issue and in two of the magazine's three digest issues. The magazine covers modems at three levels: the integrated circuit, board, and box. It is directed toward the following three classes of purchasers, all of whom qualify for a free subscription: original equipment manufacturers, including manufacturers of computers; value-added resellers; and sophisticated end users who can buy subsystems and configure them. Published monthly; the cost to nonqualified subscribers is $45 a year.

• *Business Computer Systems*—this magazine is for business users of small computer systems. The monthly periodical translates technical concepts into the businessman's language. It is not directed toward engineers or experts in technology. It is available at no charge to qualified U.S. business-based personnel. It is also sold on newstands for $2.95 an issue and is available to nonqualified readers for $35 a year.

Telecommunications Magazine

Horizon House   617-326-8220
610 Washington Street
Dedham, MA 02026

Telecommunications Magazine is a monthly publication covering a full range of topics on modems. Topics include applications, technical aspects, and new products. The magazine is directed toward people employed at a management level with some responsibility for communications systems. A free subscription is available to such people. Others can subscribe for $36 a year.

Additional Experts and Resources

The Science and Electronics Group
Office of Telecommunications   212-377-2006
International Trade Administration
U.S. Department of Commerce
14th Street and Constitution Avenue, NW
Washington, DC 20230

William J. Sullivan and Arthur Pleasants are the contact persons in this Group. They have a file on modems and can answer some general questions.

George Clark
Manager, Data Communications and Interfaces Group   301-921-3723
Division 652—Computer System Components Division
Building 225, Room A216
National Bureau of Standards
Washington, DC 20234

George Clark can answer questions about modems, interconnection of modems, and interface standards, and can refer people to other experts.

Rixon
2120 Industrial Parkway   301-622-2121
Silver Spring, MD 20904

Rixon is a manufacturer of modems. Staff members can answer technical and other types of questions. The company has literature on the products it sells.

The following is a listing of major modem manufacturers. Details about their products can be obtained by contacting them directly.

Anchor Automation
6624 Valijean Street
Van Nuys, CA 91406
818-997-6493

Hayes Microcomputer Products, Inc.
5923 Peachtree Industrial Boulevard
Norcross, GA 30092
404-449-8791

Anderson Jacobson
521 Charcott Avenue
San Jose, CA 95131
408-263-8520

Lexicon Corporation of Miami
1541 NW 65th Avenue
Ft. Lauderdale, FL 33313
305-792-4400

Bizcomp
532 Mercury Drive
Sunnyvale, CA 94086
408-733-7800

The Microperipheral Corp.
2743 151st Place, NE
Redmond, WA 98052
206-881-7544

Multi-Tech Systems
82 Second Avenue, SE
New Brighton, MN 55112
612-631-3550

Novation
20409 Prairie Street
Chatsworth, CA 91311
818-996-5060

Quest Electronics
PO Box 4430E
Santa Clara, CA 95054
408-988-1640

Racal-Vadic
222 Caspian Drive
Sunnyvale, CA 94086
408-744-0810

TNW Corp.
3351 Hancock Street
San Diego, CA 92110
619-296-2115

US Robotics, Inc.
1123 Washington West
Chicago, IL 60607
312-733-0497

Universal Data Systems
5000 Bradford Drive
Huntsville, AL 35805
205-837-8100

Multipoint Distribution Service

Organizations and Government Agencies

MDS Industry Association
655 15th Street, NW, Suite 320
Washington, DC 20005

202-639-4410—Bonnie Guthrie

This trade association follows legislative and regulatory issues in the MDS industry. Bonnie Guthrie is a good source of information on general MDS questions, and she can refer you to other experts in the field.

Federal Communications Commission (FCC)
1919 M Street, NW
Washington, DC 20554

202-632-7566—Public Reference Room
202-634-1798—Charles Gratch

The Public Reference Room (#331) maintains MDS station files and an inventory of pending applications, by location, service area, and so on.

Mr. Gratch, an electronics engineer, is one of the MDS experts at the FCC. He handles all the MDS applications and knows all the FCC rules, as well as the technology behind MDS. He will answer questions on all aspects of MDS.

Publications

Paul Kagan Associates, Inc.
26386 Carmel Rancho Lane   408-624-1536
Carmel, CA 93923

Paul Kagan Associates publish the following:

● *Multicast*—is a twice-monthly newsletter geared toward people in the MDS industry, including station owners, pay TV operators, program suppliers, and equipment manufacturers. It is read by communications attorneys, as well as federal and state regulators. The publication covers all aspects of the industry, including regulation, equipment manufacture, and systems operation. The annual subscription is $375 a year, for 24 issues.

● *MDS Data Book*—a complete directory of companies, markets, and statistics covering all aspects of the MDS industry. The cost is $35.

New Electronic Media
Broadcast Marketing Company   415-777-5400
450 Mission Street
San Francisco, CA 94105

The major five-volume reference guide covers different aspects of the new electronic media. Volume I, *The Competitive Scramble for the Pay Television Market,* deals with the following areas: pay cable, subscription television, multipoint distribution service, satellite master antenna television systems, low-power television, satellite subscription television, and so on. The annual subscription rate is $50.

Multichannel News
300 South Jackson Street, Suite 450   303-393-6397
Denver, CO 80209

This weekly industry newspaper is geared toward new electronic media company executives, but is of interest to anyone in the industry. It covers the new electronic media industry, primarily pay TV (cable, MDS, STV, DBS, SMATV). It deals with programming, systems, services, equipment, and finance. To order, send $27.50 for a one-year subscription to PO Box 1124, Dover, NJ 07801.

Additional Experts and Resources

Richard L. Vega
Vega and Associates, Inc.   301-437-7000
PO Box 191
Pasadena, MD 21122

Mr. Vega is an engineering consultant actively engaged in multipoint distribution service design and construction. He is currently president of the Microwave-Common Carrier Association, a trade association representing the MDS industry.

Music, Instructional Computing

See also: Computer Music

Organizations and Government Agencies

Music Educators National Conference (MENC)

1902 Association Drive   703-860-4000
Reston, VA 22091

A national association of music teachers, MENC is involved in exchanging information, learning about new technological advances, and promoting music education. It sponsors national and regional conferences, the proceedings of which are available to the public in the form of tapes and transcripts. The Conference is developing an on-line data base relating to instructional computing music. The staff can answer questions and make referrals. The organization publishes:

● *Music Educators Journal*—published monthly during the academic year, this periodical covers music education. Each January issue is devoted to high-tech devices and applications in music education, including the use of computers in teaching music. Instructional computing articles range from the use of electronic keyboards in elementary classrooms to profiles of specific computer programs and their applications. The *Journal* is available free to members; nonmembers can purchase individual issues for $4.

● *Journal of Research in Music Education*—this quarterly covers both experimental and theoretical research. It contains articles about computer systems and applications, investigative studies in the psychology of music, and information about varied musical applications and methodologies. The annual subscription is $8 for members and $15 for nonmembers.

National Association of Music Merchants

500 North Michigan Avenue   312-527-3200
Chicago, IL 60611

This trade Association represents the interests of music merchants. The staff in the Information Department can answer questions relating to products, producers, distributors, and various aspects of the computer music industry. The Association publishes several reports pertaining to the management and operational aspects of the music retail market. It also publishes:

● *Resource Directory*—a listing of computer software programs relating to music instruction. The 30-page publication is available for free.

Publications

Symposium Magazine
College of Music Society   303-449-1611
14444 15th Street
Boulder, CO 80302

Published twice a year, this *Magazine* covers music in higher education. It occasionally features articles about the use of computers in teaching music, as well as computer applications in music. The 200-page magazine is available for $16 a year or $10 an issue.

Additional Experts and Resources

David Peters, Ph.D.
Chairman, Music Education Department   217-333-0675
University of Illinois
Urbana, IL 61801

Dr. Peters develops computer music instructional programs for instrumental music. He is founder of the National Consortium for Computer Based Music and is chairman of the Music Education Department at the University of Illinois. Dr. Peters is available to answer questions and make referrals.

The following companies produce 95 percent of instructional computing music software. The staff may be able to provide you with brochures, catalogs, and information about instructional music computing.

Alfred Publishing Company
I5335 Morrison Street
Sherman Oaks, CA 91403
800-292-6122

Electronic Courseware Systems
309 Windsor Road
Champaign, IL 61820
217-359-7099

Silver Burdett Publishing Company
5203 Madison Road
Cincinnati, OH 45227
800-542-2007

Temporal Acuity Products, Inc.
Building I, Suite 200
300-120th Avenue, NE
Bellevue, WA 98005
800-426-2673
206-462-1007

Wenger Corporation
PO Box 441
Owatonna, MN 55060
800-533-0393

William C. Brown Publishing Company
2460 Kerper Boulevard
Dubuque, IA 52001
319-588-1451

National Security

See: Smuggling of High Technology

NMR

See: Magnetic Resonance Imaging

Noise Control

See: Acoustics

Non-Invasive Medical Technology

See: Magnetic Resonance Imaging; Medical Technology; Ultrasound in Medicine

Nuclear Energy

See: Nuclear Technology

Nuclear Magnetic Resonance Imaging

See: Magnetic Resonance Imaging

Nuclear Medicine

Organizations and Government Agencies

American College of Nuclear Medicine (ACNM)

8215 Westchester, Suite 135   214-987-1222
Dallas, TX 75225

ACNM is an association of physicians and scientists specializing in nuclear medicine. It holds semiannual scientific and educational meetings covering various topics in nuclear medicine. The proceedings are published and are available from the above address. Also published is a quarterly membership newsletter covering news in the field of nuclear medicine.

American College of Nuclear Physicians (ACNP)

1101 Connecticut Avenue, NW, Suite 700   202-857-1135
Washington, DC 20036

ACNP, an association of 1,500 members, disseminates information on the latest advances in nuclear medicine. An annual three-day conference is held, which consists of seminars and meetings on various topics in nuclear medicine. The proceedings are available on audiotape. ACNP also publishes:

● *ACNP Scanner*—a monthly newsletter that provides information on advances in research, diagnosis, treatments, and legislative issues pertaining to nuclear medicine. The newsletter, six to seven pages long, is available free of charge to members.

Medical Radioisotopes Research Group

Los Alamos National Laboratory   505-667-4675
Mail Drop J 519
Los Alamos, NM 87545

The Group maintains a laboratory with the most powerful linear accelerator, which produces isotopes for use in nuclear medicine. The Group is exploring the use of the isotopes in both diagnostic and therapeutic applications for tumors. The Laboratory is also involved in radiochemical separations, quality assurance testing, analytical testing, radiopharmaceutical research, biomedical generator developments, and animal distribution studies. A free listing of medically useful radioisotopes is available.

Dr. H. A. O'Brien, an expert in the field of nuclear medicine, is available to answer questions and make referrals.

Nuclear and Plasma Science Society (NPSS)
Institute of Electrical and Electronics Engineers (IEEE)   619-450-9811
Gamma-Metrics
5550 Oberlin Drive
San Diego, CA 92121

NPSS consists of engineers who work with doctors in developing, using, and interpreting the results of nuclear medicine as it pertains to detectors and instrumentation. The Society sponsors a three-day seminar each fall, covering many issues in the field of nuclear science. NPSS publishes:

• *IEEE Transactions on Medical Imaging*—a quarterly publication specializing in the covering equipment and instrumentation used in detecting images within organs and tissues of the body. The publication, 50-100 pages long, is available only to members.

• *Transactions on Nuclear Science*—a bimonthly publication that features articles on tomography, instrumentation, and detectors used in nuclear medicine. A wide range of nuclear science topics is covered, with an occasional emphasis on nuclear medicine. The 100-page publication is available only to members.

Nuclear Medicine Department
National Institutes of Health (NIH)   301-496-5675
Bldg. 10, Room 1C 401
9000 Rockville Pike
Bethesda, MD 20205

NIH's Nuclear Medicine Department is actively involved in the research and development of scintillation imaging, consisting of gamma rays and x-rays, in an effort to further the use of nuclear medicine. The Department oversees hospitals to ensure that the proper specifications and guidelines are followed. Papers with its research findings are submitted to other medical journals. NIH staff members are available to answer questions and make referrals in the field of nuclear medicine.

Publications

Clinical Nuclear Medicine
J. B. Lippincott Publishers   215-238-4200
East Washington Square   800-638-3030
Philadelphia, PA 19105

This monthly journal provides current technical information in the field of clinical nuclear medicine, with an emphasis on scanning, imaging, ultrasound studies, and related subjects. Issues include original papers, clinical case reports, reviews, and announcements of upcoming meetings and conferences. The annual subscription to the 50-page publication is $41; the publication can be ordered from 2350 Virginia Avenue, Hagerstown, MD 21740, 301-824-7300.

European Journal of Nuclear Medicine
Springer-Verlag
Medicine Editorial Department
Post Fach 105280
D 6900 Heidelberg 1
Federal Republic of Germany

U.S. office   201-460-1500
44 Hartz Way
Secaucus, NJ 07094
 A monthly, this *Journal* deals with nuclear medicine, providing information on
the latest research findings and treatment techniques. The 100-page publication is
available from the German office for $126 a year.

Seminars in Nuclear Medicine
Grune and Stratton, Inc.   212-614-3000
111 Fifth Avenue
New York, NY 10003
 A quarterly, this publication consists of a collection of papers covering research
and treatment in nuclear medicine. The 100-page publication is available for $54 to
physicians and for $62 to hospitals.

Additional Experts and Resources

**Most hospitals and medical centers in the U.S. have a nuclear medicine laboratory. Often the
staff at these centers can answer technical questions and direct you to additional
information sources.**

Dr. R. Carroll Maninger
Lawrence Livermore National Laboratory   415-422-6905
PO Box 808
Livermore, CA 94550
 Dr. Maninger is a senior staff scientist involved in the fusion program at the
Laboratory. He works with instrumentation and management of nuclear waste, and
is available to answer questions and make referrals to other experts in the field.

Dr. Eugene Vinciguerra
145 West 58th Street   212-582-3919
New York, NY 10019
 Dr. Vinciguerra is the director of the American College of Nuclear Medicine
and the Secretary of the American Board of Science in Nuclear Medicine. He is a
noted authority in the field, and is available to answer questions and make referrals
to other experts in the field.

Nuclear Power and Power Plants

See: Nuclear Technology

Nuclear Power Systems, Applications in Space

See: Space Nuclear Power

Nuclear Technology

Organizations and Government Agencies

American Nuclear Energy Council (ANEC)
410 First Street, SE   202-484-2670
Washington, DC 20003

A trade association interested in the peaceful application of nuclear energy, ANEC supports the development of nuclear power as an energy source. It represents the interests of the U.S. nuclear industry before the federal government. ANEC publishes a weekly newsletter, for members only, which covers legislative issues and reviews bills pertaining to nuclear energy and power plants.

American Nuclear Society (ANS)
555 North Kensington Avenue   312-352-6611
La Grange Park, IL 60525

ANS was established to advance science and engineering in the nuclear power industry. Its aim is to promote the peaceful use of the atom. ANS disseminates information, promotes research, conducts scientific and technical meetings, and works with government and regulatory agencies, educational institutions, and other organizations dealing with nuclear power issues. It publishes the following:

• *Nuclear News Monthly*—a monthly publication similar in format to *Time* magazine. It covers political developments and provides updates and articles about all aspects of nuclear energy, such as radio isotopes and reactors. The annual subscription fee is $130.

• *Nuclear Science and Engineering*—this monthly focuses on nuclear science and engineering. It provides news about the field and ongoing research. The annual subscription rate is $280.

 • *Nuclear Standard News*—a monthly, four to six pages long, providing the latest information about standards and regulations in the nuclear field. The annual subsciption rate is $765.

 • *Nuclear Technology*—this monthly publication focuses on the engineering aspect of nuclear technology. The annual subscription fee is $380.

 • *Fusion Technology*—this bimonthly publication covers fusion technology and fusion energy. The annual subscription fee for this publication plus one supplement is $240.

 • *Transactions*—this semiannual publication contains abstracts of papers read at ANS meetings devoted to nuclear technology. The cost for two volumes is $180.

Atomic Industrial Forum (AIF)

7101 Wisconsin Avenue   301-654-9260
Bethesda, MD 20814

 This organization is involved in the development and utilization of nuclear energy for constructive purposes. It monitors government regulations for nuclear energy and plant construction. AIF publications are provided free of charge to members. It publishes:

 • *Nuclear Industry*—a monthly that provides in-depth coverage of developments in the field of nuclear technology. Regulatory actions, plant construction, and planning are covered. Available to members.

 • *INFO*—this monthly newsletter covers the latest regulatory action in the field of nuclear technology. It explores licensing problems, the latest developments in nuclear technology, and the planning and construction of the plants being built. Available to members.

Center for Defense Information

Suite 303 West   202-484-9490
600 Maryland Avenue, SW
Washington, DC 20024

 This nonprofit organization consists of retired military personnel involved in research and analysis of military defense and nuclear weapons. Members deal with defense-related matters, including MX missiles, B-1 bombers, and the overall aspects of defense and nuclear warfare. The Center can provide you with information in these areas. It publishes:

 • *Defense Monitor*—issued ten times a year, this covers the latest technology and defense strategies. A contribution of $25 a year is requested.

Edison Electric Institute (EEI)

1111 19th Street   202-828-7582
Washington, DC 20036

 A trade association representing investor-owned electric companies, EEI provides its members with statistical information, research and support services, and forecast data pertaining to nuclear technology. It covers all aspects of the energy field, including nuclear technology. A listing of EEI's publications, which are available to the public, can be obtained from the Institute.

Federal Emergency Management Agency (FEMA)
500 C Street, SW
Washington, DC 20472

202-646-2500

Information about what to do in the event of a nuclear, chemical, or natural disaster is available from FEMA. The agency can be contacted 24 hours a day. Staff members are available from 8 A.M. to 4 P.M. to handle inquires; after hours you can leave a message and someone will get back to you the next business day. FEMA's publications include:

● *In Time of Emergency...A Citizen's Handbook*—provides information and guidance about what families can do to enhance survival in the event of natural or man-made disasters such as nuclear power plant accidents or nuclear attacks. It is available free of charge.

Fusion Energy Foundation (Nuclear)(FEF)
PO Box 1438
Radio City Station
New York, NY 10101

212-247-8439

The Foundation is involved in educating the U.S. about the effects of nuclear energy and laser technology. FEF received national recognition for its investigation of the Three Mile Island nuclear plant accident, and its work in the high-tech field has earned it the nickname "Star Wars." The Foundation publishes:

● *Beam Defense*—a comprehensive analysis of the use of directed energy laser beam particles for defense purposes, as well as for practical everyday uses. The publication furnishes examples of everyday uses in the fields of medicine, transportation, industry, and science. Available directly from FEF or your area bookstore for $10.

● *The Physical Principle of Thermal Nuclear Explosive Devices*—this publication covers the technology of hydrogen bombs and hydrogen fusion, as well as physical principles of thermal nuclear explosive devices. The publication is currently out of print, but copies are available for $27.

Ground Zero
806 15th Street, NW, Suite 421
Washington, DC 20005

202-638-7402

A public-interest group, Ground Zero is concerned with the effects of nuclear war. The staff will answer questions, recommend literature, and refer you to government agencies, organizations, and experts in the field of nuclear technology. Ground Zero's library is open to the public. The organization publishes:

● *Nuclear War: What's in It for You?*—a nuclear war primer, this is a thorough, readable guide to the nuclear war issue. Available in most bookstores or from Ground Zero. The cost is $5.

• *What About the Russians—and Nuclear War?*—an introduction to Russian history, culture, and politics, with a special emphasis on Soviet attitudes toward national security and the role of U.S./Soviet relations in the problem of nuclear war. Available in most bookstores or through Ground Zero. The cost is $5.

• *Thinking About Preventing Nuclear War*—a 20-page booklet proposing the broad range of strategies necessary to prevent war. It outlines the possible routes to nuclear war and the kinds of steps that must be taken to avert it. Available from Ground Zero for $.25.

• *Ground Zero Curriculum Guide for Secondary School Teachers*—this provides five consecutive lesson plans covering the history of nuclear weapons, the effect of nuclear weapons, and the history of U.S.-U.S.S.R. attempts to limit such weapons. The 50-page publication is available for $2.50.

• *Introductory Slide Shows*—Ground Zero offers two introductory slide shows, one covering nuclear war and weapons issues, and the other covering the Soviet Union and U.S./Soviet relations. Each is designed to serve as a stimulus for group discussions, and contains a written text and an optional audiotape. The slide show costs $10, and the audiotape costs $15.

Institute of Nuclear Power Operations (INPO)

1820 Water Place   404-953-3600
Atlanta, GA 30339

INPO's members are electric utilities that either have a nuclear power plant in operation or under construction. The Institute evaluates plants, conducts seminars, and publishes specifications, reports, and analyses relating to operations of nuclear power plants. Its services are primarily for members.

National Energy Information Center (NEIC)

Room 1F-048   202-252-8800
1000 Independence Avenue, SW
Washington, DC 20585

NEIC can provide statistical and analytical energy data and referrals. It sells several publications pertaining to various aspects of nuclear technology, such as investor perceptions, nuclear power prospects for the U.S. and the world, and power plant construction costs. NEIC also publishes an energy information directory, a publications catalog, and a bimonthly newsletter, all available free of charge.

National Nuclear Data Center (NNDC)

Building 197D   516-282-5205
Brookhaven National Laboratory
Upton, NY 11973

NNDC acquires, compiles, stores, retrieves, and disseminates experimental neutron data and bibliographic information on neutron and nuclear structure data. It evaluates neutron and nonneutron data for fission and fusion reactors, shielding, biomedical, and other applications; and it coordinates nuclear structure and neutron reaction data evaluation. The Center has access to several data bases: Cross Section Information Storage and Retrieval System (CSISRS), Bibliography of Neutron Data (CINDA), Bibliography of Nuclear Structure Information (NSR), Bibliography of Charged Particle Information (CPBIB), Evaluated Nuclear Data File (ENDF/B), and Evaluated Nuclear Structure Information (ENDSDF). NNDC issues several publications and a newsletter. Its services and materials are restricted to users from the U.S. and Canada.

Nuclear and Alternative Fuels Division

Energy Information Administration   202-252-6363
Department of Energy
Mailstop 2F021-EI53
1000 Independence Avenue, SW
Washington, DC 20585

The Division collects data about activity in the uranium, mining, and milling industries, including data about nuclear power generation. Data and forecasting publications are available from this office. The staff can answer questions and provide copies of reports.

Nuclear Data Project

Oak Ridge National Laboratory   615-574-4699
PO Box X
Oak Ridge, TN 37830

The Nuclear Data Project is an evaluation center, which, as part of the U.S. Nuclear Data Network, is responsible for collecting and evaluating nuclear structure information. The Project maintains a complete computer-indexed library of published works in experimental nuclear physics. In addition, the Project maintains a computer-based system for evaluated nuclear structure data (ENSDF) and a bibliographic file for nuclear structure references. The Project coordinates its activities with other national and international data centers, and has responsibility for the scientific editorship of the monthly journal *Nuclear Data Sheets*, where the evaluation work of all these centers is published.

Nuclear Regulatory Commission (NRC)

Washington, DC 20555   202-492-7715—Public Affairs Office
                                                            202-634-3273—Public Documents Room
                                                            202-492-9530—Publication Ordering Office

A federal agency, NRC has several functions. It licenses and regulates the uses of nuclear energy to protect the environment and public health and safety; licenses persons and companies to build and operate nuclear reactors and to own and use nuclear materials; makes rules and sets standards for these types of licenses; and fully inspects the activities of the persons and companies licensed, to ensure that they do not violate the safety rules of the Commission. NRC maintains a library, open to the public, which contains copies of licensing documents and plans, as well as NRC publications. The Commission publishes more than 3,000 materials, and its publications catalog, *Citizen's Guide to the Nuclear Regulatory Commission*, is available free of charge.

Nuclear Safety Information Center (NSIC)

Oak Ridge National Laboratory   615-574-0391
PO Box Y
Oak Ridge, TN 37830

NSIC was established as a focal point for nuclear safety information. It has been especially beneficial to those concerned with the analysis, design, licensing, construction, and operation of nuclear facilities in defining and solving nuclear safety problems. NSIC publishes continuing and single-issue topical bibliographies, technical reports, and fact sheets on subjects of particular interest to the safety community. Responses to specific user requests for information are another disseminating function of NSIC. The NSIC Data Base contains summaries of all licensee event reports on unusual operating experiences of U.S. reactors. The NSIC acts as the collection and distribution point for all foreign documents received in the NRC Light-Water Reactor Foreign Exchange Program, advises NRC concerning the need for an English translation, microfiches all foreign reports, and distributes them according to an NRC distribution list. The Center has access to the DOE/RECON retrieval system. NSIC's services are available free of charge to governmental organizations and their prime contractors, and on a cost-recovery basis to other users.

Nuclear Standards Program Information Center (NSPIC)

Nuclear Standards Management Center   615-574-7886
Oak Ridge National Laboratory
Mail Stop 10, Building 9204-1
PO Box Y
Oak Ridge, TN 37830

The NSPIC acquires, compiles, develops, stores, retrieves, and disseminates nuclear standards information. The NSPIC maintains a data base on the status of program standards and standards-related activities. Documents maintained and issued by NSPIC include *Nuclear Standards Master Index*, *Nuclear Standards Program Newsletter*, *Conversion of NE Standards to National Consensus Standards*, *First Steps to Standards Development*, *Guide to the Nuclear Standards Program*, *What Is the Nuclear Standards Program?*, *Monthly Summary of Unusual Occurrence Reports*, *Semiannual Index of Unusual Occurrence Reports*, and *Requirements for Preparation and Management of DOE Nuclear Energy Program Standards*. The Center also publishes a newsletter. NSPIC's services and publications are not available to parties representing a foreign organization or activity or for distribution by others to foreign interests.

Office of Reactor Deployment

Office of Nuclear Energy   301-353-3773
Department of Energy
DOE-NE-40
Washington, DC 20545

The Office manages nuclear energy programs in the following areas: high-temperature reactor technology; light water reactor safety research and development; light water reactor power plant system technology evaluation; light water reactor component evaluation, development, and testing; and light water reactor fuel fabrication technology, component evaluation, development, and testing. It publishes:

• *United States Central Station Nuclear Electricity Generating Units*—a yearly statistical report consisting of tables and charts of capacity and comparisons of utilities by state and region. Available free of charge from National Energy Information Center, 1000 Independence Avenue, SW, Washington, DC 20585, 202-252-5575. Report #DOE-NE-0030 should be referred to when ordering.

Office of Fusion Energy

Energy Research   301-353-3347
Department of Energy
ER-50 Room J204
Washington, DC 20545

This Office conducts research and develops technology for the efficient and safe use of fusion power. It oversees nuclear plant construction and operations. The Office publishes:

• *Planned, Built and Being Built*—a statistical report about reactors. It covers construction efforts pertaining to reactors that have been built, are being constructed, or are in the planning stage. The report costs $10 and is available from National Technical Information Center (NTIS), Department of Commerce, 5285 Port Royal Road, Springfield, VA 22161, 703-487-4688. Refer to publication number PB83-903001.

Office of Nuclear Energy

Department of Energy   202-252-8728
1000 Independence Avenue, SW
Washington, DC 20585

The Office provides information to the general public about nuclear energy and nuclear energy programs. The staff can provide you with information about available materials, including pamphlets, fact sheets, and booklets relating to all aspects of nuclear energy. The staff can also answer general questions or refer inquiries to appropriate program offices. The Office's basic brochure is:

• *Atoms to Electricity*—this 61-page publication describes the fission process—that is, how electricity is obtained from the atom. It also covers radiation, types of nuclear reactors, and so on. Single copies are available free of charge by writing to ENERGY/DOE, PO Box 62, Oak Ridge, TN 37830.

Uranium Enrichment Advanced Technology Project

U.S. Department of Energy   301-353-5969
NE-32
Washington, DC 20545

This federal program is responsible for the research and development of gas centrifuge technology and advanced isotope separation techniques. It publishes:

• *Uranium Enrichment 1983 Annual Report*—this covers statistical information as it pertains to the research and development of uranium. The report also summarizes the ongoing reports published during the year. It is available free of charge.

Uranium Enrichment Expansion Project

Operations and Facility Reliability   301-353-5832
Nuclear Energy
Department of Energy
Washington, DC 20545

This federal office develops program plans and directs program activities associated with the operation, construction, and maintenance of gaseous diffusion plants. It has compiled more than 400 publications on the subject. These reports are available from the Energy Information Administration, 1000 Independence Avenue, SW, Room 1F-048, Washington, DC 20585, 202-252-8800.

Publications

Nuclear Power in an Age of Uncertainty

Office of Technology Assessment (OTA)   202-226-2160—Information
600 Pennsylvania Avenue, SE   202-226-8996—Publications
Washington, DC 20510

Produced by OTA, this publication reviews research directions that could improve light water reactor technology, as well as opportunities to develop other types of reactors. It examines the crucial role of utility management in constructing nuclear power plants, and analyzes nuclear safety regulations and proposals for regulatory reform. The report describes policy approaches for Congress to consider in the future. A report summary is available, at no cost, from OTA, listed above. The full text of the report can be purchased for $10 from the Superintendent of Documents, U.S. Government Printing Office (GPO), Washington, DC 20402, 202-783-3238. When ordering, refer to GPO stock number 052-003-00941-2.

Nuclear Power Plant Innovation for the 1990's

Massachusetts Institute of Technology   617-253-3806—Elizabeth Parmelee
77 Massachusetts Avenue, Room 24-109
Cambridge, MA 02139

This report is an ongoing study of reactor options, which would increase desirability of nuclear power plants. It reviews reactor technology as well as water technology. The report is available for $7.

The Second Nuclear Era
Institute for Energy Analysis   615-576-3192—Documents Librarian
Oakridge Associate Universities
PO Box 117
Oak Ridge, TN 37830

The report studies the decline of the present nuclear era in the U.S. and examines the characteristics of a second nuclear era, which might be instrumental in restoring nuclear power to an appropriate place in the energy options of the U.S. The study has determined that reactors are much safer today than those of the Three Mile Island era. It cites reactors and specifications. It is available free of charge.

Additional Experts and Resources

David Rose, Ph.D.
Department of Nuclear Engineering   617-253-3807
Massachusetts Institute of Technology
Cambridge, MA 02139

Dr. Rose is an expert in the field of nuclear energy technology and policy development. He is available to answer questions and make referrals.

Nuclear War

See: Nuclear Technology

Nutrition

See also: Food Technology

Organizations and Government Agencies

American Institute of Nutrition (AIN)
9650 Rockville Pike   301-530-7050
Bethesda, MD 20814

The American Institute of Nutrition is a professional organization of nutrition research scientists who aim to develop and disseminate information, facilitate personal contact among specialists in nutrition and related fields, and recognize excellence in nutrition research with its awards. AIN is a member of the Federation of American Societies of Experimental Biology, and it holds its annual meeting in conjunction with the other societies that make up this Federation. Publications include:

• *Journal of Nutrition*—a monthly publication covering all phases of experimental nutrition. Articles are submitted by foreign as well as U.S. authors. The annual subscription prices are $95 for institutions and $70 for individuals.

The American Society for Clinical Nutrition (ASCN)

9650 Rockville Pike   301-530-7110
Bethesda, MD 20814

The Society's objectives are to foster research in human nutrition in health and disease; to encourage undergraduate and graduate nutrition education, particularly in medical schools; and to promote the proper application of the findings of nutrition research in the practice of medicine and related health professions. ASCN has an annual scientific meeting, which gives the opportunity to researchers to present their work, as well as to earn Continuing Medical Education (CME) credits. If you are a clinical investigator or a student with a paper to present, you may be eligible for cash and travel awards. ASCN annually offers a course in clinical nutrition, for which CME credits can be earned. Publications include:

• *The American Journal of Clinical Nutrition*—a monthly publication that discusses the progress in the field of clinical nutrition. Regular features include original research communications, perspectives in nutrition, editorials, special articles, and letters to the editor. The 200-page publication is available for $60 a year to institutions, for $45 a year to individuals, and for $17.50 a year to students.

Cooperative State Research Service (CSRS)

U.S. Department of Agriculture   202-447-3426—Mary Heltsley
West Auditor's Building
15th Street and Independence Avenue
Washington, DC 20251

The CSRS is responsible for administering and coordinating funds to 54 state agricultural experiment stations, 16 land-grant schools, and Tuskegee Institute for the purpose of carrying out research on food and agricultural issues, including human nutrition research. These projects often focus heavily on nutrient bioavailability and the composition of foods; determination of nutrient requirements; metabolic functions of nutrients and interactions; the dietary and nutritional status of special populations; dietary patterns; and alterations in the nutritional value of food supply resulting from changes in production, processing, or marketing practices.

Food and Nutrition Information Center (FNIC)

U.S. Department of Agriculture   301-344-3719
National Agricultural Library
10301 Baltimore Boulevard
Beltsville, MD 20705

FNIC acquires books, journals, and audiovisual materials dealing with human nutrition, food service management, and food science. Although FNIC resources primarily serve scientists, food service managers, and other professionals, the Center is open to the general public. The following organizations, agencies, and individuals may borrow materials from the Center free of charge: U.S. Congress, federal and state government agencies; libraries; information centers; faculty members of universities and colleges; national officers of professional societies; day-care personnel; schools, including food service personnel; nutrition education and training programs; cooperative extension programs (federal, state, and county levels); and research institutions. These organizations, agencies, and individuals may also request to have journal articles photocopied and sent to them.

Other services include a reference service and a 24-hour telephone monitor. The reference staff consists of two or three full-time nutritionists, who answer questions and/or make referrals. The reference service also offers computer searches of major data bases. The 24-hour monitor enables callers to leave messages and have their calls returned.

Human Nutrition Information Service (HNIS)

U.S. Department of Agriculture
6505 Belcrest Road
Hyattsville, MD 20782

301-436-8457—Nutrition Monitoring Division
301-436-8507—Survey Statistics Division
301-436-8491—Nutrient Data Research Branch
301-436-8474—Nutrition Education Division
301-436-8470—Diet Appraisal Research Board
301-436-5194—Guidance and Education Research Branch

The purpose of HNIS is to conduct and interpret research in human nutrition to improve professional and public understanding of the nutritional adequacy of diets and food supplies as well as the nutritive value of food, and to develop knowledge needed to improve the nutritional quality of diets. The Service also collects and disseminates information, and consults on technical and educational materials on food use, food management, and human nutrition problems. Staff members are knowledgeable about published papers, and can answer questions and make referrals. An important public service is the Nutrient Data Research Branch's Nutrient Data Bank. Its publications and magnetic tapes contain data in summary form. Publications include:

● *Agriculture Handbook No. 8 (The Composition of Foods)*—continually updated, it is the major Nutrient Data Bank publication. It covers all food groups and gives information on the nutrient composition of foods. It is available from the Government Printing Office, Washington, DC 20402, 202-783-3238. Prices range from $6 to $9.50.

Society for Nutrition Education (SNE)

1736 Franklin Street, Suite 900
Oakland, CA 94612

415-444-7133

A national and international nonprofit association, SNE is dedicated to promoting nutritional well-being. SNE holds an annual meeting, which consists of presentations of papers, screenings of audiovisuals, exhibits of educational materials, and so on. Optional workshops are offered before and after the meeting. The Society offers a free catalog of publications and films for consumers and professionals. Publications include:

• *Journal of Nutrition Education*—a quarterly publication containing new teaching ideas and activities. It also includes research and methods assessment and has an educational materials section. The annual subscription price is $30, for four issues.

• *Inventory of Nutrition Education Evaluation Instruments*—this reference describes more than 60 available research instruments designed to measure food and nutrition knowledge, attitudes, preferences, and practices among children, teachers, food service personnel, homemakers, consumers, and others. Consult the catalog for the price of the 60-page booklet.

U.S. Department of Agriculture

Agricultural Research Service (ARS)   301-344-3216
National Program Staff
Building 005
BARC—West
Beltsville, MD 20705

The ARS human nutrition research is conducted primarily at five separate Human Nutrition Research Centers and at Regional Utilization Laboratories. Each Center has a different research thrust and provides its unique contribution.

Beltsville Human Nutrition Research Center
Beltsville, MD

The purpose of the Center is to define human requirements for the essential nutrients—proteins, carbohydrates, lipids, vitamins, and minerals—for optimal health and performance, and to identify, through study of nutrient composition, the foods that meet those requirements. This Center also studies the metabolic role of nutrients, the many interactions of nutrients with other food components, and the effects of bioavailability. Emphasis at this Center is on the nutritional requirements of adults and on development of food composition analysis methodology. Call Walter Mertz, Director, at 301-344-2157.

Children's Nutrition Research Center at Baylor College of Medicine
Houston, TX

This Center is the only one to deal exclusively with research on the nutrient needs and the nutritional status of mothers, infants, and children. Its aim is to define the nutritional requirement that will ensure optimal nutritional status in pregnant and lactating women, and in infants and children through adolescence. Emphasis is given to protein and energy requirements. Call Buford Nichols, Director, at 713-799-6006.

Grand Forks Human Nutrition Research Center
Grand Forks, ND

The purpose of this Center is to develop recommendations for nutrient intakes in humans and to identify useful nutrient forms, with particular emphasis on mineral requirements. The nutrients are studied in collaboration with several other scientists at various universities and at other ARS locations. Call Leslie Klevay, Director, at 701-775-8353.

Human Nutrition Research Center on Aging at Tufts University
Boston, MA
The Center's goal is to determine the nutrient needs of the elderly and the relationship of dietary factors to the aging process. Call Harold Sandstead, Director, at 617-956-0302.

Western Human Nutrition Research Center
San Francisco, CA
The purpose of the Center is to improve methods for assessing human nutritional status and to study the factors that lead to malnutrition. This Center also conducts studies on human nutritional requirements and on factors that influence them, with emphasis on vitamin requirements. Call James M. Iacono, Director, 415-556-9699.

Plant, Soil, and Nutrition Laboratory
Ithaca, NY
This Laboratory investigates the cause-and-effect relationships among plants, soil, and nutrition. Call Darrell Van Campen, Laboratory Director, at 607-256-5480.

Regional Laboratories
Other ARS Research Centers are involved in research important to achieving a healthy human nutritional status. These include the Eastern Regional Research Center, Philadelphia, PA; Northern Regional Research Center, Peoria, IL; Southern Regional Research Center, New Orleans, LA; Western Regional Research Center, Berkeley, CA; and the Richard Russell Research Center, Athens, GA. These Centers focus on specific areas of research: food production, food processing, food storage, distribution and marketing, and food safety. Call Gerald F. Combs, Administrator for Human Nutrition, at 301-344-3216.

Publications

Directory of Food and Nutrition Information Services and Resources
Oryx Press   800-457-6799
2214 North Central at Encanto   602-254-6156
Phoenix, AZ 85004
This fully indexed paperbound guide details specialized information services, print and nonprint resources, and data bases provided by a wide variety of organizations in this country. It aids food and nutrition professionals, information providers, and consumers in identifying reliable sources of information and materials. Sections cover organizations, data bases on food and nutrition microcomputer software, jour-

nals and newsletters, abstracts, indexes, and current health awareness publications. The 256-page directory is available for $65.

Directory of Human Nutrition Activities

Agricultural Research Service   301-344-3216
U.S. Department of Agriculture
Building 005
BARC—West
Beltsville, MD 20705

This free 21-page booklet lists human nutrition services, programs, information sources, research centers, and other activities of the U.S. Department of Agriculture.

Data Bases

NutriQuest

Capital Systems Group, Inc.   301-881-9400—Lilly Gardner
11301 Rockville Pike
Kensington, MD 20895

A system for nutritional analysis, NutriQuest primarily assists dietitians, nutritionists, physicians, and technicians in evaluating dietary intakes, managing food service systems, and educating patients or clients. It can also be used in connection with weight-reduction and physical-fitness programs. The computer package consists of a data file, the software to operate the file, and a user's manual. The file contains 900 food items, identified by USDA/CFEI survey data as the foods most frequently consumed in U.S. households. The data file also contains the 1980 table for Recommended Dietary Allowances (RDA). There is a series of programs designed to generate food-intake reports and dietary data. The system is designed to operate on the Apple II and Apple II-compatible microcomputers, Apple II Plus, and the IBM personal microcomputers using Pascal UCSD or DOS 2.0. NutriQuest is updated regularly with a new handbook of information. It is sold at three different levels, with prices ranging from $100 to $700.

Nutrition Analysis System

Honeywell Data Network   800-328-5200
6400 France Avenue, South   612-870-6350
Edina, MN 55435

Sources of information for this data base are the U.S. Department of Agriculture *Handbook 456* and other USDA sources. The data base contains nutritional values for 2,500 food items. For each item, data are given for 21-22 nutrients. Users enter a recipe, and the nutrient breakdown of the ingredients in the recipe is provided.

BIOSIS Previews

Bio-Sciences Information Service

2100 Arch Street

Philadelphia, PA 19103

800-523-4806

215-587-4800

This data base contains citations from *Biological Abstracts*, *Biological Abstracts/Reports, Reviews, Meetings*, and *Bioresearch Index*. These publications provide comprehensive worldwide coverage of research in the life sciences, including nutrition. Material scanned for *Biosis* includes periodical literature, monographs, books, textbooks, technical reports, published theses, meetings, nomenclatural rules, notes, letters, annual reports, bibliographies, and guides. Available through DIALOG Information Services in Palo Alto, CA, at 800-227-1960, or through BRS in Latham, NY, at 800-833-4707.

Occupational Health and Safety

See: Industrial Hygiene

Occupations in High Technology

See: Careers and Employment

Ocean Energy

See also: Fuels, Alternative; Renewable Energy Technology; Solar Technology

Organizations and Government Agencies

Environmental and Energy Systems Division

Argonne National Laboratory

9700 South Cass Avenue

Argonne, IL 60439

312-972-8071

Argonne has been doing work in three areas in the support of Ocean Thermal Energy Conversion (OTEC) power systems development: (1) materials and biofouling, (2) heat exchanger development, and (3) system performance modeling. Mr.

Anthony Thomas is available to answer questions and/or make referrals in these areas. He is a mechanical engineer who has been with Argonne National Laboratory for 25 years. The division can also supply a list of publications.

Marine Sciences Group

Department of Paleontology   415-642-6535
University of California
Berkeley, CA 94720

The Marine Sciences Group conducts research in the physical, chemical, biological, and geological aspects of the ocean environment. Its major support is in ocean energy. A list of publications of the Marine Sciences Group is available upon request.

NOAA Special Projects Staff

Ocean Thermal Energy Conversion (OTEC)   301-443-8655
Ocean Engineering Technology Development
11400 Rockville Pike
Rockville, MD 20852

The Special Projects Staff identifies critical research needs for OTEC ocean structures and conducts a research program. Mr. Bob Taylor is the technical manager of the group. He has an MS degree in mechanical/ocean engineering and is available to answer technical questions related to ocean thermal energy conversion and ocean engineering problems, and to make referrals.

Ocean Mineral and Energy Division

National Ocean Service   202-254-3483
National Oceanic and Atmospheric Administration (NOAA)
2001 Wisconsin Avenue, NW, Room 105
Washington, DC 20235

The Division is concerned with ocean thermal energy conversion and deep seabed hard minerals. They have a licensing and regulatory program and also write environmental impact statements. Staff can answer questions and/or make referrals.

Ocean Energy Technology

Solar Electric Technologies   202-252-5517
Conservation and Renewable Energy, DOE
Forrestal Building, CE-332, Room 5H032D
1000 Independence Avenue, SW
Washington, DC 20585

This office at the Department of Energy (DOE) is responsible for designing the Ocean Thermal Energy Conversion (OTEC) proof-of-concept experiment, a 40-megawatt OTEC pilot plant in Hawaii. It manages the long-term, high-risk research and development of ocean energy systems. The office also encourages advances in use of various forms of ocean energy, such as OTEC and waves.

Ocean Engineering and Ocean Mining

See: Marine Technology

OCR

See: Optical Character Recognition

Office Automation

See also: Microcomputers; Minicomputers; Productivity and Innovation; Teleconferencing

Organizations and Government Agencies

American Federation of Information Processing Societies (AFIPS)
1899 Preston White Drive   703-620-8900
Reston, VA 22091

AFIPS, comprised of national organizations engaged in the design and/or application of computers and information processing systems, sponsors an Annual Office automation conference. The meeting encompasses automation, communications, and integration, with activities for each conference centering around a theme. It is a scientific and educational event that usually occurs in February and features speeches, more than 175 exhibitors, and 45 technical sessions, as well as the granting of awards and recognition for advances in the field. The Federation also seeks papers and reports for publication in the *OAC Digest*. This publication consists of conference papers; the price is $31.50 from the AFIPS.

Automated Office of the Future (SIG/AOF)
American Society for Information Science   202-659-3644
1010 16th Street, NW
Washington, DC 20036

A forum for sharing research about office automation, SIG/AOF is concerned with how an office can be organized to enable decision makers instant access to information, and how the use of technology can eliminate delays associated with the handling of paper-based information. The Society sponsors a conference each fall as well as several meetings throughout the year for the purpose of sharing information about office automation. SIG/AOF also offers job placement services.

Commerce Productivity Center

U.S. Department of Commerce   202-377-0940
The Assistant Secretary for Economic Affairs
14th Street and Constitution Avenue, NW, Room 7413
Washington, DC 20230

An information clearinghouse for techniques and resources pertaining to productivity improvement and quality control, the Center can provide you with information about office automation. Upon request and at no cost staff will evaluate your office systems and recommend steps for improving automation, productivity, and quality control. Staff can refer you to government agencies, private firms, and literature pertaining to office automation. Bibliographies, publications, articles, and reading lists are all available free of charge from the center.

DataPro Research, Inc.

McGraw-Hill Information Systems   609-764-9406 (in NJ)
1805 Underwood Boulevard   800-257-9406
Delran, NJ 08075

DataPro conducts research and testing of all office automation products. Its findings are published in a variety of forms including comparison listings of products for effectiveness, productivity, and price. The company provides a telephone hot line which subscribers can call to obtain answers to their questions about products and developments in the field. In addition to materials about office automation, DataPro publications also cover industry automation, communications, microcomputers, EDP systems, and much more. Specific publications of interest are:

• *Automated Office Solutions*—updated monthly, this is a solution-oriented reference work covering office methods, equipment, technology, and management. It shows how to plan for, and capitalize on, the transition to an automated office. Coverage is provided for topics such as the evolving office of the future, the transition to automation, office systems development, equipment evaluation and selection, facilities management, automated office software, systems management, the current office overview, and personnel management. Available in two looseleaf binders. The annual subscription fee is $455.

• *Office Automation*—updated monthly, this provides detailed information about the full spectrum of office products, systems, techniques, and companies. It includes product comparison tables, user ratings, and in-depth product profiles. The publication covers such topics as word processing systems, dictation, telephone and voice communications, facsimile, office facilities, forms and supplies, microcomputers, electronic copiers and printers, office automation software, executive/professional workstations, and much more. Available in three looseleaf binders, the annual subscription rate is $745.

• *Word Processing*—updated monthly, this provides practical information and guidelines to help you maximize the benefits offered by the newest word processing systems, products, and services. It includes product comparison tables, in-depth product profiles, and user ratings to help minimize the cost and risk in word processing planning, evaluation, selection, and implementation. Covers such topics as concepts and planning, time-shared word processing services, word processing software, auxiliary equipment and systems, dictation systems, industry applications, communicating word processors, and electronic mail. Available in two looseleaf binders, the annual subscription rate is $600.

• *Copiers and Duplicators*—updated monthly, this provides over 100 up-to-date reports to help you select the most cost-effective copying and duplicating equipment. Included are product comparison tables, user ratings, in-depth product reports, and how-to hints that will save time and money on office reproductions equipment. Available in two looseleaf binders, the annual subscription rate is $550.

IEEE Office Automation Technical Committee

Institute of Electrical and Electronics Engineers (IEEE)   301-589-8142
1109 Spring Street, Suite 300
Silver Spring, MD 20910

A Technical Committee (TC) of IEEE, this group of 300 members is involved in all aspects of using computer technologies to support and automate office activities. Specific topics of interest include word processors, electronic mail, electronic calendars, teleconferencing, facsimile, office information processing models, office information requirements, and comparisons of office systems. The TC emphasizes the integration of software and hardware as well as social and human engineering issues related to office systems. It sponsors an international conference every two years covering all of these topics.

• *Office Automation*—a quarterly newsletter that provides information about new products and technology for office automation. It consists of technical papers submitted by committee members. A calendar providing information about important events and conferences is also included. The newsletter is available free of charge to members.

Management Information Corporation (MIC)

140 Barclay Center   609-428-1020
Cherry Hill, NJ 08034

This company conducts market analysis of applications, products, and companies associated with office automation, data processing, and data communications. A telephone information service is provided to answer callers' questions in these areas. Examples of relevant MIC publications are:

• *Officemation Product Reports*—this monthly provides evaluations of office automation products, word processing systems, professional work stations, office automation equipment, and software available worldwide. The annual subscription is $345.

● *Officemation Management*—a monthly publication covering the status of the automated office, financial aspects of computer acquisition, electronic mail scheduling and calendaring, videotex systems, teleconferencing, executive workstations, portable computers, micrographics, time management, making meetings productive, organizing an automated office, selecting the right person for the job, productivity measurment techniques, and telephone systems. Annual subscription rate is $345.

Office Technology Research Group

PO Box 65   213-796-2675
Pasadena, CA 91102

An association of high-level managers concerned about the present and future impact of office automation, this Group is user-, not vendor-oriented. It holds meetings, seminars, and conferences. Members are provided with a ten-volume library, updated monthly, containing studies and current information about office automation and technology. The group also publishes a newsletter for members.

Society of Office Automation Professionals (SOAP)

233 Mountain Road   203-431-0029
Ridgefield, CT 06877

This Society is a grass roots organization of professionals from all fields of office automation including data processing, word processing, data and voice communication, teleconferencing and records management. SOAP's emphasis is on management effectiveness, and it promotes research, standards development, public policy, and the use of automation to serve office workers and increase productivity. Chapters exist in most major cities. The group publishes a quarterly newsletter and membership directory, both available to members only.

Publications

Automated Office Systems

The Office Systems Consulting Groups, Inc.   617-492-3300
PO Box 352
Cambridge, MA 02139

Published monthly, this publication focuses on the latest technology in office automation. It covers all aspects of the field including data processing and microcomputers. Product comparisons are provided, as well as information about new products and experts in the field. Annual subscription rate is $125.

Hearst Business Communication, Inc.

645 Stewart Avenue   516-222-2500
Garden City, NY 11530

This company offers several publications pertaining to office automation:
● *Today's Office*—a monthly publication that covers office automation technology as well as management skills necessary for running an efficient office. Special articles appear about new products, with evaluations of effectiveness and productiv-

ity. Comparison charts and consumer information is also provided. Annual subscription rate is $30.

   • *Electronic Engineers Master*—is an annual catalog of automated office equipment manufacturers. Contact and product information is provided for each entry. The cost is $62.50.

   • *Electronic Products*—this biweekly publication provides information about new electronic office products. Examples of products covered include: components, integrated circuits, diodes, and assemblers. Annual subscription rate is $44.

   • *Office Work News*—published monthly, this contains news about worldwide developments in office automation. It focuses on new products and new applications in the work area. News is also provided about people in the field. The annual subscription rate is $25.

   • *Integrated Circuits*—this catalog covers circuits used in office automation. Information is provided about the latest technology and products. Detailed listings of circuits' specifications are included. It costs $95.

Information Management

PTN Publishing Corporation   516-496-8000
101 Crossways Park West
Woodbury, NY 11797

   This monthly publication covers all phases of office automation, featuring special articles about the latest technology and applications. It also highlights new products and provides information about people in the field. The annual subscription rate is $10.

Modern Office Technology

Penton/PC, Inc.   216-696-7000
1111 Chester Avenue
Cleveland, Ohio 44114

   Issued monthly, this publication covers research in micrographics, word processing, data processing, communications, records management, and office design. Information is also provided about management and personnel relationships. The January issue contains a buyer's guide listing companies and equipment available. The annual subscription rate is $30.

Additional Experts and Resources

Mary Murphree, Ph.D.

Women's Bureau, Room 3307   202-523-6652
U.S. Department of Labor
200 Constitution Avenue, NW
Washington, DC 20210

   A sociologist, Dr. Murphree studies the effects of new office technologies on clerical workers, specializing in the secretarial occupation. Her particular concern is the impact of word processing technologies on women workers. Dr. Murphree works as a consultant to government and industry. She can answer questions in her area of expertise or refer you to colleagues in her office.

On-Line Data Base Resources

Publications

Computer Data and Data Base Source Book
Avon Books   800-238-0658—Customer Service
1790 Broadway
New York, NY 10019

A complete encyclopedia of many commercial and public information sources for use with any computer, this book provides an up-to-date guide to data available. It includes more than 1,000 commercial data bases, listed by subject and including names, addresses, phone numbers, and the fees for their use; all public data bases, most of which are accessible for a cost recovery fee; data series, including studies of such subjects as average professional salaries, stock prices, and tax rates; and much more useful information. The 900-page paperback is available for $14.95 at most bookstores.

Computer-Readable Data Bases: A Directory and Data Sourcebook
American Library Association (ALA)   312-944-6780
Publishing Services
50 East Huron Street
Chicago, IL 60611

These two volumes cover approximately 2,000 data bases in technology, engineering, and consumer information, as well as data bases in business, law, medicine, social sciences, and humanities. There is comprehensive coverage of word-oriented and numeric data bases. Entries specify basic information (data base name, former name, update frequency, and so on); subject analysis and indexing; data elements; data base services; and available user aids. Access is provided through subject, producer, on-line vendor, and name indexes. The publication is available for $87.50 per volume, with science, technology, and medicine contained in a single volume. The two volumes together are $157.50.

Data Base Informer
Information USA   301-657-1200
Computer Data Service
4701 Willard Avenue, Suite 1707
Chevy Chase, MD 20854

This monthly newsletter is designed to show decision makers where to find external data and how it can be used for computerized decision making. Items discussed (1) identify existing, unique, free or low-cost computerized data bases, which can be accessed directly or indirectly, (2) pinpoint sources of information that can be used to build a computerized data base, or (3) provide advice and helpful hints for designing and managing computerized data bases. The 12-page publication is available for $78 a year.

Database Monthly

Database Publications   512-250-1255
11754 Jollyville Road, #104
Austin, TX 78759

This monthly is an independent newspaper for users of Data General and compatible computers. In addition to presenting the news and trends affecting the Data General marketplace, it also includes feature stories, general business news, new product announcements, and user-oriented advice columns. It is available for $38 a year.

Database Newsletter

Database Research Group, Inc.   617-631-6296
23 Sewall Street
Marblehead, MA 01945

This bimonthly newsletter contains interviews with industry leaders and researchers, reports on new investigations, and covers other issues such as information centers, distributed data base designs, and new products. The subscription cost is $85 a year.

Database Update

Newsletter Management   305-483-2600
10076 Boca Entrada Boulevard
Boca Raton, FL 33433

Published monthly, the newsletter provides business people with current information on everything that's happening on-line, and how to make the most out of the material available. Subscription price is $97 a year.

Directory of Online Databases

Cuadra Associates, Inc.   213-829-9972
2001 Wilshire Boulevard, Suite 305
Santa Monica, CA 90403

A comprehensive directory that describes over 2,700 data bases of all types available to users through over 400 international on-line services. It covers bibliographic, referral, numeric, full text, textual, and software data bases in all fields. It is also available on-line through DATASTAR and Questel. Subscription to the directory includes two issues and two updates and costs $95 a year in the United States and $110 elsewhere.

Federal Data Base Finder

Information USA   301-657-1200
4701 Willard Avenue, Suite 1707
Chevy Chase, MD 20815

A directory of free and fee-based data bases and files available from the federal government, the *Federal Data Base Finder* describes more than 3,000 data bases and it identifies data sources that are not available anywhere else. Examples of the data bases included are: The Effects of Coaching on Scholastic Aptitude Tests; How Many and What Kind of People Visit Doctors; The Demographics of Divorced Individuals; and Television Programming Data Master File. This guide covers some 20 different major categories, including health, education, demographics, energy, finance, economic indicators, and general business. For each file and data base you get information on how it can be used, the name, address, and telephone number of the office in charge, and the fee, if any. The 368-page publication is available for $95.

Knowledge Industry Publications, Inc.

701 Westchester Avenue   914-328-9157—in New York
White Plains, NY 10604   800-431-1800

Knowledge Industry puts out the following publications that deal with data bases:

• *IDP Report*—this bimonthly newsletter covers every aspect of informational data base publishing with analytical articles about publishing, marketing, and using data base information. It includes surveys and charts of retrieval services by customer count, type and price, market indicators, comparisons of bibliographic utilities, network usage, on-line information experts, and videotex and teletext services. It is available for $225 a year.

• *Data Base Directory*—contains detailed descriptions of more than 1,800 numeric, full text, textual/numerical, property, bibliographic, and referral data bases available in the U.S. and Canada. The data bases are arranged alphabetically and indexed by subject, producer, and vendor. Also available on-line on the Data Base User Service, which gives current year-round information on the same data bases. The 600-page directory is available for $120.

Modem Notes

Modem Notes, Inc.   312-764-7407
PO Box 408472
Chicago, IL 60640

This monthly newsletter gives information on everything that modems can access with a computer and a telephone. It presents news about equipment, new data bases or related products, and the on-line industry.

Omni Online Database Directory

Macmillan Publishing Company
866 Third Avenue
New York, NY 10022

This directory describes over 1,000 data bases and vendors. The paperback is $10.95 and is available in most bookstores.

Online, Inc.

11 Tannery Lane   203-227-8466
Weston, CT 06883

Online, Inc. offers many different services to keep clients up to date on the latest in data bases. These services are:

- *Online*—a bimonthly journal that keeps abreast of what's new in the field and gives practical advice on using current products and services. Topics covered include: microcomputer hardware and software; new data bases; on-line searching tips; new trends and technologies. The annual subscription rate is $78.

- *Database*—complements *Online* with practical articles on data bases, including those aspects of microcomputing that relate to data bases. The subscription rate is $56 a year.

- *File SOFT*—is an on-line version of the company's *Online Software Guide and Directory*, which is a directory of over 3,200 software programs of interest to the information professional. Updated monthly, this file is available on BRS for $40 an hour.

- *The Online Chronicle*—is an electronics newspaper serving the on-line industry. It is also a source of current listings of job openings in the on-line industry. It is available on DIALOG for $35 an hour.

- *Database Search Aids*—a six-volume series of data base/system-specific tools that helps you search by showing how each data base is constructed and mounted. The first four volumes are $12.50 each and include business and management, education, government documents, and electronics. The last two, patents and health science, are $25 each.

- *Online Terminal/Microcomputer Guide and Directory*—is an extensive guide to terminals, microcomputers, printers, and modems that are useful for the information professional. The three main sections are: (1) how to choose a terminal or microcomputer, (2) a guide containing 16 lists, including equipment, specifications, manufacturers, brokers, glossary, abbreviations, and a bibliography, and (3) a directory of sales and service offices in the U.S and abroad. Available for $40, plus $30 for two supplements.

Optical Character Recognition

See also: Image Processing; Recognition Technology

Organizations and Government Agencies

American National Standards Institute (ANSI)
1430 Broadway   212-354-3473
New York, NY 10018

ANSI is a private, nonprofit federation of standards-developing organizations and standards users. Its major functions include coordinating the voluntary development and approval of American standards. ANSI can refer people to technical experts outside the organization who either write the standards or are otherwise knowledgeable in the field. The following OCR standards are available from ANSI:

- *Character Set for Optical Character Recognition (OCR-A), ANSI X3.17-1981, (FIPS 32-1)*—establishes a machine-readable standard character set, designed for optimizing OCR readability. Cost is $11.

- *Character Set for Handprinting, ANSI X3.45-1982*—provides the shapes and sizes of a set of hand-printed characters and supporting specifications and recommendations for its use in optical character recognition (OCR) systems and in interpersonal communications. Cost is $8.

- *Character Set for Optical Character Recognition (OCR-B), ANSI X3.49-1975, (FIPS 32)*—provides an optical character recognition (OCR) character set that is an alternative to ANSI X3.17-1981, (OCR-A). OCR-B is a character set in which conventionality of appearance has not been comprised for the sake of machine readability. The standard specifies the shapes, sizes, and printing positions of OCR-B alphanumeric characters and symbols. Available for $13.

- *Paper Used in Optical Character Recognition (OCR) Systems, ANSI X3.62-1979*—establishes optical and physical requirements and test procedures for paper to be used in optical character recognition (OCR) systems. This standard also revises and expands the paper specification contained in ANSI X3.17-1981. Available for $6.

- *Optical Character Recognition (OCR) Inks, ANSI X3.86-1980, (FIPS 85)*—defines the spectral band for readable inks and provides curves for the red and blue nonreadable inks, each at three levels of reflectance. Manufacturers and users can then determine the best balance between legibility to the human eye and reliability of the optical reader. Cost is $7.

- *Optical Character Recognition (OCR) Character Positioning, ANSI X3.93M-1981, (FIPS 89)*—specifies the location of OCR-A and OCR-B characters in relationship to other characters on a document or page and to reference points of the document or page. Available for $6.

• *Guideline for Optical Character Recognition (OCR) Print Quality, ANSI X3.99-1983*—describes the print quality parameters and measuring techniques for determining the quality of machine-printed characters to maximize the likelihood that they can be read by electro-optical means. Cost is $6.

ANSI Secretariat—X3A1 Committee
Computer and Business Equipment Manufacturers Association   202-737-8888
311 First Street, NW
Washington, DC 20001

The X3A1 Committee is the standards-developing panel operating under ANSI procedures. They can be contacted for information on standards currently being developed. The Committee has developed the following OCR standard: *X3.111-1983 Matrix Character Sets for Optical Character Recognition, 1983*. It is available for $9.

SPIE—International Society for Optical Engineering
PO Box 10   206-676-3290
Bellingham, WA 98227-0010

SPIE is a nonprofit, technical society dedicated to advancing engineering and scientific applications of optical, electro-optical, and optoelectronic technology, and communicating this through its publications and symposia. Members are individual engineers and corporations. A dozen symposia are held during the year. SPIE publishes:

• *Optical Engineering*—a bimonthly technical journal. Each issue of this periodical focuses on a specific topic. Other articles cover engineering and scientific applications in optical, electro-optical, and optoelectronic technology. The cost is included in the membership fee of $46 a year. The nonmember subscription price is $60 a year.

• *Optical Engineering Report*—a monthly tabloid newspaper distributed at no charge which specializes in news and employment opportunities.

Publications

Modern Office Technology
Penton/IPC   216-696-7000
1111 Chester Avenue
Cleveland, OH 44114

Published monthly, this technical magazine is directed to information systems directors or managers at the middle, upper, and senior management levels, all of whom qualify for a free subscription. The magazine covers developments in all recognition technologies and has features or news stories on OCR a few times a year. Occasionally it advertises new OCR products in the magazine's business products section. The regular subscription price is $30 a year, or $3 for a single issue.

Today's Office
Hearst Business Communications   516-222-2500
645 Stewart Avenue
Garden City, NY 11530
  This monthly covers business operations, including articles on OCR, and management topics. It also has product comparison charts and a new products section. This publication is free to individuals in management positions with buying influence. The regular subscription price is $30 a year, or $3 an issue.

Optical Fiber Communication

See: Optoelectronics; Recognition Technology

Optical Memory and Storage

Organizations and Government Agencies

Information for Graphic Communications (IGC)
375 Commonwealth Avenue   617-267-9425
Boston, MA 02115
  IGC is a private organization that disseminates information and offers conference programs on optical memory and imaging technology. It also develops multiclient research studies in these areas. Information about the studies and conferences can be obtained by contacting IGC's office.

Materials Research Society (MRS)
9800 McKnight Road, Suite 327   412-367-3003
Pittsburgh, PA 15237
  This organization provides a forum for communication between industry and educational institutions as well as interaction among scientific researchers and engineers. Members are scientists, engineers, and physicists, including many dedicated to optical and magnetic media. Tutorial lectures, technical short courses, and several symposia are offered periodically.
  A free brochure, containing information about the Society and the lectures, courses, and symposia, is available upon request. Specific technical information and referrals are provided on a limited basis.

Optical Society of America (OSA)
1816 Jefferson Place, NW   202-223-8130
Washington, DC 20036

OSA is a prominent publisher and disseminator of optical research and data. The society's membership includes more than 8,500 scientists, engineers, and technicians worldwide, as well as a large number of corporations with interest in optics. The society sponsors an annual meeting that covers a broad range of optical interests, an annual conference on lasers and electro-optics, an annual conference on optical fiber communication, a biannual conference on applied optics, and a variety of other conferences on special topics, as well as workshops and other information programs. OSA supports 18 technical groups, each on a special area of interest. Its staff of specialists can answer specific technical questions from members and nonmembers. OSA publishes the following:

• *Journal of the Optical Society of America*—issued monthly in two parts, Journal A emphasizes image science and is the general journal in the field; Journal B emphasizes spectroscopy and modern quantum optics. The cost of subscription to both parts is $225 a year for nonmembers and $16 for members.

• *Applied Optics*—a bimonthly journal geared to scientists and engineers engaged in work of an applied nature. It includes reviews of optical patents and emphasizes papers on lasers, electro-optics, information processing, atmospheric optics, and optical engineering. The cost is $28 a year for members, and $280 a year for nonmembers.

• *Optics Letters*—a monthly journal with short articles on new important developments in all areas of optics. The subscription cost is $12 a year for members, and $96 for nonmembers.

• *Journal of Lightwave Technology*—published jointly with IEEE, it covers all aspects of guided-wave science, technology, and engineering. The cost is $10 a year for members, and $96 for nonmembers.

• *Optics News*—a membership publication of the Society, issued in alternate months containing news of interest to the optical community and brief articles on areas of optical science and technology. This publication is free for members, and costs $40 for nonmembers.

• The Society also publishes English translations of two Soviet journals, *Optics and Spectroscopy* and *Soviet Journal of Optical Technology*, both containing reports of optical science advances in the U.S.S.R. The cost of the first is $44 a year for members, and $345 for nonmembers; the second is $40 a year for members, and $300 a year for nonmembers.

Rothchild Consultants
PO Box 14817   415-621-6620
San Francisco, CA 94114-0817

This group is the best source of information on optical memory, and it publishes the only reports dedicated entirely to the topic. It sponsors several conferences every year, and publishes the results of the various studies it does on the industry. Rothchild Consultants also publishes the following:

• *Optical Memory Newsletter*—a bimonthly newsletter devoted exclusively to the computer-related read-write optical memory and read-only interactive videodisc technologies. The subscription rate is $296 a year.

• *The Optical Memory Report*—a sourcebook of statistical information on the optical memory industry, including an explanation of the technology and glossary of terms. The cost is $1995 yearly including periodic updates.

Society of Photographic Scientists and Engineers (SPSE)

7003 Killworth Lane   703-642-9090
Springfield, VA 22151

SPSE is a technical society dedicated to the photographic and optical memory industry. Its members are scientists and engineers in the field of photography. The Society's executive director can be contacted by those seeking technical information and referrals. Six conferences are held by the Society every year, and some of the proceedings are published. The Society also publishes the following:

• *Journal of Photographic Scientists and Engineers*—a very technical bimonthly journal that covers topics related to photography and optical memory, including papers presented at conferences. The cost is $70 a year.

• *Journal of Imaging Technology*—a bimonthly journal that deals with technical articles on imaging technology. The cost is $70 a year.

SPIE—International Society for Optical Engineering

PO Box 10   206-676-3290
Bellingham, MA 98227-0010

SPIE is a nonprofit technical society dedicated to advancing engineering and scientific applications of optical, electro-optical, and photoelectronic technology, and communicating these applications through its publications and symposia. It holds over 140 conferences and short courses on various optical topics, including optical memory. The proceedings of all its optical memory conferences are published in full. SPIE has a large data base containing the proceedings of its conferences. The staff of the Society can answer specific technical questions, and give referrals to the general public.

SPIE's monthly journal, *Optical Engineering*, features articles on optical memory. It is available for free to members and for $60 a year to nonmembers.

Publications

Information Gatekeepers

214 Harvard Avenue   617-787-1776
Boston, MA 02134

This company publishes the following that deal with optical memory:

• *Fiber Optics Communications*—a monthly newsletter dedicated to domestic and international news, covering technical innovations, new products, calendars of events, people in the industry, and applicability of new technology. The cost is $215 a year. A condensed weekly version of the monthly edition of *Fiber Optics Communications* is also available for $128 a year.

● Conference proceedings are published after each annual meeting. More details are available from Information Gatekeepers.

Optoelectronics

Organizations and Government Agencies

Center for Laser Studies
University of Southern California   213-743-6418
Denney Research Building
Los Angeles, CA 90089-1112

The Center for Laser Studies is a research organization within the School of Engineering at USC. Research is being conducted in laser materials processing, the optical properties of matter, integrated and nonlinear optics, optical visability, laser chemistry, laser/interface interactions, and optical and thermal properties of semiconducting and insulating solids. Staff can answer your questions, and send you the following publication:

● *Technical Reports and Publications with Selected Abstracts*—a biannual publication listing technical papers generated by the School of Engineering. Copies are available from Tony Tortorice, Industrial Associates Program, 213-743-2502.

Department of Electrical and Computer Engineering (ECE)
University of California   805-961-4486
Santa Barbara, CA 93106

The Department of ECE is conducting research in optoelectronics, with particular emphasis on semiconductor lasers and integrated optics. The staff here can answer questions as well as send you copies of technical reports.

IEEE Electron Devices Society
Institute of Electrical and Electronics Engineers   212-705-7890—Technical Activities
345 East 47th Street   212-705-7866—Public Information
New York, NY 10017

The Electron Devices Society is a professional organization involved in all applications of semiconductor and other solid-state devices as they are applied to optoelectronics. The Society sponsors workshops, seminars, and conferences, including the annual International Electron Devices Meeting and the Device Research Conference. The proceedings are published after each conference and available for sale to the public. The Society also publishes:

● *IEEE Transactions on Electro Devices*—a monthly journal that focuses on the theory, design, and performance of electron devices, including electro tubes, solid-state devices, integrated electron devices, energy sources, power display, and device reliability. Each issue has approximately 160 pages, and an annual subscription rate is $159.

• *IEEE Electron Device Letters*—a monthly journal that contains news briefs on the most recent developments in the field of electron devices. Annual subscription to the approximately 40-page publication is $70.

• *Journal of Lightwave Technology*—a bimonthly journal containing articles on current research, applications, and methods used in lightwave technology and fiber optics. Topics covered include optical guided-wave technologies, fiber and cable technologies, active and passive components, integrated optics, and optoelectronics. Subscription to the 150-page journal is $96 a year.

IEEE Quantum Electronics and Applications Society

Institute of Electrical and Electronics Engineers   212-705-7405
345 East 47th Street
New York, NY 10036

The Quantum Electronics Society is a professional organization concerned with the research, development, design, manufacture, and applications of devices and systems in fields related to quantum electronics and optoelectronics. The Society cosponsors, with the Optical Society of America, the Annual Conference on Lasers. Also sponsored are the International Quantum Electronics Conference, and in cooperation with OSA, the annual Optical Fiber Communications Conference. Proceedings from these conferences are available upon request. The Society also publishes:

• *IEEE Journal of Quantum Electronics*—a monthly journal covering the latest research results in quantum electronics and its applications, including optoelectronics theory and techniques, lasers, and fiber optics. The subscription rate for this 150-page journal is $150 a year.

• *Journal of Lightwave Technology*—a bimonthly journal that contains articles on current research applications and methods used in lightwave technology and fiber optics. Topics covered include optical guided-wave technologies, fiber and cable technologies, active and passive components, integrated optics, and optoelectronics. The subscription rate for this 150-page journal is $96 a year.

• *Circuits and Devices Magazine*—a nontechnical periodical that includes feature and review articles, news, book reviews, conference reports, and research news on various subjects including quantum electronics and optoelectronics.

National Research and Resource Facility for Submicron Structures

Knight Laboratory   607-256-2329
Cornell University
Ithaca, NY 14853

This center, which is sponsored by the National Science Foundation, is a research facility for microstructures science and engineering. The facility includes a state-of-the-art public research laboratory, which is available to all qualified users.

Specific requests for information should be made by mail, and published literature about optoelectronics will be sent to you upon request.

Optical Electronics Metrology Group (OEM)

Mail Code 724.02   303-497-5341
National Bureau of Standards (NBS)
325 Broadway
Boulder, CO 80303

The OEM Group conducts work in the areas of measurements and standards for lasers, the characterization of optical fibers, and device research including detectors, modulators, and wave guide samplers. Aaron Sanders, the group leader, and his staff will respond to inquiries, or can tell you about one of the numerous research reports available. Also published is:

• *NBS Special Publication 250 and Appendix*—a source for very detailed measurement information regarding optical fibers and lasers. This is published periodically and the appendix is updated twice a year. The main publication has 114 pages, and the appendix 35 pages, and both are available free of charge from the Office of Physical Measurement Services, NBS, Physics Building, RB362, Gaithersburg, MD 20899, 301-921-2805.

Optical Science Center

University of Arizona   602-621-6997—Professor Robert Shannon, Director
Tucson, AZ 85721

The Center is comprised of 25 faculty members and 120 students who are engaged in teaching and researching many areas of optics. Specific areas include detectors, optical computers, image processing, optical design, fabrication, medical optics, and optical properties of thin films, solids, and integrated optics. Professor Shannon is available to answer inquiries and refer you to other information sources.

Optical Society of America (OSA)

1816 Jefferson Place, NW   202-223-8130
Washington, DC 20036

OSA is a membership organization composed of physicists and engineers in the field of optics and electro-optics. OSA staff can respond to questions and make referrals. Electro-optical Systems, one of OSA's technical groups, plans and sponsors technical meetings, including the Annual Conference on Lasers and Electro-optics. OSA later publishes the proceedings from these meetings, as well as:

• *Applied Optics*—a professional semimonthly journal which presents original papers on modern optics, including lasers, electro-optics, optical engineering, thin films, holography, quantum electronics, adaption optics, image processing, space optics, optical probing, and remote sensing. The subscription rate for the 150-200-page journal is $310 a year.

• *Journal of the Optical Society of America A&B*—Journal A, the general journal for basic material, also emphasizes image science. It presents articles of experimental and theoretical research regarding the principles and methods of optical phenomena. Journal B emphasizes spectroscopy and modern quantum optics. Coverage includes optical physics, optical properties of solids and interfaces, laser physics, and nonlinear optics. Subscription to the 150-page journal is $340 a year.

• *Optics Letters*—a monthly journal that covers new results in optics research. Articles cover such topics as atmospheric optics, quantum electronics, Fourier optics, integrated optics, and fiber optics. The subscription rate for the 50-page journal is $140 a year.

• *Optics News*—a monthly periodical containing news of interest to the optical industry. The articles cover many topics in the field of optics, ranging from optics theory to instrumentation and systems applications. It also includes listings of published papers and a comprehensive meetings calendar. Each issue is approximately 50 pages long, and is available free of charge to members. The annual subscription rate for others is $120.

• *Soviet Journal of Optical Technology*—a translation of a monthly Soviet journal that contains articles about theoretical and experimental details of many phases of optical, space, and astronomical engineering. Coverage includes design details of Soviet optical instruments as well as information about the physical properties of many optical materials. The subscription rate for the 70-page journal is $350 a year.

Quantum and Optical Electronics Research Laboratory

University of California   415-642-1030
Berkeley, CA 94720

The Electronics Research Laboratory coordinates research conducted by the faculty and graduate students at the University. In the area of optoelectronics, research is being done in optoelectronic switching, short pulse formation, detection, semiconductor lasers, far infrared devices, and nonlinear optics. The staff can respond to inquiries, and reprints of research papers are available upon request at no charge.

SPIE—International Society for Optical Engineering

PO Box 10   206-676-3290
Bellingham, WA 98227-0010

SPIE is a nonprofit technical society involved in all areas of electro-optics. The Society's members include scientists, engineers, and users who are interested in advancing engineering and scientific applications of optical and electro-optical instrumentation, systems, and technology. The Society's staff will answer questions and make referrals to other sources of information. The Society sponsors numerous conferences and symposia including the Annual International Symposium on Optics and Electro-optics, the Symposium on Optical and Electro-optical Engineering, the Symposium on Microlithography, and an annual conference of medical image production processing, display, and archiving. A complete listing of the proceedings are available to the public. The Society publishes the following:

● *Optical Engineering*—a technical bimonthly journal that contains original papers about applications of new optical technology and known optical technology being used in new and inventive ways. Each issue is approximately 125 pages and an annual subscription is $60.

● *Optical Engineering Report*—the monthly newsletter of the SPIE that contains employment information, announces upcoming conferences, and presents technical articles. The 30-page newsletter is available free of charge.

Additional Experts and Resources

The following are some of the major companies involved in optoelectronics. Many of them publish brochures and pamphlets that describe their products and the technology used to create them, and these are usually available for free. Public relations offices within these companies can also be a good source of information.

AT&T Bell Laboratories
150 John F. Kennedy Parkway, Room 3H234
Short Hills, NJ 07078
201-564-4244

Batelle Memorial Institute
Public Relations Department
505 King Avenue
Columbus, OH 43201
614-424-5544

Hughes Research Laboratories, Inc.
Optoelectronics Department
3011 Malibu Canyon Road
Malibu, CA 90265
213-456-6411

TRW Optoelectronics
1207 Tappan Circle
Carrollton, TX 75006
214-323-2200

Joseph T. Boyd, Ph.D.
M.L. 30
University of Cincinnati
Cincinnati, OH 45221

513-475-3066

Dr. Boyd is a faculty member of the Electrical and Computer Engineering Department and is the Director of the Solid State Electronics Lab at the University. He is involved in optoelectronics research and experimentation including guided-wave optics, layer processing of semiconductor devices, and optical characterization of materials. Research papers are available through several technical journals. Dr. Boyd and his staff can respond to inquiries and refer you to other information sources.

Gregory E. Stillman, Ph.D.
University of Illinois
Electrical Engineering Building
1406 Green Street, Room 155
Urbana, IL 61801

217-333-3097

Dr. Stillman is a professor of electrical engineering and is noted for his research in compound semiconductor materials and devices with particular application to optoelectronics. Dr. Stillman has had many of his research findings published. He and his staff are available to answer questions about optoelectronics and make referrals to other information sources.

Oral Contraception

See: Reproductive Technology

Organ Transplants

See: Transplantation Technology

Passive Solar Technology, Use of

See: Solar Technology

Patents, Trademarks, and Copyrights

Organizations and Government Agencies

American Patent Law Association (APLA)
2001 Jefferson Davis Highway, Suite 203   703-521-1680
Arlington, VA 22202

APLA tries to promote better understanding of the patent, trademark, and copyright systems. Its law library, which is open to the public, holds a wide range of material on patents. Publications include:

- *APLA Bulletin*—current information about activities relating to patents, trademarks, copyrights, and related matters.
- *APLA Quarterly Journal*—contains articles of current interest on various aspects of intellectual property law.
- *What, Why, and How of Patents*—a brochure that explains what a patent is, and why and how to obtain a patent, trademark, or copyright. Each pamphlet is 25 cents.

Commissioner of Patents and Trademarks
Public Service Center   703-557-5168
Office of Patents and Trademarks
2021 Jefferson Davis Highway
Washington, DC 20231

The Center publishes a newsletter that includes a listing of publications and brochures put out by the Patent and Trademark Office. The staff tries to answer all questions, such as finding the right office, locating publications, etc. Patent specifications and drawings, as well as trademarks, are $1 each. You must have the patent or trademark number. You can also contact the following offices.

Search Room—Patents
2021 Jefferson Davis Highway
Washington, DC 20231
703-557-2276
703-557-3158—General Information About
   Patents
703-557-2955—Scientific Library

This service will search for any patent, and give you the vendor's name, the issue date, and the title. You can also go in and search all patents in any field. The Scientific Library contains all U.S. and foreign patents and is open to the public.

Search Room—Trademarks
2021 Jefferson Davis Highway
Washington, DC 20231
703-557-3881

To search a trademark, you must either visit this office or call the Trademark Library at 703-557-3268. It is open to the public.

Trademark Information Office
703-557-3883

This office will answer questions on different aspects of trademarks, and send you a booklet describing trademarks and what the Patents and Trademarks office does.

Copyright Office
Reference and Bibliographic Section
Library of Congress
Washington, DC 20559
202-287-8700

This office will research the copyright you need and send you the information by mail. Requests must be in writing and you must specify exactly what it is you need to know.

Intellectual Property Owners, Inc. (IPO)
1899 L Street, NW   202-466-2396
Washington, DC 20006

IPO is a nonprofit trade association representing people who own patents, trademarks, and copyrights. It gathers and disseminates information on legislative and regulatory matters, and monitors international events and intellectual property developments. It publishes:
   ● *IPO News*—it keeps members up to date on developments in the field.

National Patent Council

Crystal Plaza One   703-521-1669—Vicky Schellin
2001 Jefferson Davis Highway, Suite 301
Arlington, VA 22202

The Council provides general information about patents, trademarks, and copyrights. It will answer just about any question you may have or refer you to other sources of information. The Council publishes:

● *Intellectual Property Notes*—a monthly newsletter that informs readers about the constant changes in rules, fees, and so on, at the Patent and Trademark Office. It is available for $87 a year.

The Patent Office Society

PO Box 2089   703-557-2767—Stanley Miller
Arlington, VA 22202

This professional society for patent examiners promotes the patent system to the general public. It publishes:

● *Journal of the Patent Office Society*—a journal for the dissemination and exchange of ideas in the fields of patents, trademarks and copyrights. It is available for $18.

● *Unofficial Gazette*—in-house newsletter which provides articles on court decisions, interviews, feature articles, and commentary. This is available free to members only.

The United States Trademark Association (USTA)

6 East 45th Street   212-986-5880
New York, NY 10017

The USTA keeps abreast of all aspects of the trademark field. It sponsors forums (educational meetings) and an annual meeting. USTA operates a placement service that keeps resumes on file. It maintains a comprehensive library that offers access to source material on all aspects of trademarks. The following publications are available at no cost to USTA members:

● *The Trademark Reporter*—a bimonthly law journal in the field of trademark and unfair competition law.

● *The Executive Newsletter*—offers feature articles on trademark topics.

● *Bulletins*—over 50 are published a year, reporting on general news, publications, and events related to trademark law, advertising, marketing, and design.

Publications

Superintendent of Documents

U.S. Government Printing Office   202-783-3238
Washington, DC 20402

The GPO sells several publications dealing with patents and trademarks. These are:

• *Patent Official Gazette*—the official weekly journal of the Patent and Trademark Office. It contains selected drawings and an abstract of each patent granted, indexes of patents, a list of patents available for license or sale, and other general information. Available for $250 a year.

• *Trademark Official Gazette*—the weekly journal covering trademarks. Available for $205 per year.

• *Annual Indexes*—an index of the patents issued each year is published in two volumes: an alphabetical index to patentees, and an index by subject matter of inventions. Price varies from year to year.

• *General Information Concerning Patents*—information on the application for and granting of patents, expressed in nontechnical terms. Available for $3.

• *General Information Regarding Trademarks*, $3.25.

• *Story of the Patent & Trademark Office*, $4.75.

• *Attorneys and Agents Registered to Practice Before the U.S. Patent Office*, $9.

Patent and Trademark Review

Clark Boardman Company, Ltd.   212-929-7500
435 Hudson Street
New York, NY 10014

The *Review* publishes the text of the new laws and regulations, proposed legislation, and important court rulings affecting intellectual and industrial property worldwide. Each issue contains special articles on specific patent and trademark issues and practices. Available for $33 a year, 11 issues.

World Patent Information

Pergamon Press, Inc.   914-592-7700
Maxwell House
Fairview Park
Elmsford, NY 10523

This quarterly journal for patent information and industrial innovation regularly features new regulations pertinent to patent information and documentation, advance notices and short reports on meetings and conferences on patent information, bibliographies, and book reviews. Annual subscription rate is $65 a year.

Bureau of National Affairs

1231 25th Street, NW   202-452-4500
Washington, DC 20037

The Bureau publishes:

• *BNA's Patent, Trademark and Copyright Journal*—provides weekly notification, interpretation, and analysis of significant current developments, covering a wide range of intellectual property issues such as: congressional activity, court decisions, government contracting, international development, and so on. Subscription rate is $479 a year.

• *United States Patents Quarterly*— is a full-text reporter for decisions of the federal courts and of the Patent and Trademark Office concerning patent, trademark, copyright, and unfair competition cases. An *Annual Digest* is published every year, as well as *Cummulative Digest*s every five years. This quarterly publication is available for $664 a year or $618 a year for two years.

Office of Technology Assessment (OTA)

U.S. Congress   202-226-2115—Public Affairs
Washington, DC 20510   202-224-8996—Publications

The OTA provides analytical assistance to Congress on issues which involve science and technology, and *OTA Reports* are the results of these studies. A study on patents and technology resulted in the following publication:

• *Patents and the Commercialization of New Technology*—assesses the impact of the patent system on the generation and stimulation of new technologically based firms to the future economic vitality of the U.S. A brief summary of this report is available at no cost from the OTA.

Data Bases

CLAIMS

DIALOG Information Services, Inc.   415-858-3785
3460 Hillview Avenue
Palo Alto, CA 94304

CLAIMS is a general title for several different patent data bases, including:

• *CLAIMS/US Patent Abstracts 1950-Present*—contains patents listed in the chemical, electrical, general, and mechanical sections of the *Official Gazette* of the U.S. Patent Office. The cost is $95 per on-line hour.

• *CLAIMS/US Patent Abstracts Weekly*—includes the most current weekly update and records from the current month. It also costs $95 per on-line hour.

LEXPAT™

Mead Data Central   800-227-4908
PO Box 1830
Dayton, OH 45401

LEXPAT is an on-line service that includes the full text of all utility, plant, and design patents issued by the U.S. Patent and Trademark Office from 1975 to the present.

Patent Data Base (PDB)

U.S. Department of Commerce   703-557-9236
Patent and Trademark Office
2021 Jefferson Davis Highway
Washington, DC 20231

The Patent and Trademark Office makes available its patent and trademark information on several data bases.

• *Patent Full-Text Data File*—available on Mead Data Central's LEXPAT system, 202-785-3550.

• *Patent Bibliographic Data File*—available from the Bibliographic Retrieval Service, 800-833-4707.

• *Patent Bibliographic Data File with Exemplary Claim(s)*—available from DIALOG Information Services, 800-227-1960; or Pergamon International Information Corporation, 703-442-0900.

• *Patent Bibliographic Data File With All Claims*—available through SDC Information Services, 800-421-7229.

• *Patent Master Classification File (MCF)*—available from Pergamon International Information Corporation, 703-442-0900; DIALOG Information Services, Inc., 800-227-7229; or SDC Information Services, 800-421-7229.

Pergamon Info Line, Inc.
1340 Old Chain Bridge Road   800-336-7575
McLean, VA 22101   703-442-0900—in Virginia

Several patents-related data bases are offered by Pergamon Info Line, Inc. These include:

• *INPADOC*—an international patent data base, providing access to patent documents published by 51 national and regional patent offices. Cost is $125 per hour.

• *PATLAW*—a bibliographic data base, based on the *U.S. Patents Quarterly* (USPQ) published by the Bureau of National Affairs in Washington, DC. Cost is $115 per hour.

• *INPANEW*—contains the latest 15 weeks' data from the *INPADOC Patent Gazette*. The cost is $125 per hour.

• *PATSEARCH*—a comprehensive file of U.S. patents from 1971 to date. Each file includes a complete bibliographic description and the full abstract. Cost is $95 per hour.

• *COMPUTERPAT*—contains records for all U.S. digital data processing patent documents as classified by the U.S. Patent and Trademark Office in subclasses 364/200 and 364/900 since 1942, the date that the first U.S. patent for this technology was issued. Cost is $95 per hour.

Pay Television Services

See: Subscription Television

People in Technology

Publications

American Men and Women of Science

R. R. Bowker Co.   800-521-8110
PO Box 1807
Ann Arbor, MI 48106

This seven-volume set contains 130,000 biographical sketches of experts in the physical and biological sciences. It is categorized into 65 major disciplines and 1,800 subdisciplines, and it is updated annually. The directory is also available on-line on DIALOG (800-227-1960, or 800-982-5838 in CA) and on BRS (800-833-4707, or 800-553-5566 in NY).

McGraw-Hill's Modern Scientists and Engineers

McGraw-Hill, Inc.   212-997-1221
Professional and Reference Books
1221 Avenue of the Americas
New York, NY 10020

A three-volume set, this provides 130,000 biographical sketches with photographs of over 11,000 of the most prominent experts in science and engineering. It contains mostly information on an individual's professional career and is updated approximately every ten years. It is available for $150.

Who's Who in Technology Today

Research Publications, Inc.   203-397-2600
12 Lunar Drive
Woodbridge, CT 06525

The publication provides bibliographic information about 32,000 leading scientists and technologists. Information given about each individual includes: employment and home address, current position, previous position, honors, awards, and technical achievements. The publication is organized according to 50 disciplines and divided into five volumes: Volume 1—*Electronics and Computer Science*; Volume 2 —*Physics and Optics*; Volume 3—*Chemistry and Biotechnology*; Volume 4—*Mechanical Engineering, Civil Engineering, and Earth Sciences*; and Volume 5—*Index*. The 4,086-page publication is updated every two years. The entire set can be purchased for $425, or you can purchase individual volumes for $85.90 each.

Additional Experts and Resources

Government

If an individual testified before a congressional committee or worked for the federal government, he/she may leave a trail of biographical information at either or both of the following places:

● Capitol Hill: Anyone who testifies before a congressional committee is likely to have a detailed bio filed with the committee staff. This resume is available to the public. To investigate relevant committees see listing under *Legislation*.

● Executive Branch: Anyone who has worked for the government will have a copy of his/her professional history on file. If you know the agencies he/she worked for, their personnel office will provide you with the information.

Peripherals

Organizations and Government Agencies

American National Standards Institute (ANSI)

1430 Broadway
New York, NY 10018

212-354-3345—Information System Department
212-354-3473—Sales

ANSI, a private, nonprofit corporation composed of organizations, companies from private industry, and government professionals, coordinates the development of voluntary standards by establishing a set of procedures and requirements for the development of a consensus. The standards are developed by independent organizations, technical associations, and committees that have been accredited following strict guidelines. Staff can answer technical questions and make referrals to experts in the field. It publishes reports on the standards developed, and a full directory is available for $10, covering a wide scope of topics such as data communication, software, flexible media, electronic business data interchange, programming languages, character recognition, and peripherals and codes.

Computer Business and Equipment Manufacturers (CBEM)

311 First Street, NW, Suite 500
Washington, DC 20001

202-737-8888

CBEM is a trade association dealing with regulations and proposed legislation pertaining to all computer hardware or software, including peripherals. The association has 42 members including most of the largest computer and communication manufacturers. Staff can make referrals to other manufacturers and trade associations in the field. A career brochure on the industry is published for high school students, as well as hundreds of trade publications and statistical reports on international trade, sales, and exports. CBEM also publishes:

• *Industry Marketing Data Book*—an annual publication that provides a historical look at the industry. It projects sales, revenues, and employment trends for the next five years, with a focus on news in the industry. The 200-page publication is available for $195 a year.

• *Visual Displays*—a one-time publication that deals with visual displays and focuses on various visual peripherals available in the computer industry. The 16-page publication is available for $200.

Computer Communications Industry Association (CCIA)

1500 Wilson Boulevard, Suite 512   703-524-1360
Arlington, VA 22209

CCIA is a trade association that monitors legislative issues in the computer industry for its 70 members representing computer and communication companies and manufacturers in all areas of the computer field. It maintains an information resource center for members and legislators on computers and the communications field.

Electronic Industries Association (EIA)

2001 Eye Street, NW   202-457-4900
Washington, DC 20006

EIA, a trade association for the electronics industry, sponsors two consumer electronics shows with over 200 exhibits, including many on peripherals. It also hosts two conventions made up of seminars and workshops covering computers, peripherals and consumer electronics. EIA publishes:

• *1984 Electronic Market Data Book*—an annual publication that provides sales and production figures covering computers and peripherals, and manufacturing profiles on consumer electronics. The 200-page publication is available for $60.

Office of Trade Information Service (OTIS)

Department of Commerce   202-377-2988
PO Box 14207
Washington, DC 20044

OTIS provides many reports and services on the marketing of products with an international scope, including a contact service that puts you in contact with distributors, manufacturers, retailers, and wholesalers in the country of your choice that can represent your product. OTIS publishes the following:

• *Export Statistical Profile (ESP)*—an annual two-volume publication that provides a five-year overview of specific products and export histories. The report shows the percentage of total dollar values and allows for a historical perspective of specific products. A data base is available for more specific information on products, providing a five-year export history on certain products. Volume I is $30 and Volume II is $70.

• *Market Research*—provides a five-year projected analysis on the needs for specific computer products in a particular country. The cost is $10-$250 for the reports, depending on the scope of the products and countries profiled.

• *Annual Worldwide Industry Review*—this publication augments the *Market Research Reports*, providing updated statistical information on products and country profiles. It provides a competitive and market assessment by market officers in the country being profiled. Three different reports are available: Western Hemisphere and Africa, Europe, and the Near East and Asia. Each report is $200, or all three for $500.

• *Trade List*—provides a listing by country, by product and by Standard Industry Classification (SIC) number for easy classification of specific products. It gives a list of agents, manufacturers, distributors, retailers, and wholesalers of computers and computer-related products on a worldwide scope. The 126-page publication is available for $40.

• *Trade Opportunities Bulletin*—the weekly bulletin, produced jointly by the Department of Commerce and the Department of State, consists of trade leads acquired throughout the world on all types of products. The 50-page publication is available for $175.

• *Top Notch Service*—this publication is a supplement to the *Trade Opportunities Bulletin* (described above), which provides trade leads on specific countries. The cost is $62.50 for the initial block of 50 leads.

• *Export Mailing List*—provides a custom mailing list of manufacturers of specific products for a country. It can be obtained for computer peripherals. Prices vary according to the services required.

• *World Traders Data Report*—provides financial reports on companies in foreign markets, presenting information on company history, size, production statistics, credit standings, and overall evaluation. Customized reports are available for $75 a report.

Publications

Computer Age
7620 Little River Turnpike, Suite 414   703-354-9400
Annandale, VA 22003

Computer Age is a publishing firm that produces many publications in the computer field. A free publications list is available. Some of these publications are:

• *Computer Daily*—is a daily intelligence news service available on-line which provides daily reports on hardware and software news and analytical views and expert analysis from the computer industry. The service profiles legislative issues and computer litigation suits. Available for $750 a year.

• *Peripheral Digest*—this monthly newsletter presents an information management analysis of the computer peripherals industry, covering both the domestic and foreign markets, trends, sales opportunities, and in-depth reports on peripheral topics. The 12-page publication is available for $144 a year.

• *Mini/micro Computer Report*—covers both mini- and microcomputers. It also tracks the microcomputer peripheral industry. Annual subscription to the monthly 12-page publication is $144.

• *World Trade*—this monthly publication covers the foreign market for computer hardware and peripherals. It provides an overview of world trade trends and news plus highlights on specific foreign markets. Annual subscription to the 12-page publication is $144.

Data Communications Report

Auerback Publishers	609-662-2070
6560 North Park Drive	800-257-8162
Pennsauken, NJ 08109	

The three-volume *Report* provides information on data communications. Volume I covers transmission equipment and facilities, data communication software, multiplexers, modems, and network control systems. Volumes II and III cover intelligent terminals, graphic terminals, terminal attachments and peripherals, and digital plotters and teleprinters. This report is updated monthly, and it, along with a telephone updating service, is available for $780 a year.

Datapro 70

Datapro Research Corporation	609-764-0100
1805 Underwood Boulevard	800-257-9406
Delran, NJ 08075	

This three-volume publication provides a comprehensive reference service describing the broad base of EDP hardware, software, services, and suppliers. It covers such topics as computers, terminals, graphics, memory and storage, printers, software, and data communications; and provides in-depth product profiles, and a directory of suppliers and product comparisons. The publication is updated monthly, and a *Newscom Newsletter* is also included. The annual subscription is $875.

Disk Trend

Disk/Trend, Inc.	415-961-6209
5150 El Camino Real, Suite B20	
Los Altos, CA 94022	

Disk Trend is an annual two-volume statistical market study covering both rigid and flexible disk drives. The data are broken down into categories and provide details about unit shipments and revenues on each type of disk drive. It provides a section that lists the standards and specifications for each disk drive produced in the world, and provides profiles on each manufacturer and outlines their roles in the industry. The *Rigid Disk Report* (Volume I) is 250 pages and costs $1,015; the *Floppy Disk Report* (Volume II) is 150 pages and is available for $775. Both volumes can be obtained for $1,530 for the 1984 editions.

Kahner Publishing Company

221 Columbus Avenue   617-536-7780

Boston, MA 02116

Kahner publishes the following:

• *Minimicrosystems*—this monthly computer publication provides information on personal computers, disk drives, terminals, system integrators, and many aspects of mini/microcomputer systems. Annual subscription to the 250-300 page publication is $55 a year.

• *Peripherals Digest*—a twice-yearly publication that is included with a subscription to *Minimicrosystems* (described above). It provides a directory listing peripherals and who manufactures them, as well as their specifications, such as terminals, disk drives, printers, and modems. The 230-page publication can be purchased separately for $15.

• *Computer Digest*—the annual publication is included with a subscription to *Minimicrosystems* (described above) and lists single-board computers, single-user computers, multi-user microcomputers, and minicomputers. The 150-page publication is available separately for $15.

Personal Computer

See: Minicomputers

PET

See: Medical Imaging; Nuclear Medicine

Pharmaceuticals

Organizations and Government Agencies

Academy of Pharmacy Practice (APP)

2215 Constitution Avenue, NW   202-628-4410

Washington, DC 20037

A subsidiary of the American Pharmaceutical Association (APHA), the APP consists of 4,000 members, mainly pharmacists. The Academy has several interest groups covering specific issues of concern. APP holds its annual meeting with the APHA in the spring, and it publishes:

• *Pharmacy Practice*—a monthly covering the latest developments in medical science and the retail aspect of pharmaceuticals. It covers such topics as long-term care, nuclear pharmacies, clinical practice, federal pharmaceutical regulations, and management. Available for $15 a year.

Academy of Pharmaceutical Science (APS)

2215 Constitution Avenue, NW   202-628-4410
Washington, DC 20037

A division of American Pharmaceutical Association, APS membership consists of scientists and other experts involved in the field of pharmacology and health service. Members deal with the discovery, testing, development, control, and economics of drugs. An annual conference is hosted in the fall with seminars and workshops. Papers submitted at the conference are published and available in abstract form. APS collaborates with the APHA convention in the spring. It publishes the following:

• *Academy Reporter*—a bimonthly publication that focuses on scientific and research developments in the field. It also covers the organizational news, meetings, and awards of the association. The 10-page publication is available for $15 a year.

American Pharmaceutical Association (APHA)

2215 Constitution Ave, NW   202-628-4410
Washington, DC 20037

This professional association is involved in the improvement of pharmacy practices and oversight of legislative issues. Members work closely with pediatricians, dentists, and doctors in all fields of medicine to help ensure a thorough understanding of pharmaceutical needs and uses. The 56,000 members include licensed pharmacists, scientists, chemists, students in the field of pharmacology, and retired pharmacists. The APHA is made up of two academies: the Academy of Pharmacy Science and the Academy of Pharmacy Practice, which are described earlier. The APHA holds an annual convention each spring with exhibits, seminars, and workshops. The Association publishes books, periodicals, training manuals, standards and guides, and drug updates. A free publications catalog is available. Some of the monthly periodicals are:

• *Journal of Pharmaceutical Science*—a monthly publication of reports and articles selected by a committee for publication covering such topics as: research in the field, clinical studies, advances in patient care, and practical and informative updates. There are color graphs and charts, and molecular formulas. The 200-page publication is available for an annual charge of $40 to individuals and $75 to institutions.

• *American Pharmacy*—a monthly publication covering such topics as legislative issues, research and development in the field, patient care, a news section, and practice innovations. This is the official publication of the APHA, and it is available for $30 a year to individuals, and $40 to institutions.

• *Pharmacy Student*—a quarterly publication for pharmacology students covering the activities and job opportunities in the field. It highlights the latest developments and growth potentials for university students. The 50-page publication is available for $10 a year.

• *Pharmacy Weekly*—a weekly publication covering legislative issues, committee meetings, membership information, news from the U.S. Food and Drug Administration, and health issues such as high blood pressure. The 5-page newsletter is available for $15 a year.

• *Handbook of Nonprescription Drugs*—a book listing information and warnings pertaining to nonprescription drugs. Available for $45.

• *Manager's Guide to Third-Party Programs*—a onetime publication covering different varieties of payment by third parties such as HMOs, and Blue Cross/Blue Shield insurance companies. The 100-page publication is available for $18.

American Society of Hospital Pharmacy (ASHP)

4630 Montgomery Avenue 301-657-3000
Bethesda, MD 20814

ASHP is a professional organization of 20,000 members specializing in hospital pharmacology. It holds two meetings a year consisting of seminars and workshops, after which the proceedings are published. It also holds continual educational programs and seminars for members and interested nonmembers for a nominal cost. Several data bases are available for easy access to current information. Some of the publications are:

• *American Journal of Hospital Pharmacy*—a monthly membership publication that presents papers submitted, and reviews of books in the field. It also provides the latest news on technology and patient care. The 200-page journal is available for $60 a year.

• *International Pharmacy Abstracts*—a semimonthly publication that summarizes drug reviews and therapy reviews, and provides abstracts on international medical literature. Annual subscription to the 320-page publication is $300.

• *Clinical Pharmacy*—a bimonthly publication that covers the latest research and developments in the clinical studies of pharmacology. Annual subscription to the 320-page bimonthly is $35 for members, and $45 for nonmembers.

• *ASHP Newsletter*—a monthly publication available only to members. It covers the business and news pertaining to ASHP.

• *Drug Information*—an annual that provides a detailed listing of drugs and their effectiveness, side effects, dosage requirements, and availability. The one large volume and three supplements are available for $40 a year.

American Society for Pharmacology and Experimental Therapeutics (ASPET)

9650 Rockville Pike 301-530-7060
Bethesda, MD 20814

ASPET is a society involved with research in the fields of pharmacology and therapeutics. It sponsors a five-day meeting two times a year, providing seminars, symposia, abstracts, discussion groups, and slide presentations. Some of its publications are listed here. They can be ordered from William & Wilkins, Publishers, 428 East Preston Street, Baltimore, MD 21202, 301-528-4117.

- *Journal of ASPET*—a monthly publication covering reports on research and experiments conducted in the field. The 330-page monthly is available for $120 a year.
- *Journal of Molecular Pharmacology*—a bimonthly publication covering the latest advances in the field of molecular pharmacology. The 330-page publication is available for $75 a year.
- *Journal of Drug Metabolism and Disposition*—a bimonthly publication covering the latest research and developments in the field. The 150-page publication is available for $60 a year.
- *Pharmacology Review*—a quarterly publication that highlights the latest articles and research innovations in the field. The 200-page publication is available for $30 a year.
- *Clinical Pharmacology and Therapeutics*—a monthly publication covering the latest research developments. The 150-page publication is available for $45 a year from C. V. Mosby, 11830 West Line Industry Drive, St. Louis, MO 63143.

National Center for Toxicological Research (NCTR)

Jefferson, AK 72079   501-541-4000

The Center is involved in research in most areas of toxicology and carcinogenic substances, teratology, and mutigenesis dealing with biochemical mechanism, and molecular changes from chemicals. NCTR studies immunology, biodegradation of microorganisms to break down compounds into less harmful or innoculous compounds, detoxification to develop methods to render toxic compounds into harmless substances. It publishes reports on all the research projects, which total about 200 publications a year. A listing of the individual research reports is available from NCTR, as well as the following:

- *NCTR Ongoing Research Report*—an annual report that covers the research done by the Center, including analytical methods development, method molecular biology, chemical toxicology, pathology, and microbiological studies on toxicological chemicals and substances. The 400-page publication is available free of charge.

Pharmaceutical Manufacturers Association (PMA)

Science and Technology Division   202-835-3400, ext. 540
1100 15th Street, NW
Washington, DC 20005

PMA is a national association representing more than 135 major manufacturers of pharmaceuticals that can provide directory information on any of the drug manufacturers. It keeps close contact with the industry and technological developments, specializing in the guidelines to follow in the manufacturing of pharmaceuticals. PMA puts out more than 70 publications in the field, and a complete listing is available free of charge. The publications cover such topics as regulatory issues of the FDA, deionized water for the water industry and water treatment, clinical guidelines and manufacturing guidelines, and the proceedings of seminars. Some of the publications are in a report format of 4 or 5 pages, others are 200 pages in length. Some of these publications are:

• *1980 Fact Book by PMA*—a onetime fact book on prescription drugs which serves as a diagnostic tool for doctors and pharmacists with detailed charts and tables covering new drugs introduced to the U.S. from 1940 to 1978, noting the drug origin by country. The 85-page publication is free.

• *Research and Development in Pharmacy Industry, 1977*—a onetime publication about the cost of research and how much time is needed to study and test drugs. It explores how new medicines are tested and provides the sales and research expenditures of the drug industry. The 32-page publication is available free of charge.

National Institute of General Medical Sciences (NIGMS)
Building 31, Room 4A52   301-496-7301
9000 Rockville Pike
Bethesda, MD 20205

NIGMS is a federal agency involved in drug research in synthetic chemicals, basic biological and biochemical compounds, and molecular pharmacology. It does comparative studies using laboratory animals and controlled clinical investigations of patients and normal volunteers. NIGMS provides grants for these studies, and supports anesthesiologists, chemists, biologists, and pharmacologists. It has fact sheets available on the programs sponsored, as well as the guidelines for the grants. The following are two of their publications:

• *Pharmacological Sciences Program*—an annual report providing highlights on scientific research sponsored by NIGMS. The 30-page publication is free.

• *NIGMS Annual Report*—covers research programs sponsored by the Institute. The 100-page publication is free.

Parenteral Drug Association (PDA)
1346 Chestnut Street, Suite 1407   215-735-9752
Philadelphia, PA 19107

PDA represents manufacturers of parenteral drug products and focuses on the standards used in the manufacturing process, such as validation, sterilization, and quality control. It represents 160 major pharmaceutical companies and materials suppliers. PDA sponsors an annual meeting consisting of seminars, workshops, and exhibits. Scientific papers are submitted at the conference, and a theme is usually followed. Two other meetings are also held consisting of 30 exhibits, and scientific papers are submitted during the one-day symposium. Intermittent technical reports and a listing of them are available free of charge, examples of which are: *Sterile Packaging*, and *Validation of Dry Heat Processes Used for Sterlization and Depyrogenation*. They also publish:

• *Journal of Parenteral Science and Technology*—a bimonthly that presents scientific papers, submitted at the annual meeting, that deal with ongoing research in the field. Annual subscription to the 50-page publication is $40.

World Health Organization (WHO)

Publications Center   518-436-9686

49 Sheraton Avenue

Albany, NY 12000

   Based in Geneva, Switzerland, the WHO assists countries researching their health needs. Many publications are available including over 50 on pharmacology. A free publications directory is available upon request. Some titles are:

   ● *Manpower Development in Toxicology: Report on a WHO Consultation* (1979)—$2.

   ● *International Nonproprietary Names of Pharmaceutical Substances*

   ● *International Pharmacopia*

   ● *Specifications for Reagents*

   ● *Control and Testing of Plastic Containers for Pharmaceuticals*

   ● *Expert Committee on Antibiotics*

   ● *Specifications for Pharmaceutical Preparations*

   ● *Report of the WHO Expert Committee on Specifications for Pharmaceutical Preparations*

   ● *Selection of Essential Drugs*

   ● *Principals for Preclinical Testing of Drug Safety*

   ● *Principals for the Testing of Drugs for Teratogenicity*

   ● *Principals for Testing and Evaluation of Drugs for Carcinogenicity*

   ● *Opiates and their Alternatives for Pain and Cough Relief*

   ● *Bioavailability in Drugs*

Publications

Aster Publishing

320 North A Street   503-726-1200

PO Box 50

Springfield, OR 97477

   Aster Publishing sponsors a 3-day conference in the fall with many workshops, seminars, and keynote speakers, covering such topics as chromatography, bio/tech sessions, solid dosage, validation, and quality control. The papers and articles presented at the conference are published and made available for sale. Some of Aster's publications are:

   ● *Pharmaceutical Technology*—this monthly publication covers the latest developments in the technology, production, and manufacturing of pharmaceuticals. The July issue is a buyer's guide and lists the manufacturers in the field. The 120-page publication is available for $80 a year.

   ● *Pharmaceutical Executive*—this monthly publication profiles one executive in the field, and covers marketing strategies and the latest news and developments in the industry. The 100-page publication is available for $51 a year.

   ● *Liquid Chromatography*—a monthly trade publication dealing with the latest research and development in the field of chromatography. It presents test results on absorption rates and equipment use. The 100-page publication is available for $40 a year.

Micromedex

2750 South Shoshone Street
Englewood, CO 80110

800-525-9083
303-781-7813

Micromedex is a publisher, affiliated with the Rocky Mountain Poison and Drug Center, that specializes in medical and drug information and pharmacology. They publish the following:

- *Drugdex*—a quarterly publication that provides monographs covering dosage information; pharmocokenetics; absorption, distribution, metabolism, and excretion; cautions of drugs and adverse reactions; drug interactions; and therapeutic indications. Available in microfiche for $2,075 a year. It is also available on computer tape and videodisc.

- *Poisondex*—this quarterly publication draws from a toxicology data base that deals with the identification of 350,000 substances. Indexed by trade name, brand name, generic name, and chemical name, each entry provides specific ingredients data and detailed information on how to treat any of these products that have been ingested, absorbed, or inhaled. Available in microfiche for $1,785 a year, and in computer tape and videodisc format.

Pharmaceutical Manufacturing

Canon Communications
2416 Wilshire Boulevard
Santa Monica, CA 90403

213-829-0315

This 30-page monthly publication deals with the production of pharmaceuticals and reports on the latest developments in manufacturing techniques with emphasis on quality control. Available free to qualified subscribers, or for $18 a year to others.

Additional Experts and Resources

Paul Davignon

Chief, Pharmaceutical Resources Branch
National Cancer Institute
Landow Building, Room 5C-25
Woodmont Avenue
Bethesda, MD 20205

301-496-8774

An expert in the field, Mr. Davignon oversees the development, production, and distribution of investigational drugs in the division of cancer treatment within the National Cancer Institute, most products of which are injectable dosage forms. He may be contacted on a limited basis for referrals to other experts in the field of pharmacology. Some publications of the NCI are:

- *Chemical and Analytical Information*—a onetime publication that deals with the analytical and chemical information on investigative drugs for laboratory personnel. The 200-page publication (#842654) is available free of charge.

• *Pharmaceutical Data*—an annual publication that describes data of products for the handling of drugs by pharmacists and nurses. The 200-page publication (# 842141) is available free of charge.

Leo McIntire

Office of Chemical and Allied Products   202-377-0128
U.S. Department of Commerce
14th Street and Constitution Avenue, NW, Room 4065
Washington, DC 20230
Mr. McIntire is a drug information specialist and is available on a limited basis for referrals to other agencies and experts in the field of pharmacology. He is involved in legislative efforts to reduce the tariffs on import and export expansion and promotes the sale of generic products in other countries and in the U.S. He writes the pharmacology industry outlook in the annual *U.S. Industrial Outlook*, available from the Government Printing Office.

Phobias

See: Technophobia

Photography

See: Optical Memory and Storage

Photovoltaics

Begins next page. See also: Fuels, Alternative; Renewable Energy Technology; Solar Technology

Organizations and Government Agencies

Association for the Advancement of Photovoltaic Education (AAPE)
1 Denver Place   303-292-5013
999 18th Street, Suite 1000
Denver, CO 80202

AAPE is a membership organization that provides information on many aspects of the technology and applications of photovoltaics. Programs are directed toward elementary, junior high, high school, and college-level students. The Association offers literature, workshops, seminars, and conferences.

Collector Research and Development Branch
Photovoltaics Energy Technology   202-252-1725
Solar Electric Technologies
Conservation and Renewable Energy, DOE
Forrestal Building, CE-333, Room 5E066
1000 Independence Avenue, SW
Washington, DC 20585

This office plans, implements, and evaluates research, development, and testing programs leading toward potentially low-cost, advanced photovoltaic materials and devices.

Southwest Residential Equipment Station (SWRES)
New Mexico Solar Energy Institute   505-646-1049
New Mexico State University
Box 3 SOL
Las Cruces, NM 88003

The New Mexico Solar Energy Institute operates the Southwest Residential Experiment Station located on the NMSU campus. At the experiment station photovoltaic systems mounted on prototypes of residences are tested for the southwestern United States. The SWRES is funded by the U.S. Department of Energy through its Albuquerque office. Sandia Lab provides technical advice.

Information gained through solar testing, evaluation, research, and demonstration projects is made available to the public, business and professional groups, and to target groups in industry and government through workshops, publications, presentation of papers, information services, technical assistance, and mass media programs.

Systems Research and Technology Branch

Photovoltaics Energy Technology   202-252-1724
Solar Electric Technology
Conservation and Renewable Energy
Department of Energy
Forrestal Building, CE-333, Room 5E066
1000 Independence Avenue, SW
Washington, DC 20585

This office develops and manages research programs for the design, prototype fabrication, and testing of photovoltaic systems that will be low in cost, reliable, safe, and environmentally acceptable.

Publications

Photovoltaics International

PVI Publishing Company   303-292-5013
1 Denver Place
999 18th Street, Suite 100
Denver, CO 80202

Issued monthly, the magazine contains both technical and nontechnical information. It is an application-oriented periodical aimed at the customers and consumers of photovoltaics and photovoltaic-related products, as well as a source of current information on the industry and its activities. The cost is $12 a year.

PV News

PV Energy Systems, Inc.   703-780-9236
2401 Childs Lane
Alexandria, VA 22308

This monthly newsletter screens and highlights all current activities in photovoltaics. The February issue has a world market summary and the March issue has a current-year forecast. The cost is $60 a year.

Additional Experts and Resources

Paul Maycock

PV Energy Systems, Inc.   703-780-9236
2401 Childs Lane
Alexandria, VA 22308

Mr. Maycock is a resource person in photovoltaics, and he can answer questions and/or make referrals. He has worked for Texas Instruments and was director of the PV Division at the Department of Energy. He formed PV Energy Systems in 1981 and is its current president.

Physics, Solid-State

See: Materials Science

Plant Gene Expression

See: Agriculture

Plasma Science

Organizations and Government Agencies

American Physical Society (APS)
335 East 45th Street   212-682-7341
New York, NY 10017
    The Plasma Physics Division of APS concentrates on the latest developments in the field. The Society publishes:
    • *APS Newsletter*—a quarterly newsletter for members that covers the news of the Society, as well as recent findings in the field. The newsletter runs 3 to 4 pages in length and is available free to members.
    • *Bulletin of the APS*—published ten times a year, it has information on APS and covers the activities of all the different divisions. The publication, 80 to 200 pages long, is free to members and costs $100 for nonmembers.

Publications

IEEE Transactions on Plasma Science
Group XI Mail Stop E 531   505-667-5995
Los Alamos National Laboratory
PO Box 1663
Los Alamos, NM 87545
    This bimonthly periodical is a collection of papers submitted to the IEEE, covering such topics as controlled fusion, space plasma, plasma diagnostics, plasma

circuitry, arcs, and plasma theory. The 60-page publication is available by subscription for $6 to members, and $12 to nonmembers.

Additional Experts and Resources

Dr. Egor Alexeff
Electrical Engineering Department   615-974-5467
University of Tennessee
Knoxville, TN 37916

Dr. Alexeff, a professor of electrical engineering at University of Tennessee, was previously at the Oak Ridge National Laboratory Fusion Center. He is currently the chairman of the IEEE Electrical Engineering Plasma Science Special Interest Group and Secretary/Treasurer of the American Physical Society's Plasma Group. He is a noted authority in the field of plasma science, and can answer questions and give further information about either association.

Dr. Steven Gitomer
Group XI Mail Stop E 531   505-667-5995
Los Alamos National Laboratory
PO Box 1663
Los Alamos, NM 87545

Dr. Gitomer is the editor of the *IEEE Transactions on Plasma Science* and is an expert in the field of plasma science. He is available to answer questions and/or make referrals to other information sources. He welcomes articles for publication in the *Transactions* journal.

Dr. Edward J. Powers
Department of Electrical and Computer Engineering   512-471-1430
University of Texas
Austin, TX 78712

Dr. Powers was the editor of the *IEEE Transactions on Plasma Science*, and is now involved in analyzing and interpreting fluctuation and turbulence data of plasma science research. He is a noted authority in the field and is available to answer questions and make referrals to other information sources.

Plastics

Organizations and Government Agencies

Center for Materials Science
National Bureau of Standards   301-921-3336
National Measurement Laboratory
Washington, DC 20234

     The Center is involved in the field of plastics including standards, specifications, and characterization techniques for plastics materials. Specialists at the Center cooperate with the industry in developing national and international standards. The staff can answer technical inquiries and make referrals.

     Annual reports covering work in progress, as well as technical papers published in professional journals, can be obtained by contacting the Center.

Office of Chemicals and Allied Products
U.S. Department of Commerce   202-377-0128
Bureau of Industrial Economics, Room 4045
Washington, DC 20230

     This office monitors and analyzes economic data on approximately 40 commercial materials and products, with special emphasis on international trade and the promotion of exports. Specialists are available to answer inquiries and make referrals on economic information. Contact David Rosse for information on plastic materials, coatings, adhesives, and allied products and David Blank for information on miscellaneous plastic products.

Plastics Technical Evaluation Center (PLASTEC)
Department of the Army   201-724-2778—Library Service
U.S. Army Armament, Munitions and Chemical Command   201-724-3189—Technical Specialists
Building 351 North
Dover, NJ 07801

     PLASTEC develops and compiles technical information and data on plastic materials, adhesives, and composites of interest to the Department of Defense, its suppliers, and contractors. The work done at PLASTEC covers various areas within the field of plastics research. The specialists can answer inquiries of a technical nature, provide technical advice and consulting services, and make referrals.

     The Center has a library containing 40,000 documents, primarily reports of U.S. government agencies and their contractors, and technical conference papers. A free listing of publications can by obtained from their library. PLASTEC's data base can be searched through DIALOG and ORBIT. Staff will provide literature searches and other services for the general public on a cost recovery basis. Federal agencies and their contractors can obtain searches and other services free of charge.

Polymer, Science and Standards Division

National Bureau of Standards
Washington, DC 20234

301-921-3112—Public Information Division
301-921-2318—Technical Information and Publications Division
301-921-3734—Dr. R. Eby

A division of NBS, this office is involved in the following aspects of plastics: phase behavior and characterization behavior for polymer blends; processing and relation of composites; long-term research for mechanical reliability in low-bearing applications; chemical durability; lifetime prediction; standards for electric composition; performance and dialectic properties; and reliable data in dental and medical materials.

The staff examines on a scientific basis the material properties which are of concern to the people who design, produce, and use polymer objects. If you are seeking technical advice, contact NBS's Public Information Division for referral to an appropriate expert. To obtain a list of reports, studies, and other literature produced by the Bureau's experts on plastics, contact the Technical Information and Publications Division.

Society of the Plastics Industry, Inc.

355 Lexington Avenue
New York, NY 10017

212-573-9400

This trade organization of more than 1,600 members includes resin manufacturers, distributors, machinery manufacturers, plastics processors, mold makers, and other industry-related categories. It has 50 divisions in different areas of plastics. Its information services include answering inquiries, giving referrals about other sources of information, and providing literature and on-site use of reference materials. Those seeking technical information should contact the Society's Information Center.

The Society conducts many meetings and seminars. Its main events are a National Plastics Exposition conducted every three years, the Reinforced Plastics and Composites Conferences offered annually, and an Annual Structural Foam Conference and Parts Competititon. These events are open to both members and nonmembers. The Society publishes:

• *Membership Directory and Buyer's Guide*—an annual publication listing the Society's members involved with products and processes. The cost is $45 for members and $60 for nonmembers.

• *Facts and Figures of the Plastic Industry*—an annual publication that reports on consumption and production of plastics resins.

• *Literature Catalog*—a free listing of the manuals, proceedings, surveys, and technical reports issued by the Society is available from the Information Center.

Materials Research Society

9800 McKnight Road, Suite 327   412-367-3003
Pittsburgh, PA 15237

    The Society is a forum for scientific researchers and engineers in industry and educational institutions. Several symposia are offered annually, as well as tutorial lectures and short courses. The Society calls for papers to be presented at the symposia by both members and nonmembers, after which the proceedings are published. Specific technical information and referrals are provided on a limited basis. A free brochure containing information about the Society and the lectures, courses, and symposia is available by contacting the office.

Publications

Communications Publishing Group, Inc. (CPG)

PO Box 383   617-566-2373
Dedham, MA 02026

    CPG is a data base publisher of high-technology publications, on-line data bases, and research services. CPG's products and services are based on worldwide patent literature, and are available in both print and on-line through NewsNet at 800-345-1301. Their publications dealing with plastics are the following:

    • *Plastics Materials*—a newsletter that covers new polymer-based thermoplastic and thermoset films and materials, structural composites, fiber reinforcements, flame retardant and antistatic additives, laminates, adhesive formulations, and so on. Subscription rate is $217 for 24 issues.

    • *Plastics Processing*—a newsletter covering equipment and techniques used to process all kinds of thermoplastic and thermoset materials. It presents developments in extrusion, blow molding, injection molding, heat sealing, adhesive bonding, materials handling, laser cutting, quality control, and more. Subscription is $197 for 24 issues.

    • *Industry Patent Reports*—each of these reports includes patent numbers, descriptive titles, and assignees of all U.S. patents granted in the field from 1963 to 1983. Selected patents are reproduced in their entirety. One such report is *Structural and Flame Resistant Plastics*, available for $125.

Journal of Applied Polymer Science

John Wiley & Sons   212-850-6000
605 Third Avenue
New York, NY 10158

    This journal reports on results of fundamental research worldwide in all areas of high polymer, chemistry, and physics. Each issue covers chemistry, physics, and letters. The journal is published three times a month, and the annual subscription rate is $828.

The Kline Guide to the Plastics Industry

Charles Kline and Co., Inc.   201-227-6262
330 Passaic Avenue
Fairfield, NJ 07006

This guide describes the plastics industry and covers the production of basic plastics; plastics materials; plastic processing, forms, and markets; information sources; and the 500 leading plastic companies in the U.S. The 360-page guide is revised every three years, and the last update was in 1982. The author, Mary Deitsch, works at Kline and can be contacted by those seeking technical information.

Modern Plastics

McGraw-Hill, Inc.   212-512-2000
1221 Avenue of the Americas
New York, NY 10020

Modern Plastics is a monthly magazine reporting the latest technology available for plastics materials and equipment. This is the oldest magazine covering the entire field of plastics, and is oriented toward plastic manufacturers and users in the upper management levels. The magazine maintains a staff of experts in the Reader's Services division who can answer inquiries and give referrals. Annual subscription rate is $28 for companies in construction, manufacturing, and engineering, as well as for government agencies and schools. For other companies and individuals, the cost is $32.

The 13th extra issue each October is the *Modern Plastics Encyclopedia*, a hardcover book with more than 800 pages divided into four sections, included in the subscription price. The first section is a textbook with general information on plastics materials, processes, and chemicals. It contains about 160 articles written by experts in the industry. The second section is a design guide, containing practical considerations for various uses. The third section is a data bank that covers specifications and properties of plastics and chemicals. The final section is a directory of suppliers, listing over 4,900 companies in the U.S. and Canada.

Plastics World

Cahners Publishing Co., Inc.   617-536-7780
221 Columbus Avenue
Boston, MA 02116

Oriented toward managers in the plastics industry and markets, *Plastics World* describes and interprets the significance of new products, the latest technologies, market trends, business strategies, and business conditions. In addition to its 12 monthly issues, subscribers receive a 13th issue, an annual directory of companies in the field of plastics. The magazine is available free of charge to managers in the plastics industry. The rate for others is $40 a year.

Polymers

See: Plastics

Positron Emission Tomography

See: Medical Imaging; Nuclear Medicine

Power Systems in Space

See: Space Nuclear Power

Prenatal Diagnostic Techniques

See: Reproductive Technology

Printers

See: Peripherals

Procurement

See: Research and Development

Productivity and Innovation

Organizations and Government Agencies

American Institute of Industrial Engineers (AIIE)

25 Technology Park   404-449-0460
Norcross, GA 30092

AIIE is a nonprofit international professional organization with a membership of 43,000 industrial engineers. The Institute annually sponsors two major conferences on productivity and engineering, as well as numerous symposia, seminars, trade shows, and continuing education courses and workshops. Another major annual event is the "National Productivity Improvement Campaign," which groups together many associated functions across the country. Annual membership is $60 for individuals and $12 for students. Membership includes a subscription to one of their publications:

• *Industrial Management*—a monthly magazine designed to provide timely and accurate information to business managers. Featured articles include productivity measurement, cooperative approaches to improving productivity and quality of work life, technology reports, executive profiles, and general coverage of business and economic trends. About 100 pages per issue, annual subscription rate for nonmembers is $22.

• *Industrial Engineering*—a monthly trade journal for industrial engineers that covers productivity. Each issue focuses on one of the many topics considered to be an industrial engineering challenge. Features have included improving productivity corporation-wide, productivity projects in service and support industries, high technology projects in manufacturing, the automated factory, and quality control and reliability engineering. About 125 pages an issue, annual subscription rate for nonmembers is $35.

The American Productivity Center

123 North Post Oak Lane   713-681-4020
Houston, TX 77024

This nonprofit organization serves business, labor, government, and universities by offering information and assistance about productivity and quality of working life. The Center offers a series of public seminars, presentations, and workshops, in addition to the individualized service programs which are created to meet specific company needs. The Information Services department will provide research and reference services for its members, business people, academicians, and students. The library contains some 3,000 books, 100 subscription periodicals, and 7,000 selected reports. Updated bibliographies are available for many productivity subjects. The

Center can make referrals to experts and organizations knowledgeable in the field, and will also access computerized data bases upon request. It has developed a computer network which allows participants to exchange information, ideas, and solutions on productivity topics. Acting as a clearinghouse, the Center contributes its productivity expertise, trains participants on computer use, sets guidelines and agendas for the computer discussions, and maintains the conferencing software. Publications include:

• *Productivity Digest*—an annual compendium of productivity and quality of working life literature offering authoritative reviews of the best articles, books, and conference proceedings issued in recent years. The 60-page bibliography covers seven productivity topics. Issued in each year, the single-copy price is $25.

• *Productivity Perspectives*—an annual chartbook of key facts about U.S. and worldwide productivity trends. It contains current pertinent facts, statistics, charts, and commentary in one readable document. The report costs $25 for a single copy.

• *Multiple Input Productivity Indexes*—a quarterly statistical series which provides growth rate data and indexes of trends in total factor productivity for the U.S. domestic economy, major sectors of the economy, and 20 manufacturing industry groups. In addition, historical data from 1948 for output, labor input, capital input, total factor input, total factor productivity, labor productivity, capital productivity, and labor/capital ratios are updated annually for major sectors. Annual subscription rate is $50.

Bureau of Industrial Economics (BIE)

U.S. Department of Commerce   202-377-4356
14th Street and Constitution Avenue, NW
Washington, DC 20230

BIE is responsible for the collection and analysis of both domestic and international data in all categories of industry classifications. Organized into some 300 industries covering producer and consumer goods and services, BIE has staff experts to cover each industry. A listing of each industry and its specialists is available upon request. Individuals and organizations have access to the data and analyses, and all information is available upon request at no charge.

Bureau of Labor Statistics

U.S. Department of Labor   202-523-9294
441 G Street, NW
Washington, DC 20212

The Bureau is responsible for the compilation, publication, and dissemination of a vast assortment of statistical information on U.S. industry. The productivity and technology studies section frequently issues reports and updates on the impact of technological changes and manpower trends of various industries. The following publications are available from the Superintendent of Documents, U.S. Government Printing Office, Washington, DC 20402, 202-783-3238:

- *Technology and Labor in Four Industries* (Bulletin 2104)—covers meat products; foundries; metalworking machinery; electrical and electronic equipment. The publication is available for $3.50.
- *Technology and Labor in Five Industries* (Bulletin 2033)—covers bakery products; concrete; air transportation; telephone communication; insurance. The publication is available for $2.50.
- *Technological Change and its Labor Impact in Five Energy Industries* (Bulletin 2005)—covers coal mining; oil and gas extraction; petroleum refining; petroleum pipeline transportation; electric and gas utilities. This bulletin costs $2.40.
- *Technological Change and Its Labor Impact in Five Industries* (Bulletin 1961)—covers apparel; footwear; motor vehicles; railroads; retail trade. The bulletin is available for $2.20.
- *Technological Change and Manpower Trends in Five Industries* (Bulletin 1856)—covers pulp and paper; hydraulic cement; steel; aircraft and missiles; wholesale trade. The cost for this bulletin is $1.20.
- *Technical Change and Manpower Trends in Six Industries* (Bulletin 1817)—covers textile mill products; lumber and wood products; tires and tubes; aluminum; banking; health services. The booklet is available for $1.35.

Commerce Productivity Center (CPC)

U.S. Department of Commerce   202-377-0940
14th Street and Constitution Avenue, NW
Washington, DC 20230

Part of the Office of Productivity, Technology, and Innovation (OPTI), the CPC serves as a clearinghouse for information on all aspects of productivity. With a staff of about 35 professionals with diverse areas of expertise, and access to the information resources of the entire Department of Commerce, CPC can easily provide businesses and other organizations with information on how to improve productivity, quality, and competitiveness. The CPC can provide publications, articles, reference and referral services, bibliographies, and reading lists on a variety of productivity-related topics. All services of the CPC are without charge and a publication list is available upon request.

Office of Productivity and Technology Studies

Bureau of Labor Statistics   202-523-9244
U.S. Department of Labor
441 G Street, NW
Washington, DC 20212

An inventory of the statistics available from this Office is available upon request. The data centers on three major research programs: (1) the productivity research program provides comprehensive statistics for the U.S. economy, its major component sectors, and individual industries; (2) the technological studies program investigates trends in technology and their impact on employment and productivity; and (3) the international program compiles and analyzes statistics on productivity and related factors in foreign countries for comparison with similar U.S. statistics. Also available as an additional information source is Report 671, *BLS Publications on Productivity and Technology*, free upon request from this office.

Office of Productivity, Technology and Innovation (OPTI)

U.S. Department of Commerce   202-377-1581
14th Street and Constitution Avenue, NW
Washington, DC 20230

The main purpose of OPTI is to aid development of new federal policies to create an economic climate conducive to productivity growth. OPTI assists in creating innovative research and development mechanisms to foster large-scale research and development projects, and aids in accelerating the transfer of federal technology to the private sector. Additionally, OPTI operates outreach and information programs, chiefly through the Commerce Productivity Center (CPC), to provide productivity information upon request, and conduct productivity improvement seminars and workshops for the private sector.

Productivity Improvement Research Section (PIR)

Division of Industrial Science and Technological Innovation   202-357-9805
National Science Foundation
1800 G Street, NW, Room 1237
Washington, DC 20550

The major goal of the PIR program is to provide information and sponsor research that will assist private and public sector management in enhancing U.S. technological growth. Research ranges from highly focused studies which resolve questions about a specific aspect of technological innovation and productivity to more generalized and fundamental studies. A review of the literature in the field of innovations process research is available upon request:

● *The Process of Technological Innovation: Reviewing the Literature*—focuses on the organizational context of technological change, including material/physical aspects and the social/behavioral implications. The 250-page review is available from NSF for no charge.

Science Research and Technology Subcommittee

House Committee on Science and Technology   202-225-8844
U.S. House of Representatives
Washington, DC 20515

This subcommittee has jurisdiction over productivity and innovation legislation under consideration by the U.S. House of Representatives. It has held hearings on the subject, free copies of which are available while its supply lasts. Staff members are knowledgeable about productivity and innovation issues and they can refer you to specialists, programs, and printed materials. The subcommittee publishes:

● *Congressional Innovation News Notes*—issued every other month, these notes keep track of bills, hearings, laws, and other congressional activities pertaining to innovation and productivity. The five- to six-page newsletter is available free of charge.

Technical Information and Publications Division

National Bureau of Standards   301-921-2318
U.S. Department of Commerce
Route 270
Gaithersburg, MD 20234

For expertise and assistance in scientific and technical services to aid productivity and innovation, contact the above division of NBS. Information is available in such diverse areas as computer systems engineering, computer programming science and technology, applied mathematics, building technology, chemical engineering, electronics and electrical engineering, manufacturing engineering, analytical chemistry, chemical physics, and materials science. The Publications Division can provide lists of publications on various topics available through NBS and other agencies of the federal government.

U.S. Conference of Mayors

1620 Eye Street, NW   202-293-7330
Washington, DC 20006

An organization representing cities, the U.S. Conference of Mayors is a good source for information about innovation occurring in cities. It has access to a variety of resources including a computer network providing information about innovation. Staff can answer questions.

Work In America Institute (WIAI)

700 White Plains Road   914-472-9600
Scarsdale, NY 10583

The Institute is concerned with promoting quality working life and productivity in the U.S. It disseminates information resulting from research and analysis in all aspects of productivity and working conditions. The organization participates in a national productivity network newly established to link productivity centers and information resources across the U.S. Staff will answer questions and refer you to information sources.

WIAI maintains a library collection of more than 20,000 journals, texts, periodicals, reports, and articles dealing with productivity and related topics. The library is open to the public. The Institute sponsors the Productivity Forum, a memberhip organization comprised of Fortune 500 companies, unions, and government organizations. The Forum sponsors major conferences, seminars, and visits to state-of-the-art organizations. Its publications listing contains over 60 of their own publications as well as hundreds of other abstracted articles and texts on the subject of productivity. The staff has also compiled a series of 36 individual productivity studies that contain a summary of major research and literature, abstracts from the principal publications, and a bibliography of related reference resources on each individual topic. Publications include:

• *World of Work Report*—a monthly newsletter offering both national and international coverage of contemporary productivity issues. The publication contains brief articles on new trends in the workplace, developments in technology, and experimental programs affecting productivity. There are also citations and notices on new texts and periodicals in the world productivity study. The subscription rate to the 8-page newsletter is $72 a year.

Publications

Executive Productivity

Nadel Communications   305-483-2600
10076 Boca Entrada Boulevard
Boca Raton, FL 33431

This monthly newsletter serves as a business trends forecaster, containing brief articles highlighting methods for improving productivity, including insights from experts, tips from various professionals, and analysis studies from numerous associations and institutions. In addition, the paper addresses relevant news items and offers an informative analysis on economic trends in general. The subscription rate to the 8-page paper is $97 a year.

Manufacturing Productivity Frontier

Manufacturing Productivity Center   312-567-4800
Illinois Institute of Technology
10 West 35th Street
Chicago, IL 60616

A monthly, the magazine is the official publication of the Manufacturing Productivity Center, and it covers all the meetings of the Center, including their national conferences on improving productivity. Feature articles focus on productivity improvements including the tools and technologies to implement such programs. There are also book reviews and abstracts of other relevant articles, meeting announcements, and brief news items. The subscription rate to the 50-page publication is $100 a year.

Guide to Innovation Resources and Planning for the Smaller Business

The Small Business Technology Liaison Division   202-377-1093
Office of Productivity, Technology, and Innovation
U.S. Department of Commerce, Room 4816
14th Street and Constitution Avenue, NW
Washington, DC 20230

This guide identifies more than 50 federal and 85 state government offices that assist smaller businesses bringing new technologies to market. It is written for individuals, companies, and state and local government economic development planners.

The *Guide* has two basic sections. The first examines the many steps in the innovation process and the skills and resources needed. The second section identifies a wide range of resources (federal, state, and private) available to assist the smaller business in areas such as financing, information gathering, and management. The capabilities of these resources are summarized and contact phone numbers and addresses are given. Each resource has been identified by the stage of the innovation process to which it applies.

The 85-page publication is available for $13.50 from the National Technical Information Service, U.S. Department of Commerce, 5285 Port Royal Road, Springfield, VA 22161, 703-487-4650.

National Productivity Report

1110 Greenwood Road   312-653-4424
Wheaton, IL 60187

Each issue of this biweekly newsletter contains a case study. After a brief review of the facts and conditions, various improvement techniques are offered and analyzed, with an emphasis on productivity improvement. There are also technology reports and reviews of new publications. The subscription to the 8-page newsletter is $78 a year.

National Productivity Review

Executive Enterprises Publications Co., Inc.   212-489-5912
33 West 60th Street
New York, NY 10023

A quarterly journal containing approximately 12 articles on issues related to productivity. The articles cover such topics as problems, improvement techniques, technology, and organizational structure. There are also book reviews of recent productivity publications and summaries of current articles in the field, and a comprehensive calendar of workshops, symposia, and conferences on related topics. The *Review* has an accurate list of quality working life and productivity centers throughout the U.S. and Canada, giving addresses, contact persons, and emphasis of study for each center. The subscription rate to the 100-page journal is $96 a year.

National Technical Information Service (NTIS)

5285 Port Royal Road   703-487-4650
Springfield, VA 22161

Two productivity reports available from NTIS are:

• *White House Conference on Productivity Growth: A Better Life for America* (Pub #PB84-159144)—covers the key areas of the White House conference on productivity, held September 21-23, 1983. The conference focused on four main areas: capital investment, human resources, the role of government, and the role of the private sector. The softcover report is available for $5.

• *Restoring Productivity Growth in America: A Challenge for the 1980s* (Pub #84-217603)—is the final report of the National Productivity Advisory Committee dated December 30, 1983, and it outlines 46 specific recommendations for improving productivity growth. The report is available for $5.

Productivity

Productivity, Inc.   203-322-8388
PO Box 16722
Stamford, CT 06905

A monthly, this publication focuses on the white collar work force and on methods of improving productivity in American business. The newsletter covers the major conferences sponsored by Productivity, Inc., including abstracts of the key presentations and subject matter listings of the many workshops comprising the conferences. In addition, there are reports of new technology, reviews of current literature, interviews and advice from experts. The subscription rate to the 12-page publication is $126 a year.

Science Indicators

National Science Foundation (NSF)   202-357-9498
1800 G Street, NW
Washington, DC 20550

This 1982 document is the sixth in a series prepared by the National Science Board, the policy-making body of the NSF. It is a quantitative assessment of U.S. science and technology, demonstrating the priority for investment and reflecting their importance to the economy and to national security. Statistics covered in the report include amounts spent nationally on research and development (both industrial and federal), patent applications, stock offerings and venture capital investments in high technology companies, amounts spent on basic and applied research in the education sector, science and engineering education statistics, and public reactions to status and expenditures in the science and technology fields. Copies of the report (#038-000-0538-3) are available from the Superintendent of Documents, U.S. Government Printing Office, Washington, DC 20402, 202-783-3238. Each copy is $9.50.

Programming Languages (Computer)

Organizations and Government Agencies

ADA Information Clearinghouse (ADA IC)

3D139 (400 AN)   703-685-1477
The Pentagon
Washington, DC 20301

Sponsored by the Department of Defense (DOD), the primary user of the ADA computer language, this Clearinghouse acts as a vehicle for the dissemination of current information about ADA. The office coordinates the collection, integration, and distribution of documents on all aspects of the ADA language and associated aspects of DOD's software initiative. The Clearinghouse also coordinates the development and introduction of the language, serving as a link between the government, the user community, and universities. The office is also responsible for maintaining the standardization and consistency of the ADA language as well as promoting use of language throughout the software community. It does this several ways: by offering training programs on ADA; by keeping up on who are the experts and who is working on what project; by knowing where to go to get documentation on ADA; and by providing information about validated compilers. The Clearinghouse also maintains an on-line data base called *ADA Information* that contains announcements of recent ADA-related activities as well as comprehensive information about the ADA program. Access is available through GTE Telenet. In addition to free information packets about ADA, the Clearinghouse also publishes:

• *ADA IC Newsletter* (and special updates)—a bimonthly newsletter providing program updates and information about available ADA documentation and education.

Computer Languages Technical Committee (TC)
IEEE Computer Society   301-589-8142
1109 Spring Street, Suite 300
Silver Spring, MD 20910

A technical committee within the overall Institute of Electrical and Electronics Engineers, this TC extends its interest beyond conventional high-level programming languages to include other areas such as specification, design, test, and query languages. The TC sponsors workshops on language-related issues and plans for an international conference on computer languages. Through committees within the TC, individual languages and special topics receive detailed coverage.

Information Systems Engineering
Institute for Computer Sciences and Technology   301-921-2431
National Bureau of Standards
Building 225, Room A 265
Route 270
Gaithersburg, MD 20899

This govenment agency administers programming language standards within the federal government. The division represents federal interests within both national and international standards bodies. Staff can tell you what's happening within the federal government and where to go for information about language programming standards.

Federal Software Testing Center
5203 Leesburg Pike, Suite 1100   703-756-6153
Falls Church, VA 22041

The Center tests language compilers used by the Government to verify their performance in Cobol, Fortran, Basic, ADA, and Pascal. Reports are published on the research, and cover such topics as compilers, software engineering, tests for acceptance and productivity, and software tool selection and evaluation. The reports vary from 10 to 200 pages in length and are available from NTIS. George Baird, Director of the Federal Software Testing Center, is an expert in the field and is available to answer questions and make referrals.

SIGADA

Association for Computing Machinery (ACM)   212-869-7440
PO Box 12115
Church Street Station
New York, NY 10249

SIGADA, an ACM special interest group for the ADA computer language, is involved in all aspects of software development related to ADA, including development methodology, language compilers, and tools. Members can answer questions, make referrals, and provide you with bibliographic listings. SIGADA publishes:

• *ADA Letter*—a bimonthly publication covering all aspects of ADA. Articles appear on training and education methodologies used to develop ADA programs; design languages related to ADA; and the compilers, tools and technologies developed for ADA. Information is also provided about conferences and literature available. Nonmembers can subscribe to the 100-page publication for $23 a year.

SIGAPL

Association for Computing Machinery (ACM)   212-869-7440
PO Box 12115
Church Street Station
New York, NY 10249

A special interest group within ACM, SIGAPL's 2,200 members share ideas and techniques about the development and application of the APL programming language. The group is active in promoting the standardization of the language and sponsors conferences and publications to help users. SIGAPL publishes:

• *APL Quote Quad*—a quarterly newsletter dedicated to APL programming language which contains programming examples, articles, algorithms, technical papers, and recreational puzzle configurations. The publication also contains SIGAPL conference proceedings. Each issue is approximately 24 to 30 pages, and the annual subscription fee is $30.

SIGPLAN

Association for Computing Machinery (ACM)   212-869-7440
PO Box 12115
Church Street Station
New York, NY 10249

A special interest group within ACM, SIGPLAN's 10,000 members are involved in all aspects of programming languages and programming language processors. The group covers both practical and theoretical approaches to topics of interest to programming language users, developers, implementers, and theoreticians. SIGPLAN sponsors two conferences a year, and the office listed above can refer you to an appropriate SIGPLAN member. The following is published:

- *SIGPLAN Notices*—a monthly newsletter containing articles about programming languages, design, and implementation. The publication also contains conference proceedings. Annual subscription fee is $22 for members and $36 for nonmembers.

- *Fortran Forum*—a quarterly newsletter with articles describing Fortran implementation and Fortran 8X design. Annual subscription is $6 for members and $16 for nonmembers.

X-3 Secretariat

American National Standards Institute (ANSI)   202-737-8888
Computer and Business Equipment Manufacturers Association (CBEMA)
311 First Street, NW, Suite 500
Washington, DC 20001

CEMBA serves as the Secretariat for ANSI's X-3 Committee, a portion of which is involved in developing voluntary standards for programming languages. These standards are important as they are developed by experts in the field and, although voluntary, the standards are adhered to by the computer industry. Twenty-four of the X-3 Technical Committees cover programming languages, with committees on specific languages, reference models, computer graphics, data bases, and more. Names of the Technical Committee chairpersons, X-3 reports, and standards are all available from the ANSI Secretariat at CBEMA.

Publications

TOPLAS

Association for Computing Machinery (ACM)   212-869-7440
11 West 42nd Street
New York, NY 10036

A quarterly publication, *TOPLAS* is the acronym for ACM's Transactions on Programming Languages and Systems. It contains papers covering the design, implementation, evaluation, and use of programming systems, as well as the methodologies, techniques, and applications of programming languages. Annual subscription is $18 for members and $66 for nonmembers.

Computer Language

2443 Filmore Street, Suite 346   415-957-9353
San Francisco, CA 94115

Published monthly, this publication is geared toward programmers and engineers who write their own software. It covers the latest programming languages in operating systems, with emphasis on programming techniques and tools for programmers' daily operations. The 84-page publication is available for $24 a year.

Structured Language World

PO Box 5314   203-288-0283
30 Mowry Street
Mt. Carmel, CT 06518

This quarterly contains information about products, people, and markets relevant to the languages covered. Among the programming languages and systems given coverage are Pascal, Modula-2, ADA, and P-System. The subscription rate is $12 a year.

Additional Experts and Resources

George Baird

Director   703-756-6153
Federal Software Testing Center
5203 Leesburg Pike, Suite 1100
Falls Church, VA 22041

George Baird is a software expert and director of the Federal Software Testing Center. He is available to answer questions about programming languages and can refer you to other experts and information sources.

Joseph Urban

Associate Professor   318-231-6304
Computer Science Department
University of Southwestern Louisiana
PO Box 44330
Lafayette, LA 70504

Responsible for the software engineering program at the University, Dr. Urban is involved in teaching and research in the area of software engineering. He is also chairman of IEEE's Computer Language Technical Committee. Dr. Urban can answer questions about computer programming and refer you to other experts and information sources.

Mary S. Van Deusen

34 Archer Street   617-384-2526
Wrentham, MA 02093   914-945-2394

Ms. Van Deusen's major area of interest is programming language design and human engineering. She has been very active in programming language organizations for the past 11 years, and can give you insight into the field and refer you to other experts and information sources.

Public Perceptions and Attitudes Toward High Technology

See: Labor and Technological Change; Social Impact of Technology; Technophobia

Public Policy on Scientific and Technological Change

See: Technology Assessment

Publishing, Electronic

See: Electronic Publishing

Radiation Therapy

See: Medical Imaging

Radio

Organizations and Government Agencies

Electronic Industries Association (EIA)
2001 Eye Street, NW   202-457-4919
Washington, DC 20006

      EIA is an association of manufacturers of radio, television, video systems, electronics parts, and components. Members submit data and EIA compiles the data into statistical reports. This is a good source of information for statistics on radio equipment. The Association also sponsors consumer electronics trade shows during the winter and summer, international telecommunications seminars and exhibits, and an annual Design Engineers Electronic Component Conference. Information is provided to members and nonmembers by their library staff. Trade show and seminar information can be obtained from the Public Affairs Division. The EIA publishes:

●   *Electronic Market Data Book* (annual)—a compendium of statistical information on the electronics industry. Available for $55.

Federal Communications Commission (FCC)

1919 M Street, NW   202-632-7000—Consumer Assistance
Washington, DC 20554   202-632-7100—Library Branch

The FCC regulates interstate and international communications by radio, and oversees the development and operations of radio broadcasting services nationwide. These services include AM, FM, commercial and noncommercial, educational, and developmental services. Other radio services include aviation, marine, public safety, industrial, land transportation, amateur, personal disaster, and experimental.

Those interested in more details about the FCC may obtain publication lists from the Office of Public Affairs. The FCC also has books, reprints, journals, newsletters, publications lists, bibliographies, regulations, and historical background information. The FCC experts specializing in radio include:

Audio   **FM**
Larry Eads   Ray Laforge
202-632-6485   202-632-6908

AM
John Morgan
202-254-9570

National Association of Broadcasters (NAB)

1771 N Street, NW   202-429-5333
Washington, DC 20036   800-368-5644

This organization is dedicated to the U.S. broadcasting industry on issues affecting the electronic media. It holds an annual convention, an annual radio programming conference, and several seminars. The NAB library maintains a wide-ranging collection of books, reports, and periodicals on all aspects of broadcasting, and the staff of the science and technology department will answer questions and help you find the information you need. The NAB Employment Clearinghouse is a job bank for minority and women members. NAB also publishes:

●   *Radio Active*—a monthly magazine on various aspects of radio.

●   *Quarterly Labor Review*—a review of all reported labor relations and collective bargaining activities and developments in the broadcast industry.

●   *Radio Financial Report*—an annual, it contains statistical information on median revenues, expenses, and profit margins of radio stations. Available for $20.

- *Broadcasting Bibliography*—a guide to useful and available books in broadcasting. The cost is $2.
- *FCC Telephone Directory*—lists more than 70 of the FCC departments and the people working in them. Available for $2.

National Radio Broadcasters Association (NRBA)

1705 De Sales Street, NW, Suite 500   202-466-2030
Washington, DC 20036

NRBA provides information on many aspects of radio such as research, program engineering, management, and sales. A library is planned, and the public inquiries department will answer any questions. Publications, which are available to members only, include the following:

- *Radio and New Technology*—a monthly magazine with updates on the latest satellite technology, and new technology developed by manufacturers of radio equipment.
- *Monday Morning Memo*—a weekly newsletter on the latest Congressional developments, FCC activities, and news about members of the Association.
- *SCA Action*—a monthly newsletter about the broadcasting industry covering radio paging, background music, and data transmission, as well as interviews with engineers on the technology used in the broadcasting field.
- *Radio Q&A*—a monthly publication that presents interviews with personalities in the field.
- *Radio Pro-Gram*—a newsletter published monthly covering radio formats and different kinds of programming.

Radio Advertising Bureau (RAB)

485 Lexington Avenue   212-599-6666
New York, NY 10017

RAB deals with the advertising aspects of the radio industry. It holds an annual conference, and publishes a twice-monthly newsletter. The Member Service department will answer any questions from members as well as make referrals to other sources of information. They maintain a sound library of more that 30,000 radio commercials indexed by advertiser category. They also publish:

- *Radio Co-op Sources*—a directory of manufacturers' current co-op programs, published twice a year.
- *Radio Facts*—annual overview of radio advertising trends. Available free of charge.

Publications

Broadcasting Publications

1735 De Sales Street, NW   202-638-1022
Washington, DC 20036

This company puts out many important publications in the field of radio. Their major publications are:

• *Broadcasting Magazine*—a weekly journal of the communications industry that covers radio, television, cable television, and satellites. Available for $60 a year.

• *Broadcasting Cablecasting Yearbook*—this standard reference volume, lists the rates for radio stations, major radio advertisers and advertising agencies, the addresses of major programming companies, and courses available in the broadcasting field. The yearbook also includes a brief history of FCC regulations affecting radio broadcasting. Annual subscription is $80 a year.

Electronics
McGraw-Hill, Inc.   212-512-2000
1221 Avenue of the Americas
New York, NY 10020

Electronics is a weekly magazine in the field, covering electronics technology and electronics in the marketplace. Annual subscription is $24 a year, or $59 for three years.

Radio News
Phillips Publishing, Inc.   301-986-0666
7315 Wisconsin Avenue, Suite 1200N
Bethesda, MD 20815

This biweekly newsletter covers the latest techniques for improving sales and profitability, new industries, ways to improve station management, programming analysis techniques, views on markets, syndication and technology, and related issues. Subscription is $197 a year for 26 issues. Also available on-line via NewsNet at 800-345-1301 or 215-527-8030.

Television/Radio Age
Rockefeller Center   212-757-8400
1270 Avenue of the Americas
New York, NY 10020

A biweekly broadcast journal, this covers many aspects of the radio and television industry. Each issue includes: FCC information, viewpoints, the Wall Street Report, programming and production, commercials, Business Barometer, and different special studies. Available for $40 a year.

Weekly Radio Action Update
Television Digest, Inc.   202-872-9200
1836 Jefferson Place, NW
Washington, DC 20036

This weekly newsletter reports on FCC actions that affect radio stations. The three main sections are: station sales and transfers, AM stations, and FM stations.

U.S. Industrial Outlook

U.S. Department of Commerce   202-377-2872
Bureau of Industrial Economics (BIE)
Washington, DC 20042

This is an excellent source of information about the radio and television industries. Two chapters are devoted to "Radio and Television Communications Equipment" and "Broadcasting." Along with the industry forecast are definitions, explanations and descriptions of new technological breakthroughs, referrals to other sources of information, and the name of the BIE industry specialist for radio and television. Published annually, it costs $14 and is available from the Government Printing Office, Washington, DC 20402, 202-783-3238.

Additional Experts and Resources

Arthur Pleasants

U.S. Department of Commerce   202-377-2872
International Trade Administration
14th Street and Constitution Avenue, NW, Room 1015B
Washington, DC 20230

Mr. Pleasants, an electronics industry specialist, will answer questions on consumer-oriented products and on all military equipment such as sonars, satellites, and the like.

Radiology

See: Magnetic Nuclear Resonance; Medical Imaging

Railroad Transportation

See: Supertrains; Transportation

Recognition Technology

See also: Bar Code Technology; Image Processing; Magnetic Ink Character Recognition; Optical Character Recognition; Voice Recognition

Organizations and Government Agencies

Association for Information and Image Management (AIIM)

AIIM Resource Center   301-587-8202
1100 Wayne Avenue
Silver Spring, MD 20910

AIIM's Resource Center is open to the public, and it contains books, trade journals, guides, equipment evaluations, directories, reports, standards, and case histories in the following subject areas: micrographics, optical character recognition, word processing and information processing, computer assisted retrieval, optical and videodiscs, office automation, and CAD/CAM. Statistical information contained in market studies on the micrographics industry can be obtained over the phone. Article reprints from the many trade journals are available from the Center at $4 a reprint for members and $5 a reprint for nonmembers.

An on-line bibliographic data base of all sources in the Resource Center is available to nonmembers for $5 a search, and at no charge to members. An index is available on microfiche and can be purchased by nonmembers for $25 a year and by members for $20.

Association of Records Managers and Administrators (ARMA)

PO Box 8540   913-341-3808
Prairie Village, KS 66208

ARMA membership consists of professionals in the records and information management field. Annual international conferences as well as individual association chapters cover all of the information and records management elements and technologies. A standards committee develops and publishes standards for records and information management operations. Louis Snyder, the executive director and an excellent resource person, can answer questions and/or make referrals. ARMA publishes:

• *Records Management Quarterly*—a professional journal that encompasses the total records and information management field, and contains articles on trends in the field and specific how-to information. Subscription price is $30 a year. Price of the journal is included in the cost of membership, which is $45.

Graphic Communications Association (GCA)

Printing Industries Association   703-841-8160
1730 North Lynn Street, Suite 604
Arlington, VA 22209

This special industry group within the Printing Industries Association tries to introduce computers and new electronic technologies into the manufacturing and distribution processes of the publishing industry. Association services include a resource person who can answer questions and/or make referrals. GCA sponsors about 50 meetings a year, which are open to the public. These include six major conferences relating to the prepress area, with other conferences relating to the press area, output end, and distribution via the mails. Proceedings of meetings and other publications are available for purchase. The following standards are available: *PASS Manual—Package and Stock Sequencing for Publication Mailings*; *Gen Code GCA Standard 101-1983*; and *Document Mark Up Metalanguage*. Cost is $48 a single copy, with discounts for volume purchases.

Optical Society of America (OSA)

1816 Jefferson Place, NW   202-223-8130
Washington, DC 20036

OSA is a source of information on OCR and image processing, publishing and disseminating information about optical research. The Society's membership includes more than 8,500 scientists, engineers, and technicians worldwide, as well as a large number of corporations with interest in optics. OSA sponsors an annual meeting that covers a broad range of optical interests, an annual conference on lasers and electro-optics, an annual conference on optical fiber communications, a biennial conference on applied optics, and a variety of other conferences on special topics, as well as workshops and other informational programs. OSA supports 18 technical groups, each dealing with a special area of interest.

Their staff of specialists can answer specific technical questions from members and nonmembers. OSA publishes the following:

• *Journal of the Optical Society of America*—issued monthly in two parts, Journal A emphasizes image science and is the general journal in the field for basic material. Journal B gives emphasis to spectroscopy and modern quantum optics. The cost of subscription to both parts is $225 a year for nonmembers, and $16 for members.

• *Applied Optics*—a bimonthly journal geared toward scientists and engineers engaged in work of an applied nature. It includes reviews of optical patents, and emphasizes papers on lasers, electro-optics, information processing, atmospheric optics, and optical engineering. Cost is $28 for members and $280 for nonmembers.

Recognition Technologies Users Associations (RTUA)

PO Box 2016   802-362-4151
Manchester Center, VT 05255

This membership organization provides information on recognition technologies such as OCR, OMR, OBR, MICR, image and voice, as a means of automated data entry. It conducts educational, publishing, and research activities, and provides up-to-date recognition technology and applications information. Staff will answer questions and provide referrals. RTUA publishes:

• *Recognition Technologies Today*—a free bimonthly journal that covers applications areas such as airline, banking, chemical, credit card, data entry services, education, government, health care, graphic arts, insurance, military, energy, postal, publishing, research, retailing, service bureaus, transportation, utilities, office automation and word processing.

• *Annual Buyer's Guide*—a free corporate listing of a segment of the recognition technology industries, including manufacturers of hardware and those offering related supplies and services.

• *The History of OCR*—a complete study of OCR's beginnings, and what makes it an integral part of business and industry today. Also included is a comprehensive subject, author, and manufacturing company index, as well as a glossary with contemporary optical character recognition language. Available for $16.95.

Publications

Information Management

PTN Publishing Corporation   516-496-8000
101 Crossways Park West
Woodbury, NY 11797

A nontechnical monthly, the publication is management oriented and deals with the systems, technology, software, and media for managing information. It is directed toward corporate executives responsible for their organization's information management function, all of whom qualify for a free subscription. Otherwise the price is $10 a year or $2 an issue.

Modern Office Technology

Penton/IPC   216-696-7000
1111 Chester Avenue
Cleveland, OH 44114

This monthly technical magazine monitors developments in all recognition technologies and has feature articles and news stories on OCR several times a year. It also occasionally advertises new OCR products in their business product section. The magazine is directed to information systems directors or managers at management levels, all of whom qualify for a free subscription. The regular subscription price is $30 a year, or $3 an issue.

The Office

Office Publications, Inc.   203-327-9670
1600 Summer Street
Stamford, CT 06904

The monthly trade journal is free to those responsible for purchasing directly or indirectly any kind of office equipment. It contains articles dealing with trends in the industry and specific applications of OCR and other recognition technologies in a particular environment.

Additional Resources and Experts

National Computer Systems (NCS)
Scanning Systems Division   800-447-3269—Susan Wicks
4401 West 76th Street   612-830-7600
PO Box 9365
Minneapolis, MN 55440
   NCS designs, markets, and sells optical mark reading (OMR) equipment and provides several brochures describing its products and the technology behind OMR.

Recombinant DNA

See: Biotechnology

Record and Information Management

See: Office Automation; Recognition Technology

Recycling

See: Waste Utilization

Remote Sensing

See: Satellite Communication

Renewable Energy Technology

See also: Biomass/Bioconversion; Fuels, Alternative; Geothermal Power and Hydro-power; Ocean Energy; Photovoltaics; Solar Technology; Waste Utilization; Wind Energy

Organizations and Government Agencies

Alternative Sources of Energy, Inc.
107 South Central Avenue
Milaca, MN 56353

612-983-6892

The primary purpose of this nonprofit scientific and educational organization is to disseminate information on the development and use of renewable energy sources. It places special emphasis on information concerning the renewable energy power industry. Staff of the energy information service can answer questions and/or make referrals, with fees depending on the information requested. Publications include:

● *Alternative Sources of Energy*—a bimonthly industry-oriented magazine that covers developments in the renewable energy power production field, including photovoltaics, hydropower, wind power, and biomass cogeneration. Available for $25 a year.

American Solar Energy Society (ASES)
2030 17th Street
Boulder, CO 80302

303-443-3130

This technical society provides information services that further the work of scientists and practitioners in the renewable energy field. The Society holds two conferences a year, as well as local chapter meetings to which the general public is invited. Conference proceedings and books written by the Society's members are available through the ASES. Publications include:

● *International Directory of New and Renewable Energy Information Sources and Research Centers*—a guide that offers 2,955 entries with details of national governmental organizations, research centers, information resources, professional/trade associations, and publications for 138 countries. Indexes include publications, subjects, and organizations. Typical entries include contact names as well as addresses and brief descriptions of activities and affiliates. Available to members for $20 and to nonmembers for $35.

California Energy Commission (CEC)

1516 Ninth Street
Sacramento, CA 95814

916-324-3298—General Public Information Office
800-952-5670—Solar Information (California only)

CEC has programs on renewable financing in the following areas: solar, wind, biomass, geothermal, photovoltaics, and methanol. It maintains programs on contingency planning for oil shortages or emergencies, and responsibility for siting both nuclear and nonnuclear power plants. CEC also does forecasting for energy supply and needs. The information office makes available a free brochure, a publications catalog, and energy efficiency standards for appliances, residential buildings and nonresidential structures. The Wind Information Center staff can answer questions and make referrals.

Conservation and Renewable Energy Inquiry and Referral Service

PO Box 8900
Silver Spring, MD 20907

800-523-2929
800-462-4983—in Pennsylvania
800-233-3071—in Hawaii and Alaska

This free service provides packets of information on all short-term renewable energies: solar, wind, alcohol, fuels, biomass, wood heating, small-scale hydro, photovoltaic, and energy conservation. The service will also answer questions, and can refer callers to organizations or groups for help with other types of questions.

The Energy Library

Department of Energy
Forrestal Branch:
Forrestal Building, MA-232.2, Room GA138
1000 Independence Ave, SW
Washington, DC 20585

202-252-9534

The Energy Library
Department of Energy
Germantown Branch
Route 270
Germantown, MD 20874

301-353-4301

This library system provides the Department of Energy headquarters staff with a wide variety of bibliographic and reference services, and it is a major repository of energy-related information.

The Forrestal Branch collects materials on administrative and regulatory matters, nonnuclear research and development, energy conservation, and renewable energy sources.

The Germantown Branch collects materials on nuclear energy and fossil fuels, energy research, environmental protection, safety, and emergency preparedness.

Hawaii Natural Energy Institute (HNEI)

University of Hawaii at Manoa   808-948-8890
2540 Dole Street
Holmes Hall
Honolulu, HI 96822

HNEI was established to provide focus and support for natural energy research, development, and demonstration. It publishes an annual report that provides information about the type of work done and sponsored. Staff is available to answer questions relating to liquid fuel production from biomass, and to make referrals to other sources of information. Publications include:

- *HNEI Newsletter*—a free publication, usually issued quarterly, that describes the Institute's activities in renewable energy.
- *Wind Bulletin*—a free quarterly.

Institute for Gas Technology

3424 South State Street   312-567-3650
Chicago, IL 60616

The Institute does research for the gas utility industry and for other utility and engineering companies, and provides industrial education and information services for industry and the public. It does contract research for anyone who wants to support a project. Members are gas utility companies, oil companies, and engineering companies. A technical library is maintained, consisting of textbooks on gas technology, patents, periodicals, and government and industry reports, and an information specialist is available to answer questions. Interlibrary loans on the OCLC system are also made. Publications include:

- *Gas Abstracts*—a monthly journal covering planning, policy, supply, distribution, production, processing, transmission, storage, utilization, instrumentation, and so forth. Each issue has about 150 abstracts from about 200 journals. The cost is $120 a year.
- *Energy Abstracts*—a quarterly publication covering reserves, production, and consumption; imports and exports; and prices of energy, natural gas, oil, coal, electricity, and uranium. It also presents business indicators. The cost is $100 a year.
- *International Gas Technology Highlights*—a biweekly four-page newsletter of current events concentrating on the gas utility field. An insert, "Energy Topics," reviews one specific subject and is included in every two or three issues. The cost is $90 a year.

Kenneth E. Johnson Environmental and Energy Center

University of Alabama in Huntsville   205-895-6257
Huntsville, AL 35899

The Johnson Center's major field of study involves the southeastern U.S.'s climate. Applied research is being conducted in the areas of robotics, electric vehicles, heat transfer, biomass, and photovoltaics. The Center maintains a data base, a library, and audiovisuals. Staff is available to answer questions and/or make referrals.

Los Alamos National Laboratory Library

PO Box 1663-MSP362   505-667-4448

Los Alamos, NM 87545

This library primarily serves the laboratory, but is open to the public, who are allowed to use library materials only in the library. The staff also honors interlibrary loan requests.

National Center for Appropriate Technology (NCAT)

Assistance Service   800-428-2525

PO Box 2525

Butte, MT 59702

NCAT is a nonprofit organization established to research, develop, and transfer small-scale technologies. It has gone on to promote the application of conservation and renewable energy technology. Energy-efficient home construction and home weatherization are major research areas for NCAT. Professional staff can provide technical assistance and training, publications, research, and referrals. NCAT has more than 60 publications in stock on topics ranging from the use of solar mobile homes to small-scale fuel alcohol research.

National Energy Information Center (NEIC)

Information Referral Division   202-252-8800

Energy Information Administration

Department of Energy

Forrestal Building, EI-22, Room 1F048

1000 Independence Avenue, SW

Washington, DC 20585

NEIC provides statistical and analytical energy data, general information, and referral assistance to the government and private sectors, as well as the general public. It distributes EIA publications and single copies of blank data survey forms. EIA publications sold through the Government Printing Office (GPO) may be purchased at NEIC. The division provides search assistance for the Data Resources Directory (DRD), a data base of all EIA survey forms, files, tables, models, and data systems, and the Public Use Energy Statistics Data Base.

New Mexico Energy Research and Development Institute

Pinon Building, Room 358   505-827-5886

1220 South Saint Francis Drive

Santa Fe, NM 87501

The Institute's primary purpose is to assist in the economic development of New Mexico by attracting new industries into the state and helping existing industries. Funding is provided for energy projects in New Mexico and is available to anyone who has a well-defined research and development project. The Institute's Seed Capital for Energy-related Projects program seeks entrepreneurs who have inventions or products well along in the research and development process and are looking for money to help them over the last hurdles of research and development in order to commercialize their product. Staff can answer questions, disseminate information, and make referrals. A free monthly newsletter, *Energy Source*, covers energy-related topics in New Mexico.

Office of Business Loans

Small Business Administration (SBA)   202-653-6470
Washington Central Office
1441 L Street, NW
Washington, DC 20416

This government agency provides, through its regional and local offices, loan guarantees of up to $500,000 to help small businesses market inventions in any of the following categories: solar cells and related equipment; solar thermal energy equipment; any device or technique that helps existing equipment, methods of operation, or systems to conserve fossil fuels; bioconversion of wood, biological waste, grain, and similar fuels; industrial waste; hydroelectric power equipment; and wind energy equipment.

Renewable Energy Institute (REI)

1516 King Street   703-683-7795
Alexandria, VA 22314

A publicly supported nonprofit organization, REI conducts policy research and education activities concerning the development of all renewable energy technologies. One of REI's primary functions is to provide policy research support for trade associations and public policymakers. An extensive library is open to the public with resource materials on various renewable technologies. Policy reports are available on the utility industry, tax policy, trade and investment policies in developing countries, and agriculture programs. A write-up of proceedings from the annual symposium is available for $35.

Renewable Technology

Conservation and Renewable Energy   202-252-5347
U.S. Department of Energy
Forrestal Building, CE-32, Room 5F059
1000 Independence Avenue, SW
Washington, DC 20585

This office manages the research, development, and proof-of-concept activities of biomass, energy technology, energy from municipal waste, and geothermal and small hydropower energy technology. Emphasis is placed on development of renewable technology with potential to increase significantly the supply of fuels, heat, and electricity.

Solar Energy Institute of North America

1110 Sixth Street, NW   202-289-4411
Washington, DC 20001-3687

This organization offers inquiry, referral, and consulting services in all areas of renewable energy with an emphasis on solar thermal and solar electric. No charge is made for questions over the phone requiring less than five minutes to answer. Staff will either refer callers to other sources of information or provide more extensive research services for a fee.

Technical Information Center (TIC)

U.S. Department of Energy (DOE)   615-576-6837—General
PO Box 62   615-576-1305—Public Inquiries
Oak Ridge, TN 37830   615-576-1194—Information Services
                                            615-576-1301—Technical Reports
                                            615-576-1155—Energy Data Base

TIC manages the technical information program of DOE. It collects, evaluates, analyzes, stores, and disseminates energy information resulting from DOE-funded research and development, as well as relevant technical literature produced world-wide for use by the DOE community. It is a central processing and distribution point for scientific and technical reports generated by DOE programs, including reports with classified and limited distribution. TIC maintains the DOE Energy Data Base (EDB), with over 1.8 million citations of technical energy literature; maintains the central DOE Research-in-Progress (RIP) data base; publishes abstract journals and bibliographies; provides on-line retrieval through DOE/RECON; offers microfiche publication services for technical reports to DOE and its contractors; works with technical program offices to assist them with special technical information needs; represents DOE in international technical information exchanges; and responds to public requests for energy information on behalf of program offices.

Tennessee Valley Authority (TVA)

Power Information Staff   615-751-2864
Chattanooga, TN 37401

The Tennessee Valley Authority is an independent federal agency charged by Congress to oversee the development of natural resources and human resources. Its greatest involvement is in power production; it is the largest electric utility in the U.S. TVA is actively involved in numerous energy research and development projects, including several in the renewable energy area. Staff is available to answer questions and/or make referrals. Brochures containing specific information on the different areas of renewable energy are available.

Publications

Solar Engineering & Contracting

Business News Publishing Co.   313-362-3700
PO Box 3600
Troy, MI 48007

Issued bimonthly, this industry-oriented journal contains both technical and nontechnical information in the areas of wind, geothermal, solar, and photovoltaics. It serves people who design, sell, install, service, or maintain solar and other alternative energy systems for profit. The cost is $24 a year.

Solar Age

Solar Vision, Inc.   603-827-3347
Church Hill
Harrisville, NH 03450

This monthly periodical covers the total energy problem with an emphasis on solar energy. Subscription rate is $24 a year. Solar Vision also has a brochure listing books available. Cost is $4.

Data Bases

DOE/RECON (Department of Energy's Remote Console Information System)

Technical Information Center   615-576-1272
Department of Energy (DOE)
PO Box 62
Oak Ridge, TN 37830

Western Regional Information Service Center   415-486-6307
Building 50, Room 130
Lawrence Berkeley Laboratory
One Cyclotron Road
Berkeley, CA 94720

DOE/RECON comprises one of the largest and most comprehensive files of worldwide energy information. Its numerous data bases cover such subject areas as energy conservation; storage and conversion; coal, petroleum, natural gas, oil shale, and tar sands; solar, geothermal, and nuclear energy; fusion and reactor technology; other synthetic and natural fuels; wind and tidal power; and energy management and policy. The system's Energy Data Base alone contains more than two million bibliographic references to worldwide literature ranging from technical research and development reports to practical how-to articles and nontechnical information on energy-related social, economic, and environmental concerns. DOE/RECON also contains data bases with information about completed and in-progress research projects and a data base about sources of federal data. DOE contractors can access the data base directly. Others should contact the Western Regional Information Service

Center for searches. Also contact DIALOG Information Services, Inc., 3460 Hillview Ave, Palo Alto, CA 94304, 800-227-1927 (800-982-5838), 415-858-3785.

Reproductive Technology

Organizations and Government Agencies

American College of Obstetricians and Gynecologists (ACOG)
600 Maryland Avenue, SW   202-638-5577
Washington, DC 20024-2588

This nonprofit organization of 24,000 physicians specializing in obstetric and gynecological care developsand sponsors continuing education programs for its members, and creates standards of care and guidelines for evaluating and improving medical care. The college publishes technical bulletins with the results of latest research and clinical findings. These bulletins are approximately six pages long and all deal with some aspect of gynecology and obstectrics. A free listing of the 40 technical bulletins is available from the ACOG.

American Fertility Society (AFS)
2131 Magnolia Avenue, Suite 201   205-251-9764
Birmingham, AL 35256

AFS is an educational membership organization of professionals in the field of fertility. An annual meeting is held, and six regional postgraduate courses on reproductive medicine are sponsored. The Society publishes a quarterly newsletter for its members and puts out reports on such topics as infertility, artificial insemination, ethical guidelines, and in vitro fertilization. It also publishes:

• *Fertility and Sterility*—a highly technical journal on reproductive medicine and endocrinology. The journal covers many issues such as how to cause and prevent pregnancies, and in vitro fertilization. The 175-page journal is available to nonmembers for $70, and is free to members.

International Society of Reproductive Medicine
2 Jack Lynn Court   314-991-3304
St. Louis, MO 63132

A membership organization of physicians and scientists, the Society is involved in reproductive medicine and research and development in this area. An annual meeting is held, and the proceedings are published in:

• *The Journal of Reproductive Medicine*—a monthly journal directed toward physicians in private practice, with the goal of providing the latest information and treatment techniques in reproductive medicine. The publication covers such topics as oral contraception, pediatric and adolescent gynecology, in vitro fertilization, artificial insemination, and surrogate embryo transfer. Annual subscription to the 100-page journal is $45.

National Institute of Child Health and Human Development (NICHHD)

Office of Research Reporting   301-496-5133
Department of Health and Human Services
Building 31, Room 2A-32
9000 Rockville Pike
Bethesda, MD 20205

NICHHD is involved in conducting clinical research on the reproductive and behavior processes of human development. It supports research programs and informs scientists and the public about research advances. Staff can answer questions and make referrals to other information sources. Research findings are published in reports, and a free listing of these reports is available from the office. Also published is:

• *Diagnostic Ultrasound Imaging in Pregnancy*—a free 215-page report on the proceedings of a conference held on ultrasound.

National Institute of Environmental Health Sciences (NIEHS)

Public Affairs Office   919-541-3345
Research Triangle Park, NC 27709

NIEHS can supply you with information about the effects of toxic substances on reproduction. The staff can refer you to appropriate experts. The Institute has published:

• *Research Program Booklets*—a free annual publication that profiles ongoing research at the Institute.

• *Environmental Health Perspective*—a quarterly that profiles specific toxic substances in the environment and their toxic effects on health. Back issues are free, and issues after Volume 60 cost $11.

Resolve

PO Box 474   617-484-2424
Belmont, MA 02178

The national office of Resolve and its 44 affiliated chapters offer medical information on infertility. The staff can make referrals to infertility specialists, therapists, adoption services, in vitro fertilization clinics, and artificial insemination programs. Besides maintaining support groups and educational programs, they publish fact sheets, articles, and books on infertility. A free listing of these publications is available upon request. Also published is a national newsletter that comes out three times a year and provides medical updates on new technologies in the area of infertility.

Publications

Consumer Information Center
PO Box 100
Pueblo, CO 81002

Upon written request, this government publications distribution center will send you the free publications described below. You may want to order its free quarterly publications catalogue to see if new literature is also available.

- *Contraception: Comparing the Options (554N)*—in a fold-out chart format, this lists the nine common methods of birth control along with the pros and cons of each.
- *Infertility and How Its Treated (578N)*—a two-page pamphlet that discusses the causes of infertility and methods of treatment.

Journal of In Vitro Fertilization and Embryo Transfer
Plenum Press   212-620-8000
233 Spring Street
New York, NY 10013

This quarterly journal reviews and updates human and animal data relevant to the process of in vitro fertilization and embryo replacement. It contains research reports as well as papers delivered at symposia. Annual subscription fee for the 90-page publication is $45 for individuals and $80 for institutions.

Additional Experts and Resources

Listed below are the major medical centers involved in research and development in the area of reproductive technology. The staff generally can supply information about the latest technology.

Department of OB/GYN
Eastern Virginia Medical School
600 Medical Tower
Norfolk, VA 23507
804-628-3370

Department of OB/GYN
Women's Hospital
1240 North Mission Road
Los Angeles, CA 90033
213-226-3421

University of Texas
Health Science Center
Department of OB/GYN
Division of Human Reproduction
7703 Floyd Curl Drive
San Antonio, TX 78284
512-691-7270

Dr. Marian D. Damewood, Director

In Vitro Fertilization Program   301-955-2016

Johns Hopkins Hospital

600 North Wolff Street

Baltimore, MD 21205

Dr. Damewood is Assistant Professor of OB/GYN and Director of the In Vitro Fertilization Program at the university hospital. Her main interests are techniques of infertility surgery, reproductive physiology, and in vitro fertilization. She is available to answer questions and make referrals to other sources of information.

Dr. Gary Hodgen

Scientific Director, Eastern Virginia Medical School   804-446-5600

Jones Institute for Reproductive Medicine

Lewis Hall, Room 2047

700 Olney Road

Norfolk, VA 23507

Dr. Hodgen is director of the Eastern Virginia Medical School and is a leading expert in the field of in vitro fertilization and infertility techniques. The Norfolk Institute was responsible for the first test-tube baby birth in the U.S. Dr. Hodgen is interested in research involving human birth technologies, including the development of procedures to freeze human eggs and embryos for later use with infertile couples. The School is an excellent source of information on the latest techniques used for fertility.

Research and Development

See also: Cooperative Programs; Entrepreneurism; Small Business Incubators; State High-Tech Initiatives; Technology Transfer; Venture Capital

Organizations and Government Agencies

The Conference Board

845 Third Avenue   212-759-0900

New York, NY 10022

This nonprofit research and membership organization is concerned with economics and management. Conferences are open to the public, but Conference Board researchers are only available to answer members' questions. Two research and development reports available to the public are:

● *Research and Development: Key Issues for Management, Report 842*—a one-time publication consisting of an edited collection of about 18 papers presented at the April 1983 Research and Development Management Conference sponsored by the Conference Board. Major topics covered include: top management looks at research and development; perspectives from Washington; organizing for research and development; exploiting outside resources; management's role in the innovative process; research and development and corporate strategy. The 102-page publication is available to nonmembers for $125 and to members for $25.

• *Technology Imports into the United States, Research Bulletin 129*—a one-time report providing a survey of U.S. companies' imports and use of foreign technology in production in the U.S. It lists the countries from which such technology is imported and industries where it is significant. The 15-page publication is available for $50 to nonmembers and $10 to members.

Federal High Technology Conference

Foresight Science and Technology   202-833-8844
2000 P Street, NW, Suite 305
Washington, DC 20036

Sponsored annually by the National Science Foundation, this Conference is designed to plug small high-technology companies into the federal and national research and development system. The Conference is run by Foresight Science and Technology, a company that conducts conferences for government agencies, and provides policy research, marketing, and government-relations services for clients. Generally held in the fall at different locations throughout the U.S., the Conference consists of workshops and one-on-one meetings with representatives from participating government agencies, prime contractors, and professional associations.

Foundation Center

79 Fifth Avenue   212-620-4230 in New York City
New York, NY 10003   800-424-9836 elsewhere

This independent clearinghouse is the only national source of factual information about philanthropic giving, and it can help you find out where to apply most appropriately for funding. The Center operates offices in New York (listed above), Cleveland, San Francisco, and Washington, D.C. It also maintains 140 collection centers, holding publications and records, throughout the country. Using the Foundation Center's publications and its nationwide network of library reference collections, you will be able to identify foundation programs that correspond with your need. You can visit any of the centers and use the resources available there free of charge. Staff will answer telephone inquiries, assist visitors with research, and conduct computer searches, often at no cost. Some publications are available free of charge; others must be purchased.

• *Foundation Directory*—a publication of particular interest is this standard reference work, which describes more than 4,000 of the largest foundations in the United States representing 90 percent of the grant dollars given. The Directory provides the following information for each foundation it lists: name, address, and phone number; assets; total grants; high grant; low grant; description of the foundation's purpose; application procedures; contact person; and names of donors, trustees, and officers. The 775-page directory is published every other year and a supplement is issued between publications of the basic edition. The Directory is available for $60 and the supplement for $30. If purchased as a package, the cost is $75.

National Institute for Entrepreneur Technology
2000 P Street, NW, Suite 305   202-833-2326
Washington, DC 20036

The Institute is a membership organization of 1,000 representatives predominately from small high-technology businesses, universities, and state and federal government agencies involved in promoting innovations. Focused on science and technology, the Institute keeps its members informed about federal research and development funding and provides support to people engaged in entrepreneurial activities. The Institute's services, primarily for its members, include a telephone hotline service through which callers can obtain information about the federal government's science and technology activities, legislation, regulations, telephone numbers, and so on. The public can obtain free copies of back issues of its newsletter:

• *Institute Insight*—a monthly newsletter containing feature articles on topics of interest to members, research reports, a monthly column highlighting a company that has provided money or services to further technology, a legislative update section, and a publications section containing abstracts of recent publications.

National Institutes of Health (NIH)
U.S. Department of Health and Human Services   301-496-5126
Building 1, Room 105
9000 Rockville Pike
Bethesda, MD 20205

Each of NIH's Institutes promotes research pertaining to health-related issues. The office above can refer you to the Institute and office most likely to sponsor research in your field of interest. Funding, research reports, and technical assistance are often available from NIH. A report that is helpful for determining research and development projects supported by NIH is:

• *The Research Award Index*—an annual catalog of NIH grantees and contractors with descriptions of the research being done. This two-volume set is heavily indexed and cross-indexed by grant or contract number and name of the principal investigator. The publication (stock #017-040-00493-1) is available for $31 from the Superintendent of Documents, U.S. Government Printing Office, Washington, D.C. 20402, 202-783-3238.

National Technical Information Service (NTIS)
5285 Port Royal Road   703-487-4600
Springfield, VA 22161

NTIS is the central source for the public sale of U.S. government-sponsored research, development, and engineering reports, as well as computer software and data files. The NTIS collection exceeds one million titles with 60,000 new reports being published. As the government's central technical information clearinghouse, NTIS provides a variety of information awareness products and services to keep the public abreast of the technologies developed by the federal government's multibillion-dollar annual research and engineering effort. For a detailed description of NTIS publications and services, see "National Technical Information Service" write-up under the *Technology Transfer* section of this directory.

Office of Productivity, Technology, and Innovation (OPTI)

U.S. Department of Commerce   202-377-1093
14th Street and Constitution Avenue, NW, Room 4816
Washington, DC 20230

OPTI reviews issues of industrial innovation and productivity enhancement in the private sector. It provides a variety of information services to the private sector and state and local governments. Its Small Business Technology Liaison Division provides information about the innovation process, resources available, and a financial sensitivity model available for purchase. It links smaller firms, state and local governments, and other organizations involved in innovation, and provides federal policy input on smaller firms and technology. This division also produces the *Guide to Innovation Resources and Planning for the Smaller Business*, which is described separately later in this section of the directory.

OPTI's Industrial Technology partnership program provides information and training on the Research and Development Limited Partnership (RDLP), which is a financing mechanism, prepares feasibility packages to identify high-technology projects suitable for RDLPs, assists in the establishment of cooperative research and development projects, and provides federal policy input on research and development tax policy.

OPTI's Productivity Center answers productivity inquiries with off-the-shelf publications. Its staff can provide you with bibliographies, reading lists, and articles, and refer you to other sources.

Office of Technology Assessment (OTA)

U.S. Congress   202-224-9241—Public Affairs Office
Washington, DC 20510

A nonpartisan analytical agency, OTA prepares objective analyses of major policy issues related to scientific and technological change. Some of its reports have implications for research and development funding. While these reports are primarily prepared for the U.S. Congress, OTA has published more than 100 reports available to the public. Its staff consists of experts in a wide range of fields, including productivity and innovation, engineering, energy, health and life sciences, communication and information technologies, and space. OTA's staff are available to answer questions and provide information about the subjects they have researched. Upon request, OTA will send you a free catalog of its publications, as well as descriptions of new projects under way. Summaries of most OTA reports are available, at no cost, directly from the agency. Full text of the reports must be purchased from the Superintendent of Documents, U.S. Government Printing Office, Washington, D.C. 20402, 202-783-3238. Described below are OTA reports relating to research and development funding:

● *Census of State Government Initiatives for High Technology Industrial Development*, (stock #052-003-00912-9, Publication Code OTA-BP-STI-21)—this is a directory of 153 state government programs that encourage high-technology development. It includes names and phone numbers of program directors, as well as descriptions of program services. Cost is $4.50.

● *Encouraging High Technology Development #1, Background Paper #2* (Publication Code OTA-BP-STI-25)—this publication provides a discussion of high-technology development initiatives by state and local governments, universities, and private sector groups. It includes a directory of 54 local high-technology initiatives with names and phone numbers of program directors. Cost is $4.75.

● *Technology, Innovation, and Regional Economic Development* (Publication Code OTA-STI-238)—this report assesses the potential for local economic growth offered by high-technology industries, the types of programs used by state and local groups to encourage development of these industries, and the implications of these programs for federal policy. Cost is $4.50.

Science Research and Technology Subcommittee
House Committee on Science and Technology   202-225-8844
U.S. House of Representatives
Washington, DC 20515

The Subcommittee has jurisdiction over science, research, and technology legislation under consideration by the U.S. House of Representatives. It holds hearings on these topics and publishes testimony received. Free copies of its hearing minutes are available while the committee's supply lasts. Staff members are knowledgeable about research and development issues and they can refer you to specialists, programs and printed materials. The Subcommittee publishes:

● *Congressional Innovation News Notes*—issued every other month, these notes keep track of bills, hearings, laws, and other congressional activities including those pertaining to research and development opportunities. The newsletter, five to six pages long, is available free of charge.

Science Resources Studies (SRS)
National Science Foundation   202-634-4634—Charles E. Falk, Division Director
1800 G Street, NW
Washington, DC 20550

This office can provide information and publications about research and development funding and personnel resources. Staff conduct economic surveys and studies designed to provide national research and development funding totals by economic sector. Surveys are carried out in industry, universities, and government. Separate reports are available for each sector. A free publications catalog listing more than 100 materials is available from the above office. All publications are available free of charge from National Science Foundation as long as the initial printing lasts. When the NSF supply is exhausted, the publications must be purchased from the U.S. Government Printing Office. SRS's primary publications are:

● *Funds for Research and Development*—published each fall, this report presents data and analysis based on a survey of all federal agencies supporting research and development programs. The data cover such categories as basic research, applied research and development, the research performing sectors, fields of science, and geographic distribution. The report is available in two sections, Volume I being 200 pages of tables and Volume II, 80 pages of text.

● *Research and Development in Industry*—published each summer, this report provides data based on an annual survey conducted for NSF by the Bureau of the Census. Companies surveyed are from a sample consisting of about 1,500 companies. The data represent historical trends, as well as current survey year data. The report is available in two sections, Volume I being 175 pages of tables and Volume II, 50 pages of text.

● *Academic Science Research and Development Funds*—this annual report is based on data collected each year from academic institutions spending over $50,000 for separately budgeted research and development. The data elements include sources of support and breakdowns by basic/applied/development research and discipline. Data are displayed for individual institutions. The report runs approximately 100 pages.

● *National Patterns of Science and Technology*—published each spring, this report assembles and analyzes data collected in the sectional studies. It provides a concise and comprehensive summary and estimates for national totals. The publication complements the National Science Board's *Science Indicators* and NSF's *Science Engineering Personnel: A National Overview.*

● *Science Indicators*—a biannual publication of the National Science Board, this reports on the overall state of the U.S. for all facets of science and technology. *Science Indicators* also provides comparisons of the scientific and technical activities conducted in the U.S. and other major industrialized countries, primarily Japan, Western Europe, and the Soviet Union.

U.S. Small Business Administration

Office of Innovation, Research, and Technology   202-653-6458
1441 L Street, NW, Room 500-A
Washington, DC 20416

Small Business Innovation Research Program (SBIR)

The SBIR Program stimulates technological innovation, encourages small science and technology-based firms to participate in government-funded research, and provides incentives for converting research results into commercial applications. Twelve federal agencies with research and development budgets greater that $100 million are required by law to participate: the Departments of Defense, Health and

Human Services, Energy, Agriculture, Commerce, Transportation, Interior, and Education; the National Aeronautics and Space Administration; the National Science Foundation; the Nuclear Regulatory Commission; and the Environmental Protection Agency.

Businesses of 500 or fewer employees that are organized for profit are eligible to compete for SBIR funding. They must be in business for profit by the time they receive the award. Nonprofit organizations and foreign-based firms are not eligible to receive awards. (See the next entry for description of SBIR information services for small businesses.)

SBIR representatives listed below can answer questions and send you materials about their agency's SBIR plans and funding:

Department of Agriculture
Ms. A. Holiday Schauer
Office of Grants and Program Systems
Department of Agriculture
1300 Rosslyn Commonwealth Building, Suite 103
Arlington, VA 22209
703-235-2628

Department of Defense
Mr. Horace Crouch, Director
Small Business and Economic Utilization
Office of Secretary of Defense
Room 2A340 Pentagon
Washington, DC 20301
202-697-9383

Department of Education
Dr. Ed Esty
The Brown Building
1900 M Street, NW, Room 722
Washington, DC 20208
202-254-8247

Department of Energy
Mrs. Jerry Washington
SBIR Program
U.S. Department of Energy
Route 270
Gaithersburg, MD 20545
301-353-5867

Department of Health and Human Services
Mr. Richard Clinkscales, Director
Office of Small and Disadvantaged Business Utilization
Department of Health and Human Services
200 Independence Avenue, SW, Room 513D
Washington, DC 20201
202-245-7300

Department of Interior
Dr. Thomas Henrie, Chief Scientist
Bureau of Mines
U.S. Department of the Interior
2401 E Street, NW
Washington, DC 20241
202-634-1305

Department of Transportation
Dr. Robert Ravera, Acting Director
Transportation System Center
Department of Transportation
Kendall Square
Cambridge, MA 02142
617-494-2222

Environmental Protection Agency
Mr. Walter Preston
Office of Research Grants and Centers
  (RD-675)
Office of Research and Development
Environmental Protection Agency
401 M Street, SW
Washington, DC 20460
202-382-5744

**National Aeronautics and Space
  Administration**
Mr. Carl Schwenk
National Aeronautics and Space
  Administration
SBIR Office—Code A
600 Independence Avenue, SW
Washington, DC 20546
202-755-2306

National Science Foundation
Mr. Ritchie Coryell
Mr. Roland Tibbetts
SBIR Program Managers
National Science Foundation
1800 G Street, NW
Washington, DC 20550
202-357-7527

Nuclear Regulatory Commission
Mr. Francis Gillespie, Director
Administration and Resource Staff
Office of Nuclear Regulatory Research
Nuclear Regulatory Commission
Washington, DC 20555
301-427-4301

Technology Assistance Program

The Small Business Administration, Office of Innovation, Research, and Technology, in cooperation with the University of Connecticut and the University of Southern California, will provide a fast-reaction technology information service for small businesses interested in participating in the Small Business Innovation Research (SBIR) Program. The service provides, within five days, state-of-the-art information useful in preparing SBIR proposals or in guiding SBIR research efforts. The output is a comprehensive bibliography (often with abstracts) derived from a computerized search of a wide variety of data bases. The cost of this service to small businesses is $125 an inquiry. SBA provides supplemental funding to offset actual costs which are significantly higher. Documents can also be ordered for an additional fee and can typically be delivered within three weeks. To obtain this service or additional information, contact one of the following university-based centers:

**University of Southern
  California**
Western Research Applications Center
  (WESRAC)
3716 South Hope Street, #200
Los Angeles, CA 90007
213-743-6132

The University of Southern California provides service for small firms in the states of Alaska, Arizona, California, Colorado, Hawaii, Idaho, Montana, Nevada, North Dakota,

Oregon, South Dakota, Utah, Washington, and Wyoming.

For firms in all other states:

University of Connecticut
New England Research Application Center
(NERAC)
Mansfield Professional Park
Storrs, CT 06268
203-486-4586

For all states not served by WESRAC.

Triservice Industry Information Centers

The Centers below provide information on Air Force, Army, and Navy acquisitions, research and development requirements, plans, and future needs. The Centers were established to encourage industry and university participation in solving DOD problems. All centers have reading rooms where documents can be reviewed.

The Centers' services are available to current or potential DOD contractors who are registered for access to DOD information services.

Triservice Industry Information Center
5001 Eisenhower Avenue
Alexandria, VA 22333-0001
Attn: (Specific branch)
202-274-9305—Air Force Office (AFIFIO)
202-274-8948—Army Office
202-274-9315—Navy Office
202-284-9305—AUTOVON (Intragovernment)

Triservice Industry Information Center
1030 East Green Street
Pasadena, CA 91106
Attn: (Specific branch)
818-792-3192—Air Force Office
818-792-7146—Army Office
818-792-5182—Navy Office
818-360-2456—AUTOVON (Intragovernment)

Triservice Industry Information Center
Air Force Wright Aeronautical Laboratories
   (AFWAL/TST)
Wright-Patterson AFB, OH 45433
Attn: (Specific branch)
513-258-4259—Air Force Office
513-258-4260—Army Office
513-258-4261—Navy Office
513-785-5572—AUTOVON (Intragovernment)

Taft Group
5125 MacArthur Boulevard, NW   800-425-3761
Washington, DC 20016   202-966-7086—in Washington Metropolitan Area

The philosophy of this consulting and publishing organization is "profit thinking for nonprofit organizations." Staff can answer questions about the Group's publications and services. Publications include two directories of givers, monthly newsletters to supplement the information in the directories, and a series of short books. Described below are the directories:

• *The Taft Corporate Directory*—this annual publication lists more than 500 corporations, all of which donate more than $100,000 per year. Information provided for each company includes: names, address, phone number, and contact person; how much the company earned during the year covered by the directory; giving priorities; philosophy; how and when to approach the company; biographical information about company board members; and recent grants given. The directory is heavily indexed. Cost is $267.

• *The Taft Foundation Reporter*—this annual publication lists private foundations. Information provided for each foundation includes: name, address, phone number, and contact person; how much the foundation earned during the year covered by the report; giving priorities; philosophy; how and when to approach the foundation; biographical information about board members; and information about recent grants given. The report is heavily indexed. It costs $267.

Technology Utilization Division (NASA TU)
Code I   202-453-8415
NASA Headquarters
National Aeronautics and Space Administration
400 Maryland Avenue, SW
Washington, DC 20546

This Division is involved in transferring NASA-developed technology to industry for the develoment of useful commercial products. The Technology Utilization Division can provide you with information about technologies developed under NASA-funded projects. A variety of services and publications are available from this office. Staff will answer questions and make referrals. (NASA TU information and services are restricted to U.S. citizens and U.S.-based industries.) For further information about the program, see entry for NASA Technology Utilization and Industry Affairs Division under the *Technology Transfer* section of this directory.

Publications

Budget of the United States Government
Office of Management and Budget   202-395-3000
Old Executive Office Building
Washington, DC 20503

Published each January, this document contains the President's proposed budget to the U.S. Congress. Information is provided for all federal expenditures, including past and proposed research and development funding. Staff of the Office of Management and Budget can answer questions regarding the budget. The publication (stock #041-001-00270-7) is available for $12 from the Superintendent of Documents, U.S. Government Printing Office, Washington, D.C. 20402, 202-783-3238.

Catalog of Federal Domestic Assistance
General Services Administration (GSA)   202-673-5302
Federal Program Information Branch
1825 Connecticut Avenue, NW, Room 804
Washington, DC 20405

This book describes all federally funded domestic assistance programs. These include seven financial types—for example, formula grants, project grants, and direct loans—as well as eight nonfinancial types, including provision of specialized services. The programs span 20 functional categories and 176 subcategories, such as agriculture, community development, consumer protection, education, employment, labor and training, and health. The *Catalog* contains program descriptions, appendixes, and indexes, as well as information about application procedures, eligibility of applicants, and forms required to apply for assistance. Each program description contains the following information: authorization that funds the program; federal agency that administers the program; objectives and uses of the program; eligibility requirements; applications and awards process; financial information for past, present and future years; and information contacts. The 1,000-page publication is published annually in May, with a supplement issued in December. The latter covers new programs, revisions of existing programs, deleted programs, changes in program identification numbers, and similar information. The *Catalog*, plus supplement, is available for $40 from the Superintendent of Documents, U.S. Government Printing Office, Washington, D.C. 20402, 202-783-3238.

Commerce Business Daily

Superintendent of Documents   202-783-3238
U.S. Government Printing Office
Washington, DC 20402

The publication contains notices of opportunities and awards for federally sponsored research and development. Published Mondays through Fridays (except on federal legal holidays) by the U.S. Department of Commerce, in cooperation with the federal agencies, it provides a daily list of U.S. government RFPs (Requests for Proposals), invitations for bids, contract awards for possible subcontracting leads, sales of surplus property, and foreign business opportunities. Generally, this list is accompanied by a brief description of the proposed procurement action and information about how to get a copy of the formal solicitation or announcement. Copies of the publication are available for reference purposes at Department of Commerce field offices and in most public libraries. It may be obtained on a subscription basis for $100 a year via second class mail or $175 via first class mail from the Superintendent of Documents (address listed above.) A purchase order must be accompanied by payment.

This publication is also available in an electronic, on-line edition. See below under *Data Bases*.

The Foundation 500

Douglas M. Lawson Associates, Inc.   212-759-5660
39 East 51st Street
New York, NY 10022

Updated every three years, this publication provides information about 500 foundations. It includes: name, address, phone number, and contact person; number of grants given; the kinds of organizations the foundation gave to; and total amount and average amount donated. The directory is broken down into seven major categories which are further subdivided into a total of 66 grants categories. Relevant major categories include science, health, education, the arts and humanities, and international areas. A state-by-state breakdown showing which foundations are giving in a particular geographic area is also provided. The 76-page publication is available for $32.50.

Getting Yours
Viking Penguin, Inc.
40 West 23rd Street
New York, NY 10010

A condensed version of the *Catalog of Federal Domestic Assistance* described earlier in this section, this book provides a capsule description of funding opportunities available from the federal government. Directed toward the general public, it gives examples of people and projects that have received funding as well as advice about how to select federal programs that would be most receptive to your idea. The 292-page book, which is updated periodically, is available from most bookstores for $7.95.

Grants for Scientific and Engineering Research
National Science Foundation (NSF)   202-357-9498—Office of Legislative and Public Affairs,
1800 G Street, NW   Public Affairs and Publications Group
Washington, DC 20550

This publication contains instructions and applications for submitting a proposal for NSF funding. It goes hand-in-hand with the Foundation's guide to its programs, which is a helpful tool for identifying the various subject areas for which NSF funds research projects. These and other NSF publications are available free of charge from NSF while its supply lasts. After that, publications must be purchased from the Superintendent of Documents, U.S. Government Printing Office, Washington, D.C. 20402, 202-783-3238.

Guide to Innovation Resources and Planning for the Smaller Business
The Small Business Technology Liaison Division   202-377-1093
Office of Productivity, Technology, and Innovation
U.S. Department of Commerce
14th Street and Constitution Avenue, NW, Room 4816
Washington, DC 20230

The *Guide* identifies more than 50 federal and 85 state government offices that assist smaller businesses in bringing new technologies to market. It is written for individuals, companies, and state and local government economic development planners. The *Guide* has two basic sections: the first examines the many steps in the innovation process and the skills and resources needed; the second section identifies a wide range of resources (federal, state, and private) available to assist the smaller business in areas such as financing, information gathering, and management. The capabilities of these resources are summarized and contact phone numbers and addresses are given. Each resource has been identified by the stage of the innovation process to which it applies. The 85-page publication is available for $13.50 from the National Technical Information Service, U.S. Department of Commerce, 5285 Port Royal Road, Springfield, VA 22161, 703-487-4650.

The R&D Annual

Technical Insights, Inc.   201-568-4744
PO Box 1304
Fort Lee, NJ 07024

This yearly publication is divided into three parts: the first of these is a lengthy introduction that reviews technological advances of the past year and previews what may be coming next year; the second section is a compilation of hundreds of technical advances that the company's researchers have uncovered during the year; the last part contains research and development statistics, including what companies, governments, and other entities spend, with historical comparisons of the numbers. The 235-page publication is available for $315.

Selling to the Military

Superintendent of Documents   202-783-3238
U.S. Government Printing Office (GPO)
Washington, DC 20402

This is a guide for people wanting to do business with the Department of Defense and all its military departments. It is a listing of items purchased by Defense agencies and locations of military purchasing offices. Updated irregularly, the 138-page publication (stock #008-000-00345-9) is available for $6 from GPO.

Small Business Guide to Federal R&D Funding Opportunities

Office of Small Business R&D   202-357-7464
National Science Foundation
1800 G Street, NW
Washington, DC 20550

The purpose of the *Guide* is to provide scientifically and technically oriented small businesses with information about opportunities for obtaining federal funding for research and development activities. It contains information about the substantive priorities of federal research and development programs; the criteria that companies must meet in order to do business with the federal government; the procedures used by federal research and development programs to publicize funding opportunities, to solicit ideas from the private sector, and to fund extramural research and development activities; certain of the federal laws, regulations, and policies that affect small business participation in federally funded research and development activities; and points of contact for obtaining additional information about each research and development program. The 136-page publication (stock #038-000-00522-7) is available for $6 from the Superintendent of Documents, U.S. Government Printing Office, Washington, D.C. 20402, 202-783-3238. It is also available for $9.50 from the National Technical Information Service, 5285 Port Royal Road, Springfield, VA 22161, 703-487-4650; refer to number PB83-192401.

Technology Review

Massachusetts Institute of Technology   617-253-8292—Subscriptions
77 Massachusetts Avenue, Building 10, Room 140
Cambridge, MA 02139

Directed towards leaders in the fields of science, engineering, business, and policy-making, this magazine focuses on current and future developments in technology policy. Articles touch on the ways technology policy and research and development funding impact upon each other, with emphasis on the implications for society and the environment. Annual subscription price is $24 for 8 issues.

Data Bases

The Commerce Business Daily

An electronic, on-line edition of the *Commerce Business Daily* is available from several Department of Commerce contractors. Interested parties may contact the contractors for full details:

DIALOG Information Systems, Inc.   **United Communications Group**
3460 Hillview Avenue   8701 Georgia Avenue
Palo Alto, CA 94304   Silver Spring, MD 20910
800-227-1927   800-638-7728—Joanne Gionnola
                                              301-589-8875—Call Collect

DMS/ONLINE
100 Northfield Street
Greenwich, CT 06830
800-243-3852
203-661-7800

Additional Experts and Resources

Philip Speser, J.D., Ph.D.
President of Foresight Science and Technology   202-833-2323
2000 P Street, NW, Suite 305
Washington, DC 20036

Dr. Speser is a good resource for information about where to go in the federal government to find out about research and development opportunities. His special interests include marketing, high technology, strategic planning, and policy research on science and technology innovation.

William L. Stewart
Deputy Director, Division of Science Resources Studies   202-634-4625
Head, R&D Economic Studies Section
National Science Foundation (NSF)
1800 G Street, NW
Washington, DC 20550

Employed in NSF's Division of Science Resources Studies since 1962, Mr. Stewart's work involves the collection and analysis of research and development funding and employment data for four major sectors of the economy including industry, government, universities and colleges, and other nonprofit organizations. He conducts surveys within these sectors in order to find out who's doing the research and who is paying for it. Mr. Stewart has been involved in the publication of some 20 reports.

Dr. Albert H. Teich
Director, Office of Public Sector Programs   202-326-6600
The American Association for the Advancement of Science (AAAS)
1333 H Street, NW
Washington, DC 20005

Dr. Teich's work at the AAAS involves preparation of an annual cycle of reports about research and development in the federal budget, including a congressional action report, annual research and development budget book and proceedings from the annual research and development colloquium held in Washington, D.C. He has managed this conference, as well as workshops and seminars devoted to research and development. He is available to answer questions.

Professional societies sometimes publish surveys and reports covering opportunities and trends in research and development. Often staff members can give you leads to research being done or funding available. Check with associations listed under topics of interest to you in this directory.

Retraining Programs

See: Careers and Employment; Labor and Technological Change

Reviews of Software

See: Software Guides, Directories, and Reviews

Risk Analysis and Management

See: Social Impact of Technology; Technology Assessment; Technophobia

Robotics

See also: Artificial Intelligence

Organizations and Government Agencies

Communications Publishing Group, Inc. (CPG)
PO Box 383   617-566-2373
Dedham, MA 02026

CPG is a data base publisher of high-technology publications, on-line data bases, and research services. CPG's products and services are based on worldwide patent literature and are available in both print and electronic form (from NewsNet, 800-345-1301). Their publication dealing with robotics is:

- *Robotics*—a biweekly newsletter covering components, equipment, systems, and processes of industrial automation as well as visual systems, sensors, controllers, grippers, and hydraulic and pneumatic devices. The annual subscription rate is $197.

Robot Industries Association (RIA)
1 SME Drive   313-271-7800, ext. 585
PO Box 1366
Dearborn, MI 48128

The Association is a trade group representing corporations working in the robotics field. RIA maintains an updated library of educational materials, technical papers, books, films, videotapes, and journals to keep members abreast of the latest developments in robotic technology. It sponsors research in all areas of robotics, and holds many conferences, seminars, and workshops. Staff members will answer inquiries and make referrals to other sources of information. They also publish the following:

- *Industrial Robots: A Survey of Foreign and Domestic U.S. Patents*—highlights of this book include a detailed list of 212 United States patents along with a summary of robotic patents from foreign countries. Cost is $115.
- *Worldwide Robotics Survey and Directory*—provides economic projections, applications information, a directory of manufacturers, and charts that detail licensing and joint venture agreements. Also includes key details on robot manufacturers around the world. This annual costs $30.

Robotics International of SME (RI)
1 SME Drive   313-271-1500, ext. 416
P.O. Box 930
Dearborn, MI 48128

RI is an educational and scientific society which covers all phases of robot research, design, installation, and operation. CARIC, the Computerized Automation and Robotics Information Center, a service of SME, gathers and disseminates technical data on robotics, and staff members will answer inquiries, perform limited literature-searching services for a fee, make referrals to other sources of information, and permit on-site use of their literature collection. Many searches are done on their in-house data base, IN-TIME, which is not available on-line to the public. Call CARIC at extension 203 for further details. RI publishes the proceedings from various conferences on robotics. Papers are available for $3.50 to $6 each. The procedings are also available on microfiche for $2.75 per copy. RI also publishes the following journals:

• *Today*—a bimonthly journal of applications-oriented industrial robotics information. It is one of the top periodicals in the field. A subscription costs $36 per year.

• *North American Directory of Robotics Education and Training Institutions*—this directory, published every two years, lists colleges, universities, and technical schools in the United States and Canada that offer robotics education and training programs. It provides names, addresses, contact persons, and tells whether the school has a robotics library. The publication is available free of charge.

Publications

High-Tech Industries: Profiles and Outlooks—The Robotics Industry
U.S. International Trade Commission   202-523-0377
Machinery-Equipment Division
701 E Street, NW
Washington, DC 20436

This report on the robotics industry is published by the International Trade Administration and is available for $4 from the Superintendent of Documents, U.S. Government Printing Office, Washington, DC 20402, 202-783-3238. The order number is 003-009-0036-6.

Robomatix Reporter (Worldwide Robotics Intelligence Network)
EIC/Intelligence   800-223-6275
48 West 38th Street
New York, NY 10018

This monthly magazine summarizes important technological innovations and applications, as well as new patent listings, market highlights, corporate monitors, and previews of conferences. EIC/Intelligence, the clearinghouse that publishes the magazine, also maintains a comprehensive microfiche collection on robotics literature since 1973. Also available in data base form.

Robotics Age

PO Box 358   603-924-7136
Peterborough, NH 03458

A monthly magazine oriented toward hobbyists in the field of robotics. Articles cover all aspects of robotics, such as artificial intelligence, CAD/CAM, industrial robots, sensors, and speech synthesis. The annual subscription rate is $21. (This publication is also available via the computerized system NewsNet, 800-345-1301).

Robotics Industry Directory

Technical Data Base Corporation   409-539-9688
PO Box 720
Conroe, TX 77305

Covers industrial robot models and components with special emphasis on specification information. Features over 200 computer-generated, full-page robot model listings. Other sections cover special application systems, controllers, mechanical components, consultants, systems houses, research institutes, and names, addresses, and phone numbers of marketing contacts. Also published is a bimonthly update, *Computerized Manufacturing* which reports on new products, new applications, and other developments in industrial robot technology. The directory is $35 and the update is $15 a year.

Robotics Technology Abstracts

Cranfield Press   011-44-234-752727
Management Library
Cranfield, Belford
MK43 OAL
United Kingdom

A monthly publication providing worldwide coverage of scientific and technical journals, technical reports, patents, and conference proceedings relevant to both the manufacture and use of robots.

Robotics Tracer Bullet

Reference Section   202-287-5670
Science and Technology Division
Library of Congress
Washington, DC 20540

The Science and Technology Division of the Library of Congress has put together a *Tracer Bullet* on Robotics. It provides a listing of organizations and literature which are good resources regarding robotics. These "Bullets" are available free of charge.

Robotics World

Communications Channels, Inc.   404-256-9800
6255 Barfield Road
Atlanta, GA 30328

A leading monthly periodical in the flexible manufacturing industry, *Robotics World* is oriented toward manufacturers and users of industrial robots. Topics covered include news of the industry, international news, events in robotics, machinery and equipment, trade literature, and patent reviews. Cost is $25 a year.

Technical Insights, Inc.

PO Box 1304   201-944-6204
158 Linwood Plaza
Fort Lee, NJ 07024

Technical Insights publishes several reports on various aspects of robotics, including:

● *Robots in Industry*—six individual reports on the applications of robots for foundries, assembly, metal fabrication, the plastics industry, general applications, and the electronics industry. The first five reports sell for $65 each, the last report sells for $125.

● *Intelligent Robots*—a 200-page technical report that presents the technologies that provide the intelligence for a new generation of robots. Available for $285.

● *The Intelligent Robot*—a 258-page report with information on various technologies—artificial intelligence, voice systems, vision, tactile sensing—as they pertain to robotics. It sells for $285.

● *Machine Vision for Robotics and Automated Inspection*—a three-volume, 350-page report that explains the successful application of machine vision systems. Available for $185.

● *High Technology Manufacturing for the Food Industry*—a three-volume, 300-page report that applies robotics to food processing plants. It costs $225.

Data Bases

RBOT—Robotics Information

BRS   800-833-4707
1200 Route 7   800-553-5566—in New York
Latham, NY 12110   518-783-7251

This data base collects and summarizes data from the latest industry research, conference papers, journals, government reports, newsletters, and books. It provides access to the full range of robotics information, from sensor systems to machine intelligence. Coverage dates back to 1970, and data is updated monthly. Photocopies of most articles and conference papers in the RBOT data base may be obtained from MILDOCS of Cincinnati Milacron for a fee. Contact Corporate Information Center (MILDOCS), Cincinnati Milacron Industries, Inc., 4701 Marburg Avenue, Cincinnati, OH 45209, 503-841-8110.

Robotronics Age Newsletter
NewsNet, Inc.   800-345-1301
945 Haverford Road
Bryn Mawr, PA 19010

> Full text of the newsletter of the same name. Contains information on the manufacture and use of robots worldwide. Produced by Twenty-first Century Media Communications, Inc. (604-688-7103).

Additional Experts and Resources

Many universities run research centers for robotics, and staff are often willing to answer questions about the more technical aspects of robotics. The following is a list of some of the centers:

University of Maryland
Robotics Lab
TERH Building, Room 1240
College Park, MD 20742
301-454-7694

Carnegie-Mellon University
Robotics Institute
Schenley Park
Pittsburgh, PA 15213
412-578-2597

University of Michigan
Center for Robotics & Integrated
    Manufacturing
Ann Arbor, MI 48190-1185
313-764-7494

University of Florida
College of Engineering
Center for Intelligent Machines & Robotics
Gainesville, FL 32611
904-392-0814

University of Rhode Island
Robotics Research Center
Kelley Hall
Kingston, RI 02881
401-792-2514

For further assistance identifying sources of information in the field of robotics, contact any of the following government agencies:

National Bureau of Standards
Industrial Systems Division 739
Building 220, Room A123
Washington, DC 20234
301-921-2381

Office of Producer Goods, Room 4042
Bureau of Industrial Economics
U.S. Department of Commerce
14th Street and Constitution Avenue, NW
Washington, DC 20230
202-377-0314—Thomas Gallogly

Machinery-Equipment Division
U.S. International Trade Commission
701 E Street, NW
Washington, DC 20436
202-523-0377

Satellite Communications

Organizations and Government Agencies

American Institute of Aeronautics and Astronautics (AIAA)
1633 Broadway   212-518-4300
New York, NY 10019

AIAA is a professional society that represents the engineering community. Its Technical Information Library collects and translates information and data from international sources. The society produces educational programs, seminars, and short courses, and also sponsors exhibits. They publish:

● *AIAA Communications Satellite Systems Conference Proceedings*—includes the papers presented at the annual AIAA conference. Available to members for $95 and to nonmembers for $110.

● *AIAA Journal*—includes technical articles on satellite communications, and reprints of papers presented at AIAA conferences, all with indexes. Annual subscription rate is $19 for AIAA members, and $185 for nonmembers.

Electronic Industries Association (EIA)
2001 Eye Street, NW   202-457-4900
Washington, DC 20006

This national trade organization represents the electronics manufacturing community, including satellite communications manufacturers. EIA is involved in such services as technical standards development, petition filing, and market statistics development. Publications include:

● *Telecommunications Trends and Directions*—presents papers given at the EIA Telecommunications Conference. Available for $20.

● *Electronic Market Data Book*—this annual publication contains text and over 100 statistical tables, graphs, and charts based on production and sales figures provided to the EIA by several hundred companies and selected government and private sources. Also covered are developments in the communications and industrial electronics markets, including satellite communications. Available for $8.

European Space Agency (ESA)

955 L'Enfant Plaza, SW, Suite 1404   202-488-4158
Washington, DC 20024

ESA, representing eleven European governments whose combined space program is similar to NASA's, was founded to promote European cooperation in research and development. ESA has prepared special reports, conference proceedings, bulletins, a journal, and other documents. Its Washington office will answer questions from the public, and send copies of any ESA publications it may have on hand. All publications are available free of charge, and should be ordered from: ESA Scientific and Technical Publications Branch, ESTEC, 2200 AG Noordwijk, Netherlands. Their publications include:

● *ESA Bulletin*—a quarterly magazine covering all aspects of space programs and research, with a status report on major developments.

● *ESA Journal*—this quarterly journal provides detailed technical information about space activities and is of particular interest to scientists and engineers.

● *Development and Application of New Technologies in the ESA Olympus Program*—a conference paper presented at the AIAA annual Communications Satellite conference.

Institute of Electrical and Electronics Engineers (IEEE)

Aerospace and Electronic Systems Society   212-705-7600
c/o TAB Office
345 East 47th Street
New York, NY 10017

The IEEE organizes and sponsors technical meetings related to satellite communications, including the "Position, Location, and Navigation Symposium," the "National Aerospace Electronics Conference," and the "National Telecommunications Conference." It also publishes the papers presented at these meetings, which are available for $35-$40 a publication. IEEE also publishes other literature on the subject, including indexes, books, and tutorial publications. The bimonthly *IEEE Transactions on Aerospace and Electronics Systems* presents papers covering all aspects of satellite communications, and is available to members for $15 a year, and to nonmembers for $75.

International Trade Administration (ITA)

U.S. Department of Commerce   202-377-2872
14th Street and Constitution Avenue, NW, Room 1015B
Washington, DC 20230

This agency reports on Bureau of the Census data for industries as classified by the Standard Industrial Classification (SIC), including that of the satellite communications systems industry. It gathers, analyzes, and presents its findings on product shipment for that SIC number, and these findings are published as a chapter in the annual *U.S. Industrial Outlook*, available for $14.

Society for Private and Commercial Earth Stations (SPACE)

1920 N Street, NW, Suite 510   202-887-0605
Washington, DC 20036

SPACE is the trade association of the satellite earth station industry. It holds two annual conventions, and sponsors nationwide seminars providing analysis of the private cable (SMATV) industry. Annual membership is $35, and includes the newsletter and other services. Their publications include the following:

- *Q&A About Satellite Earth Stations*—a basic outline of the earth station industry. Available for $4.95.
- *The Satellite Earth Station Zoning Book*—answers most zoning questions and covers many zoning issues. Available for $25.
- *The Business of Private Cable (SMATV)*—$115.
- *SATVISION*—this monthly membership magazine is $30 a year.
- *Inside SPACE*—a newsletter for members.

Publications

SAT Guide

PO Box 1048   208-788-9531
Hailey, ID 83333

A magazine published for the cable television industry and its alternative distribution systems, *SAT Guide* is devoted exclusively to the domestic satellite industry. It addresses the issues, problems, and changing technologies. Regular feature articles include programming, reviews, hardware, trends, a yellow pages reference, and reviews of regulatory matters.

Satellite Communications

Cardiff Publishing Co.   303-694-1522
6530 South Yosemite Street
Englewood, CO 80111

A monthly magazine devoted to the news, application, and technology of space telecommunications, *Satellite Communications* centers on the issues and development of the satellite communications industry through feature articles on regulatory and technical news, satellite business and applications, new products and systems, service applications, and case history features. Available for $25 a year.

Satellite News

Phillips Publishing, Inc.   301-986-0666
7315 Wisconsin Avenue, Suite 1200N
Bethesda, MD 20814

This weekly newsletter covers the latest news in marketing, management, regulation, finance, technology, programming, NASA, DBS, international developments, remote sensing, cable, DTS, video-conferencing, launch schedules, common carriers, professional conferences, and so forth. The annual subscription is $397 for

50 issues. Also available on-line via NewsNet, Inc. (800-345-1301 or 215-527-8030).

Satellite Week

1836 Jefferson Place, NW   202-872-9200
Washington, DC 20036

A weekly, the newsletter covers the satellite industry, with special emphasis on communications. It follows satellite and satellite communications policymaking in the U.S., the U.S.S.R., and European nations. In addition, it reports news on military satellites. The annual subscription cost is $462.

Data Bases

COMPENDEX

Engineering Information, Inc.   202-705-7600
345 East 47th Street   800-221-1044
New York, NY 10017

The data base provides the engineering and information communities with abstracted information on engineering developments from engineering and related technological literature worldwide. Indexed publications include approximately 4,500 journals, publications of engineering societies and organizations, works from conferences and symposia, technical reports, and monographs. The data base is accessible through nine commercial vendors, including DIALOG, SDC ORBIT, BRS, and ESA. Engineering Information may be contacted directly if a search is needed.

Additional Experts and Resources

Roger Herbstritt

Federal Communications Commission (FCC)   202-653-8106
Manager of Satellite Systems Branch
Office of Science and Technology
2025 M Street, NW
Washington, DC 20554

Mr. Herbstritt is an expert on satellite developments being studied by the FCC. He is knowledgeable about the overall satellite industry and can refer you to experts in industry and government.

Roger Cochetti

Director of the Office of Public and Investor Relations   202-863-6800
Communications Satellite Corporation
950 L'Enfant Plaza
Washington, DC 20024

Mr. Cochetti serves as a resource for information on international, domestic, and maritime satellite communications systems, services, and related equipment.

C. Louis Luccia

Manager of Advanced Programs   202-453-1522
Communications Division
NASA Headquarters
600 Independence Avenue, SW
Washington, DC 20546

> Mr. Luccia is a scientist who has been involved actively in international satellite programs for the last twenty years. He has an expertise in digital communications by satellite.

Arthur Pleasants

International Trade Administration   202-377-2872
U.S. Department of Commerce, Room 1015B
Washington, DC 20230

> Mr. Pleasant's work concerns product shipments within Standard Industrial Classification 3662, which includes satellite communications systems. He can answer questions and/or make referrals.

Raul R. Rodriguez

Office of International Affairs   202-377-1866
National Telecommunications and Information Administration
U.S. Department of Commerce
14th Street and Constitution Avenue, NW
Washington, DC 20230

> Mr. Rodriguez is the Director of International Satellite Policies of the Office of International Affairs. His particular expertise relates to two functions: overseeing of INTELSAT and INMARSAT, and conducting an instructional process regarding COMSAT policy.

Carl H. Schmitt

Assistant to the Director for Communications   202-863-7842
COMSAT General Corporation
950 L'Enfant Plaza, SW
Washington, DC 20024

> Mr. Schmitt is an educational engineer who specializes in international satellite coordination and systems design. He has compiled geostationary satellite data for both in-orbit and planned communications, broadcasting, and maritime satellite systems.

Phil Tremper

Federal Communications Commission   202-653-8163
2025 M Street, NW
Washington, DC 20554

> Mr. Tremper is an expert in space-related data bases available to the public. He is knowledgeable about data bases on satellite communications, transmission, and

FCC licensure. Mr. Tremper can help you find technical information and he will refer you to appropriate sources within the field.

Scanners, Medical

See: Medical Imaging; Medical Technology

Security, Computer

See: Computer Security

Semiconductors

See also: Gallium Arsenide; Microelectronics

Organizations and Government Agencies

Communications Publishing Group, Inc. (CPG)
PO Box 383   617-566-2373
Dedham, MA 02026

> Besides publishing sundry high-technology publications, CPG offers on-line data bases and research services. Its products and services are based on worldwide patent literature, and are available in both print and electronic form through News-Net (800-345-1301). Publications dealing with semiconductors include:
>
> • *Semiconductors/ICs*—a newsletter covering devices such as discrete devices, integrated circuits, logic circuits, and all LSI, and VLSI circuits. Available for $247 a year for 24 issues.

Electronic Industries Association (EIA)
2001 Eye Street, NW   202-457-4900
Washington, DC 20006

> The Association has available 14 standards on semiconductors. These documents, written by committees consisting of association members, vary in price. Information regarding their cost can be obtained from the Sales Office at 202-457-4966. They also publish:

• *Electronic Market Data Book*—contains information on semiconductors. The more than 150-page annual publication costs $60 an issue, and is available from the Marketing Services Department at 202-457-4955.

National Bureau of Standards

Semiconductor Materials and Processes Division   301-921-3786—Robert Scace, Chief
Technology Building, Room A305
Gaithersburg, MD 20899

National Bureau of Standards

Semiconductor Devices and Circuits Division   301-921-3541
Building 225, Room B344
Gaithersburg, MD 20899

The Semiconductor Materials and Processes Division will handle materials-related questions. The secretary can refer callers to an expert who will be able to answer questions or make referrals. Staff at the Semiconductor Devices and Circuits Division can answer highly technical questions. NBS publishes materials on semiconductors including:

• *Semiconductor Measurement Technology*—a list of publications containing reports of work performed at the National Bureau of Standards in the field of semiconductor measurement technology. The publications are grouped by subject matter and within groups are arranged in chronological order. Free copies are available from NBS.

• *Semiconductor Technology for the NonTechnologist*—the properties of semiconductor materials, the methods of processing them, and the solid-state products made from them are described in terms intended to be understandable by the lay person. Available free from Robert Scace.

Semiconductor Equipment and Materials Institute (SEMI)

625 Ellis Street, Suite 212   415-964-5111
Mountain View, CA 94043

SEMI represents the equipment and materials manufacturers to the semiconductor industry. It sponsors trade shows worldwide, and also offers the following publications:

• *Membership Directory*—$50.
• *Japanese Semiconductor Industry Report*—$100.
• *SEMI Standards* (five volumes: Chemicals, Equipment, Materials, Packaging, and Photomask)—$150 for the set.
• Transcripts of all the technical programs can be purchased for $50 each.
• Genealogy chart—of all semiconductor companies in Silicon Valley is available for $10.

Semiconductor Industry Association (SIA)

4320 Stevens Creek Boulevard, Suite 275   408-246-1181
San Jose, CA 95129

This association represents U.S. semiconductor manufacturers. They publish the following:

• *SIA Yearbook/Directory*—published every two years, this includes a directory of semiconductor manufacturers. The following information is provided: names and addresses of companies; phone and telex numbers; names of senior executives; key products; products manufactured; and annual sales for members of SIA. Available for $50 a copy.

• *World Semiconductor Trade Statistics*—this monthly publication provides information about the U.S., European, and Japanese semiconductor manufacturers. The cost is $1,200 a year.

Publications

Electronics

McGraw-Hill, Inc.   609-426-5989
Princeton-Hightstown Road
Hightstown, NJ 08520

Published every two weeks, *Electronics* reports and analyzes new technological trends in the electronics industry. It is directed toward electrical engineering designers, particularly of solid-state devices, and managers in the field. The annual subscription price is $24.

Institute of Electrical and Electronics Engineers (IEEE)

Publications Sales Department   201-981-1393
445 Hoes Lane
Piscataway, NJ 08854

The IEEE publishes the following literature pertinent to semiconductors:

• *Electron Device Letters*—a monthly that presents new results in the field of electron devices. Items are restricted to two pages and appear usually two months after receipt.

• *Journal of Solid-State Circuits*—this bimonthly presents device coverage of circuits and related areas, with emphasis on practical applications.

• *Transactions on Electron Devices*—this monthly covers theory, design, and performance of electron devices, including electron tubes, solid-state devices, integrated electron devices, energy sources, power devices, displays, and device reliability.

Semiconductors International

Cahners Publishing Co.   312-635-8800
1350 East Touhy Avenue
Des Plaines, IL 60018

This monthly journal deals with processing, testing, and assembling in the semiconductor industry and integrated circuits. The annual subscription rate is $45, but is available free of charge to qualified individuals. Contact Cahners at 270 St. Paul Street, Denver, CO 80206, 303-388-4511, to find out if you are eligible for a free subscription.

Semiconductor Industry and Business Survey

HTE Management Resources   408-684-0125
PO Box 60276
Sunnyvale, CA 94086

The journal is published 18 times a year and reports on the semiconductor business for electronics management. The annual subscription rate is $400.

Sewing Systems

See: Apparel

Ship Building

See: Marine Technology

Silicon

See also: Ceramics; Chips and Integrated Circuits; Materials Science; Metals; Micro-electronics; Plastics; Semiconductors

Organizations and Government Agencies

General Electric Silicon Products Division (GE)

Waterford, NY 12188   518-237-3330—Customer Service

One of the major producers of silicon, GE will send catalogs for each of their products, and data sheets that describe in detail the technology behind the product. Their Customer Service Department can handle general questions and refer you to other information sources within the company or outside.

Fairchild Materials Center

33 Healdsburg Avenue   707-433-7000—Marketing Department
Healdsburg, CA 95448

The semiconductor company produces its own silicon, and it will send you a brochure that describes the work being done at the Materials Center. Marketing Department personnel will answer questions and refer you to other sources of information.

Hemlock Semiconductor Corporation

PO Box 187   517-642-5201
12334 Geddes Road
Hemlock, MI 48626

This subsidiary of Dow Corning is a major producer of silicon. Their annual report contains information about the company and what types of work they are doing. Several free descriptive brochures are available, as well as a product guide.

Institute of Electrical and Electronics Engineers (IEEE)

345 East 47th Street   212-705-7866—Public Information
New York, NY 10017   212-705-7890—Technical Section

A major electronics professional society, IEEE has published many papers on the topic of silicon and its applications. It publishes a directory of papers with a key word index. The engineering society's library, located in the same building, carries all the IEEE papers, and the staff will search their data base for a fee.

Monsanto Electronics Materials Company

755 Page Mill Road   415-493-3300
Palo Alto, CA 94303

Monsanto is one of the major producers of silicon and components using silicon. Upon request, it will send you a product information catalog that lists all of its products. Its customer relations personnel can answer questions about the company and its work.

Motorola Semiconductor Product Sector

725 South Madison Drive   602-994-6316—Lothar Stern, Manager of Technical Communications
Tempe, AZ 85281

Motorola is a semiconductor manufacturing firm that produces its own silicon. Informational brochures are available for each of its products. Mr. Stern can answer specific questions about the products.

SEH America

4111 Northeast 112th Avenue   206-254-3030—Marketing Department
Vancouver, WA 98662

This Japanese silicon-producing firm has a base in Washington state. Through its marketing department you can receive brochures describing its products, and personnel is on hand to answer questions regarding the company and its products.

SEMI

625 Ellis Street, Suite 212
Mountainview, CA 94043

415-964-5111—Public Relations
Carol Seaborn—Standards Department

SEMI is a trade association of 3,200 members that serves the semiconductor equipment and materials industry. It has a subcommittee on standards for silicon, and it publishes a materials volume ($75). The Standards Department also publishes a newsletter.

Additional Experts and Resources

Robert Scace

National Bureau of Standards
Building 225, Room A305
Route 270
Gaithersburg, MD 20899

301-921-3786

Mr. Scace, an expert at the National Bureau of Standards, can answer questions about what is going on in the field of silicon, and he can make referrals to other sources of information.

Small Business Incubators

See also: Cooperative Programs; State High-Tech Initiatives

Organizations and Government Agencies

American Planning Association (APA)

1313 East 60th Street
Chicago, IL 60637

312-955-9100

A professional organization of land-use planners, APA represents local and federal governments, consulting firms, educators, and private citizens. The Association is involved with issues relating to economic development including business incubators. APA sponsors an annual conference that includes seminars and presentations. Workshops have been offered which deal with the incubator concept. The Association, through its Planning Advisory Service, can provide specific information packets in your interest area. APA has several publications including:

● *Planning Magazine*—a monthly illustrated periodical that includes articles about urban land-use planning. This sometimes features business incubators. The 44-page magazine is $29 a year.

Ben Franklin Partnership (BFP)

Pennsylvania Department of Commerce   717-787-4147
Forum Building, Room 463
Harrisburg, PA 17120

A bureau within Pennsylvania's Department of Commerce, BFP provides the largest state-funded program for the advancement of technology in the United States. Ben Franklin has served as a model for other states pursuing similar efforts. Under this program, Pennsylvania sponsors four incubator centers, each hosting a number of high-tech incubator businesses. Each center draws on its own network of resources including industry, universities, and technical facilities. This bureau also operates a small loan program for incubators that includes low-cost space, business services, and entry/exit procedures. BFP will send you reports about incubators and the program, and staff members will answer questions.

Corporation for Enterprise Development (CED)

1211 Connecticut Avenue, NW   202-293-7963
Washington, DC 20026

A small, nonprofit research organization, CED conducts research and offers technical assistance in the area of small business development. The corporation assists local and state legislatures in creating economic programs and policies. CED is involved with business incubator programs, particularly those projects involving depressed areas and women entrepreneurs. The staff will respond to inquiries. CED has several publications, including reports and surveys on business incubators that can be sent to you upon request.

Council for Community Development (CCD)

10 Concord Avenue   617-492-5461
Cambridge, MA 02138

This for-profit consulting firm provides information about how states and localities can set up business incubators. CCD can provide advice about the availability of financing and recommend policies and programs. It specializes in coordinating the resources of an incubator with current state and local financial plans. The staff can answer your questions and direct you to other sources. Some publications are available at minimal cost.

Industrial Science and Technological Innovation Division (ISTID)

National Science Foundation   202-357-7527
1800 G Street, NW
Washington, DC 20550

ISTID has funded incubator centers for the past ten years. A majority of these centers have supported high-tech enterprises, many of which have matured into successful businesses. ISTI staff can provide information and reports about the incubators.

National Council for Urban Economic Development (NCUED)

1730 K Street, NW, Suite 1009   202-223-4735
Washington, DC 20006

    This national, private, nonprofit membership organization is involved in research concerning business incubators. Areas of interest regarding both public and private incubator facilities include funding, location, management, and types of businesses involved. A report by Mia Purcell, *The Industrial Incubator*, is available, and Ms. Purcell can respond to your questions and make referrals. NCUED publishes:

    • *Incubator Resource Kit*—a packet of materials from various incubator facilities including documents, tenant information on adaptive reuse of facilities, and high-tech and university-affiliated incubators. The cost is $18.

National Development Council (NDC)

1025 Connecticut Avenue, NW, Suite 515   202-466-3906
Washington, DC 20036

    A nonprofit corporation, NDC serves as a consultant in structuring business financing for projects such as incubators. NDC primarily packages loans for entrepreneurs on behalf of states, communities, and nonprofit organizations. Mark Barbash can answer your questions. Some general information describing specific incubator programs is available.

National Economic Development & Law Center

1950 Addison Street   415-548-2600
Berkeley, CA 94704

    The Center provides technical assistance to low-income communities regarding economic development, which sometimes involves incubator programs. Assistance primarily includes legal and tax services. Staff can provide you with some basic information about business incubators and can refer you to other sources.

Office of Advocacy

Office of Private Sector Initiatives (OPSI)   202-634-9841—Jackie Wodard
U.S. Small Business Administration
1411 L Street, NW
Washington, DC 20416

    The Office of Private Sector Initiatives follows state and local activities regarding business incubators. Staff can provide you with information about legislative activities in this area. The office sponsors conferences and educational programs on how to develop and manage successful incubator facilities. Technical assistance is available to communities and organizations sponsoring small business incubators projects. Through SBA's Office of Advocacy and Local Affairs you can get current information on state and legislative initiatives to promote small business incubators. OPSI has established a national data base on incubator projects around the country. Numerous publications are available, including:

● *Small Business Incubators*—a resource summary prepared in cooperation with the National Council for Urban Economic Development. The publication includes an overview of what an incubator is and how it functions. It also provides a listing of SBA publications relating to incubators as well as relevant publications produced by other organizations. A contact list of people and organizations that can supply information is also included. The eight-page publication is available free of charge.

● *Starting a Small Business Incubator*—a handbook for sponsors and developers. The majority of the text concerns state programs available to businesses in Illinois. The book describes the role and characteristics of an incubator and the steps of development including where to get help in starting and managing an incubator. The 60-page publication is available free of charge.

Public Works Program

Office of Public Works   202-377-5113—Public Affairs Office
Economic Development Administration
U.S. Department of Commerce
14th Street and Constitution Avenue, NW, Room 7800B
Washington, DC 20230

The Public Works Program funds incubator facilities through grants to cities, counties, and nonprofit corporations. Generally, the facilities house high-tech industries. The staff can provide general information about existing programs, tell you about eligibility requirements and designated areas for incubators, and refer you to Regional Offices where an Economic Development Representative can provide more specific information.

Publications

Business Incubator Profile: A National Survey

Hubert H. Humphrey Institute of Public Affairs
909 Social Sciences Tower
267 19th Avenue South
Minneapolis, MN 55455

This survey, conducted between 1983 and 1984, describes fifty business incubators, including high-tech companies. Each company is discussed according to when it was established, size, rent, number of jobs created, funding sources, business assistance, job training, and development. The report is expected to be updated at regular intervals. The 130-page publication is available at a cost of $5 for community-based nonprofit organizations, and $15 for all others.

Additional Experts and Resources

David N. Allen, Ph.D.
Assistant Professor of Public Administration   814-865-2536
Institute of Public Administration
205 Burrows Building
Pennsylvania State University
University Park, PA 16802

Dr. Allen's specialty is advanced technology development policies, at both the state and local level. He has conducted workshops and published papers about business incubators.

Candace Campbell
Cooperative Community Development Program   612-376-9798
Hubert H. Humphrey Institute of Public Affairs
909 Social Sciences Tower
267 19th Avenue South
Minneapolis, MN 55455

Ms. Campbell has conducted extensive research on incubators. She coauthored a report, *Business Incubator Profile: A National Survey*, which describes fifty incubators, including high-technology companies. Ms. Campbell is familiar with the planning and establishment of business incubators in both the public and private sectors. Her special interest is in incubators which are set up to benefit local economies, and she is willing to answer questions and refer you to other sources of information.

Peat, Marwick, Mitchell and Co.
303 East Wacker Drive   312-938-1000—Michael Lavin (high tech business
Chicago, IL 60601   incubators), Conrad Van Hazebroek (other
                                                                                   business incubators)

This accounting firm serves in an advisory and consulting capacity regarding incubator programs, especially with financing, general business advice, setting up necessary services, and referrals to lawyers and marketing consultants. The firm has also sponsored and participated in conferences dealing with incubator programs. Michael Lavin's specialty is in high tech incubators, and Conrad Van Hazebroek's is in all other types of business incubators; both of them are willing to answer questions in their areas.

James Roberts, President
Pryde, Roberts and Co.   202-659-9054
1150 17th Street, NW, Suite 607
Washington, DC 20036

Mr. Roberts has experience in the organization, development, financing, and management of incubator facilities. He and his associates are experienced in financing centers and raising seed capital.

Loren Schultz

Technology Centers International (TCI)   215-646-7800
1060 Route 309
Montgomeryville, PA 18936

Mr. Schultz, president of TCI, is an entrepreneur and a business activist who has put together programs to assist small businesses grow through incubators. He has implemented and managed a network of incubators, both in the public and private sectors, and has sponsored seminars about business incubators, about which literature is available.

Mark Weinberg

Ohio University   614-594-5617
Department of Political Science   614-594-5626
Bentley Hall
Athens, OH 45701

Professor Weinberg is a member of the faculty at Ohio University and Director of the Institute for Local Government Administration for Local Development. He has conducted research concerning incubators in rural areas across the country, incubator financing, and government-related incubators. He can answer your questions and direct you to other information sources.

At present there are nearly 100 incubator programs nationwide, and as the concept is catching on, new programs are frequently being started. Described below are five programs representative of high-technology incubators.

Advanced Technology Development Center

Georgia Institute of Technology (GIT)   404-894-3575
O'Keefe Building, Suite 42
Atlanta, GA 30332

The Center, funded by the state of Georgia, specializes in early stage high-tech companies. It provides companies with business planning services, marketing assistance, information on where to go for financial backing, where to get low-cost housing, resources from GIT, and an introduction to the business community. High-tech incubator companies have included veterinary, pharmaceutical, digital communications, software, and medical engineering firms. Staff can answer your questions and send you publications describing the Center and the incubators.

Ben Franklin Technical Center

South Mountain Drive   215-861-0584
Bethlehem, PA 18015

Created by the Pennsylvania State Challenge Grant Program, this Technical Center coordinates comprehensive assistance for entrepreneurs and very small companies. The primary function of the center is to sponsor research and development projects through joint efforts by industry and technical institutions and universities. The incubator companies benefit from the guidance of academic and research experts and free business assistance. Many high-tech companies have been accepted as incubators and have been involved in advancing technological research and applications, including computer-aided design and manufacturing, microelectronics, and biotechnology. Louis Robinson can answer your questions and send you literature about the Center.

Ohio University Innovation Center (OUIC)

1 President Street   614-594-6682
Athens, OH 45701

OUIC is a university-affiliated incubator facility that receives public funds. The clients of the center are generally high-tech companies. Dinah Adkins, Assistant to the Director, can assist you with your inquiries. Publications are available upon request.

University City Science Center

3624 Market Street   215-387-2255
Philadelphia, PA 19104

The Center has sponsored high-tech incubators for several years, with participating companies coming from varied fields: data processing, information processing, computer software, and biomedical products. The Center assists these companies with financial arrangemnts, housing, and business services provided by consulting, accounting, and legal firms. Conferences are sponsored on behalf of incubators. The Center makes available information about the incubators as well as its other services.

Utah Innovation Center

417 Wakara Way   801-583-4600
Salt Lake City, UT 84108

The Center is a private, for-profit corporation that creates high-tech manufacturing companies. It is unique among incubation facilities because it provides risk capital for the companies it assists. Each company also receives space, equipment, management, legal aid, accounting, and technical assistance. Companies have been involved in telecommunications, biomedical products, computer peripherals, and laser imaging. Staff will respond to your questions, as well as send you literature about the Center.

Small Business, Funding for

See: Research and Development; Small Business Incubators; Venture Capital

Smuggling of High Technology

See also: International Competition; Technology Transfer

Organizations and Government Agencies

Computer and Communications Industry Association (CCIA)

1500 Wilson Boulevard   703-524-1360—Edward Black
Arlington, VA 22209

A trade association, CCIA promotes the needs of manufacturers and suppliers of computer and communications equipment. The Association can respond to inquiries concerning U.S. export policy, laws, and regulations. Staff can assist you with any difficulties you might experience in regard to export regulations.

U.S. Department of Defense (DOD)

Public Correspondence Division   202-697-5737
OASD (Public Affairs-PC)
Room 2E777 The Pentagon
Washington, DC 20301-1400

This office responds to public inquiries regarding the activities of the Department of Defense. Staff can answer your questions and can refer you to the most current unclassified DOD publications. DOD has prepared several reports about U.S. interest in, and efforts to prevent, the illegal flow of critical high technologies, particularly those which have military implications. Under Congressional mandate, DOD has been submitting an annual report to Congress entitled *The Technology Transfer Control Program*, which summarizes DOD's efforts to control high-tech smuggling. The report is available free of charge as long as the supply lasts.

Electronic Industries Association (EIA)

2001 Eye Street, NW   202-457-4924—Alan Spurney, Director of International Business
Washington, DC 20006   202-457-4940—Publications

This trade association represents U.S. manufacturers of electronic components and equipment. It helps companies establish, within their own organization, an export compliance system. EIA also provides companies with information about how to avoid inadvertant stoppages by the State Department's Office of Munitions Control. An EIA publication of particular interest is:

• *A Practical Guide on Export Licensing Procedures Under Munitions Control*—this 37-page book serves as a supplement to the International Traffic and Arms Regulations. It explains, in lay terms, licensing requirements and procedures as well as other regulations. It was prepared by EIA in cooperation with the Office of Munitions Control. The cost is $10 for members and $20 for nonmembers.

Office of Munitions Control (OMC)
PM/MC, Room 800, SA6   202-235-9756
U.S. Department of State
Washington, DC 20520

OMC administers export control of defense articles and services originating in the U.S. It provides licenses and other forms of approval for export. The office can answer questions about whether a product or service is considered to have military implications. Staff will provide companies with advice, in the form of written evaluations, about licensing approval of their products or services. This office recommends two publications:

• *International Traffic in Arms Regulations* (22CFR, parts 121-128)—the operational guide for OMC, which contains all the regulations, policies, and procedures for the functioning of OMC. It also explains how to prepare for licenses and agreements. The cost of this guide (order #044-000-01603-6) is $4.50 and can be ordered from the Government Printing Office, Superintendent of Documents, Washington, DC 20402, 202-783-3238.

• *OMC Newsletter*—published irregularly, this newsletter contains information about regulations and policy changes. It is available to all manufacturers and exporters registered with OMC as well as organizations involved in munitions exportation. The eight-page publication is available free of charge from OMC. Back issues are available.

Operation Exodus
U.S. Customs Service   202-566-5286—Public Affairs Office
Office of Enforcement
1301 Constitution Avenue, NW
Washington, DC 20229

Operation Exodus is an enforcement program designed to stop illegal high-tech exports, including munitions, to Eastern bloc countries. Staff can answer your questions concerning the illegal transfer of high technology. They can also refer you to other information sources in the government and elsewhere. Press releases, factsheets, business advisories, and statistics are available from this office.

Office of Export Enforcement
International Trade Administration (ITA)   202-377-3618
U.S. Department of Commerce
14th Street and Constitution Avenue, NW
Washington, DC 20230

This office is responsible for administering export regulations, as cited in the Export Administration Act, as well as protecting domestic industries from unfair trade practices by other nations. ITA's Office of Export Administration regulates the flow of high-technology commodities, services, and technical data applicable to civilian and military uses. The office can give you information about regulations and licensing procedures, as well as product and service export eligibility status. ITA offers seminars and consultations to help businesses become aware of the indications of potential illegal exports. A publication of special interest is:

● *Business America—ITA Export Services*—a special reprint of four magazine articles about the International Trade Administration. It offers information about export regulation, antiboycott compliance, support services, and fair trade. Available free of charge upon request.

Publications

The Bureau of National Affairs, Inc.
1231 25th Street, NW   202-452-4286—Deanne E. Neuman, Managing Editor
Washington, DC 20037

The Bureau publishes reports, analyses, and briefings on international trade developments as well as relevant text and reference materials. ITR's four publication services are:

● *International Trade Reporter Current Reports*—a weekly publication that provides news of the important developments during the week concerning all aspects of international trade including export and import policy, and legal actions. It averages 30 pages an issue, and an annual subscription is $388.

● *International Trade Reporter Reference File*—bimonthly guide to statutes, regulations, and related reference materials that covers the entire import process. It includes analyses of the import procedure, duties, import competition, regulated trade administration and enforcement, and negotiations. Each issue is approximately 231 pages, and an annual subscription is $296.

● *International Trade Reporter Decisions*—this biweekly publication includes digests and full texts of judicial and administrative decisions in major fields of import law. These include rulings of the International Trade Commission and Commerce Department. It averages 50 pages per issue and an annual subscription is $296.

● *International Trade Reporter Export Shipping Manual*—A comprehensive, three-volume loose-leaf reference that is updated weekly. It covers foreign import regulations and documenting requirements as well as U.S. export controls. The manual is designed to facilitate export shipments to every country in the world. An annual subscription is $296.

Additional Experts and Resources

Steve Hirsch
Bureau of National Affairs   202-452-4445
1231 25th Street, NW
Washington, DC 20036

Mr. Hirsch is the senior staff editor of the *International Trade Reporter*, published by the Bureau of National Affairs and described above. As a journalist, he has covered U.S. export controls and issues having to do with the Export Administration Act. He can answer your questions about U.S. policy regarding high-tech smuggling, as well as refer you to other sources.

Mr. Eduardo Lachica
Wall Street Journal   202-862-9200
1025 Connecticut Avenue, NW, Suite 800
Washington, DC 20036

Mr. Lachica is a reporter who covers the export control policies of the U.S. for the *Wall Street Journal*. He has written many articles about high-tech smuggling, and can refer you to other information sources on the topic.

The Export Administration Act, which contains provisions affecting high technology, has been under congressional consideration for extension and amendment. Listed below are the House and Senate subcommittees that have jurisdiction for reviewing the Act. Committee staff can respond to your inquiries concerning export policy and current legislative activity. They can also send you copies of congressional hearings and reports dealing with export legislation. Since the two subcommittees differ greatly in their stand on the issues, it would be wise to contact both for information.

Subcommittee on International Economic Policy and Trade
Committee on Foreign Affairs
702 House Annex #1
Washington, DC 20515
202-226-7820

Subcommittee on International Finance and Monetary Policy
U.S. Senate Committee on Banking, Housing, and Urban Affairs
SD-534
Dirkson Senate Office Building
Washington, DC 20510

Social Impact of Technology

Begins next page. See also: Bioethics; Environmental Impact Assessment; Ergonomics; Futures; Technology Assessment; Technology Transfer; Technophobia

Organizations and Government Agencies

Computers and Society (CAS)

Association for Computing Machinery (ACM)   212-869-7440
11 West 42nd Street
New York, NY 10036

CAS is a special-interest group within ACM whose members are computer programers, analysts, engineers, and others interested in computers and their impact upon society. CAS publishes:

● *Computers and Society*—a quarterly newsletter containing articles on such subjects as computer crime, computers in the workplace, the economic effects of computers, computers in the classroom, and computer ethics. Annual subscription is $8 for members, and $18 per year for nonmembers.

IEEE Society on Social Implications of Technology (SIT)

345 East 47th Street   212-864-5046
New York, NY 10017

SIT is a transnational organization of scientists and engineers who are concerned with the impact of technology on society. Its interests include both the positive and negative effects of technology, the impact of society on the engineering profession, the history of societal aspects of electrotechnology, social and economic responsibility, and professional ethics. Staff can answer questions and make referrals. The Society's principle publication is:

● *IEEE Technology and Society Magazine*—a 36-page quarterly journal with articles on the societal impact of computers, telecommunications, and electrotechnology. The annual subscription is $28.

International Association for Impact Assessment (IAIA)

Industrial and Systems Engineering   404-894-2330—Alan Porter
Georgia Institute of Technology
Atlanta, GA 30332

IAIA is an international association of 500 people involved in studying technology assessment, including the social and environmental impact of technology. It is an excellent resource for information about the impact of technology upon society. Its members include scientists, academicians, and business executives involved in planning activities. Staff will answer public inquiries and make referrals. IAIA holds meetings and annual conferences, and publishes the following:

● *Impact Assessment Bulletin*—a 100-page quarterly geared primarily toward technology assessment practitioners and academicians. The bulletin covers technology assessment, including risk assessment, social impact assessment, and environmental assessment. Articles range from technical papers to speeches. Book reviews, a resource section, news of professional opportunities, and a calendar of events are also included. Subscription to the bulletin is included in the membership fee of $18 a year for individuals and $38 a year for institutions. The annual subscription for nonmembers is $38.

Society for Risk Analysis (SRA)

1340 Old Chain Bridge Road, Suite 300   703-790-1745
McLean, VA 22101

SRA was established to promote knowledge and understanding of risk analysis technology. It is a diverse group involved in the question of risks to health, safety, and environment. SRA represents health sciences, engineering, physical sciences, social sciences, economics, and many other fields. SRA members will answer questions and/or make referrals. Publications include:

• *Risk Analysis: An International Journal*—a 100-page quarterly covering risk analysis. Topics have included how democratic societies make decisions, indoor air pollution, highway safety issues, and how people behave in the presence of risk. Membership fee of $30 includes a complimentary subscription to the journal. Nonmember subscription rate is $60 a year.

Additional Experts and Resources

Rachelle Hollander, Ph.D.

Ethics and Values in Science and Technology   202-357-7552
National Science Foundation (NSF)
1800 G Street, NW
Washington, DC 20550

Dr. Hollander directs NSF's program supporting research and related activities that examine the ethical and value issues associated with current scientific, engineering, and technical activities. Funded projects involve scientists, engineers, scholars, consultants, and groups affected by specific technological developments. Dr. Hollander and her staff can provide information about ongoing social impact research and experts in the field.

Rowan Wakefield

American Family Council   202-635-5487
Cardinal Station
Washington, DC 20064

Rowan Wakefield is an expert on the changes technology has brought to family life and society, as well as how family use of home computers is shaping the development of new technologies. He writes a monthly column on the topic, has published articles, and is currently compiling a book on the subject. His pubications, available from the American Family Council, include:

• *The Impacts of Home Computer Use on the Professions Serving Families*— a monograph series, the first of which focuses on the mental health profession. Material was gathered from a survey of 36 health professionals, including psychiatrists, psychologists, and social workers. This 36-page publication is available for $12.50. Individual monographs are planned for each profession.

• *Families and Telematics*—a bound volume published each September, this contains articles and annotated bibliographies about the impact of families and computers on each other. The bulk of the volume is a collection of the monthly family and telematics column Rowan Wakefield writes for the *American Family Newsletter*. The cost is $10.

Many large established technical societies are starting to form committees to study the impact of their particular technology on society. Often these committees are good information sources. Many produce reports, newsletters, and journals, as well as answer questions and make referrals.

Software Engineering

Organizations and Government Agencies

Goddard Space Flight Center
National Aeronautics and Space Administration (NASA)   301-344-6846
Code 702
Greenbelt, MD 20771

The Center maintains several different ongoing research software engineering projects, some of which are developed in conjunction with the Computer Science Department of the University of Maryland. The personnel can be helpful and refer you to other sources of information.

Office of Computer Software and Systems (CSS)
Office of the Under Secretary of Defense for   202-694-0208
Research and Advanced Technology
U.S. Department of Defense (DOD)
The Pentagon
Washington, DC 20301

CSS is responsible for the management of the ADA language and Software Technology for Adaptable Reliable Systems (STARS). ADA is a modern, high-order computer programming language created to be the standard language for software for computer applications at DOD. It was designed to help reduce the costs of military software systems and to improve the quality of software engineering. The ADA Information Clearinghouse gathers and disseminates information, and manages the collection and distribution of documentation on all aspects of the ADA language. It maintains a public data base file to announce recent activities and general information on the program. For instructions on obtaining access to the file through MIL-NET or TELENET, contact Cynthia Hillman at 202-694-0210.

Rocky Mountain Software Engineering Institute

1670 Bear Mountain Drive　　　　　　　　　　　　　　　　　　303-499-4783
Boulder, CO 80303

This nonprofit educational and research organization is involved in software engineering. A small newsletter is published on a monthly basis. Mr. Riddle, a founder of the Institute, can answer many questions and refer you to other good sources of information. The Institute also sponsors an annual two-week tutorial program in software engineering.

SIGSOFT—Special Interest Group for Software Engineering

Association for Computing Machinery　　　　　　　　　　　212-869-7440
11 West 42nd Street
New York, NY 10036

SIGSOFT is devoted to the collection of techniques and principals involved in the production of high quality software for practical identification of, and search for solutions to, problems inherent in program development. It holds an international meeting every 18 months, and 10 to 20 meetings and symposia yearly. Membership in SIGSOFT is $6 a year. SIGSOFT publishes:

● *Software Engineering Notes*—a quarterly publication of 200-300 pages devoted to high-quality software. Relevant areas of interest include methodologies for design and implementation, program techniques, tools, software economics, portable programs, and program validation. It includes two special research and development issues a year. Annual subscription rate is $6 a year for members, $4 for students, $12 for educational institutions and libraries, and $14 for nonmembers.

Software Engineering Technical Committee

IEEE Computer Society　　　　　　　　　　　　　　　　　　301-589-8142
1109 Spring Street, Suite 300
Silver Spring, MD 20910

The Software Engineering TC encourages the application of engineering methods and principles to the development of computer software, and works to increase professional knowledge of techniques, tools, and empirical data to help improve software quality. The Committee sponsors a major conference, the International Conference on Software Engineering, and several informal workshops every year. A quarterly newsletter, *Software Engineering*, informs all members of current events and technical results.

Publications

IEEE Computer Society

1109 Spring Street, Suite 300   301-589-8142
Silver Spring, MD 20910

The Computer Society publishes the following that deal with software engineering:

- *IEEE Transactions on Software Engineering*—this bimonthly journal carries fundamental and original research papers on all aspects of the specification, design, development, testing, maintenance, and documentation of software systems. Specific areas of interest include programming methodology, software testing and validation, performance and design evaluation, programming environments, and hardware and software monitoring. Annual subscription rate is $100 a year.

- *Software Magazine*—this quarterly magazine presents tutorials and surveys on current techniques and new products in software design and development. It focuses on such topics as software tools, measuring program reliability, designing software tests, PCs as programming workstations, localization of bugs, and making programs readable. Available for $50 a year.

Journal of Systems and Software

Elsevier Science Publishing Co., Inc.   212-867-9040
52 Vanderbilt Avenue
New York, NY 10017

Issued quarterly, the *Journal* that presents technical papers covering all aspects of program methodologies, software engineering, and related hardware/software issues. Topics of interest include software tools; multiprocessing; real-time and distributed systems; and techniques for developing, validating, and maintaining software. It also includes research papers, surveys, and representations of practical papers. Annual subscription cost for the four issues is $77.

Software Practice & Experience

c/o Expediters of the Printed Word, Ltd.   212-838-1364
515 Madison Avenue
New York, NY 10022

This is a monthly journal, published by John Wiley & Sons, which is oriented toward the software writer. Articles cover software design and implementation, case studies, and critical appraisals of software systems. The emphasis is on practical experience, although articles of a theoretical or mathematical nature are also included. The subscription price is $225 a year.

Data Bases

Software Engineering Laboratory Data Base Maintenance System (SEL/DBAM)

COSMIC—NASA's Computer Software Management and Information   404-542-3265
  Center
112 Barrow Hall
University of Georgia
Athens, GA 30601

SEL/DBAM is a data base for evaluating software development methodologies. It has over 17,000 records. The data base is accessible only through COSMIC, and the program can be purchased for $2,400.

Additional Experts and Resources

Several universities and institutes across the country conduct research in software engineering:

University of Massachusetts at Amherst
COINS Department
Lederle Graduate Research Center
Amherst, MA 01003
413-545-1328—Professor Laury Clark

University of Colorado
Department of Computer Sciences
Campus Box 430
Boulder, CO 80309
303-492-6361—Prof. Leon Ostereell

University of Maryland
Computer Science Department
Computer and Space Science Building
College Park, MD 20742
301-454-7690

Wang Institute of Graduate Studies
Tyng Road
Tyngsboro, MA 01879
617-649-9731—Dr. Dick Fairley

USC Information Sciences Institute
4676 Admiralty Way
Marina Del Rey, CA 90292
213-822-1511—Dr. Joel Goldberg

Carnegie-Mellon University
Department of Computer Sciences
Pittsburgh, PA 15213
412-578-2565—Dr. Mario Barbacci

University of California at Irvine
Department of Information & Computer
  Sciences
Irvine, CA 92717
714-856-7403

Stanford Research Institute
(SRI International)
Menlo Park, CA 94025
415-859-2859—Dr. Peter Neuman

William Adrion
National Science Foundation   202-357-7375
1800 G Street, NW
Washington, DC 20550

Mr. Adrion, a member of the software engineering special interest group at the Association for Computing Machinery, is an excellent source of information on software engineering. He can answer questions, and can tell you where to get further information.

Professor David R. Hanson
Department of Computer Science
University of Arizona
Tucson, AZ 85721

Professor Hanson is the U.S. editor of the journal *Software Practice and Experience*. He will answer technical inquiries.

Software Guides, Directories, and Reviews

Organizations and Government Agencies

COSMIC—NASA's Computer Software Management & Information Center
112 Barrow Hall   404-542-3265
University of Georgia
Athens, GA 30601

This Center has produced more than 1,350 computer programs for NASA's projects. The programs are divided into 75 subjects, including data base management, optics, structural mechanics, thermo-analysis, fluid flow, and others in engineering and other high-technology topics.

All the programs are listed in the Center's annual publication, *Cosmic Software Catalog*, which is available in microfiche for $10, on microcomputer tape for $50, and in printed copy for $25.

If you are interested in particular computer programs or applications, contact the Center and specialists will check to see if they have the type of programs you need. Prices for the programs range from $100 to $4,000.

National Energy Software Center
Argonne National Laboratory   312-972-7250
9700 South Cass Avenue
Argonne, IL 60439

The Argonne National Laboratory produces a listing of about 1,000 software packages from the National Energy Software Center. Contact the Center for further information about subscription rates and plans.

PC Telemart, Inc.

11781 Lee-Jackson Highway
Fairfax, VA 22033

703-352-0721
800-368-4422
800-552-4422—in Virginia

This company runs the PC Software Resource Center, a reference and research facility that houses a collection of thousands of microcomputer software programs and associated documents. Users have access to the Center's software and its 30 computers. Members can use the Center free of charge; others must pay $100 a day.

PC Telemart produces an on-line service for retail stores and corporations. It provides on-line access to over 30,000 software packages. Access is handled through Tymnet (703-827-9110).

Telemart publishes the following directories:

• *PC Telemart Software Directory*—an annual containing over 34,000 listings of microcomputer software and products. Available for $34.95.

• *PC Telemart Van Loves's Apple Software Directory*—an annual directory of Apple software. The 965-page directory is $24.95.

Publications

DataPro Research, Inc.

McGraw-Hill Information Systems
1805 Underwood Boulevard
Delran, NJ 08075

609-764-0100

DataPro publishes four major directories on software. Each runs about 200 pages, and includes a monthly 4-page newsletter with the subscription cost. The directories are:

• *DataPro Directory of Software*—a two-volume directory oriented toward mainframe computers. It compares and makes recommendations for over 6,000 software packages. It is updated monthly, and costs $495 a year.

• *DataPro Microcomputer Software*—this two-volume monthly publication is oriented towards microcomputer users, and covers over 4,000 software packages. Available for $525 a year.

• *DataPro On-line Services*—a monthly, two-volume publication that covers in detail more than 100 on-line software programs, as well as 800 data base profiles. Available for $420 a year.

• *DataPro Reports on Microcomputers*—this two-volume monthly presents detailed information on microcomputers, including in-depth coverage of the top 30 hardware products and the top 30 software packages. Reviews of peripherals and data communications equipment are also provided. Available for $600 a year.

Software Catalog

Elsevier Science Publishing Co., Inc.   800-223-2115
52 Vanderbilt Avenue   212-867-9040
New York, NY 10017

This directory, which covers over 40,000 software packages, is a thorough and comprehensive guide to published software. It provides information on availability, price, applications, and compatibility of packaged software. The *Catalog* is published in two series, each twice a year, one for microcomputers ($75 each), and one for minicomputers ($115 each).Updates are published twice during the year. The *International Software Catalog* is available on-line through the Lockheed DIALOG Information Service. Call Lockheed at 800-227-1927 for more information.

Federal Software Exchange Catalog

National Technical Information Service (NTIS)   703-487-4650
U.S. Department of Commerce
5285 Port Royal Road
Springfield, VA 22161

NTIS publishes the *Federal Software Exchange Catalog*, which contains the most up-to-date listings and descriptions of all the software available through the Exchange (a federal software clearinghouse) and NTIS. The catalog presents an abstract and a description of each package, as well as a chapter on software tools, and listings of all the software approved by the General Services Administration for purchase by government agencies. The directory costs $50 an issue, and is accompanied by a supplement available for $25.

List Magazine

Redgate Publishing Company   305-231-6904
3381 Ocean Drive
Vero Beach, FL 32963

A monthly publication, this magazine covers microcomputer software that deals with business solutions. The magazine contains partial directories of software products. Each year it publishes a compilation of the directories presented in each monthly issue. The subscription price is $24.95 a year.

Meckler Publishing

520 Riverside Avenue   203-226-6967
Westport, CT 06880

This company produces:

● *SR Software Review*—published quarterly, *SR* provides the education and library community with information about prewritten software. Each issue features articles on software concepts, evaluation, and selection; reports on available software products suitable for education and library applications; reviews of books and other recent publications pertaining to software; and provides a listing of software developed by library and educational organizations. The annual subscription is $58.

• *Software Publishers' Catalog Annual*—this microfiche directory lists software publishers' catalogs. It is accompanied by a printed index that lists each software publisher, in alphabetical order, and also the 40 different categories covered. Available for $97.50.

Personal Software

Hayden Publishing Company   201-393-6165
10 Mulholland Drive
Hasbrouck Heights, NJ 07604

The monthly magazine reviews 25 software products in each issue. These reviews are like software demos in print, with color photos of screens, documentation, and even packing. Each issue includes coverage of major software categories (educational, home financial/home control, recreational), productivity, coverage of all new software releases, and a comprehensive monthly buyer's guide. The annual subscription rate is $24.

SoftKat Educational Computer Software Guide

SoftKat, Inc.   213-781-5280
15015 Oxnard Street   800-641-1057
Van Nuys, CA 91411

This directory provides complete reviews of the best educational software programs available. It is a quick reference guide listing programs by subject area and grade level, with all the information needed to make a buying decision. It is available for $75.

Software Merchandising

Eastman Publishing Company, Inc.   213-995-0436
15720 Ventura Boulevard, Suite 222
Encino, CA 91436

This monthly trade magazine is written for current and potential retailers of home computer software and accessories. It provides information on software selling, and covers the new trends involved in the marketing of house computer software. The December issue includes a software directory. Available free of charge to retailers, manufacturers, and distributors, or for $18 a year for others.

Software Reports

Allenbach Industries   619-438-2258
2101 Las Palmas Drive, Department TJ   800-854-1515
Carlsbad, CA 92008

The publication evaluates educational programs for Apple, Atari, the IBM PC, Commodore, and TRS 80 microcomputers, for all grade levels, preschool through college. Twenty subject areas are covered including mathematics, science, language arts, history, computer literacy, and foreign languages. The manual is regularly updated and is available for $59.95.

Swift's Educational Software Directory

Apple II Edition or IBM Edition   512-282-6840
Sterling Swift Publishing Company
7901 South IH-35
Austin, TX 78744

This directory contains over 400 pages of software listings specifically suited for the educational market. It is organized by subject, publisher, and title. Subjects covered include administration, basic living skills, business education, college preparatory and career search, computer literacy, courseware development, early childhood education, language arts, library skills, special education and statistics, and graphing utilities. Curriculum subjects covered include fine arts, foreign languages, geography, history and political science, mathematics, and science. The spiral-bound directory is $18.95.

Whole Earth Software Catalog

Quantum Press/Doubleday   516-294-4561
Doubleday and Company
501 Franklin Avenue
Garden City, NY 11530

The catalog identifies and comparatively describes personal computer products, with emphasis on software. The reviews are based upon actual experience with the products, including hardware, magazines, books, accessories, suppliers, and on-line services. The catalog is 200 pages long and costs $17.50.

Data Bases

MENU—International Software Database (ISD)

International Software Database Corporation   303-482-5000
1520 South College Avenue   800-843-6368
Fort Collins, CO 80524

MENU-ISD is a computerized, comprehensive data base of computer software available through on-line systems. It includes computer systems with their compatible programming languages, operating systems, and microprocessors. The ISD also includes an evaluative data base that stores comparative information about both software and hardware. The data base is also available in printed copy.

Solar Energy

See: Solar Technology

Solar Technology

See also: Appropriate Technology; Biomass/Bioconversion; Fuels, Alternative; Geothermal Power and Hydropower; Photovoltaics; Renewable Energy Technology; Waste Utilization; Wind Energy

Organizations and Government Agencies

Florida Solar Energy Center
300 State Road, Suite 401   305-783-0300
Cape Canaveral, FL 32920

The Center conducts research in low-energy building design,photovoltaics, solar domestic water heating, and swimming pool heating. It also does testing and rating of solar equipment. Workshops are held on solar installation, marketing, and other topics of interest to professionals in the industry.

The Public Information Office disseminates free publications, including design notes, energy notes, installation manuals, professional papers, and fact sheets. The Center also has an information referral service and reference library to answer questions.

Materials and Technology Development Branch
Passive and Hybrid Solar Energy   202-252-8110
Solar Heat Technologies
Conservation and Renewable Energy
U.S. Department of Energy (DOE)
Forrestal Building, CE-312.2, Room 5H047
1000 Independence Avenue, SW
Washington, DC 20585

This office manages high-risk and high-payoff research and development, from concept generation through prototype field tests, for solar passive and hybrid materials and components. Research and development activities are conducted in cost-shared projects with industry, DOE national laboratories, private sector companies, universities, and other federal agencies in order to identify and mitigate system technical barriers and to improve technical performance and cost-effectiveness. Activities include research in the thermal sciences and materials and components concerned with the collection, thermal storage, thermal transport, rejection/dehumidification, and appropriate controls associated with the use of passive and hybrid technologies in buildings. Areas of interest with respect to buildings include apertures/building envelopes, thermal storage in the interior structural elements, and the transfer of energy among these elements.

Passive Solar Industries Council

125 South Royal Street   703-683-5003
Alexandria, VA 22314

The primary purpose of the Passive Solar Industries Council is to promote the use of passive solar technology in buildings. Members are large industry organizations; large and small corporations, usually manufacturers of building products; utilities, and individual building professionals. Services include a newsletter, information series, and slide presentations. The Council publishes:

- *Passive Solar News*—a bimonthly newsletter directed toward the building industry. It provides an industry review, including articles on projects, products, resources, and industry news. The cost is $9.95 a year.

- *Passive Solar Trends*—a free information series consisting of reports on specific topics within the passive solar area.

- A slide presentation on passive solar technology, available at a cost of $55 for members, and $75 for nonmembers.

Solar Electric Technologies

Conservation and Renewable Energy   202-252-5540
U.S. Department of Energy
Forrestal Building, CE-33, Room 5E080
1000 Independence Avenue, SW
Washington, DC 20585

This office is responsible for the research and development activities of wind energy systems, ocean energy systems, and photovoltaic energy systems, serving as the solar energy program's focal point and interfacing with the electric utility industry.

Solar Energy and Energy Conservation Bank

U.S. Department of Housing & Urban Development   202-755-7166
451 7th Street, SW, Room 7110
Washington, DC 20410

Many states and U.S. territories now give federally-funded low-interest loans or grants to homeowners who want to install solar energy devices. This program is for low- and moderate-income families owning single- and multi-family dwellings. For specific information about currently available assistance, contact the Solar Energy and Energy Conservation Bank.

Solar Energy Industries Association

1156 15th Street, NW, Suite 520   202-293-2981
Washington, DC 20005

The Solar Energy Industries Association is a national industry trade group including 14 state and regional chapters. Members are manufacturers, dealers, installers, and component suppliers. Services and publications include:

- *Technical Briefs*—these have been written for eight different kinds of solar technologies.
- *Buyer's Guide*—a listing by technology of companies in the solar energy industry. It has descriptions of technologies and covers all kinds of solar energy systems including heating, cooling, and electricity. Each entry includes contacts, phone numbers, and addresses. The cost is $10.
- *SEIA News*—a monthly newsletter covering the state of research and development and all kinds of technical and legislative issues. The cost is $1.50 an issue.

Solar Energy Research Institute (SERI)

1617 Cole Boulevard   303-231-1181—Public Affairs
Golden, CO 80401   303-231-7303—Technical Inquiry Services

SERI is a government-owned, contractor-operated research institute funded by the Department of Energy. It will respond to technical questions from researchers, scientists, engineers, and educators on the state-of-the-art in the solar technologies. Services of SERI include an interlibrary loan service for researchers and scientists in the field as well as a data base which is described separately in this section. The Institute also issues subcontracts based on competitive bidding to qualified persons and organizations. Publications include:

- *In Review*—a monthly technical digest describing recent discoveries and work in research. It lists publications and conferences sponsored, is designed for scientists and engineers in the field of solar energy, and is available free of charge to these professionals.
- *Technical Information Guides* are available in all areas that SERI covers.
- *Catalog of Publications*—presents a selected list of documents published by the Institute since its inception. Available free of charge to scientists in the field.

System Research and Development Branch

Passive and Hybrid Solar Energy   202-252-8121
Solar Heat Technologies
Conservation and Renewable Energy
U.S. Department of Energy
Forrestal Building, CE-312.1, Room 5H047
1000 Independence Avenue, SW
Washington, DC 20585

This office develops integrated passive and hybrid solar heating, cooling, and lighting systems for residential and nonresidential buildings. The branch is responsible for the following activities: concept generation, schematic design, development, and system analysis of residential and commercial building systems; energy management, system control, and performance studies of large-scale systems; and development of large-scale hybrid solar systems for a wide range of building applications in different climatic regions.

Systems Test and Evaluation Branch
Solar Thermal Technology
Solar Heat Technologies
Conservation and Renewable Energy
U.S. Department of Energy
Forrestal Building, CE-314, Room 5H041
1000 Independence Avenue, SW
Washington, DC 20585

This office directs the design, characterization, and operations of solar thermal pilot plants, experimental systems, and field tests. It assesses the technical feasbility of advanced solar thermal systems.

Data Bases

National Climatic Data Center (NCDC)
User Services Branch   704-259-0682
Federal Building
Asheville, NC 28801

The NCDC collects statistical information on solar energy and wind, and other meteorological data. The Center can supply data tapes and publications. As the Center is user-oriented, it frequently prepares custom tapes to meet requestors needs. You do not need specific ordering information about a tape when contacting NCDC; just specify the type of data you need, the time period, and the geographic location. Standard tapes are $99, and custom tapes are $185 for the first reel and $125 for additional tapes in set. Below is a sampling of data tapes relating to solar energy.

• *SOLMET—Hourly Solar Radiation Plus Surface Meteorological Observations* (TD-9724)—a common tape designed to provide solar energy users with quality-controlled hourly solar information and collateral meteorological data in a single Fortran-compatible tape. Each tape consists of an identification portion followed by solar radiation data and then surface meteorological data. Data are primarily from July 1952 to December 1976.

• *Typical Meteorological Year (TMY)* (TD-9734)—solar radiation and surface meteorological data, recorded on an hourly basis, were used as input to form this data file. The TMY is comprised of specific calendar months selected from the entire period of record span for a given station as the most representative, or typical, for that station month. The major parameters that make up this file are solar time; extraterrestrial radiation, including direct, diffuse, net, and global; ceiling height; skycondition and visibility; weather; sea level and station pressure; dry-bulb and dew-point temperatures; cloud amounts; and total and opaque sky cover.

• *POST 1976 SOLMET* (TD-9736)—each tape consists of an identification portion followed by the solar radiation data and then surface meteorological data. The major parameters that make up this file are local standard time; extraterrestrial radiation; direct, diffuse, and net radiation; global radiation on a tilted surface; global radiation on a horizontal surface including observed data, Engineering Corrected data, and Standard Year Corrected data; sunshine; sky conditions; visibility; weather; pressure; temperature; and snow cover indicator. Data are from January 1977 through December 1980.

- *SOLDAY* (TD-9739)—the data in this file are meteorological and are metric conversions of all parameters currently available in the TD-9750 file. The major parameters in this file are WBAN station number, year, month, day, sunrise/sunset, extraterrestrial radiation, global radiation, sunshine, temperature, precipitation, snow, day with weather, and sky cover. Data are from July 1952 through December 1976.

- *Input Data for Solar Systems*—the sources of data in this file are heating–degree days and cooling–degree days extracted from *Climatology of the United States No. 81* (1941-1970) and solar radiation data from the SOLMET magnetic tape TD-9724. The data in this file are monthly and annual averages only based on the entire period of a record. Data are from 1941 through 1976.

- *Historical Sunshine Data* (TD-9788)—the primary source of this file is from historical publications. Data in each of the three files are by station identifier, year, and WBAN Station number, followed by the monthly and annual information. File 1 contains monthly and annual percentages of possible sunshine computed from the data in files 2 and 3.

National Solar Technical Audience File (NSTAF)

Solar Energy Research Institute (SERI)
Technical Information Branch
1617 Cole Boulevard
Golden, CO 80401

The Institute (described earlier in this section) maintains a data base of people who have indicated their interest in solar energy. The data base can be searched by particular solar technologies, i.e., alcohol fuels, biomass energy, photovoltaics, and wind energy. Staff uses the data base to notify people about conferences and new publications in the field of solar energy. Upon request, SERI staff will add your name free of charge to their data base. Staff will also search the data base and supply a printout (on mailing labels) for a charge of $.07 a name.

Additional Experts and Resources

Joan Moody

Solar Lobby   202-466-6350
1001 Connecticut Avenue, NW, Suite 530
Washington, DC 20036

Ms. Moody is an expert in federal and state legislative activities in the area of solar energy.

Solid-State Chemistry and Physics

See: Materials Science

Solid-State Electronics

See: Chips and Integrated Circuits; Microelectronics; Semiconductors

Space Exploration

See: Aeronautics

Space Nuclear Power

Organizations and Government Agencies

DARPA/STO
Defense Advanced Research Projects Agency (DARPA)   202-694-1703
1400 Wilson Avenue
Arlington, VA 22209
    DARPA Program SP-100's objective is to develop the technology for power systems in space for both military and civilian applications. The program is a joint effort of NASA and the Department of Defense, and by the 1990s the system should be operational. William E. Wright is the director of the program and an expert in the area of space nuclear power. He is able to answer questions and make referrals to other experts in the field. Publications are available.

G. A. Technologies
PO Box 85608   619-455-2997—Collin Fisher
San Diego, CA 92138
    This company is under contract with the Department of Energy, the Jet Propulsion Laboratory, and the Defense Advanced Research Projects Agency for the

development of thermionics power systems programs. Collin R. Fisher is project manager of the systems engineering study and an expert in the field.

General Electric Space Systems Division

Product Information   215-962-1260
PO Box 8555
Philadelphia, PA 19101

General Electric, under the sponsorship of DARPA, the Department of Energy, and NASA, examines the options of nuclear reactor-based space power systems. The SP-100 program base line system is scheduled for partial space operations by 1991 to meet smaller energy requirements. A free one-page brochure is available that provides an overview of the program and the developmental objectives.

Jet Propulsion Laboratory (JPL)

4800 Oak Grove Drive   818-354-5011
Pasadena, CA 91109

The JPL is working in conjunction with NASA on the SP-100 program to study the civilian and commercial uses of a manned space station. They are developing dynamic power conversion systems able to produce hundreds of kilowatts of power. The work is classified, but staff may be able to refer you to other information sources.

NASA's Lewis Research Center

21000 Brookpark Road   216-433-4000
Cleveland, OH 44135

NASA is involved in a cooperative program with DARPA and the Department of Energy in the development of the SP-100 program. The Jet Propulsion Laboratory directed NASA's role in the development of the dynamic power conversion system. Dr. Joe Sovie is the director of the NASA program and an expert in the field.

Rockwell International

Atomic International Programs   818-700-4267
8900 DeSoto Avenue
Canoga Park, CA 91304

Rockwell International is involved in the research, development, testing, manufacturing, and operations of nuclear power systems for terrestrial and space applications. It works in collaboration with the Department of Energy and the Department of Defense to develop advanced reactors for space applications. J. O. Gylfe, the Director of Business Development and External Affairs, is an expert in the field and is available to answer questions and make referrals to other experts in the field. Publications are available describing the project, but they should be requested from the Department of Energy (described below).

U.S. Department of Energy (DOE)
Office of Nuclear Energy   301-353-3321—Earl Wahquist
NE-54, GTN
Washington, DC 20545

The DOE is involved in a triagency agreement with DARPA and NASA to develop the SP-100 program. Earl Wahquist is the director of the program and oversees the research and development aspects. He can answer questions, and refer you to other sources of information.

Spacecraft and Rocketry

See: Aeronautics

Special Education

See: Computers for the Handicapped

Standards and Measurement

Organizations and Government Agencies

American National Standards Institute (ANSI)
1430 Broadway   212-354-3300—General Information
New York, NY 10018   212-354-3471—Sales Office

A private, nonprofit organization operating in the public interest, ANSI is the coordinating organization for the U.S.'s standards system. The ANSI federation consists of 900 companies, as well as technical and professional organizations. Consumers are represented through ANSI's Consumer Council, which works to foster strong and direct communication between individual consumers and standards-writing committees. Standards are handled in several fields, including construction, electrical and electronics; heating, air conditioning, and refrigeration; information systems; medical devices; mechanical; nuclear; physical distribution; piping and processing; photography and motion pictures; textiles; and welding. ANSI is the clearinghouse and information center for all approved American National Standards. Major ANSI publications include:

• *ANSI Catalog of American National Standards*—an annual catalog of all approved American National Standards. The catalog, as well as the periodic supplements, inform the public of the availability of new and revised domestic standards. The price is $10 a copy plus handling and shipping charges.

• *Standards Action*—a biweekly periodical that reports on standards being considered for approval, final actions, and newly published standards. A package consisting of this publication as well as the *ANSI Reporter* is available for $30 a year.

• *ANSI Reporter*—a biweekly newsletter that informs about policy-level actions of ANSI and the international organizations to which it belongs. It also covers standards-related actions and proposals of the U.S. government.

American Society for Testing and Materials (ASTM)

1916 Race Street   215-299-5400
Philadelphia, PA 19103

A nonprofit organization, ASTM has 140 technical committees that concentrate on writing standards for various fields. The Society also sponsors about 40 symposia on technological subjects each year. The ASTM committees are managed by staff managers who can answer questions and/or make referrals to technical experts on the committees. Major publications include:

• *The Annual Book of ASTM Standards*—this publication contains voluntary consensus documents developed by the 140 committees and 30,000 members around the world. The 66 volumes divided into 16 sections include test methods, specifications, practices, definitions, and guides. Prices vary according to the number of volumes and sections purchased.

• *Special Technical Publications*—these are collections of papers given at ASTM symposia. They are state-of-the-art publications covering a variety of technical subjects. Prices vary from $15 to $75 depending on the number of pages.

Association for the Advancement of Medical Instrumentation (AAMI)

1901 North Fort Meyer Drive, Suite 602   703-525-4890
Arlington, VA 22209

The Association has more than 20 standing technical committees that develop voluntary standards and technical information reports. Areas in which they develop standards include cardiovascular surgery; monitoring and therapy; sterilization and infection control; gastroenterology/urology; neurosurgery; equipment maintenance and design; and anesthesiology. Standards are available for public review and comment as well as for sale in the form of completed publications. AAMI staff can answer general and procedural questions, and refer technical inquiries to the appropriate committees. The *AAMI Resource Catalog of Publications and Services* lists standards, and is available free of charge. Another publication is:

• *AAMI News*—this bimonthly newsletter contains a Standards Monitor section, which is a status report and announcement of standards available for review and sale. The publication is available for $70 a year.

Association for Computing Machinery (ACM)
11 West 42nd Street   212-869-7440
New York, NY 10036

An international educational and scientific society serving the computing community, ACM has recently established standards-developing procedures. Many of the members of ACM's special interest groups (SIGs) have designed standards, and many of the SIGs' representatives serve on standards committees of other organizations. Members involved with standards can be a good source of information on standards. The ACM staff can refer inquiries to the current chairman of ACM's Standards Committee and/or to members of special interest groups.

National Bureau of Standards (NBS)
Laboratories and Centers   301-921-1000
Route 270
Gaithersburg, MD 20899

NBS provides advisory and consulting services to assist government and industry in the development of standards. As the national reference for physical measurement, NBS produces measurement standards data necessary to create, make, and sell U.S. products and services at home and abroad. Staff members work closely with industry and consumers at every level; therefore, they are a good resource for information about the latest technological developments in their specialty area. Generally, a staff member can lead you to major companies, research centers, experts, and literature. Listed below are NBS's laboratories and centers, which can provide scientific and technological services as well as measurement, instrumentation, and standards information.

The National Engineering Laboratory
Technology Building, Room B119
National Bureau of Standards (NBS)
Route 270
Gaithersburg, MD 20899
301-921-3434

The Laboratory provides technology and technical services to the public and private sectors to address national needs and to solve national problems; conducts research in engineering and applied science in support of these efforts; builds and maintains competence in the necessary disciplines required to carry out this research and technical service; develops engineering data and measurement capabilities; provides engineering measurement traceability services; develops test methods and proposes engineering standards and code changes; develops and proposes new engineering practices; and develops and improves mechanisms to transfer results of its research to the ultimate user. The Laboratory consists of the following centers:

Center for Applied Mathematics
Administration Building, Room A438
National Bureau of Standards (NBS)
Route 270
Gaithersburg, MD 20899
301-921-2541

The Center is involved in statistical models, computational methods, math tables, handbooks, and manuals.

Center for Building Technology
Building Research, Room B250
National Bureau of Standards (NBS)
Route 270
Gaithersburg, MD 20899
301-921-3377

The Center is involved in performance criteria and measurement technology for building owners, occupants, designers, builders, manufacturers, and regulatory authorities.

Center for Chemical Engineering
Cryogenics Building, Room 2000
325 Broadway
Boulder, CO 80303
303-497-5108

The Center is involved in measurement data, standards, and services for fluids, solids, and gases.

Center for Electronics and Electrical Engineering
Metrology Building, Room B358
National Bureau of Standards (NBS)
Route 270
Gaithersburg, MD 20899
301-921-3357

The Center is involved in engineering data, measurement methods, standards, and technical services.

Center for Fire Research
Polymers Building, Room A247
National Bureau of Standards (NBS)
Route 270
Gaithersburg, MD 20899
301-921-3143

The Center is involved in engineering data, methods and practices, measurement, and test methods for fire safety.

Center for Manufacturing Engineering
Metrology Building, Room B322
National Bureau of Standards (NBS)
Route 270
Gaithersburg, MD 20899
301-921-3421

The Center tests basic metrology, automation, and control support for discrete part manufacturers and others.

Law Enforcement Standards Laboratory
Physics Building, Room B157
National Bureau of Standards (NBS)
Route 270
Gaithersburg, MD 20899
301-921-3161

The Laboratory in involved in technical data for standards used by law enforcement officials to evaluate commercial products.

The National Measurement Laboratory

Physics Building, Room A363
National Bureau of Standards (NBS)
Route 270
Gaithersburg, MD 20899
301-921-2828

This Laboratory provides the national system of physical, chemical, and materials measurement; coordinates the system with measurement systems of other nations and furnishes essential services leading to accurate and uniform physical and chemical measurement throughout the nation's scientific community, industry, and commerce; conducts materials research leading to improved methods of measurement, standards, and data on the properties of materials needed by industry, commerce, educational institutions, and government; provides advisory and research services to other government agencies; develops, produces, and distributes Standard Reference Materials; and provides calibration services. The Laboratory consists of the following centers:

Center for Analytical Chemistry

Chemistry Building, Room A309
National Bureau of Standards (NBS)
Route 270
Gaithersburg, MD 20899
301-921-2851

The Center is involved in the measurement methods and services for analysis of chemicals of importance in industry, medicine, energy, and pollution control.

Center for Basic Standards

Physics Building, Room B160
National Bureau of Standards (NBS)
Route 270
Gaithersburg, MD 20899
301-921-2001

This Center develops and maintains the scientific competences and laboratory facilities necessary to preserve and continue to refine the base physical quantities upon which the nation's measurement system is constructed. It improves, maintains, and transfers the measurement base for time, frequency, electricity, temperature, pressure, mass, and length.

Center for Chemical Physics

Chemistry Building, Room B162
National Bureau of Standards (NBS)
Route 270
Gaithersburg, MD 20899
301-921-2711

This Center is involved in the measurement methods and services for industry in surface science, molecular spectroscopy, chemical kinetics, and thermodynamics.

Center for Materials Science

Materials Building, Room B308
National Bureau of Standards (NBS)
Route 270
Gaithersburg, MD 20899
301-921-2891

The Center is involved in measurement methods and services to evaluate materials such as gasses, ceramics, metals, and polymers; and to deal with failure and substitution questions.

Center for Radiation Research
Radiation Physics Building, Room C229
National Bureau of Standards (NBS)
Route 270
Gaithersburg, MD 20899
301-921-2551

The Center develops methods, calibrations, and products essential to the measurement of radiation for health care, nuclear energy, radiation processing, and radiation safety.

The Institute for Computer Sciences and Technology (ICST)
Technology Building
National Bureau of Standards (NBS)
Route 270
Gaithersburg, MD 20899
301-921-3151

The Institute conducts research and provides scientific and technical services to aid federal agencies in the selection, acquisition, application, and use of computer technology to improve effectiveness and economy in government operations. ICST carries out this mission by managing the Federal Information Processing Standards Program, developing federal ADP standards guidelines, and managing federal participation in ADP voluntary standardization activities. It provides scientific and technological advisory services and assistance to federal agencies, and it provides the technical foundation for computer-related policies of the federal government. The Institute consists of the following centers:

Center for Programming Science and Technology
Technology Building, Room A247
National Bureau of Standards (NBS)
Route 270
Gaithersburg, MD 20899
301-921-3436

The Center is involved in standards and guidelines related to programming languages, software engineering, data management, computer security, and systems selection and evaluation.

Center for Computer Systems Engineering
Technology Building, Room A231
National Bureau of Standards (NBS)
Route 270
Gaithersburg, MD 20899
301-921-3817

The Center is involved in standards and guidelines related to computer system components, computer network protocols, local and area networks, and computer-based office systems.

National Standard Reference Data System (NSRDS)
Office of Standard Reference Data (OSRD)
A323 Physics Building
National Bureau of Standards (NBS)
Route 270
Gaithersburg, MD 20899
301-921-2228—OSRD Reference Center

This Office is responsible for managing and coordinating the National Standard Reference Data System (NSRDS), a decentralized network of Data Centers and short-term projects conducted at universities, government laboratories, industrial laboratories, and NBS technical divisions. The program's aim is to provide numerical data to the scientific and technical community. Its centers and special projects are aggregated into three application-oriented areas: (1) Energy and Environmental Data, which includes data from fields such as spectroscopy, and related research and development and environmental modeling; (2) Industrial Process Data, which primarily covers thermodynamic and transport properties of substances important to the chemical and related industries; and (3) Materials Properties Data, which includes structural, electrical, optical, and mechanical properties of solid materials of broad interest.

NSRDS's principal output is compilations of evaluated numerical data and critical reviews of the status of data in particular technical areas. It also produces annotated bibliographies and procedures for computerized handling of data. NSRDS's data is stored in a computerized system, and made available to the public via publications, magnetic tapes, and on-line retrieval systems. Staff in the above office will answer questions and/or make referrals. Publications lists, as well as information about tapes, printouts, and data base searches, are also available from this office.

Listed below are the NSRDS Data Centers throughout the United States. Staff will supply information and conduct computerized data base searches for the public. A cost-recovery fee may be charged for searches.

Alloy Phase Diagram Data Center
Dr. Kirit Bhansali
Center for Materials Science
Materials Building, Room B150
National Bureau of Standards
Route 270
Gaithersburg, MD 20899
301-921-2982

Aqueous Electrolyte Data Center
Dr. David Smith-Magowan
Center for Chemical Physics
Chemistry Building, Room B348
National Bureau of Standards
Route 270
Gaithersburg, MD 20899
301-921-2108

Atomic Collision Cross Section Data Center
Dr. Jean Gallagher
Joint Institute for Laboratory Astrophysics
University of Colorado
Boulder, CO 80309
303-492-7801

Atomic Energy Levels Data Center
Dr. W. C. Martin
Center for Radiation Research
Physics Building, Room A167
National Bureau of Standards
Route 270
Gaithersburg, MD 20899
301-921-2011

Atomic Transition Probabilities Data Center
Dr. W. L. Wiese
Center for Radiation Research
Physics Building, Room A267
National Bureau of Standards
Route 270
Gaithersburg, MD 20899
301-921-2071

Center for Information and Numerical Data Analysis and Synthesis (CINDAS)
Dr. C. Y. Ho
Purdue University
CINDAS
2595 Yeager Road
West Lafayette, IN 47906
317-494-6300
Direct inquiries to: Mr. W. H. Shafer

Chemical Kinetics Information Center
Dr. John J. Herron
Center for Chemical Physics
Chemistry Building, Room A147
National Bureau of Standards
Route 270
Gaithersburg, MD 20899
301-921-2565

Chemical Thermodynamics Data Center
Dr. David Garvin
Center for Chemical Physics
Chemistry Building, Room A158
National Bureau of Standards
Route 270
Gaithersburg, MD 20899
301-921-2773

Corrosion Data Center
Dr. Gilbert Ugiansky
Center for Materials Science
Materials Building, Room B266
National Bureau of Standards
Route 270
Gaithersburg, MD 20899
301-921-2811

Crystal Data Center
Dr. A. D. Mighell
Center for Materials Science
Materials Building, Room A221
National Bureau of Standards
Route 270
Gaithersburg, MD 20899
301-921-2950

Diffusion in Metals Data Center
Dr. John R. Manning
Center for Materials Science
Materials Building, Room A153
National Bureau of Standards
Route 270
Gaithersburg, MD 20899
301-921-3354

Fluid Mixtures Data Center
Mr. N. A. Olien
Center for Chemical Engineering
Mail Code 773.00
National Bureau of Standards
Boulder, CO 80303
303-497-3257

Fundamental Constants Data Center
Dr. Barry N. Taylor
Center for Basic Standards
Metrology Building, Room B258
National Bureau of Standards
Route 270
Gaithersburg, MD 20899
301-921-2701

Fundamental Particle Data Center
Dr. Thomas Trippe
Lawrence Berkeley Laboratory
University of California
Berkeley, CA 94720
415-486-5885

High Pressure Data Center
Dr. Leo Merrill
PO Box 7246
University Station
Provo, UT 84602
401-224-0389

Ion Kinetics and Energetics Data Center
Dr. Sharon Lias
Center for Chemical Physics
Chemistry Building, Room A139
National Bureau of Standards
Route 270
Gaithersburg, MD 20899
301-921-2439

Isotopes Project
Dr. Janis Dairiki
Lawrence Berkeley Laboratory
University of California
Berkeley, CA 94720
415-486-6152

JANAF Thermochemical Tables
Dr. Malcolm W. Chase
Dow Chemical Company
1707 Building
Thermal Research Laboratory
Midland, MI 48640
517-636-4160

Molecular Spectra Data Center
Dr. F. J. Lovas
Center for Chemical Physics
Physics Building, Room B265
National Bureau of Standards
Route 270
Gaithersburg, MD 20899
301-921-2023

Molten Salts Data Center
Dr. G. J. Janz
Rensselaer Polytechnic Institute
Department of Chemistry
Troy, NY 12181
518-266-6344

National Center for Thermodynamic Data of Minerals
Dr. John L. Haas, Jr.
U.S. Geological Survey
U.S. Department of the Interior
959 National Center
Reston, VA 22092
703-860-6911

Phase Diagrams for Ceramists Data Center
Dr. Lawrence P. Cook
Center for Materials Science
Materials Building, Room A227
National Bureau of Standards
Route 270
Gaithersburg, MD 20899
301-921-2844

Photon and Charged-Particle Data Center
Dr. Martin J. Berger
Center for Radiation Research
Radiation Physics Building, Room C311
National Bureau of Standards
Route 270
Gaithersburg, MD 20899
301-921-2685

Radiation Chemistry Data Center
Dr. Alberta B. Ross
University of Notre Dame
Radiation Laboratory
Notre Dame, IN 46556
219-239-6527

Thermodynamics Research Center
Dr. Kenneth R. Hall
Thermodynamics Research Center
Texas A & M University
College Station, TX 77843-3112
409-845-4971

Thermodynamic Research Laboratory
Dr. Buford Smith
Department of Chemical Engineering
Washington University
St. Louis, MO 63130
314-889-6011

Office of Physical Measurement Services (OPMS)

Physics Building, Room B362
National Bureau of Standards (NBS)
Route 270
Gaithersburg, MD 20899
301-921-2805—Mr. R. Keith Kirby, Chief

The principal responsibility of OPMS is to provide general administrative support for NBS calibration programs and to provide a central point of contact within NBS for resolving calibration issues. This office is the focal point for responding to customer requests for measurement assurance programs (MAPS) and calibration services. OPMS maintains a data base on all NBS measurement services and provides information to the public about these services. Their primary publication is:

- *The Calibration Services of NBS, Special Publication 250*—this book lists the services and briefly describes them. It also provides a listing of references to the tests themselves, that is, how they are performed. The appendix is a price list for calibration services. Special Publication 250 is available free of charge.

Office of Weights and Measures

Administration Building, Room A617
National Bureau of Standards (NBS)
Route 270
Gaithersburg, MD 20899
301-921-2401

This Office coordinates the weights and measures system in the United States, including state and local government programs. It sponsors the annual National Conference on Weights and Measures, which is held in a different city each year, as well as various committee meetings held in cities across the country throughout the year. The Conference is an association of federal, state, and local jurisdictions, as well as industries involved in the manufacturing, processing, and packaging of products for sale at the retail level. Manufacturers of and service companies for measurement and weighing devices all participate. A brochure describing the organization, procedures, and membership plan of the Conference is available from The National Conference on Weights and Measures, PO Box 3137, Gaithersburg, MD 20760, 301-921-3677.

Working directly with the states and industry, and through the National Conference, the Office of Weights and Measures promotes the standardization of commercial practices; model laws and regulations; and specifications for the design and inspection of measuring and weighing devices. Staff can answer questions and make referrals. The office publishes many technical materials, including handbooks and manuals for weights and measures, administration, and training. An important publication is:

• *NBS Handbook 44—Specification Tolerances and Other Technical Requirements for Weighing and Measuring Devices*—updated annually, this handbook contains the latest technology in the field. It is used by every state and is incorporated into state regulations for use by inspectors of weighing and measuring devices, for example, gas pumps at gas stations, grocery store scales, and highway truck scales. The publication is available for $6 from the Superintendent of Documents, U.S. Government Printing Office, Washington, D.C. 20402, 202-783-3238.

Standards Office

Electronic Industries Association (EIA)   202-457-4966
2001 Eye Street, NW
Washington, DC 20006

An association involved in the development and publication of electronics standards, EIA is concerned with all aspects of electronics. Examples include telecommunications, solid state products, electronic displays and tubes, industrial automation, and consumer electronics. EIA membership consists largely of electronics manufacturers, many of whose representatives serve on various standards-writing committees. EIA staff can answer questions and refer technical inquiries to committee members involved in standards development. A *Standards Catalog* is available from EIA for $5.

Federal Laboratory Consortium (FLC)

U.S. Department of Agriculture   202-447-7185
3865 South Building
Washington, DC 20250

FLC is a national network of 300 individuals from federal labs and centers, some of which deal in the areas of standards and measurements. Through the FLC, the public has access to virtually every aspect of unclassified research activity within the federal goverment. Ted Maher, the Director of FLC, can refer you to an FLC member in your specialty or geographical area. The directories listed below will help you to locate FLC members and laboratories, a number of which are involved in instrumentation, standards, and measurement.

• *Federal Laboratory Directory*—provides information about 388 federal laboratories with ten or more professionals engaged in research and development. Summary data arranged by federal agency and by state provides a broad overview of the federal lab system. For each lab, a contact for obtaining technical information is given by name, address, and phone number. The directory is available free of charge from the Office of Research and Technology, National Bureau of Standards, Room 402, Administration Building, Washington, DC 20234, 301-921-3814.

• *Directory of Federal Technology Resources*—describes the hundreds of federal laboratories, agencies, and engineering centers willing to share their expertise, equipment, and sometimes even their facilities. It also includes descriptions of 70 technical information centers. The Directory is indexed by subject, geographical location, and resource name. The Directory was prepared by the Center for the Utilization of Federal Technology, a division of NTIS. It costs $25, and can be purchased from NTIS, 5285 Port Royal Road, Springfield, VA 22161, 703-487-4650. (Order #PB84-100015/AAW).

• *The FLC*—an annual report that highlights activities of the Federal Laboratory Consortium. Examples include conferences, research projects, and new product development. This document also lists FLC Regional and National Contacts. The publication is available free of charge.

GATT Agreement on Technical Barriers to Trade (Standards Code)
Standards Code and Information Office   301-921-2092
A629 Administration Building   301-921-3200—Hotline
National Bureau of Standards
Route 270
Gaithersburg, MD 20899

The Standards Code and Information Office assists U.S. exporters and manufacturers by keeping them informed of proposed foreign regulations that may affect U.S trade opportunities. This Office has copies of proposed foreign regulations. Its staff can answer questions and make referrals. The GATT Hotline provides a recording with the most recent information on proposed foreign regulations. An important publication is:

• *GATT Annual Report*—describes the activities for the year performed by the Standards Code and Information program in support of the GATT Standards Code. Activities covered include operating the U.S. GATT inquiry point for information on standards and certification activities; notifying the Secretariat of proposed U.S. federal government standards-based rules that may significantly affect trade; assisting U.S. industry with trade-related standards problems; and responding to inquiries or foreign and U.S. proposed regulations. The publication is available free of charge.

IEEE Computer Society
1109 Spring Street, Suite 300   301-589-8142
Silver Spring, MD 20910

West Coast Office:
10662 Los Vaqueros Circle   714-821-8380
Los Alamitos, CA 90720

This large professional society promotes technical interaction through a variety of programs and activities, general interest magazines, scholarly journals, conferences and tutorials, standards, technical committees, and local chapter organizations. An important service of the society is the development of standards that benefit computer users and industry, and a number of society task groups are engaged in developing standards. Standards documents developed by the Computer Society are available from the Society's West Coast Office.

IEEE Standards Department

Institute of Electrical and Electronics Engineers

345 East 47th Street

New York, NY 10017

212-705-7960—General Information

201-981-0060, ext. 398—Publications Sales

A major standards developer in the U.S., the IEEE is involved in the areas of electrical and electronics engineering and computer technology. The standards department staff can answer questions and/or make referrals to technical committees or Standards board members. You can obtain the *IEEE Standards Listing*, a numeric listing of about 500 standards, from the IEEE Standards Department. Specific standards are available from the IEEE Service Center, 445 Hoes Lane, Piscataway, NJ 08854, 201-981-0060, ext 398.

International Electrotechnical Commission (IEC)

Central Office:

3, rue de Varembe

1211 Geneva 20

Switzerland

011-41-22-3401-50

U.S. contact:

American National Standards Institute (ANSI)

1430 Broadway

New York, NY 10018

212-354-3361—Technical Information

212-354-3379—Publication Sales

The authority for standards for electrical and electronic engineering, the IEC is composed of national committees from 42 countries. Each committee is expected to be representative of all the major electrical and electronic interests in its country. The aims of IEC world standards, which cover the whole gamut of electrical and electronic engineering, are to secure unambiguous communication of engineering information, reliability and safety of equipment made by different manufacturers worldwide, and the elimination of unnecessary diversity of components used in the construction of equipment. IEC world standards are prepared by specialized committees, and once adopted are published. Besides its standards publications, IEC publishes these books:

• *IEC Multilingual Dictionary of Electricity*—gives terms and definitions, including those used in standards development. Price is $30.

• *Letter Symbols Handbook*—covers the principles governing conventions for electric and magnetic circuits, including currently approved standard letter symbols in general use. The price is $29.

International Organization for Legal Metrology (OiML)

Office of Product Standards Policy   301-921-3287—David Edgerly
Administration Building
National Bureau of Standards
Route 270
Gaithersburg, MD 20899

   The Organization consists of member nations whose basic purpose is to harmonize legal metrology requirements and to foster international trade. Mr. Edgerly or other staff members can answer questions and/or make referrals to the committees. The Organization holds a general conference usually once every four months. An important publication is:

   ● *OIML Newsletter*—published irregularly, this periodical generally contains brief articles on the status of any harmonization effort going on. The newsletter's coverage includes reports on committee activities. The six-page publication is free.

International Organization for Standardization (ISO)

Central Secretariat   011-41-22-34-12-40
1, rue de Varembe
CH-1211 Geneva 20
Switzerland

U.S. Contact: 212-354-3319—Daniel W. Smith, Director of Operations
American National Standards Institute (ANSI)   212-354-3379—Publications Sales
1430 Broadway
New York, NY 10018

   This nongovernmental organization is the specialized international agency for standardization. Its members are standards bodies from some 90 countries representing more than 95 percent of the world's industrial production. ISO work results in international technical agreements which are published as international standards. A general assembly is held every three years. ISO also supports international collaborative work in special council advisory committees dealing with questions such as product certification, reference materials, consumer needs, and assistance to developing countries. Some of the publications issued by ISO are:

   ● *ISO Catalog*—an annual publication that lists all published ISO standards. It is updated quarterly by cumulative supplements. The price is $25.

   ● *ISO Standards Handbooks*—ISO standards are compiled into 19 handbooks covering certain technical fields. Examples of topics covered include information transfer; measurement units; statistical methods; acoustics, vibrations, and shock; and machine tools.

Publications

Standards Reference Materials Catalog
NBS Special Publication 260   301-921-2045
Office of Standard Reference Materials
B311 Chemistry Building
National Bureau of Standards
Route 270
Gaithersburg, MD 20899

Updated every two years, this free catalog lists and describes 900 real materials which are certified for their chemical or physical properties. Included under the category of standard reference materials are environmental type and clinical reference materials. This office is the source for ordering standard reference materials, as well as the catalog.

Data Bases

National Center for Standards and Certification Information (NCSCI)
A633 Administration Building   301-921-2587
National Bureau of Standards
Route 270
Gaithersburg, MD 20899

NCSCI maintains several data bases with up-to-date information about domestic and foreign standards and certification programs. Keyworded indexes retrieve information such as industry and government standards for electric toasters; test methods for determining various characteristics of fireclay brick; whether the nomenclature used in quality control has been defined; and what nationally recognized organization has developed specifications for a specific product. The data bases are used to answer public inquiries about the existence, source, and availability of standards, and to produce lists, indexes and bibliographies of standards. The system has records on more than 30,000 U.S. standards and references to over 240,000 standards, specifications, test methods, codes, and recommended practices. Searches and printouts are provided free of charge.

Additional Experts and Resources

University research centers and company research and development divisions are often involved in measuring the products and services they provide. Experts, associations, and other sources listed under specific high-technology topics in this directory may also be of assistance. They may be able to provide you with standards, measurements, and instrumentation information.

Star Wars: Antimissile Systems in Space

See also: Military Technology

Organizations and Government Agencies

Air Force High Energy Laser Program
Air Force Office of Public Affairs   202-695-5766
Washington, DC 20330

This Office can provide you with information about the Department of Defense's High-Energy Laser Program, as well as information about laser weapon technology in general. Staff members can answer questions and direct you to the correct government office for further information. Publications distributed by the office include:

• *The Department of Defense High-Energy Laser Program*—a report describing the U.S. military's efforts to develop laser and space laser weapons. It details the technology, program history, and funding of high-energy lasers. The 20-page report, which has illustrations, is available free of charge.

Public Correspondence Branch
Office of Assistant Secretary of Defense for Public Affairs   202-697-5737—Public Inquiries
Department of Defense (DOD)   202-659-0192—Press Inquiries
Pentagon, Room 2E777
Washington, DC 20301-1400

Information about the overall Star Wars defense program, the efforts of the Defense Advanced Research Projects Agency (DARPA) in this area, and the DOD Laser and Space Program can be obtained from this branch. Fact sheets, press releases, congressional testimony, and reports about strategic defense initiative programs and studies are available.

Publications

Defense 84
American Forces Information Service   202-696-5294
1735 North Lynn Street, Room 210
Arlington, VA 22209

This monthly publication, produced by the Department of Defense, covers all aspects of defense technology and applications. Each August issue is devoted to strategic defense initiatives. The publication is distributed free of charge to DOD agencies and contractors. Others can subscribe for $23 a year from the Superintendent of Documents, U.S. Government Printing Office, Washington, DC 20402, 202-783-3238.

Additional Experts and Resources

Several companies, under contract to the Department of Defense, are working on Star Wars-related projects. Staff, especially in the public relations office, may be able to answer your questions and send you literature.

BDM Corporation
7915 Jones Branch Drive
McLean, VA 22l02
703-821-5000

Under contract to the Department of the Army, BDM provides engineering and support services to the High-Energy Laser System Test Facility in White Sands, New Mexico. It works with the Air Force to design hardware and software for controlling laser beams. BDM is also developing special shells to protect U.S. warheads from laser attack.

E. G. & G.
45 William Street
Wellesley, MA 02181
617-237-5100

A design-engineering and consulting firm, E. G. & G. is involved in specialized military research and electronics for Star Wars defense.

Itek Division
10 Maguire Road
Lexington, MA 02173
617-276-2000

This company is involved in the manufacturing and development of satellites, weapon systems, and defense electronics for Star Wars initiatives.

Lockheed-Missiles and Space Company, Inc.
PO Box 504
Sunnyvale, CA 94086
408-742-4321

This company has received the largest share of DOD contracts for the development of satellite-based weapons systems. It is also developing ground-launched projectiles.

Logicon
225 West Fifth Street
San Pedro, CA 90731
213-831-0611

A software engineering company, Logicon is under contract to the Department of Air Force's High-Energy Laser Program and the Air Force's High-Energy Laser and Space Program. It is involved in developing laser-resistant materials and providing technical assistance to help DOD integrate lasers with other weapon systems.

Rockwell International Corporation
600 Grant Street
Pittsburgh, PA 15219
412-565-2000

Rockwell is active in developing Star Wars defense satellites, hardware, and defense electronic supplies.

United Technologies Corporation
1 Financial Plaza
Hartford, CT 06101
203-728-7000

This company is involved in the development of Star Wars defense technologies, weapons systems, and communications.

State High-Tech Initiatives

Publications

Guide to Innovation Resources and Planning for the Smaller Business
Office of Productivity, Technology, and Innovation   202-377-1093
U.S. Department of Commerce
14th Street and Constitution Avenue, NW, Room 4816
Washington, DC 20230

       This guide identifies 135 government offices, including 85 at the state level, that offer assistance to smaller businesses attempting to bring new technologies to market. It is written for both individual companies and state and local government economic development planners. The guide is divided into two parts. The first examines the many steps in the innovation process and the skills and resources needed. The second part was compiled from studies conducted and published by the Office of Technology Assessment and the National Governors Association. (Both studies are described in this section.) It identifies a wide range of resources (federal, state, and private) available to assist the smaller business in areas such as financing, information gathering, and management. This section contains a state-by-state listing of government resources (both state and local) and initiatives in the high-tech area. Contact information is provided. The publication is available for $13.50 from the National Technical Information Service, 5285 Port Royal Road, Springfield, VA 22161, 703-487-4650. Refer to order number PB84-176304/LAH.

Office of Technology Assessment (OTA)
Public Affairs Office   202-224-9241
U.S. Congress
Washington, DC 20510

       A nonpartisan analytical agency, OTA prepares, for the U.S. Congress, objective analyses of major policy issues related to scientific and technological change. Upon request, OTA will send you a free catalog of its publications as well as descriptions of new projects under way. Summaries of most OTA reports are available, at no cost, directly from the agency. Full text of the reports must be purchased from the Superintendent of Documents, U.S. Government Printing Office, Washington, DC 20402, 202-783-3238. Described below are OTA reports covering state high-tech initiatives:

      ● *Census of State Government Initiatives for High-Technology Industrial Development, Background Paper* (Stock #052-003-00912-9, Publication Code OTA-BP-STI-21)—this publication is a directory of 153 state government programs to encourage high-technology development. It includes names and phone numbers of program directors as well as descriptions of program services. It costs $4.50.

● *Encouraging High-Technology Development, Background Paper #2* (Publication Code OTA-BP-STI-25)—this publication provides a discussion of high-technology development initiatives by state and local governments, universities, and private sector groups. It includes a directory of 54 local high technology initiatives with names and phone numbers of program directors. It costs $4.75.

● *Technology, Innovation, and Regional Economic Development* (Publication Code OTA-BP-STI-238)—this report assesses the potential for local economic growth offered by high-technology industries, the types of programs used by state and local groups to encourage development of these industries, and the implications of these programs for federal policy. It costs $4.50.

Technology and Growth—State Initiatives in Technology Innovation
National Governors Association (NGA)   202-624-7880
Hall of the States
444 North Capitol Street
Washington, DC 20001

Based on a 1983 survey of the 50 states, this report presents current state activities that use technological innovation as a tool for economic development. It identifies ways in which states encourage technological innovation through employment and training, education, policy development, university/industry linkage, technical and management support, and economic support initiatives. The 113-page publication is available for $10.50.

Sterility

See: Reproductive Technology

Subscription Television

See also: Telecommunications; Television

Organizations and Government Agencies

Subscription Television Association (STA)
1425 21st Street, NW   202-822-9463
Washington, DC 20036

STA is a trade association for the subscription television (STV) industry. It holds an annual conference and exposition, as well as publishes a membership directory, monthly newsletters, and Washington Reports. *STA News* is the monthly newsletter that covers all aspects of the STA industry.

Publications

Broadcast Marketing Company

450 Mission Street   415-777-5400
San Francisco, CA 94105

This company has published *Changes, Challenges, and Opportunities in the New Electronic Media* by William McGee, Lucy E.Garrick, and Joseph H. Caton, a major five-volume reference guide on different aspects in the new electronic media. Volume I, *The Competitive Scramble for the Pay Television Market*, deals with pay cable, subscription television, multipoint distribution service, satellite master antenna television systems, low-power television, satellite subscription television, and other areas. It is available for $50 per volume, or $295 for the whole set of five volumes.

Broadcasting Publications, Inc.

1735 De Sales Street, NW   202-638-1022
Washington, DC 20006

This company publishes the following:

• *Broadcasting Magazine*—this weekly journal of the communications industry also covers subscription television. Available for $60 a year.

• *Broadcasting Cablecasting Yearbook*—this standard reference volume lists the rates for television stations, the addresses of major programming companies, and courses available in the broadcasting field. The yearbook also includes a brief history of FCC regulations affecting the broadcasting industry. Subscription rate is $80 a year.

Federal Communications Commission (FCC)

1919 M Street, NW   202-632-7000—Consumer Assistance Division
Washington, DC 20554   202-632-7100—Library Branch

The FCC regulates licenses and policies related to the operation of subscription television stations. The FCC has specialists in the areas of licenses and policies regarding subscription television in its television branch (202-632-6357). Those looking for information about the FCC may obtain copies of bulletins such as "Subscription Television," "The FCC in Brief," and "How to Apply for a Broadcast Station," among others. Single copies are available free of charge.

Paul Kagan Associates

26386 Carmel Rancho Lane   408-624-1536
Carmel, CA 93923

This company publishes the following newsletters on subscription television:

• *Cable Television Technology*—issued twice a month, this newsletter deals with technology advances in cable television, and the aspects of construction of new cable systems as well as the rebuilding of existing ones. The subscription rate is $495 for one year.

● *The Pay Television Newsletter*—this semimonthly newsletter on pay television services covers news and economic analysis of subscription television. The subscription rate is $525 a year.

Television Digest, Inc.
1836 Jefferson Place, NW   202-872-9200
Washington, DC 20036

TV Digest is one of the main publishing companies in the broadcasting sector. Some of its publications dealing with subscription television are:

● *Television and Cable Factbook*—is a complete compilation of subscription television stations, the services they provide, the markets they serve, and the equipment used, as well as current information on professional associations, and regulatory agencies. It is available for $165.

● *Television Digest with Consumer Electronics*—this weekly reports on the broadcasting, cable, and consumer electronics sectors. Half of the report is dedicated to broadcast television topics, with coverage of companies, legislation, new technology, conferences, and other items. The subscription rate is $406 a year.

● *Television Action Update*—a weekly addenda to the *TV and Cable Factbook* that presents news of FCC actions affecting television stations. The subscription rate is $187 a year.

● *Cable and Station Coverage Atlas*—this annual reference guide to cable and television station signal coverage and distribution information has one chapter covering subscription television. It is available for $114.

Supercomputers

Organizations and Government Agencies

ACCESS Program
Supercomputing Centers   202-357-9777
Office of Advanced Scientific Computers
1800 G Street, NW
Washington, DC 20550

The National Science Foundation (NSF) established the ACCESS Program to respond to university researchers' need for access to supercomputers. Under this program, NSF contracts with supercomputer centers to allow researchers to use the equipment. Researchers desiring to participate should contact the appropriate NSF Program for their field of research, i.e., aerospace, physics, biology. If you do not know which program to contact, the office listed above can direct you to the appropriate division.

California Institute of Technology

1201 East California Boulevard   213-356-3765
Pasadena, CA 91125

The Physics Department of this major university is currently studying a new approach to supercomputing. Under the direction of Dr. Jeffrey Fox and Dr. Steve Otto, the Institute has initiated a study and implementation of a parallel processing supercomputer linking as many as 64 processors. Each processor is assigned a specific task within a problem solution, solves that part, and then communicates the results to the other processors. Research findings and application theories treated by the Institute have been published in the form of technical journal papers appearing in issues of *Physics Today* and *Physical Review Letters*. Copies of these articles, as well as other reference materials, may be obtained upon request from staff at the Institute.

Carnegie-Mellon University

Computer Science Department   412-578-2617
Pittsburgh, PA 05213

Researchers in the Computer Science Department at Carnegie-Mellon are currently in the process of patenting a large parallel processing supercomputer. Staff members can discuss and give information about currently available hardware and software and make referrals to publications and experts in this new parallel process approach to supercomputers.

Control Data Corporation (CDC)

8100 34th Avenue South   612-853-8100
Bloomington, MN 55420

In addition to its data systems and information services, Control Data also develops and manufactures supercomputers and ultracomputing systems. Sales personnel at CDC headquarters can answer questions and send information about the specifications and capabilities of their equipment. They will also put you in contact with previous purchasers for specific requirements and performance information, and can make referrals to institutions doing research and development in the area of supercomputers.

Courant Institute of Mathematical Sciences

New York University   212-460-7100
251 Mercer Street
New York, NY 10012

The Courant Institute is operated as a research institute by senior NYU faculty members of the departments of mathematics and computer science. There are approximately 300 staff members, most of whom are scientists and have doctoral degrees. The library houses more than 35,000 books, 16,000 volumes of periodicals, 650 microfilm reels, and a large collection of technical reports. Staff at the Computer Center, the laboratory, and the library are all helpful in answering questions about current research as well as with referrals to written materials and experts in the supercomputer field.

Cray Research, Inc.
608 Second Avenue South   612-333-5589
Minneapolis, MN 55402

Cray and its subsidiaries design, develop, manufacture, market, and support large-scale, high-speed supercomputers. Cray staff members at both their manufacturing and developing facility in Chippewa Falls, WI, and their software and support facility in Mendota Heights, MN, will discuss current applications of supercomputers by Cray customers, as well as technological research done by Cray Division. Their government relations office in Washington, DC, can give locations and applications of federally purchased Cray equipment. Sales staff at their regional business centers can give technical information about Cray equipment and software. In addition to the various brochures, product releases, and technical reports, Cray also publishes:

● *Cray Channels*—a quarterly publication that contains feature articles on both Cray and other supercomputers, and their applications in research and industry. There are also announcements of technological breakthroughs by Cray and other computer manufacturers. The Corporate Register column announces new purchases of Cray Systems and new uses by previous purchasers. There is also an upcoming events section to announce related meetings, conferences, and so on. The magazine is available free of charge.

Denelcorp
17000 East Ohio Place   303-337-7900
Aurora, CO 80017

One of the major U.S. producers of supercomputers, Denelcorp provides systems for many of the major corporations in the U.S. Sales staff at Denelcorp can answer questions about their equipment, as well as that of major competitors, and can refer you to previous purchasers for specific use information. Upon request brochures and publications describing their equipment can be sent.

Institute of Electrical and Electronics Engineers (IEEE)
345 East 47th Street   212-705-7900
New York, NY 10017

Although no technical advisory committee exists specifically to deal with supercomputers, several of the established groups such as the one on computer architecture and the committee on microprocessing have members who are specialists in this field. Plans are being made for a conference in 1985 covering the supercomputing field. Staff at the administrative offices in Silver Spring, Maryland (301-589-8142), can refer you to appropriate members specializing in supercomputers, and add your name to their mailing list to ensure that you receive all future information in this field. The following is published:

• *Supercomputers—Design and Application*—a 648-page study by Dr. Kay Hwang of Purdue University addressing all the major facets of supercomputers. It reports on state-of-the-art technology, from design issues and systems architecture to applications and data flow. There are technical surveys of vector processes, multi-processing, parallel algorithms, and languages. A detailed bibliography provides resource information on many of the topics covered. The price is $36.

Lawrence Livermore National Laboratory (LLNL)
PO Box 808   415-422-1100
Livermore, CA 94550

Although research in nuclear weapons is the foundation of the work at LLNL, their work focuses on five other major programs: magnetic fusion energy, laser isotope separation, laser fusion energy, energy and resources, and biomedical and environmental sciences. Much of the lab's scientific calculations are performed at the Livermore Computer Center (LCC), one of the most powerful computer centers in the world, and also at the National Magnetic Fusion Energy Computer Center (NMFECC) located at Livermore. The Centers will respond to specific questions concerning their computers' capabilities and accessibilities and will forward brochures, pamphlets, reference lists, and photographs on the computer center work. Their research findings in various areas are available through the Departments of Energy and Defense.

Los Alamos National Laboratory (LANL)
Los Alamos, NM 87545   505-667-7000

One of the eleven multipurpose laboratories operated by the Department of Energy, LANL forms part of a collective national repository of technical information and basic research. Programs include nuclear weapons design, development, and testing; magnetic and inertial fusion; nuclear fission; nuclear safeguards and security; solar, fossil, and geothermal energy; environmental studies; and basic sciences. LANL has one of the most powerful computing facilities in the world, and has begun work in two areas of artificial intelligence. Staff members at the Laboratory's Computer Center will direct you to the appropriate publications office to obtain documents and technical reports on supercomputer research that is available through a data base called File 103, DOE Energy, available through DIALOG.

LANL sponsored a conference in 1983 called "Frontiers of Supercomputing" that centered on the government's role in the development of supercomputers and competition with Japan in this area. Copies of conference papers are available upon request directly from LANL.

Publications

Report of the Panel on Large-Scale Computing in Science and Engineering
Courant Institute of Mathematical Sciences   212-460-7100
New York University
251 Mercer Street
New York, NY 10012

This is the report of the Washington Workshop on Supercomputing held in June 1982, and the follow-up workshop held in Michigan in August. A large panel met to discuss the current U.S. position in the world of supercomputers. The report covers what was discussed in the meeting. In addition, the report contains an extensive bibliography of articles published on the different facets of the supercomputing field. There is also an index of information on selected facilities with supercomputer capability, a section addressing the current and future applications of supercomputers, and an assortment of position papers by many of the panel members.

Additional Experts and Resources

Jim Bottum
Staff Associate for Supercomputing   202-357-9776
National Science Foundation
Office of Advanced Scientific Computers
1800 G Street, NW
Washington, DC 20550

Mr. Bottum can refer you to experts, literature, and programs involved in supercomputers. His expertise is in guiding people to information sources, not supplying technical information.

Lawrence A. Lee, Ph.D.
Program Director for Supercomputing Centers   202-357-9717
National Science Foundation
1800 G Street, NW
Washington, DC 20550

As former Director of Computing at the National Center for Atmospheric Research, Dr. Lee oversaw use of two supercomputers. He is very knowledgeable about many aspects of supercomputers, and will respond to inquiries.

George Michael
Lawrence Livermore National Laboratory (LLNL)   415-422-4239
PO Box 808
Livermore, CA 94550

Dr. Michael is knowledgeable about supercomputers and research being conducted in the U.S and abroad. He is in charge of the Computing Research Department at LLL and his work entails figuring out how to compute much faster than is presently possible. Dr. Michael can answer technical inquiries and refer you to experts and other information sources.

Supercritical Fluids

Organizations and Government Agencies

Critical Fluid Systems
25 Acorn Park   617-864-5770
Cambridge, MA 02140

Critical Fluid Systems is involved in selling processes and equipment that use fluids at their critical point to extract solvents. Thomas J. Cody, vice president of the company, is available as an information source on supercritical fluids. Over 30 publications are available, and a listing of these is available for free. The reports, usually two to three pages in length, are free. Some of these are:

• *Application of Supercritical Fluids to Regeneration of Granular Activated Carbon*

• *Critical Fluid Extraction Processes for High-Volume Organic Chemicals*

• *Critical Fluids Extraction of Vegetable Oils from Oil Seeds*

• *Industrial Waste Treatment with Critical Fluid Extractions*

• *The Promise of Supercritical Fluids*

• *Supercritical Fluids Regeneration of Activated Carbon for Absorption of Pesticides* (PB 80197569)—it is available for $16.50 from National Technical Information Service, 5285 Port Royal Road, Springfield, VA 22161, 703-487-4600.

Kerr-McGee Refining Corporation
Refining Technology Sales Department   405-270-2427
PO Box 25861
Oklahoma City, OK 73125

Kerr-McGee uses ROSE (Residuum Oil Supercritical Extraction), a solvent extraction process used on heavy oil to upgrade the bottoms of refineries, which has applications in asphalt and feed stock. The company has published over 15 papers that are available free of charge.

• *Creation of the ROSE Process*—an eight-page paper that provides an explanation of how the ROSE process works, along with a commercial operations history.

• *ROSE Process Offers Energy Savings for Solvent Extraction*—the nine-page paper focuses on the advantages of the ROSE energy savings process.

• *ROSE Process Improves Residue Feed*—an eight-page paper that provides the results of hydrocarbon studies.

Pittsburgh Energy Technology Center
Department of Energy   412-675-5797
PO Box 10940
Pittsburgh, PA 15236

The Center works with projects using supercritical fluids. Robert Warzinski is the project leader and oversees the research and development at the Center. Staff members are available to answer questions and make referrals to other sources in the area of supercritical fluids.

SRI International
333 Ravenswood Avenue   415-326-6200
Menlo Park, CA 94025

SRI is involved in the conversion of coal to liquid fuel or chemical feed stock material in supercritical media. Reports are published on the research being done, and are available to the public. David Ross is the program manager and is an expert in the field. He can be contacted for assistance or referrals to other experts in the field.

Supertrains

Organizations and Government Agencies

Bombardier, Inc.
Mass Transit Division   514-655-3830
1350 Nobel Street
Boucherville, Quebec
Canada

Bombardier, Inc., is a Canadian corporation that designs, develops, manufactures, and markets transportation-related equipment and products. In cooperation with two other Canadian corporations, Bombardier has developed the only high-speed passenger train of North American design, the LRC (light, rapid, and comfortable) train. It is the only high-speed train in the world that can be operated on existing railway tracks. The staff can answer your questions about the LRC train as well as send you a booklet describing it. They can also send you several reports:

• *Financial Aspects of High-Speed Intercity Rail*—a report presented to the American Public Transit Association on June 14, 1984. It covers some of the more critical issues and necessary actions for the successful financing of high-speed rail projects. The 15-page publication is available free of charge.

• *Florida High-Speed Rail System Conceptual Proposal*—overview of Bombardier's approach in appraising market factors, technology, and solutions for an economical, high-performance rail passenger system. This proposal considers ridership projection, route structure, service, equipment technology, financing, environmental assessment, economi impact, legislative requirements, and qualifications. The ten-page publication is available free of charge.

• *Synopsis of Bombardier Presentation of the Ohio High-Speed Rail Task Force*— a recommendation for the strategic implementation of high-speed rail services in Ohio. It describes the LRC train system objectives and advantages. It considers ridership, revenue projection, the system technology, capital investment, and system economics. The 37-page publication includes graphics and is available free of charge.

Budd Company

3155 Big Beaver Road
Troy, MI 48054

313-643-3520—Paul Sichert, VP Public Affairs

Budd Company is the U.S. representative for the recent German production of the Mag-Lev, a magnetically levitated rail system with trains that travel 400 miles per hour. Mr. Sichert can send you literature about this train, which is currently in the test stage in Germany. He can also answer general questions about supertrains and refer you to other sources of information.

Division of Science, Research, and Technology

Embassy of the Federal Republic of Germany
4645 Reservoir Road
Washington, DC 20007

202-298-4000

This division serves as liaison between German research and development efforts and U.S. industries, scientific institutions, and government agencies. The office can provide you with materials about Germany's high-speed rail technology, such as publications, slides, films, and videotapes. The staff can answer questions and make referrals.

Federal Railroad Administration (FRA)

RPF-1, Room 5411
400 Seventh Street, SW
Washington, DC 20590

202-426-9660—Louis S. Thompson

FRA has sponsored and jointly funded studies regarding super-high-speed rail systems. The studies focus on the demand and economic feasiblilty of such transportation systems in U.S. corridors. Mr. Thompson can answer your questions concerning high-speed rail and provide more information about the studies.

Florida Department of Transportation (DOT)

605 Suwannee Street
Tallahassee, FL 32301

904-488-4224—Carl Huff, Staff Director, High-Speed Rail

Florida's DOT has a special committee which has been investigating the possibility of a high-speed rail system in the state. The office can send you conceptual proposals designed for prospective franchise applicants. The proposals provide general assessments of ridership, routing, cost, and environmental consequences. The office can also send you a generic study of high-speed rail, which was financed by the Federal Railroad Administration. Mr. Huff and his staff will answer questions and make referrals.

High-Speed Rail Association (HSRA)
1625 Eye Street, NW, Suite 1015   202-296-8001
Washington, DC 20006

HSRA is a trade association of organizations and individuals involved in providing products or services for high-speed rail passenger transportation systems. The purpose of the organization is to promote the design, development, construction, and use of high-speed rail transportation. HSRA sponsors an international exposition, which is open to the public, and after which the proceedings are published. HSRA publishes a monthly newsletter for its members. The staff can respond to your inquiries, refer you to other sources of information, and send you relevant articles from its newsletter.

Japanese National Railway (JNR)
45 Rockefeller Plaza   212-757-9070
New York, NY 10011

The U.S. office of the Japanese government's transportation system can provide you with information about its high-speed rail system. Upon written request, staff will send you technical information and illustrated brochures about the Japanese high-speed bullet train called *Shinkansen*. Or contact:

Japanese National Railroad Assistance Team to the Northeast Corridor Improvement Project   This office represents the Japanese National Railroad.
c/o Federal Railroad Administration
RPF-1
400 Seventh Street, SW
Washington, DC 20590
202-472-5597—Ichiroh Mitsy, Director

Railway Progress Institute (RPI)
700 North Fairfax Street, Suite 601   703-836-2332
Alexandria, VA 22314

This trade association represents manufacturers and suppliers of high-speed rail, railroad, and rail mass transit equipment. The Institute has a high-speed rail data bank that contains information about all the high-speed rail projects in the U.S. Members obtain regular updates from the data bank as well as an inquiry service free of charge. Others may receive these services on a fee basis. RPI sponsors symposia that are open to the public, after which the proceedings are published. Upon request, RPI will conduct seminars for your organization.

Texas Railroad Company (TRC)

9405 Mountain Quail Road   512-837-3880
Austin, TX 78758

TRC is involved in the planning and building of a high-speed rail passenger system to serve the Texas Triangle area of Houston, Dallas, and San Antonio. The company will be using the German ICE technology. Staff will send you a project summary, financing information, and project description documents, and will also respond to inquiries and make referrals.

TGV Company

1801 K Street, NW   202-293-5920
Washington, DC 20006

TGV is the U.S. partnership representing *Francorail MTE* and *Althom Atlantic*, the French manufacturers of TGV, a very high-speed train. The company hopes to market this high-speed train in the U.S. This office can answer questions of a general nature and staff can send you promotional literature with illustrations and information about the TGV train.

Publications

U.S. Passenger Rail Technologies

U.S. Government Printing Office   202-783-3238
Superintendent of Documents
Washington, DC 20402

This report addresses the feasibility of high-speed rail service in the U.S. It analyzes the experiences of foreign countries having such rail services and considers the criteria necessary to make this kind of rail service economically successful. The 120-page report's order number is 052-003-009-38-2, and the cost is $4.50.

Additional Experts and Resources

Marshall G. Beck

Director of Marketing LRC   514-655-3830
Bombardier, Inc.
Mass Transit Division
1350 Nobel Street
Boucherville, Quebec
Canada J4B 181

Mr. Beck oversees the marketing of LRC (light, rapid, and comfortable), the Canadian high-speed train produced by Bombardier. He is responsible for total system proposals to various localities in the U.S., Canada, and overseas. Mr. Beck is knowledgeable about technical and economic issues regarding high-speed rail, strategic planning, development of criteria for system proposal evaluations, and selection and project risk analyses. Mr. Beck's staff can send you literature about the LRC.

William W. Dickhart, III
Assistant General Manager   215-643-2950
Budd Company Technical Center
375 Commerce Drive
Fort Washington, PA 19034

Mr. Dickhart is a technical expert on Mag-Lev, the German high-speed train technology. He serves as a liaison between his company, Budd (the U.S. representative of the German Mag-Lev company) and the U.S., as well as others involved in high-speed and super-speed ground transportation systems. Mr. Dickhart and his staff will answer questions and send you literature on Mag-Lev.

Dr. Henry Kolm
EML Research, Inc.   617-661-5655
725 Putnam Drive
Cambridge, MA 02139

Dr. Kolm, along with Dr. Richard Thorton, developed in the mid-seventies the magneplane, a linear magnetically levitated transportation system. Dr. Kolm continues to conduct research in electromagnetic technology. He is also a consultant in the area of positive ground transportation systems. He can respond to your inquiries.

Lucia Turnbull
Project Director   202-226-2024
Science, Transportation, and Innovation
Office of Technology Assessment
U.S. Congress
Washington, DC 20510

Ms. Turnbull directed the 1984 study *U.S. Passenger Rail Technologies*. She can answer questions concerning the planning, decision making, and policy implications of high-speed rail systems. Ms. Turnbull can also refer you to other information sources.

Professor David Wormley
Department of Mechanical Engineering   617-253-2246
Massachusetts Institute of Technology
Cambridge, MA 02139

David Wormley is a professor of mechanical engineering. He has done extensive research in high-speed ground transportation, including conventionally wheeled, magnetially levitated, and air cushioned transportation. He has written more than 20 journal papers on this subject, and is available to respond to your inquiries.

Synthesizers

See also: Computer Music; Electronic Keyboards

Organizations and Government Agencies

American Music Conference

150 East Huron Street 312-266-7200
Chicago, IL 60611

This nonprofit organization represents 200 members of the music industry, including manufacturers, retailers, and wholesalers. It is concerned with promoting the sale of musical instruments and their use by the general public and educational systems. The Conference collects statistics and compiles reports on all different aspects of the industry. Staff are available to answer inquiries and make referrals. It publishes:

● *Music USA*—an annual statistical survey containing information about all aspects of the music industry. Data are provided on sales, imports and exports, the music industry as a whole, specific instruments, household use, and more. Profiles are provided about specific electronic instruments such as synthesizers and keyboards. The publication also includes a directory, with contact information, of all music associations and companies. The 33-page publication is available free of charge.

Audio Engineering Society (AES)

60 East 42nd Street 212-661-8528
New York, NY 10165

A professional society of 10,000 members, AES collects and disseminates information about the latest research and developments in the field of recording and reproducing sound. The Society publishes:

● *Journal of the Audio Engineering Society*—published monthly, the journal contains papers as well as research and engineering reports on audio engineering and synthesizers. It provides coverage of new products, literature, and book reviews. A membership guide and directory are also included. The 90-page publication is available free of charge to members. Nonmembers can subscribe for $60 a year.

Creative Audio and Music Electronic Association (CAMEA)

10 Delmar Avenue 617-887-6459
Framingham, MA 01701

CAMEA is an organization of audio and music electronics manufacturers and distributors catering to the creative audio market. It sponsors educational seminars and conferences on audio and musical electronics. Subjects have included synthesizers, computer music, and electromagnetic technology. The organization publishes:

● *Creative Dictionary of Audio Terms*—defines more than 1,000 audio and musical electronic terms. The 100-page publication is available for $4.95.

National Association of Music Merchants (NAMM)

500 North Michigan Avenue   312-527-3200
Chicago, IL 60611

This trade association represents manufacturers of musical instruments and equipment. The staff in the Information Department can answer questions relating to products, producers, and distributors. The organization sponsors trade shows and conferences. It publishes several reports relating to the music retail market. It also produces:

● *Program Directory*—updated frequently, this lists all companies involved in manufacturing high-tech musical instruments and equipment. It includes manufacturers of synthesizers. The 70-page publication is available free of charge.

Publications

Keyboard Magazine

20085 Stevens Creek Boulevard   408-446-1105
Cupertino, CA 95014

Published monthly, this magazine profiles all keyboard instruments and provides information about electronic music and synthesizers. It contains consumer guidelines on subjects such as buying a synthesizer or electronic keyboard instruments. The publication is geared toward musicians. An annual subscription to the 90-page magazine is $18.95.

MIX Magazine

MIX Publications   415-843-7901
2608 Ninth Street
Berkeley, CA 94710

This monthly magazine is geared toward the audio industry professional, especially audio recording engineers, studio owners, and managers of recording studios. It covers high-tech innovations in synthesizers, digital audio, and the recording studio. The magazine also provides coverage of trade shows. Each issue is approximately 200 pages, and an annual subscription is $24.

Music Trades Magazine

PO Box 432   201-871-1965
Englewood, NJ 07631

This monthly trade publication is geared toward music retailers and musical instrument manufacturers. It provides information about new products and feature articles about new musical instruments. The magazine has information about the manufacturing, production, distribution, and marketing of synthesizers. Each issue is approximately 150 pages, and an annual subscription is $10.

Upbeat Magazine
22 West Adams   312-346-7820
Chicago, IL 60606

Geared toward the music retailer, this monthly publication focuses on new products and new features of musical equipment. It covers synthesizers, electronic keyboards, computer music, and musical instruments. Each issue focuses on a specific topic or musical instrument and profiles the product's retailers. Business and trade show information is also provided. Each issue is approximately 75 pages, and an annual subscription is $8.50.

Additional Experts and Resources

Judith Corea
Consumer Goods Office, Durable Goods Division   202-377-2132
U.S. Department of Commerce
IPA Room 4312
14th Street and Constitution Avenue, NW
Washington, DC 20230

An industry trade specialist with the Department of Commerce, Ms. Corea is responsible for following developments of the musical industry as a whole. She can provide you with import/export and product shipment data, as well as an overview of the industry, its new developments and trends. Ms. Corea can also refer you to other sources.

Technology Assessment

See also: Bioethics; Environmental Impact Assessment; Social Impact of Technology; Technology Transfer; Technophobia

Organizations and Government Agencies

Commercial Development Association (CDA)
1133 15th Street, NW   202-429-9440—Diane Taylor, Administrative Assistant
Washington, DC 20005

CDA is an association of 800 professionals, many of whom are the senior entrepreneurs of technically oriented companies. Eighty-five percent of their members are in the chemical and petroleum fields, the others are in electronics, metals, agriculture, paper, and industry machinery. CDA members are highly involved in technology assessment and transfer; Diane Taylor of their staff, will refer you to an appropriate member who can help you with technology assessment questions.

Division of Policy Research and Analysis (DPRA)

National Science Foundation (NSF)   202-357-9828
Washington, DC 20550

DPRA supports professional research in technology assessment and policy analysis. The staff can refer you to NSF experts in nearly every field who can serve as resources in technology assessment. The office publishes hundreds of reports, special analyses, abstracts, and other documents relating to technology assessment. Periodically, DPRA publishes a list of recent publications and abstracts available to the public.

International Association for Impact Assessment (IAIA)

Industrial and Systems Engineering   404-894-2330—Alan Porter
Georgia Institute of Technology
Atlanta, GA 30332

This international association of 500 people is involved in studying technology assessment, and to a lesser extent, social and environmental impact assessment. Its members include scientists, academicians, and business executives involved in planning activities. The association will answer public inquiries and make referrals. They hold meetings and annual conferences, and publish the following:

• *Impact Assessment Bulletin*—a 100-page quarterly oriented primarily toward technology assessment practitioners and also academicians. The bulletin covers the general range of topical interest technology assessment, risk assessment, social impact assessment, and environmental assessment. The format is six short articles ranging from technical papers to speeches delivered at meetings. Book reviews, a resource section, news and opportunities in the field, professional practice, and forthcoming meetings are also included. Subscription is included in the membership fee of $18 a year for individuals and $38 a year for institutions. Nonmembers can subscribe for the same $38 a year.

National Academy of Sciences and Engineering (NASE)

Office of Information   202-334-2138
2101 Constitution Avenue, NW
Washington, DC 20418

The Academy serves as official advisor to the federal government on matters of science and technology. Its members are scientists and engineers selected for their distinguished achievements. The Academy has 800 committees doing ongoing studies, many of which relate to technology assessment. NASE is divided into the National Academy of Sciences, the Academy of Engineers, the Institute of Medicine, and the National Research Council. Each organization produces newsletters, holds seminars, collects technical data, and follows new and developing fields of technology. A free catalog of their publications is available to the public. Call the general information number above to order the catalog or to be referred to an appropriate Academy expert. Since the Academy is a large operation it is best to be as specific as you can about what you need.

Office of Technology Assessment (OTA)

Public Affairs Office   202-224-9241
U.S. Congress
Washington, DC 20510

This nonpartisan analytical agency supports the U.S. Congress by providing objective analysis of major public policy issues related to scientific and technological change. The office has produced more than 100 reports that are available to the public. Its staff consists of experts in a wide range of fields including law, engineering, energy, life sciences, international security, health and life sciences, communication and information technologies, natural resources, space, transportation, and innovation. OTA's staff will answer questions from the public and provide general information regarding technology assessment. While the full text of OTA reports must be purchased from the U.S. Government Printing Office, all their publications listed below are available free of charge.

• *List of Publications*—catalogs by subject area all of OTA's published reports. Upon request, you can be added to their mailing list for this publication.

• *Assessment Activities*—contains brief descriptions of assessments presently underway, with estimated dates of completion.

• *Report Briefs*—one-page summaries of OTA reports.

• *Summaries of OTA Reports*—the summaries of most OTA reports published since late 1979 are available at no cost, as separate, self-mailing booklets.

• *Annual Report*—details OTA's activities and summarizes reports published during the preceding year.

Rand Corporation

Public Information Office   213-393-0411
1700 Main Street
Santa Monica, CA 90406-2138

Rand Corporation is a private, nonprofit research organization funded by federal, state, and local governments, foundations, and other organizations. Rand publishes 200 new reports yearly on topics ranging from computing technology to satellite communications. Each year Rand prepares 50 topical bibliographies (with abstracts) listing its publications. These bibliographies are available free of charge to the public. Full text of the reports are often available in libraries and they can be purchased from Rand Corporation. The Public Information Office will answer inquiries and make referrals.

Technology Assessment and Business Analysis
IIT Research Institute (IITRI)   312-567-4609—Robert Levi
10 West 35th Street
Chicago, IL 60616

     IITRI is a $65-million contract research organization involved in science and engineering research and development for government agencies and private businesses. The Institute tracks technology and assesses it in light of market needs. Mr. Levi is an expert in this area and can refer you to helpful sources.

Technology Law

See: Computer and Technology Law

Technology Transfer

See also: Environmental Impact Assessment; Smuggling of High Technology; Social Impact of Technology; Technology Assessment

Organizations and Government Agencies

Federal Laboratory Consortium (FLC)
Ted Maher, Director
U.S. Department of Agriculture
3865 South Building
Washington, DC 20250

     This is a national network of 300 individuals from federal labs and centers representing 11 federal agencies. Each FLC member is responsible for taking inventory and assessing the technology developed at his/her facility and then passing the research information on to industry, state and local governments, and to the general public. Through the FLC, the public has access to virtually every aspect of unclassified research activity within the federal government. Ted Maher can refer you to an FLC member in your speciality or geographical area. The directories listed below will help you locate FLC members.

     • *Federal Lab Directory, 1985*—provides information about 388 federal labs with ten or more full-time professionals engaged in research and development. Summary data arranged by federal agency and by state provide a broad overview of the federal lab system. Lab lists by staff size, by state, and by agency provide a cross-reference. For each lab a contact for obtaining technical information is given by name, address, and phone number. Major mission and major scientific or testing equipment is listed for each laboratory. This directory is available free of charge from Office of Research and Technology Applications, National Bureau of Standards, Room 402, Administration Building, Washington, DC 20234, 301-921-3814.

● *Directory of Federal Technology Resources*—describes the hundreds of federal laboratories, agencies, and engineering centers willing to share their expertise, equipment, and sometimes even facilities. It includes descriptions of 70 technical information centers and is indexed by subject, geographical location, and resource name. It was prepared by the Center for the Utilization of Federal Technology, a division of the National Technical Information Center (NTIC). The cost is $25, and it can be purchased from NTIS, 5285 Port Royal Road, Springfield,VA 22161, 703-487-4650 (order number is PB84-100015/AAW).

Foundation of Osaka Science and Technology Center
International Department   011-81-6-443-5321
1-8-4, Utsubo Hommachi, Nishi-ku
Osaka, 550, Japan

This nonprofit organization undertakes various projects in close cooperation with industries, government offices, and academic circles to promote scientific technology and industrial development in the Kansai region. Its publications include: *Overseas Affiliation Opportunities*, *New Airport Review*, and *Center Report* (all published monthly in Japanese); *Technical Exchange Bulletin* (quarterly in English); and *R&D Summaries* (in Japanese). The foundation provides advisory and consulting services; conducts seminars and workshops; and distributes publications. Services are available to anyone and may be provided for a fee.

Licensing Executives Society, Inc. (LES)
PO Box 1333   203-323-3143
Stamford, CT 06904

LES tracks worldwide developments in technology transfer. The international organization, with 4,100 members throughout the world, aims to advance licensing and international technology transfer. Each year the Society sponsors three regional meetings in the U.S., a conference, and an international meeting. In conjunction with these events, LES offers mini-courses, workshops, and seminars of interest to both novices and experts involved in licensing and technology transfer. LES members will answer questions from the public, and Jim Menge of their staff can refer you to an appropriate expert. Most LES publications are available to members only, but the public can obtain the following:

● *Bibliography of Licensing*—this free publication lists periodical references from 1966 through 1982 and it explains the "how, where, why, and when" of licensing.

● *LES Nouvelles*—a quarterly publication of 40 to 60 pages that covers a multitude of topics relating to technology transfer and all aspects of licensing in the U.S.

and abroad. Examples of topics include updates on new technologies; how to transfer new software technologies; and the packaging of a technology transferred from the U.S. to a specific foreign country, with information on the country's laws, customs, business moves, and so forth. While *LES Nouvelles* is available on a subscription basis to members only, nonmembers can purchase back issues for $6 each. To do so, write to Jack Ott, Editor, 1225 Elbur Avenue, Cleveland, OH 44107.

Little People's Productivity Center (LPPC)

2580 Grand Avenue   516-623-6295
Baldwin, NY 11510

The Center's president, Thomas V. Sobczak, is an expert on technical knowledge acquisition and transfer. He can explain where to go in the government for information and refer you to appropriate experts and studies. His Center has built a computerized data base of information sources relating to technology transfer. Dr. Sobczak has authored two papers on technology transfer, which are available free of charge from the Center. Below is a description of LLPC's data base and papers.

• *Man Tech SID*—this is a computerized data base containing manufacturing technologies resulting from all research sponsored by the federal government and done by its contractors. It contains data from the Defense Technical Information Center, the National Technical Information Center, other federal agencies, and an engineering index. Man Tech also contains details about technology transfer information sources. The data base is searchable by services, technology, and project. Retrievable information includes the name of project, dollars allocated, whether anyone is working on the project, and where in the federal budget you can find the project. Searches are done for the general public, and the cost is $65 per hour of research time plus $20 per hour for connection time.

• *Sources of Information for Technology Transfer*—this six-page paper lists 25 information sources which can be of assistance to small- and medium-sized businesses. Sources range from federal clearinghouses to commercial on-line data bases. Available free of charge.

• *Small Business and the Transfer of Technology*—this paper describes technology transfer resources, with an assessment of the government activities. Available for free from the Center.

National Technical Information Service (NTIS)

5285 Port Royal Road   703-487-4600
Springfield, VA 22161

NTIS is the central source for the public sale of U.S. government-sponsored research, development, and engineering reports, as well as computer software and data files. The NTIS collection exceeds one million titles with 60,000 new reports being published yearly. As the government's central technical information clearinghouse, NTIS provides a variety of information awareness products and services to keep the public abreast of the technologies developed by the federal government's multibillion dollar annual research and engineering effort. These include:

Industry Awareness List

The Center for the Utilization of Federal Technology (CUFT) will place the name of any small business on its list. This is a free service to help businesses locate and use federal technology. CUFT alerts people on its mailing list to new federal information services, and in some instances, it forwards the company's name to other government agencies which offer additional technology-oriented assistance. To be placed on the CUFT mailing list, send your name, address, company needs and interests, as well as company size, to Center for the Utilization of Federal Technology, Room 8R, at the above address.

Current Awareness Services

● *Weekly Abstracts Newsletters*—a weekly bulletin available on a subscription basis in each of 26 different subject areas. The bulletin announces summaries of newly released government research and development reports and provides complete coverage of broad areas of government research. Prices range from $60 to $120 a year, depending on subject. For further information, order NTIS's free brochure PR 205.

● *Government Inventions for Licensing*—a weekly subscription bulletin that announces more than 1,500 federally owned patents or patent applications that are available for licensing, often on an exclusive basis. Each issue divides inventions into eleven subject disciplines and provides a summary and, when appropriate, a drawing of the invention. Subscription cost is $205 a year. For further information, order NTIS's free brochure, PR 750.

● *NTIS Tech Notes*—a monthly subscription service alerting readers to new federal technology having practical or commercial potential. *Tech Notes* provides information about newly developed technology in 12 different areas. The publication includes one- to two-page fact sheets, often illustrated, of new processes, equipment, software, and material. For further information, order NTIS's free brochure, PB 365. Yearly subscription rate is $60 a category or $250 for all 12 categories.

Special Bibliographies

NTIS publishes a free 300-page directory entitled *Published Searches: Master Catalog*, which lists subject searches, with complete citations, from the NTIS bibliographic data base. It also has citations from searches of 20 commercial data bases. Approximately 3,000 NTIS searches are listed, each of which can be ordered in full for $35 from NTIS. For further information, order NTIS's free brochure, PR 701.

Computer Software

NTIS's Software Center has a collection of more than 450 computer programs covering an array of subjects. This machine-readable information produced by the government can be purchased by the public.

On-line Data Bases

NTIS runs several on-line data bases which can be accessed directly by subscribers of BRS, DIALOG, and SDC. NTIS staff will also search their data bases for the general public; fees for NTIS searches generally start at $125. NTIS data bases include:

● *NTIS Data Base*—includes summaries of research reports prepared, since 1964, by the federal agencies, their contractors, or grantees. The data base is comprised of summarized citations of more than 950,000 technical reports. It includes 2,000 *NTIS Tech Notes* and 34,000 government inventions. Approximately 60,000 new citations are added annually.

● *Federal Research in Progress*—provides access to federal research and development information before the final reports are issued. This data base was formerly the Smithsonian Information Exchange (SSIE).

Directories

NTIS, through its Center for the Utilization of Federal Technology (CUFT), has prepared two directories which provide information helpful to people interested in technology transfer. They are:

● *Directory of Federal Technology Resources, 1984*—a 150-page directory of more than 800 federal government sources. Included are federal agencies, laboratories, and engineering centers willing to share their expertise, equipment, and sometimes even their facilities. Detailed summaries of the sources are arranged into 30 subject-oriented areas. Subject, geographic, and resource name indexes are provided. The cost is $25.

● *Federal Technology Catalog*—a 230-page directory describing practical technology selected for commercial and/or promising applications to the fields of computer technology, energy, electro-technology, engineering, life sciences, machinery and tools, manufacturing, materials, physical sciences, and testing and instrumentation. The directory describes more than 1,000 new processes, inventions, equipment, software, and techniques developed by and for federal agencies during 1983. The cost is $23.50.

● *Foreign Technology Information*—NTIS's on-line data bases, current awareness services, and other materials contain information on foreign technology.

Society of Manufacturing Engineers (SME)

1 SME Drive   313-271-1500
PO Box 930
Dearborn, MI 48121

SME has more than 75,000 members in 65 countries. It functions as a technology transfer agent, accumulating technical information and disseminating it throughout the manufacturing environment via numerous vehicles. The staff can put you in contact with SME members who are experts in a variety of fields. The Society sponsors 30 to 40 educational and state-of-the-art conferences each year, and they are open to the public. SME also holds hardware expositions, and publishes technical papers on specific topics, books, videotapes, magazines, and newsletters. In addition, SME has a research library and manufacturing information data base available.

Society of University Patent Adminstrators (SUPA)

Spencer Braylock, President   515-294-4740
315 Beardshear
Ames, IA 50011

     SUPA is an association of people involved in technology transfer for universities, teaching hospitals, and their nonprofit organizations. Its members are researchers, patent administrators, attorneys, and others, responsible for taking university research results and finding companies interested in developing the technology into a product. Most Society activities are for members only, but Mr. Baylock can refer you to appropriate members who can answer questions regarding the transfer of technology from a university, setting up a research foundation, and patent policies of a specific university. SUPA members will supply information about the technology transfer policy, but they will not get involved in the commercial aspect of locating or selling technology for a requestor.

Technology Transfer Society

N. J. Goldstone, Executive Secretary   213-874-2535
7033 Sunset Boulevard, Suite 302
Los Angeles, CA 90028

     The Society was formed in 1975 as a nonprofit international organization that could act as a catalyst to help accelerate the movement of technology out of the laboratory and into the mainstream of commerce. Its membership is open to technologists, scientists, economists, attorneys, venture capitalist, bankers, businesspeople, information specialists, consumer groups, and others. The Society publishes a bimonthly newsletter for its members and sponsors workshops, seminars, and conferences throughout the world. The staff can answer questions from the public and refer people to appropriate sources and experts. The following materials are available to the public:

     ● *Journal of Technology Transfer*—a reference journal of papers submitted to the society covering mechanisms and processes of technology transfer. The journal is published twice yearly and available for $35 a year.

     ● *Proceedings from the Society's International Symposium*—the entire conference proceedings are available in print. Cost varies according to the length of the publication.

NASA Technology Utilization and Industry Affairs Division (TU)

Walter Heiland, Director   301-859-5300, ext. 241—Baltimore Residents
NASA Scientific and Technical Information Facility   301-621-0241—Elsewhere
PO Box 8757
BWI Airport, MD 21240

     NASA's TU program offers a variety of free publications and personnel services to help transfer aerospace technology to nonaerospace applications. NASA TU information is restricted to U.S. citizens and U.S.-based industry. Their publications include:

- *NASA Tech Briefs*—published quarterly and free to engineers in U.S industry and to other domestic technology transfer agents. It is both a current awareness medium and a problem solving tool. Potential products, industrial processes, basic and applied research, shop and lab techniques, computer software, and new sources of technical data concepts can be found. The short section on New Product Ideas highlights a few of the potential new products contained in each issue. The remainder of the volume is organized by technical category to help you quickly review new developments in your areas of interest. Finally, a subject index makes each issue a convenient reference file.

- *NASA Tech Brief Journal*—a free quarterly publication covering 125 new technologies in each issue. The *Journal* has information about potential products, industrial processes, basic and applied research, shop and laboratory techniques, computer software, and new sources of technical data contacts.

- *Special Publications Series*—NASA TU publishes technology utilization notes; handbooks and bibliographies on specific technologies; conference proceedings; technology utilization surveys on anything from advanced valve technology to measurement of blood pressure in the human body; and technology utilization reports and compilations that include all types of documents about products, processes, and other topics in the technology utilization field.

- *NASA Annual Report*—produced to make the public aware of practical investments in aerospace research, this document contains a section profiling at least 35 companies that have used NASA technologies to improve their products. Issued each June or July.

NASA's Technology Transfer System

The NASA system of technology transfer personnel and facilities extends from coast to coast and provides geographical coverage of the nation's primary industrial concentrations, together with regional coverage of state and local governments engaged in transfer activities. For specific information concerning the activities described below, contact the appropriate technology utilization personnel at the addresses listed on the following pages.

Field Centers

Field Center Technology Utilization Officers manage center participation in regional technology utilization activities.

Ames Research Center
National Aeronautics and Space
   Administration
Moffett Field, CA 94035
   Technology Utilization Officer: Stan
Miller
415-965-6471

Goddard Space Flight Center
National Aeronautics and Space
   Administration
Greenbelt, MD 20771
   Technology Utilization Officer:
Donald S. Friedman
301-344-6242

Lyndon B. Johnson Space Center
National Aeronautics and Space
   Administration
Houston, TX 77058
      Technology Utilization Officer:
Marvin F. Matthews
713-483-3809

John F. Kennedy Space Center
National Aeronautics and Space
   Administration
Kennedy Space Center, FL 32899
      Technology Utilization Officer:
U. Reed Barnett
305-867-3017

Langley Research Center
National Aeronautics and Space
   Administration
Langley Station
Hampton, VA 23655
      Technology Utilization and Applications
Programs Officer: John Samos
804-865-3281

Lewis Research Center
National Aeronautics and Space
   Administration
2100 Brookpark Road
Cleveland, OH 44135
      Technology Utilization Officer:
Harrison Allen, Jr.
216-433-4000, ext. 422

George C. Marshall Space Flight Center
National Aeronautics and Space
   Administration
Marshall Space Flight Center, AL 35812
      Director, Technology Utilization Office:
Ismail Akbay
205-453-2223

Wallops Flight Center
National Aeronautics and Space
   Administration
Wallops Island, VA 23337
      Technology Utilization Officer:
Gilmore H. Trafford
804-824-3411, ext 663

Resident Office
Jet Propulsion Laboratory
4800 Oak Grove Drive
Pasadena, CA 91103
      Technology Utilization Officer:
Aubrey D. Smith
213-354-4849

Industrial Applications Centers

Industrial Applications Centers provide information retrieval services and assistance in applying technical information relevant to user needs. The Centers will do in-depth searches and provide you with everything they have on a particular topic. Prices may vary.

Aerospace Research Applications Center
PO Box 647
Indianapolis, IN 46223
      John Ulrich, Director
317-264-4644

Kerr Industrial Applications Center
Southeastern Oklahoma State University
Durant, OK 74701
      Tom J. McRorey, Ph.D., Director
405-924-6822

NASA Industrial Applications Center

701 LIS Building
University of Pittsburgh
Pittsburgh, PA 15260

Paul A. McWilliams, Ph.D., Executive Director
412-624-5211

NASA Industrial Applications Center

University of Southern California
Research Annex, 2nd Floor
3716 South Hope Street
Los Angeles, CA 90007

Robert Mixer, Ph.D., Director
213-743-6132

New England Research Applications Center

Mansfield Professional Park
Storrs, CT 06268

Daniel Wilde, Ph.D., Director
203-486-4533

North Carolina Science and Technology Research Center

PO Box 12235
Research Triangle Park, NC 27709

James E. Vann, Ph.D., Director
919-549-0671

Technology Applications Center

University of New Mexico
2500 Central Avenue, SE
Albuquerque, NM 87131

Stanley Morain, Ph.D., Director
505-277-3622

State Technology Applications Centers

State Technology Applications Centers provide technology transfer services similar to those of the Industrial Applications Centers, but only to state governments and small businesses within the state.

NASA/Florida State Technology Applications Center

State University System of Florida
500 Weil Hall
Gainesville, FL 32611

J. Ronald Thornton, Director
904-392-6626

NASA/UK Technology Applications Program

University of Kentucky
109 Kinkead Hall
Lexington, KY 40506

William R. Strong, Manager
606-257-6322

Computer Software Management and Information Center

The Computer Software Management and Information Center (COSMIC) offers government-developed computer programs adaptable to secondary use.

COSMIC

112 Barrow Hall
University of Georgia
Athens, GA 30602

John A. Gibson, Director
404-542-3265

Application Team

Application Team works with public agencies and private institutions in applying aerospace technology to solutions of public sector problems.

Research Triangle Institute
PO Box 12194
Research Triangle Park, NC 27709
    Doris Rouse, Ph.D., Director
919-541-6980

Data Bases

Government Industry Data Exchange Program (GIDEP)
Officer in Charge   714-736-4677
GIDEP Operations Center
Corona, CA 91720

GIDEP is a cooperative activity between government and industry providing participants with a means to exchange information about certain types of technical data, especially the design, development, and operational phases of the life cycle of systems and equipment. The program is managed by the U.S. government and participation is open to individuals and organizations able to exchange technical information and willing to abide by GIDEP program guidelines. GIDEP's services, data bases, and publications are available free of charge to participants. GIDEP staff will occasionally perform simple data base searches for nonparticipants, but on a one-time basis. Currently GIDEP members include 444 industry representatives, and 204 U.S. government officials, primarily from NASA and the Departments of Defense, Energy, Transportation, and Labor. Canadian businesses and government officials, mainly from the Department of Defense, also participate. GIDEP services include a bimonthly newsletter about new developments and an on-line data base. Participants can obtain free, direct, on-line access to the data base's main computers located in Washington, DC and Corona, CA. Members can also request searches or obtain the entire data base on microfilm and microfiche. GIDEP data base files are:

- *The Engineering Data Interchange*—contains engineering evaluation and qualification test reports, nonstandard parts justification data, parts and materials specifications, manufacturing processes, and other related engineering data on parts, components, materials, and processes. This data interchange also includes a section of reports on specific engineering methodology and techniques, air and water pollution reports, alternate energy sources, and other subjects.

- *The Reliability-Maintainability Data Interchange*—contains failure rate/mode and replacement rate data on parts, components, and materials based on field performance information and/or reliability demonstration tests of equipment, subsystems, and systems. This data interchange also contains reports on theories, methods, techniques, and procedures related to reliability and maintainability practices.

- *The Metrology Data Interchange*—contains metrology-related engineering data on test systems, calibration systems, and measurement technology and test equipment calibration procedures, and has been designated as a data repository for the National Bureau of Standards (NBS) metrology-related data. This data interchange also provides a Metrology Information Service (MIS) for its participants.

• *The Failure Experience Data Interchange*—contains objective failure information generated when significant problems are identified on parts, components, processes, fluids, materials, or safety and fire hazards. This data interchange includes the ALERT and SAFE-ALERT data, failure analysis, and problem information data.

• *Manufacturer Test Data and Reports*—the GIDEP data base also includes certified test reports from manufacturers detailing test results and inspections conducted on devices by their manufacturers. Test data pertain to commercial as well as military and high-reliability devices. The availability of this test data in GIDEP provides participants the opportunity to apply the data in every phase of system design, development, production, and the support process.

I. P. Sharp Associates, Ltd.

Suite 1900, Exchange Tower   416-364-5361
2 First Canadian Place
Toronto, Ontario
Canada M5X 1E3

    SITC—United Nations Commodity Trade Statistics—is a data base that allows the user to monitor the transfer of technology for 28 selected reporting countries by looking at the imports and exports of over 3,000 commodities. Examples of commodities include heavy machinery and office automation products.

TECHNOTEC

Control Data Corporation, World Tech, Inc.   612-893-4640
Technology and Information Service
7600 France Avenue South
Edima, MN 55435

    TECHNOTEC is a computerized technology file containing technologies, patents, processes, and expertise being offered by individuals or organizations for sale, license, exchange, or joint venture. The technologies available for transfer are owned by individuals, companies, and governments. The data base's Quest file contains technologies for people interested in joint ventures for locating or developing technology. Control Data staff will search their data base for the general public. There is no charge for searches of technologies offered for marketing by Control Data. Searches of other data generally cost from $50 to $200.

Technology Transfer Network

National Congress of Inventor's Organizations (NCIO)   214-528-8050—Tom E. Workman
4000 Rock Creek Drive
Dallas, TX 75204

NCIO maintains a computerized data base of inventor organizations and individuals looking for buyers to further develop their new technologies. Tom Workman will search NCIO's data base and provide referrals.

Additional Experts and Resources

Professor James A. Jolly
School of Business and Public Administration 916-454-6640
California State University 916-929-8454
Sacramento, CA 95819

Under sponsorship of the Department of Commerce's Business Development Agency, Professor Jolly's office provides a number of services to help the public learn about new technologies developed by the 680 federal laboratories across the U.S. His office serves as a clearinghouse to help interested parties locate new technologies and their producers. His staff produces three monthly newsletters containing information extracted from three corresponding computerized data bases maintained at Cal State. The newsletters and searches of the data base are all available free of charge. This office also compiles a *Technology Transfer Directory of People*, which is available free of charge while supplies last. Below is a description of the publications and data bases available:

● *Directory of Persons Interested in Technology Transfer*—is a free annual publication with contact information for more than 2,000 people worldwide who are experts in the full range of technologies, from acoustics to wholesale trade. Individuals listed have agreed to provide initial information to inquiries at no cost. Geographical and occupational indexes are provided.

● *TECTRA*—is a computerized data base containing information about new or more efficient technology that has been developed by a federal laboratory and is currently available for use by interested persons. Technologies in this data base have already been used. Retrievable data includes the name of a person and organization that has produced and used the product. Data base searches and a monthly newsletter titled *TECTRA* are both available free of charge. The two-page newsletter provides case summaries of technologies in the data base.

● *TECLAB*—is a computerized data base containing information about new technologies that have been developed by a federal laboratory. The technologies listed in TECLAB are believed to have commercial potential, but they have not yet been commercially applied. Data base input information is from the 680 member laboratories of the Federal Laboratory Consortium for Technology Transfer. Retrievable data include the name of laboratory, name of scientist or engineer who is responsible for the technology, and the person to contact. Data base searches and a monthly newsletter entitled *TECLAB* are both available free of charge. The two-page newsletter provides summaries of cases in the data base.

● *TECUNI*—this is a computerized data base of information about new technologies that have been developed by a university school of engineering and are available for commercialization. The data base input is received on a voluntary basis from engineering schools. Retrievable data include the name of school, name of scientist or engineer responsible for the technology, and the person to contact for more information. Data base searches and a monthly newsletter entitled *TECUNI* are both available free of charge. The two-page newsletter provides summaries of cases in the data base.

Technology Utilization

See: Technology Transfer

Technophobia

See also: Ergonomics; Social Impact of Technology

Organizations and Government Agencies

The Phobia Society of America (PSA)
6181 Executive Boulevard
Rockville, MD 20852

301-231-9350

This national, nonprofit organization serves as an information clearinghouse for phobic people, their families, clinicians, researchers, and the general public. The Society's goals are to stimulate the development of more effective treatment and to help phobic people find treatment. Through an annual national phobia conference, the Society provides an opportunity for experts in the field to exchange information, train others, and form support groups. PSA's staff can refer you to treatment resources and experts. The following are some of the publications available:

• *The National Phobia Treatment Directory*—a reference guide containing information about practitioners and programs throughout the country, with addresses, telephone numbers, and treatment descriptions. The *Directory* also contains a primer on how to find treatment, what questions to ask a prospective therapist, and how to evaluate treatment. The cost of the Directory is $2.50.

• *Phobia*—a comprehensive summary of new treatment approaches, offering clinical guidelines for their application to a wide range of phobic behaviors. Emphasis is placed on the active participation of the phobic in therapy, the importance of self-help groups, and the use of former phobics as paraprofessional therapists. The

252-page hardbound book is available to nonmembers for $25 and to members for $15.

• *Your Phobia*—this book was written for people who suffer from phobias. It contains information that can help phobic people understand and overcome their phobias. The publication is available to nonmembers for $18, and to members for $15.

Science Indicators Unit

National Science Foundation   202-634-4682
1800 G Street, NW
Washington, DC 20550

This office prepares the biannual *Science Indicators Report* which includes a chapter about public attitudes towards science and technology. The report is the result of a survey taken of three sectors: the general public; that portion of the public interested and knowledgeable about science and technology (20%); and nongovernmental science policy leaders. The staff at this office can answer questions regarding the public's fear of, and attitudes towards, relating to high technology.

Publications

The Analysis of Actual Versus Perceived Risks

Plenum Publishing Company   212-741-6680
233 Spring Street
New York, NY 10013

This book deals with issues relating to automobile accidents, nuclear power, cancer, cigarette smoking, and toxic wastes. It covers factors that influence public perceptions of technological risks, and how an understanding of such factors can be used to design more effective risk management policies. The 375-page publication is available for $55.

Additional Experts and Resources

Donald Buzzelli, Ph.D., SCD

Science Indicators Unit   202-634-4682
National Science Foundation
1800 G Street, NW
Washington, DC 20550

Dr. Buzzelli's primary interest is public attitudes toward science and technology, including specific technologies. He is knowledgeable about public preferences and expectations regarding scientific and technological developments as well as the public's science literacy.

Dr. Vincent Covello, Program Director

Risk Analysis Program　　　　　　　　　　　　　　　　　202-357-9828
National Science Foundation
1800 G Street, NW
Washington, DC 20550

　　　　An economist and sociologist, Dr. Covello works in the area of public perceptions of technological hazards, including responses and behavior related to them.
Hazards include nuclear power; low-probability-high-consequence accidents, including airplane crashes; oil spills; hazardous waste sites; biotechnology; and dam failures.
Dr. Covello can answer questions in his area of expertise. He is the author of two
books, *The Analysis of Actual Versus Perceived Risks* and *Low-Probability-High-
Consequence Risk Analysis*, as well as over 20 articles.

Robert L. DuPont, M.D.

6191 Executive Boulevard　　　　　　　　　　　　　　　301-468-8980
Rockville, MD 20852

　　　　A specialist in phobic fear, Dr. DuPont's main interest is people's fear of modern technology. He has written a series of articles about nuclear phobia and testified
before the U.S. Congress on the subject.

Gerald M. Rosen, Ph.D.

Cabrini Medical Tower　　　　　　　　　　　　　　　　206-343-9474
901 Boren Avenue, Suite 1910
Seattle, WA 98104

　　　　Dr. Rosen maintains a private practice as a clinical psychologist and serves as
an associate professor at the Department of Psychiatry at the University of Washington School of Medicine. He has worked with people who have fears of computers and
other new technologies, and can answer questions in his area of expertise. Dr. Rosen
is the author of more than 20 articles that have been published in professional
journals, many in the area of fear reduction. He has also written two self-help books,
one about fear reduction and the other about relaxation therapy.

Telecommunications

*Begins next page. See also: Cable Television; Electronic Mail; Fiber Optics; Radio;
Satellite Communications; Videotex/Teletext*

Organizations and Government Agencies

AID—Rural Satellite Programs Library
1255 23rd Street, NW, 4th Floor   202-862-3855
Washington, DC 20036

The library is funded by the Agency for International Development (AID), and its primary focus is telecommunications technology in Third World countries. Most publications in the library are oriented toward telecommunications applications such as rural telephony and rural communications via satellite. The library issues a quarterly newsletter. It has a wide range of books, periodicals, papers from conferences, and technical articles relating to telecommunications.

Center for Telecommunications Studies
Dr. Christopher Sterling, Director   202-676-7062
George Washington University
2000 G Street, NW
Washington, DC 20052

The Center, established in 1980, develops research on telecommunications for the public and private sectors, with emphasis on policy areas. The staff can furnish technical information and give referrals. The Center also maintains a library with technical books, journals, and other publications in the telecommunications sector.

Electronic Industries Association (EIA)
2001 Eye Street, NW   202-457-4919
Washington, DC 20006

A national trade association, EIA represents over 300 electronics companies, which include manufacturers and distributors of equipment. Its Telecommunications Group serves as liaison between government and pertinent industry groups, and it provides a wide range of technical and statistical advice and opinions on standards, spectrum usage, and legislative action. EIA's library provides technical information to both members and nonmembers. The Association publishes the following pertaining to telecommunications:

• *Electronic Market Trends*—a monthly publication containing comprehensive articles on new technologies, international developments, and important policy decisions on telecommunications and electronics. Cost is $150 a year.

• *Telecommunications—Trends and Directions*—annual seminar proceedings containing the latest information about various segments of the telecommunications industry. This publication is of particular interest to financial analysts and investors. Annual subscription rate is $20.

• A brochure describing the Telecommunications Group, its activities, as well as the EIA Publications Index, can be obtained without charge from the Public Affairs Office.

Institute for Telecommunication Sciences (ITS)

U.S. Department of Commerce   303-497-3484—Val O'Day, Executive Officer
National Telecommunications and Information Administration (NTIA)
325 Broadway
Boulder, CO 80303

The Institute is a research and engineering branch of the NTIA. The work done in its offices consists of applied research and analyses on spectrum and telecommunications. Mr. O'Day's office personnel can furnish information on technical matters regarding telecommunications. Technical reports and publications are produced by the Institute, and a list of these can be obtained from the ITS office listed above.

National Telecommunications and Information Administration (NTIA)

U.S. Department of Commerce   202-377-1551—Kenneth Robinson, Policy Advisor
14th Street and Constitution Avenue, NW
Washington, DC 20230

NTIA is an executive branch agency responsible for the development and presentation of domestic and international telecommunications policy. It is also responsible for the administration of telecommunications facility programs receiving capital grants from the federal government. NTIA manages the government use of radio telecommunications, and regulates radio spectrum within the country. Its staff of specialists develop research in support of policy, and provide services in the area of radio frequency.

Mr. Robinson's office can be contacted for technical information and referrals. Technical publications and reports are issued by the NTIA, as well as a publications listing.

North American Telecommunications Association (NATA)

2000 M Street, NW, Suite 550   202-296-9800
Washington, DC 20036

NATA is a trade association that represents manufacturers, suppliers, and distributors of telecommunications equipment. It holds an annual convention, and several workshops, and serves as a clearinghouse of information for its members. NATA publishes:

• *Washington Update*—a biweekly newsletter reporting on federal developments affecting the telecommunications sector. $100 a year for nonmembers and free of charge to members.

• *Telecommunications Sourcebook*—an annual guide that includes a comprehensive directory of telecommunications equipment and services in the U.S. and Canada. Free of charge for members, and $50 a year for nonmembers.

Telecommunications Information Center

George Washington University (GWU)   202-676-5740
Gelman Library, Room 610
2130 H Street, NW
Washington, DC 20052

The Center is a research facility for the students of GWU's telecommunications policy courses. The library can be used by those seeking technical or marketing information. Its reference room and main collection hold more than 1,000 current books and 60 journal subscriptions, as well as government documents and industry reports. Bibliographic lists can be obtained from the library.

Publications

Communications Week

CMP Publications, Inc.   516-365-4600
111 East Shore Road
Manhasset, NY 11030

A weekly tabloid, this business newspaper covers the entire telecommunications and voice industry ranging from manufacturers and distributors of equipment and services to their users. It also covers local, long distance, and international communications carriers, as well as their users. Available free of charge to the people in the industry.

Data Communications

McGraw-Hill, Inc.   212-512-2000
1221 Avenue of the Americas
New York, NY 10020

This is a monthly magazine oriented toward data communications and their users, covering technology in this field, applications, feature articles, news and business stories, and a list of new products introduced in the market, in terms of hardware and software. The subscription price is $24 a year.

Information Gatekeepers

214 Harvard Avenue   617-787-1776
Boston, MA 02134

Information Gatekeepers chairs an annual conference for the telecommunications industry. It publishes the conference proceedings along with several newsletters, handbooks, and other materials about telecommunications. A free brochure of their publications is available from the company. Examples of their publications include:

• *Local Area Network*—this monthly newsletter covers the major telecommunications networks worldwide. It deals with technology and marketing innovations, informing the public about the applicability of new products. The annual subscription fee is $250.

• *Fiber Optics Communications*—this monthly newsletter covers domestic and international telecommunications news. It is a comprehensive look at technological innovations, new products, events, and the people in the industry. The subscription cost is $215 a year. A weekly version of *Fiber Optics Communication* is available for $128 a year.

Telecommunications Magazine

Horizon House   617-326-8220
610 Washington Street
Dedham, MA 02026

Issued monthly, the magazine covers a full range of topics on telecommunications, including new products, applications, and technical experts. *Telecommunication* is directed toward management level personnel in the telecommunications sector. Subscription price is $36 a year.

Telecommunications Policy

Butterworth Scientific, Ltd.
88 Kings Way
London WC2B 68B
England

A quarterly magazine covering the main aspects of policy in telecommunications internationally. Cost is $150 a year.

Telecommunications Systems and Services Directory

Gale Research Co.   313-961-2242
Book Tower
Detroit, MI 48226

This is a three-volume directory that covers a wide range of services for suppliers of telecommunications. Covered are international and interstate long-distance telephone services; local, national and international data communications; teleconferences; electronic mail; videotex and teletext systems; and related areas.

It contains descriptions and contacts for data, voice, text, and image transmission services and systems, as well as related consultants, associations, publishers, government agencies, and research organizations. The price for the three-volume directory is $170.

Data Bases

NewsNet, Inc.

945 Haverford Road   215-527-8030
Bryn Mawr, PA 19010   800-345-1301

The on-line data base service presents over 25 newsletters in the area of telecommunications, dating back to 1982. *NewsNet* maintains the most current edition of each newsletter, as well as all back issues. *NewsNet* is accessed through any microcomputer equipped with a modem.

Additional Experts and Resources

Tom Kieffer, Director of Marketing
Mainstream Software   612-831-3030
4940 Viking Drive, Suite 344
Minneapolis, MN 55435

Mr. Kieffer is a telecommunications specialist in the areas of electronic mail, electronic access, and business communications. He is a good source of information for these topics, and can refer you to vendors of on-line data bases, as well as provide information about technology available in the areas above.

Mr. Kieffer has written a book entitled *Getting Connected*, published by Ashton-Tate in 1984, that covers many areas of telecommunications, including electronic mail, electronic access, professional services of on-line data bases, videotex, and personal computers. The book explains how one can use the technology available, and it lists the suppliers of on-line data bases. The cost of the book is $24.95.

U.S. Department of Commerce
National Telecommunications and Information Administration (NTIA)
14th Street and Constitution Avenue, NW, Room H-4725
Washington, DC 20230
202-377-1880—William J. Sullivan

William Sullivan develops industry and market analysis, and can provide information on the telecommunications sector. The Office of Producer Goods can be contacted for information and referrals on the telecommunications industry.

Federal Communications Commission

The following is a listing of office and reference rooms within the FCC that are good sources of information about various aspects of telecommunications and satellite communications.

Consumer Assistance and Small Business Division
FCC
1919 M Street, NW, Room 252
Washington, DC 20054
202-632-7000

This Division is the place to go if you have questions about FCC telecommunications services, such as the use of radio, broadcast, telephone, cable television, and satellite communications. They will provide you with assistance in locating information concerning FCC rules, policies, and procedures.

If you need general information about communications issues or background material about the FCC, the following bulletins are available free of charge:
- *Information Services and Publications*
- *How to Apply for a Broadcast Station*
- *Broadcast Services*
- *The FCC in Brief*

- *Radio Stations and Other Lists*
- *A Short History of Electrical Communication*
- *Private Radio Services*
- *Common Carrier Services*
- *Station Identification and Call Signs*
- *Regulation of Wire and Radio Communications*
- *Frequency Allocation*
- *Educational Television*
- *Memo to All Young People Interested in Radio*
- *Letter to a Schoolboy*
- *Field Operations Bureau*
- *Subscription Television*
- *Public Radio*
- *Cable Television*
- *International Communications in Amateur Radio*
- *UHF Television Comparability Pamphlet*

FCC Library
FCC
1919 M Street, NW, Room 639
Washington, DC 20054
202-632-7100

The library maintains a collection of various types of legal and technical information. The legal collection contains federal and statutory case histories, indexes, reference works, treatises, and loose-leaf services. The technical collection covers telecommunications and related subjects. The library also has approximately 400 periodicals, including scholarly research, trade journals, law reviews, and literature reviews.

Office of Public Affairs (Press Office & Consumer Assistance)
FCC
1919 M Street, NW, Room 202
Washington, DC 20054
202-254-7674—Information
202-632-0002—Recorded Message

This office publishes a *Daily Digest* that lists Commission notices, releases, texts, and decisions issued each day. The FCC recorded message number provides information about late releases. Staff will answer questions about the FCC and what it does.

Mass Media Bureau Public Reference Room
FCC
1919 M Street, NW, Room 239
Washington, DC 20054
202-632-7566

The Reference Room contains information on application processing procedures, procedures for requesting files, Federal Record Center file requests, procedures for duplication of documents, contact representatives, and reference materials.

Contact Representative (CR)
Mass Media, FCC
1919 M Street, NW, Room 248
Washington, DC 20054
202-632-6334

The CR is responsible for providing appropriate information and expertise to the public, by phone and in person, on any aspect of AM/FM radio or television.

Informal Complaints and Public Inquiries Branch
Common Carrier Bureau
FCC
2025 M Street, NW, Room 6319
Washington, DC 20054
202-632-7553

Staff will answer questions about common carriers. The Common Carrier Bureau also has several public reference rooms that are open to the public:
- Tariff Review—Room 513, 202-632-5550
- Economics—Room 535, 202-632-7984
- International Facilities—Room 537, 202-632-7265
- Domestic Facilities—Room 331, 1200 19th Street, 202-634-1512
- Mobile Services—Room 628, 202-632-6400
- Cellular Mobile—Room 650, 202-632-6400

Private Radio Bureau
PRB Operations Review Branch
FCC
2025 M Street, NW, Room 5114
Washington, DC 20054
202-632-6497
202-632-6940

Contact the PRB Branch for questions about PRB rules and issues pertaining to private radio broadcasting.

PRB Licensing Division
FCC
Route 116
Gettysburg, PA 17325
717-337-1212

Contact this division with questions about PRB license applications.

Field Operations Bureau (FOB)

FOB Public Contact Branch
FCC
1919 M Street, NW, Room 5114
Washington, DC 20054
202-634-1940

This branch handles inquiries from consumers.

Cable TV Room

FCC
1919 M Street, NW, Room 239
Washington, DC 20054
202-254-3420

The division specializes in all aspects of the cable television industry. Its cable files are kept in the Mass Media Reference Room, and can be used there.

Satellite Systems Branch

Science and Technology
FCC
2020 M Street, NW
Washington, DC 20054
202-653-8107

This technical group works on all aspects of satellite communications and related issues that come before the FCC. It conducts technical studies for the application and use of satellites including the technical details of satellite characteristics, and the use of frequencies. The following experts can be contacted directly at the above address:

● Bruno Pattan—concentrates on the technical aspects of direct broadcasting systems.

● George Sharp—specializes in calculating and predicting interference between satellite systems and in determining how new applications are compatible.

● Oleg Efremov—studies how satellite orbits can be used efficiently.

Satellite Radio Branch

Common Carrier Bureau
FCC
1919 M Street, NW
Washington, DC 20054
202-634-1624

This branch provides licenses to operate space stations (satellites), as well as earth stations for domestic use. This office can also provide information about who has applied for licenses for space and/or earth stations.

International Facilities Division
Common Carrier Bureau
FCC
1919 M Street, NW
Washington, DC 20054
202-634-1624

The Division provides licenses to operate space stations (satellites) as well as earth stations for international use. This office can also provide information on who has applied for licenses.

FCC Reports
FCC   202-632-0426
Publications Branch
Washington, DC 20054

This branch publishes approximately 600 reports a month on topics of legal significance that result from FCC meetings. These reports are available to the public.

Data Tapes
National Technical Information Service (NTIS)   703-487-4807
5285 Port Royal Road
Springfield, VA 22161

NTIS makes FCC data tapes available on magnetic tape. They publish a 13-page catalog describing more than 60 available tapes. Contact NTIS for further information.

Commission Telephone Directory
International Transcription Services, Inc.   703-352-2400
4006 University Drive
Fairfax, VA 22030

This is a telephone directory of the FCC, including telephone and room numbers of staff members, as well as functional listings and identification of contact persons by subject areas. Single copies are available for $100.

Teleconferencing

Begins next page. See also: Videotex/Teletext

Organizations and Government Agencies

Applied Business Communications (ABC)
5 Crow Canyon Court, Suite 209   415-820-5563
San Ramon, CA 94583

ABC sponsors the largest annual meeting in the field of teleconferencing. It also organizes workshops dedicated to training executives in teleconferencing systems. It publishes:

• *Teleconference*—a business communications magazine targeted to the professional video conferencing user that gives full coverage of all the news about teleconferencing. The subscription rate is $60 a year for nine issues.

Center for Interactive Programs (CIP)
University of Wisconsin   608-262-4342
975 Observatory Drive
Madison, WI 53706

This international training and resource center tracks developments in the field of teleconferencing and electronic communications and acts as a clearinghouse for information and research. Through its programs and publications, CIP informs people about current developments, and provides professional training in teleconferencing. Each year CIP presents a number of seminars, a conference, and training programs. Publications include:

• *Telcoms*—reports exclusively on teleconferencing (audio, audiographics, video) and two-way electronic communications. It covers key factors in using and marketing a system in addition to new technologies, as well as giving a concise update on the industry with the latest developments in products and services. Subscription rate for eight issues a year is $65.

• *Teleconferencing and Electronic Communications, Volumes I & II*—a collection of articles devoted to teleconferencing and electronic communications. Volume I was published in 1982; Volume II in 1983; each costs $25.

• *Teleconferencing and Interactive Media* (1983)—a publication with information on developments and applications of interactive media. The 60-page publication is available for $20.

• *The Teleconferencing Directory*—an annual comprehensive and up-to-date resource on teleconferencing. Includes information on organizations using teleconferencing systems worldwide; hardware suppliers and providers of technical, production, and information sources; a summary of market projections; and a glossary of 150 terms. Available for $35.

• *Teleconferencing Technology and Applications* (1983)—a shopper's guide to all the current teleconferencing systems, their benefits, and general development trends. The major forms of teleconferencing and their applications are discussed: audio teleconferencing, audio-graphics, video teleconferencing, and computer conferencing. The 350-page publication is available for $45.

International Teleconferencing Association (ITA)

1299 Woodside Drive, Suite 100   703-556-6115
McLean, VA 22102

      An international professional association, ITA aims to provide a noncommercial source of information and statistics about teleconferencing. Its purpose is to educate the public and to address issues involving teleconferencing before local and federal regulatory and legislative bodies.

Institute for the Future (IFTF)

2740 Sand Hill Road   415-854-6322
Menlo Park, CA 94025-7097

      This research organization is involved in three general areas of research: forecasting and planning; teleconferencing; and teletext and videotex. The focus of its work in the teleconferencing area is on the social, organizational, and policy issues raised by the use of new teleconferencing systems. IFTF has many publications relating to teleconferencing. Examples include a working paper covering ten years of teleconferencing; a paper concerned with approaches to teleconferencing justification; and an audio conferencing handbook. The Institute's free publications catalog describes approximately 40 materials available to the public.

Publications

Data Communications Special Projects Center

1221 Avenue of the Americas   212-512-4852
New York, NY 10020

      The Center publishes:

      ● *Teleconferencing and Beyond: An Exploration of Communications in the Office of the Future*—this 250-page book by Robert Johnson presents the full story behind the teleconferencing revolution. Detailed chapters cover (1) appraisal of all current forms of teleconferencing, (2) historical overview, (3) teleconferencing today, (4) users' needs and wants, (5) proposals for new methods for planning and forecasting, and (6) a literature review. Also, more than 20 teleconferencing systems are examined and evaluated. Available for $35.

      ● *Definitive Buyer's Guide to Teleconferencing Products and Services*—is based on the information stored in a computerized data base created and maintained by *Telespan* (see below). There are 44 sections covering all major categories of products and services, with information on suppliers; products and services; company addresses, contacts and phone numbers; and an up-to-date glossary. Available for $40.

The Teleconferencing Resources Directory

Knowledge Industry Publications, Inc.   914-328-9157
701 Westchester Avenue   800-431-1800
White Plains, NY 10604

This reference guide locates hundreds of teleconferencing suppliers, including permanent rentable video conference facilities; consultants and system designers; transmission services; training and information sources; and manufacturers of such specialized equipment as speakerphones and freeze-frame TV, as well as manufacturers of basic audio and video equipment. Company listings include detailed information on whom to contact and specific services provided. The 148-page book is $47.50.

Teleguide: A Handbook for Video-Teleconferencing Planners
Public Service Satellite Consortium   202-331-1154
1660 L Street, NW, Suite 907
Washington, DC 20036

This book provides a comprehensive review of the considerations that must go into planning a teleconference, from reasons for choosing to teleconference through developing and designing an effective program.

Telespan
50 West Palm Street   213-797-5482
Altadena, CA 91001

Telespan is a monthly newsletter serving as a clearinghouse for information on suppliers, new equipment, public policy and procedures, current research, and other information pertaining to teleconferencing. It features in-depth case history articles, and a regular listing of single-event conferences. Annual subscription rate is $165 a year.

Additional Experts and Resources

Several universities research the topic of teleconferencing, and they are excellent sources for up-to-date information. Some of these schools are:

University of Wisconsin
Center for Interactive Programs
975 Observatory Drive
Madison, WI 53706
608-262-4342

New York University
Center for Interactive Telecommunications
725 Broadway
New York, NY 10003
212-598-3338

University of Southern California
Teleconferencing Research Group
Annenberg School of Communications
University Park
Los Angeles, CA 90089-0281
213-743-7400

Telephones

Organizations and Government Agencies

National Telephone Cooperative Association (NTCA)
2626 Pennsylvania Avenue, NW   202-342-8200
Washington, DC 20037

NTCA is a nonprofit trade association that represents small, locally owned and controlled telephone cooperatives and companies in the U.S. It keeps up to date on all the developments in telecommunications as it applies to telephone cooperatives. The staff will answer members' questions on all aspects of telecommunications, including new technologies and services. Publications include:

- *Rural Telecommunication*—focuses on key areas of access charges, corporate planning, and rural utility financing. Also included are articles, written in nontechnical language, on industry trends, new technologies, and the legislative and regulatory environment. Annual subscription is $12 a year.

- *The NTCA Exchange*—a newsletter of association activities.

- *Washington Report*—a weekly one-sheet update on legislative and regulatory developments.

Publications

Telephone Bypass News
Telestrategies, Inc.   703-734-7050
PO Box 1218
McLean, VA 22101

A monthly newsletter covering developments in telephone network local access and bypass technologies and local access regulatory developments. Covers status, forecasts, and updates of DTS, private microwave, cellular radio, cable television, and teleports. Annual subscription is $287 a year.

Telephone News
Phillips Publishing, Inc.   301-986-0666
7315 Wisconsin Avenue, Suite 1200N
Bethesda, MD 20814

A weekly newsletter that keeps the reader up to date on the most important events in the industry, such as new technology, telco mergers, FCC and congressional actions, new products, new state strategies that may shape national trends, top executive career moves, and so forth. The annual subscription rate is $247 a year for 50 issues. It is also available electronically via NewsNet, Inc. (800-345-1301 or 215-527-8030).

Teletext

See: Videotex/Teletext

Television

Organizations and Government Agencies

National Association of Broadcasters
1771 N Street, NW   202-429-5300
Washington, DC 20036   800-368-5644

NAB, which is comprised of the U.S. broadcasting industry, monitors issues affecting the electronic media. It holds a convention and several seminars annually. The NAB library maintains a wide-ranging collection of books, reports and periodicals on all aspects of broadcasting, and the staff of the Science and Technology Department will answer questions and help find the information needed. Call the Television Department or the Science and Technology Department for information about any aspect of the television industry, technology, conventions, publications, and so forth. NAB also publishes the following:

• *Telemedia*—a bimonthly magazine on various aspects of television, with a special focus on the business side.

• *Highlights*—a weekly Washington newsletter.

• *NAB Engineering Handbook*—a handbook on technological developments in the broadcasting field. The cost is $40 for members, $120 for non-members.

• *Proceedings* of their annual conventions are available.

• *Broadcasting Bibliography*—guide to useful available books in broadcasting. Available for $2.

Television Information Office
745 Fifth Avenue   212-759-6800
New York, NY 10022

This is an educational membership organization in the communications field. Its Department of Research Services maintains an extensive information center and library specializing in the social, cultural, and programming aspects of television. It collects and disseminates information about television from all print sources. It also publishes:

- *TV Mini-File*—a pocket summary of vital statistics and public opinions on television as a social force. Forty cents a pamphlet.
- *How Free TV Works*—basic facts about television. The cost is twenty-five cents a copy.
- *Periodicals in Broadcasting and Communications*—a selected bibliography listing primary sources of information about all aspects of broadcasting. It costs twenty cents a copy.
- *Publications List*—describes available items and their prices.

Electronic Industries Association (EIA)

2001 Eye Street, NW   202-457-4919
Washington, DC 20006

EIA is composed of manufacturers of radio, television, video systems, and equipment, and of electronics parts and components. Members submit data and EIA compiles the data into statistical reports. The association is a good source of statistical information on television equipment. It also sponsors trade shows on consumer electronics during the winter and summer, international telecommunications seminars and exhibits, and an annual Design Engineers and Electronic Component Conference. Information is provided to members and nonmembers by their library. EIA publishes:

- *Electronic Market Data Book*—an annual compendium of statistical information on the electronic industry. It is available for $55 from the Marketing Services Division.

Federal Communications Commission (FCC)

1919 M Street, NW   202-632-7000—Consumer Assistance
Washington, DC 20554   202-632-7100—Library Branch
                                      202-632-6357—Ray Pendarvis, Television Branch
                                      202-632-9660—John Reiser, Technical Branch

The FCC regulates interstate and international communications by television, cable, and satellite, and oversees the development and operation of broadcasting services nationwide. These services include commercial and noncommercial educational, pay and low-power television, TV translators, and experimental and developmental services. Those interested in more details about the FCC may obtain a publications list from the Office of Public Affairs. The FCC has several publications, such as books, reprints, journals, newsletters, publication lists, bibliographies, and regulations history background.

International Television Association

3 Dallas Communications Complex   214-869-1112
6311 North O'Conner Road, Suite 110
Irving, TX 75039

This membership organization is concerned with the use of television for communication in business, industry, health care, government, education, and other fields. They hold monthly chapter meetings, regional seminars, and workshops, as well as publish reports and other publications, and newsletters from the local chapters.

National Cable Television Association (NCTA)

1724 Massachusetts Avenue, NW   202-775-3550
Washington, DC 20036

NCTA serves as a major information clearinghouse on cable television. Its members include cable channels, equipment producers, publishing companies, distributors, and interested individuals. The Association serves as a national medium for an exchange of experiences developed through research, studies, and publications. It also provides information on legal and regulatory issues affecting the cable television sector. The staff will provide orientation and answer questions of members and nonmembers.

Publications

Broadcasting Publications, Inc.

1735 DeSales Street, NW   202-638-1022
Washington, DC 20036

This company publishes some of the major sources of information in the field of television. Its main publications are:

- *Broadcasting Magazine*—this weekly journal of the communications industry covers television, cable television, radio, and satellites. Annual subscription rate is $60 a year.
- *Broadcasting Cablecasting Yearbook*—is a standard reference volume listing the rates for television stations, major television advertisers, advertising agencies, addresses of major programming companies, and courses available in the broadcasting field. The *Yearbook* also includes a brief history of FCC regulations affecting radio broadcasting. Annual subscription rate is $80 a year.

Electronics

McGraw-Hill, Inc.   212-512-2000
1221 Avenue of the Americas
New York, NY 10020

Electronics, one of the major weekly magazines in the field, covers electronics technology and electronics in the marketplace. Cost is $24 a year, or $59 for three years.

Television Digest, Inc.

1836 Jefferson Place, NW   202-872-9200
Washington, DC 20036

This company publishes several television-related materials. It offers free four-week trial subscriptions to some of these:

- *Television and Cable Factbook*—a complete compilation of television stations, cable systems, the services they provide, the markets they serve, and the equipment used, as well as current information on equipment manufacturers and suppliers, professional associations, industry consultants and specialists, regulatory agencies, and so forth. Available for $165.

● *Television Action Update*—this weekly addenda to the *Television Factbook* presents news of FCC actions affecting television stations. Annual subscription rate is $187 a year.

● *Television Digest with Consumer Electronics*—this weekly reports on the broadcast, cable, and consumer electronics fields. Half of the report is devoted to broadcast television topics, with coverage of companies, legislation, new technology, conferences, and other news items. Annual subscription rate is $406 a year.

● *Cable & Station Coverage Atlas*—a reference guide to cable and television station signal coverage and distribution information. Chapters include low-power television stations, subscription television station directory, earth station directory, MDS directory, microwaves serving cable systems, copyright law, cable rules, cable television organizations, management and technical services directory, and cable television equipment directory. The cost of each annual edition is $114.

● *Cable Action Update*—weekly for a cost of $187.

● *Satellite Week*—weekly for a cost of $442.

● *Communications Daily*—daily for a cost of $995.

● *Video Week*—weekly for a cost of $378.

● *Public Broadcasting Report*—biweekly for a cost of $176.

Television/Radio Age

Rockefeller Center   212-757-8400
1270 Avenue of the Americas
New York, NY 10020

A biweekly broadcast journal covering all aspects of the radio and television industry. Each issue includes the following features: Inside the FCC, Viewpoints, the Wall Street Report, Programming and Production, Commercials, Business Barometer, and Special Studies. Annual subscription rate is $40.

● *Television Age International*—this quarterly supplement to *TV/Radio Age* is distributed throughout the world. Feature articles are devoted to the developments in the broadcasting industry outside the U.S. Subscription rate is $20 a year.

U.S. Industrial Outlook

U.S. Department of Commerce   202-377-2872
Washington, DC 20023

This annual publication is an excellent source of industry information for the television and radio industries. Two chapters are devoted to the topics "Radio and Television Communications Equipment" and "Broadcasting." Along with the industry forecasts are included definitions, explanation of new technology, referrals to other sources of information, and the name of the industry expert for television and radio. The *U.S. Industrial Outlook* is organized by standard industrial classification

(SIC) codes. Available for $14 from the Government Printing Office, Superintendent of Documents, Washington, DC 20402.

Additional Experts and Resources

U.S. Department of Commerce
International Trade Administration   202-377-2872—Arthur Pleasant
14th Street and Constitution Avenue, NW, Room 1015B
Washington, DC 20230

Mr. Pleasant is an electronics industry specialist in military and industrial radio and television equipment within SIC 3662. He will answer questions and make referrals to other experts in the field. He does not deal with inquiries about broadcasting.

Test Tube Babies

See: Reproductive Technology

Textiles

See also: Apparel

Organizations and Government Agencies

Economic Information Division
American Textile Manufacturers Institute   202-862-0500
1101 Connecticut Avenue, NW
Washington, DC 20036

This trade association of the textile industry represents manufacturers dealing with the raw fiber and the processing of the fiber. It has an annual three-day meeting in the spring of each year, consisting of seminars, workshops, and keynote speakers covering the latest technology. The Institute provides seminars throughout the year to disseminate information on such topics as taxes, health and safety, computer automation, and the latest research and development in the field. Besides its lobbying activities in the U.S. Congress, it is also involved in the regulatory and administrative functions of the government agencies. It has a membership of 200 companies, and a weekly newsletter is published for the members. It provides the following publications:

- *Textiles High Lights*—a quarterly economic report with monthly supplements. It provides a general overview of the textiles industry and the economy, with focus on prices, international trade, fiber consumption, production, and fiber resources. The 34-page report and seven-page supplements are $50 a year.
- *Educational Brochures*—geared toward academic institutions. They cover processes involved in changing fiber to cloth, and job prospects for students in the textile/design field. They are available free of charge.

Fiber Society
Textiles Research Institute   609-924-3153
PO Box 625
Princeton, NJ 08540

This is a nonprofit organization for the advancement of scientific knowledge pertaining to fiber products or fibrous materials. Its 400 members include chemists, physicists, and engineers interested in fiber sciences and related topics. The Society provides speakers for universities and interested groups on fiber science technology. It also sponsors two technical meetings a year, one a research symposium and the other a conference. All publications are for members only, and are not available to the public.

Gurber Garment Technology Company
55 Gurber Road West   203-644-2401
South Windsor, CT 06075

Gurber manufactures computer-aided design and computer-aided manufacturing equipment for the garment industry, such as: pattern marking, pattern grading, automated cutting, and computer-automated manufacturing systems that consist of material handling and controlled production in the sewing room. Staff provide seminars for customers on the latest technological advancements. The company publishes brochures on their products, and a bimonthly newsletter on the latest technology and advances in the field of automation in the garment industry. Both the newsletter and the product brochures are available free of charge.

Institute of Textile Technology
PO Box 391   804-296-5511
Charlottesville, VA 22902

This research center and graduate school concentrates on textile manufacturing. It is funded by 40 member manufacturing companies involved in the development of raw material fibers. The Institute offers seminars and courses in management, computer-aided manufacturing, textile processing, weavings, and many other topics. It produces several publications available to the public, including:

● *Textile Technology Digest*—a monthly abstract journal covering more than 700 articles and journals devoted to the latest innovations and technological developments in textiles. Annual subscription is $300. *Textile Technology Digest* is also available as an on-line computerized data base through DIALOG.

● *Institute of Textile Technology Report*—an annual report providing comprehensive coverage of a specific topic in the textile field. A different subject is covered each year. Cost is $50.

Society for the Advancement of Material and Process Engineering (SAMPE)

PO Box 2459   213-331-0616
Covina, CA 91722

SAMPE is a nonprofit international society of 5,400 members, dedicated to the advancement of material and process engineering. Each year it sponsors two conferences, one technical conference and one symposium, and exhibits displays of the latest technical products in the field. Papers from the conference are published. Their other publications include:

● *SAMPE Journal*—a bimonthly publication that covers the latest technology advances in the field, and provides abstracts profiling the research and development achievements in the field. A copy is provided with membership. The 200-page publication is available for $31 a year.

● *SAMPE Quarterly*—a quarterly publication containing papers submitted at the two SAMPE conferences held each year. The 100-page publication is available for $25 a year.

Textile Research Institute

PO Box 625   609-924-3150
Princeton, NJ 08542

This nonprofit research institute is involved in the research and development of the materials and processes of the fibrous materials industry. The staff is involved in the research of polymer, fiber, and textile systems designed to aid the industry in product and process development, in quality and productivity enhancement, and in energy conservation and pollution control. It sponsors educational programs on fiber, textile science, and technology, and annual symposia and conferences dealing with such topics as computer applications to textile processing, textile dyeing and finishing, and textile evaluation. The Institute is also involved in product and environmental regulations. It publishes over 20 papers yearly on topics such as "Foam Wet Processing in the Textile Industry," and "Dyeability-Structure Relationships in High-Speed Spun Yarns." A complete listing of their publications is available from the Institute. A publication available by subscription is:

● *Textile Research Journal*—a monthly publication covering fundamental and applied scientific information on the physical, chemical, and engineering sciences related to the textile and apparel industry. The comprehensive journal provides illustrations, charts, and graphs. The 75-page publication is available for $100 a year.

Publications

Daily News Record
Fairchild Publications   212-741-4470
7 East 12th Street
New York, NY 10003

Published Monday through Friday, this newspaper reports on textiles, men's apparel, and retail. The paper provides information about new technological developments, and it contains legislative and financial news. Annual subscription rate is $50.

Symposium on Cotton Dust: Sampling, Monitoring and Control
American Society of Mechanical Engineers   212-705-7785
345 East 47th Street
New York, NY 10017

This onetime publication is a collection of technical papers on research and development in the field of high-tech cotton dust control that were presented at a conference in 1980. The 65-page publication (order #I00136) is $12 and can be obtained from PO Box 3199, Grand Central Station, New York, NY 10163.

Textile World
Circulation Department   609-426-5000
McGraw-Hill, Inc.
Princeton-Hightstown Road, N-1
Hightstown, NJ 08520

A monthly publication covering all phases of the textile industry from the raw fiber through the processing to fabric. It provides new product news, trade literature, automation, electronics, robotics, reports on new fibers, and special features monthly on manufacturing techniques and new machinery uses. Annual subscription rate is $35.

Textiles Products and Processes
4170 Ashford Dunwoody Road, Suite 420   404-252-0626
Atlanta, GA 30319

A small bimonthly publication that provides listings of new products, new machinery, and equipment in the textiles field. Annual subscription is $25.

Additional Experts and Resources

School of Textiles
North Carolina State University   919-737-3059—General Information
Raleigh, NC 27695-8301   919-737-3043—Library

The school offers undergraduate and graduate courses on the latest technology in the textile industry. The courses cover such topics as computer-based markers for pattern construction, the application of the computer in textile testing and quality control, textile production in the apparel industry, robotics automation, and management techniques. The school has a large library with extensive sources on the textile industry and high technology and can provide literature surveys and pathfinders listing background information on specific topics pertaining to the apparel industry. The library is available for use by the community and members in the industry. You may contact the Textile Library at 919-737-3043 for telephone resource information. All the publications by the faculty of the School of Textiles are available in the library. Available publications include:

- *Newsletter*—a quarterly publication, provides synopses of the research being done in this field, covers the new publications by the School of Textiles, and spotlights noted publications in the field. The newsletter is available free of charge.
- *The School of Textiles Research Abstract*—provides a brief description of all publications/reports on research being done by the school. It covers the sponsor and principal investigators, and their objectives and conclusions. Available free of charge.

Dr. Perry O. Grady
Associate Dean, School of Textiles   919-737-3059
North Carolina State University
Raleigh, NC 27695

Dr. Grady, Associate Dean of the School of Textiles, is a noted authority on advancements in the textile industry. He is available on a limited basis as an expert in this field. He was co-editor of a textbook covering robotics and computer automation in the field of textiles.

Frances Zeglan
Director of Textile School   215-951-2757—Hotline
Philadelphia College of Textiles and Science
School House Lane and Henry Avenue
Philadelphia, PA 19144

Frances Zeglan, Director of the Textile School, is involved in research and development in the fields of apparel, knitting, woven fabrics, and nonwoven fabrics. He is an expert on new textile developments, and can provide assistance, answer questions, and make referrals to other experts in the field of textiles. The School's major emphasis of study and ongoing research is in the area of product evaluation, and they provide a free hotline service to help callers with specific questions about textiles.

Thermal Sciences

See: Solar Technology

Three-Dimensional (3-D)

See: Computer Animation

Timesharing Services

See: Computers (General); Telecommunications

Tomography

See: Medical Imaging

Toxicology

See also: Pharmaceuticals

Organizations and Government Agencies

Agency for Toxic Substances and Disease Registry
Superfund Implementation Group   404-452-4100
Atlanta, GA 30333

This intergovernmental group is involved in coordination and planning—with other agencies such as the Food and Drug Administration (FDA), Centers for Disease Control (CDC), National Toxicological Program (NTP), and National Institutes for Health (NIH)—to ensure that the cleanup of hazardous waste dump sites and hazardous spills is carried out. The agency oversees a comprehensive environmental program involving testing of chemicals, health studies, testing equipment and clothing for safety, training programs, data base development, and coordinating adequate safety in cleanup sites and cleanup groups. It does not have any publications.

Centers for Disease Control

Clinical Chemistry Division   404-452-4176
Toxicology Branch
1600 Clifton Road
Atlanta, GA 30333

Staff members at this government agency analyze pesticides, arsenic and poisons, trace metals, and dioxins. They can provide analytical data about the clinical research performed on certain chemical compounds.

Center for Environmental Health

Centers for Disease Control   404-452-4161
Atlanta, GA 30333

This government agency publishes more than 150 free publications a year on a wide variety of topics. Subjects have included cancer, laboratory techniques, population studies, lead poisoning, and birth defects. A free publication catalog is available from the Center.

The National Academy of Sciences

Toxicology Information Center   202-334-2387
2101 Constitution Avenue, NW
Washington, DC 20418

The Academy studies specific topics and problems for various government agencies such as the FDA and EPA, and committees are set up consisting of experts from the Academy. The findings of the research, generally taking 18 months to complete, are published in reports and are available from the National Academy Press.

National Institute of Environmental Health Sciences (NIEHS)

PO Box 12233   919-541-3345
Research Triangle Park, NC 27709

This federal agency is studying the effects of chemical, physical, and bio-metabolism on health. NIEHS's pharmacology staff members conduct research on toxicological effects and metabolic pathways in the body and affected organs. They also explore harmful agents in the environment, neurological and behavioral toxicology, pulmonary functions, genetics, and reproduction and development. NIEHS publishes the following:

• *Annual Report to the Science Community*—an annual report providing an overview of the research and developments achieved in the field over the year. The 400-page publication is available free of charge.

• *Research Programs of NIEHS*—a biannual publication that outlines the research being conducted by the Institute and new findings. The 80-page publication is available free of charge.

Refer to the additional listings below for more information about NIEHS's National Toxicology Program.

National Institute of General Medical Sciences (NIGMS)
Building 31, Room 4A-52   301-496-7301
9000 Rockville Pike
Bethesda, MD 20205

NIGMS staff can provide information and/or referrals to other agencies and experts in the field of toxicology. The Institute provides grants and training in pharmacological sciences on the predoctoral and postdoctoral levels. NIGMS publishes the following:

● *Medicines and You*—provides information on how your age, genes, and diet affect the way medicines work in your body. This booklet describes what medical researchers are learning about the biological individuality of each human being's response and tolerance to medicines and other chemicals. The 62-page publication is available free of charge.

● *Inside the Cell*—this booklet describes discoveries about the cell that are leading scientists to develop new concepts of health and disease. The 96-page publication is available free of charge.

● *New Human Genetics*—provides information about how gene splicing helps researchers fight inherited disease. The 48-page publication is available free of charge.

● *Research Reports*—published several times a year, this is a collection of the research being done by NIGMS covering such topics as genetics and pharmacology, scientific instrumentation, biomedical engineering, and cell biology. A mailing list is maintained for this free publication.

National Toxicology Program (NTP)
National Institute of Environmental Health Sciences   919-541-3991—Information Office
B2-04, PO Box 12233
Research Triangle Park, NC 27709

NTP studies the effectiveness rate of chemicals and tests for carcinogenic toxins. It publishes the following (a mailing list is maintained for all publications):

● *Annual Report of Carcinogens*—available each December, it provides a detailed rundown on the effectiveness rate and metabolic rate of various chemicals based on controlled dosages. Illustrative graphs and charts are provided. Annual subscription rate is $32.50.

● *Summary of Annual Report on Carcinogens*—provides a summary of the research findings of 117 chemicals found to be carcinogenic. Covers the laboratory animals used in the testing and the dosages given. The 235-page publication is available free of charge.

● *Technical Report Series*—contains 250 reports, each covering a specific chemical. The reports are issued when final testing of the chemical is completed. Each report provides an in-depth explanation of the research and testing performed. The reports run approximately 75 to 100 pages and are available free of charge.

● *NTP Annual Plan*—a detailed listing of all ongoing and projected research studies. Volume I primarily covers NTP research underway, and plans for future projects. Volume II reviews the toxicology research being conducted by other federal agencies. Both volumes are available free of charge.

• *NTP Technical Bulletins*—published every few months, it provides updates on current research projects as well as test results. The 25-page publication is available free of charge.

• *Environmental Health Perspective*—a bimonthly publication dealing with the potential health hazards associated with particular elements in the environment. Emphasis is placed on problem areas, and toxicological information summaries are provided on metallothioein cadmium, environmental epidemiology, and target organ toxicity issues. Back issues are available from NIEHS and current issues from the Government Printing Office. The 300-page publication is available for $40 a year, or $8 an issue.

Society of Toxicology

475 Wolf Ledges Parkway   216-762-2289
Akron, OH 44311

This is an organization of 1,800 scientists practicing toxicology in laboratories, government, industry, and educational institutions. Members are involved in studying the physiological effects on the body of the environment, chemicals, drugs, and other toxic substances. The Society hosts an annual meeting in March consisting of seminars, workshops, and exhibits. Papers submitted are reviewed and published in the journals. The Society's publications are:

• *Journal of Fundamental and Applied Toxicology*—a bimonthly publication, providing information about regulatory issues, drug safety, chemical safety evaluation standards, genetic toxicology, and acute and chronic drug use. The *Journal* includes conference papers, current issues in the field of toxicology, and articles about upcoming events. The 250-page publication is available for $25 to individuals, $68 to institutions, and free to members.

• *Journal of Toxicology and Applied Pharmacology*—a monthly publication covering the cellular and subcellular levels of toxic effects of chemicals, drugs, and natural toxins such as plants and snakes. The 1,200-page publication is available free to members, or for $40 a year to others.

Toxicology Information Response Center (TIRC)

Oak Ridge National Laboratory   615-576-1743
PO Box X
Building 2024
Oak Ridge, TN 37831

The staff at the Center can provide information, extensive bibliographies, computerized data base searches, and publications on all aspects of toxicology. The staff has access to a vast library and more than 100 computerized government and commercial data bases. A free publication catalog is available from TIRC.

Publications

Elsevier Science Publishing Co., Inc.
Journal Information Center   212-867-9040
52 Vanderbilt Avenue
New York, NY 10017

The following are published by Elsevier:

● *Excerpta Medica Abstracts*—produced in Amsterdam, the Netherlands, *Abstracts* covers 45 medical and scientific topics. Some of these topics are toxic interactions, mutigens genetics, target organ toxicology, predictive toxicology, and epidemiology and toxicological methods and laboratory findings. These are published as 20 issues in two volumes, both available for $322 a year.

● *Toxicology Journal*—is published in Ireland. This international journal is devoted to that aspect of toxicology concerned with the effects of chemicals on living systems. The audience consists of toxicologists, pharmacologists, and epidemiologists. The *Journal* probes the effects of food additives and pesticides on animals and humans. The *Journal* is published as 16 issues in four volumes and is available for $454 a year.

Food and Chemical Toxicology Journal
Pergamon Press, Inc.   914-592-7700
Maxwell House
Fairview Park
Elmsford, NY 10523

A monthly journal containing original papers and reviews relating to biological and toxicological research findings, with special emphasis on food. Annual subscription rate is $370.

Journal of Toxicology
Marcel Dekker, Inc.   212-696-9000
270 Madison Avenue
New York, NY 10016

The journal consists of three separate publications:

● *Clinical Toxicology*—published six times a year, this covers clinical research studies being conducted in the field. Information is provided about research methods and results. Annual subscription rate is $195.

● *Cutaneous and Ocular Toxicology*—published four times a year, this publication covers research on toxins affecting the skin and eyes. Available for $155 a year.

● *Toxin Review*—published three times a year, the publication provides general information about toxins, highlighting the latest research and developments in toxicology. Available for $95 a year. Subscription to all three journal publications is available for $383 per year.

Micromedex

2750 South Shoshone Street   800-525-9083
Englewood, CO 80110   303-781-7813

Micromedex is a publisher, affiliated with the Rocky Mountain Poison and Drug Center, that specializes in medical and drug information and pharmacology. It publishes the following:

- *Drugdex*—this quarterly publication provides monographs covering dosage information, pharmocokenetics, absorption, distribution, metabolism, excretion, cautions of drugs and adverse reactions, drug interactions, and therapeutic indications. Available in microfiche for $2,075 a year, or on computer tape or videodisc.

- *Poisondex*—the quarterly publication draws from a toxicology data base that deals with the identification of 350,000 substances. Indexed by trade name, brand name, generic name, and chemical name, each entry provides specific ingredients data and detailed information on how to treat any of these products that have been ingested, absorbed, or inhaled. Available in microfiche for $1,785 a year, or in computer tape or videodisc format.

National Academy Press

2101 Constitution Avenue, NW   202-334-3113
Washington, DC 20418

A division of the National Academy of Sciences, it has over 900 publications available on many different topics including toxicology and pharmacology. A free publication directory is available from the Academy. *Toxicity Testing*, a onetime 395-page publication, is available for $22.50. It provides information about chemicals that have been tested and proven to be toxic.

Toxicology Newsletter

417 Mellon Hall   412-434-6380
Duquesne University
Pittsburgh, PA 15219

Toxicology is a quarterly publication providing information about jobs in the field and details about upcoming seminars nationwide. The four-page publication is available free of charge.

Data Bases

MEDLARS

National Library of Medicine   301-496-6193
9000 Rockville Pike
Bethesda, MD 20814

The Library has four on-line data bases pertaining to toxicology. For details about accessing the system, see the MEDLARS entry under *Medical Technology*.

- *CHEMLINE*—a chemical dictionary that provides registry numbers for over 600,000 chemicals, synonyms of the chemical names, and locator information as to where the registry number is used in the other Library of Medicine files. Average cost is $56 an hour.

- *TOXLINE*—a bibliographic data base that covers the literature from 1965 to the present, this is a freetext searchable data base, not an indexed file data base. Average cost is $59 an hour.

- *RTECS*—*Registry of Toxic Effects of Chemical Substances*, produced by the National Institute of Occupational Safety and Health (NIOSH). Deals with occupational chemicals and is essentially a chemical file. Each record is a chemical with information on acute and chronic toxicity, with standards and regulations. NIOSH created and maintains the data base. Average cost is $55 an hour.

- *TOXICOLOGY DATA BANK (TDB)*—each record is a chemical pair reviewed by a team of toxicologists who look at the sources of information and the data used to build the record. A comprehensive record of information on manufacturer use, environmental aspects, toxicity aspects (both human and animal), metabolism, antidote and treatment, poison potential, and pollution potential to soil, water, and air. Average cost is $75 an hour.

Cost for the different data bases varies, depending on the extent of the use. The rate is less when the data base is used during nonprime time, but there is an additional charge for printed information.

Additional Experts and Resources

Dr. Vera Glockin, Assistant Director for Pharmacology & Toxicology
Office of Drug Research and Review   301-443-4330
Center for Drugs and Biology
U.S. Food and Drug Administration (FDA)
5600 Fishers Lane
Rockville, MD 20857

An expert in the field of toxicology, Dr. Glockin is involved in the development of drug toxicology guidelines for new drugs. She reviews drug applications submitted to the FDA. She is involved with synthetic drugs (not biological drug development), and she and her staff provide guidelines for new drug production. Dr. Glockin can answer questions and refer you to other toxicology experts in the government.

Dr. Frederick Oehme
Comparative Toxicology Laboratories   913-532-5679
VCS Building
Kansas State University
Manhattan, KS 66506

An expert in comparative and veterinary toxicology, Dr. Oehme has published more than 450 scientific papers in the past 20 years. He can answer technical questions and refer you to other experts and resources.

Dr. Allen Heim, Director of Science Coordination
U.S. Food and Drug Administration (FDA)   301-443-1587
5600 Fishers Lane
Rockville, MD 20857

In an attempt to eliminate duplicated research, Dr. Heim's office is responsible for coordinating FDA's in-house science activities, as well as FDA's activities involving other federal agencies, such as EPA, NIH, CDC, and NIOSH. Dr. Heim and his staff can direct you to appropriate toxicology research programs and experts. FDA itself has several committees dealing with toxicology issues and regulatory guidelines.

Trade Shows, Computer

See: Conferences, Meetings, and Trade Shows on High Technology

Trademarks

See: Patents, Trademarks, and Copyrights

Transplantation Technology

See also: Medical Technology

Organizations and Government Agencies

American Council on Transplantation
1825 Eye Street, NW, Suite 400   202-429-2067
Washington, DC 20006

The Council is an independent, voluntary federation of organizations and individuals concerned with the transplantation and donation of both organs and tissues. It was established to educate both health professionals and the general public about donating organs and tissues, as well as to address the technical, ethical, and financial issues involved in organ recovery and transplantation. Staff members are knowledgeable about information resources pertaining to transplantation and they can answer questions about new developments in the field.

American Society of Transplant Surgeons

716 Lee Street   312-824-5700
Des Plains, IL 60016

This professional society of 350 physicians specializes in transplant surgery. It sponsors an annual meeting, the proceedings of which are published in the Society's publication:

- *Journal of Transplantation*—a monthly publication for researchers and clinicians who need to know what is going on in immunology. It covers such specialties as hematology, endocrinology, genetics, and embryology. Annual subscription fee for the journal is $55 for members, $115 for institutions, and $95 for others. Order from Williams and Wilkins Company, 428 East Preston Street, Baltimore, MD 21202, 301-528-4255.

Living Bank

PO Box 6725   800-528-2971
Houston, TX 77265   713-528-2971

This is the only national organ donor referral service that maintains a complete listing of all organ banks and transplant facilities in the United States. It provides educational services for both the health professional and the general public. Free informational brochures, a quarterly newsletter, and educational films are available from the Living Bank. The films are loaned free of charge.

National Association of Patients On Hemodialysis and Transplantation (NAPHT)

150 Nassau Street   212-619-2727
New York, NY 10038

NAPHT is a support group for patients that has regional chapters and local groups throughout the United States. It conducts educational programs about kidney transplants. In addition to informational pamphlets, the Association publishes:

- *NAPHT News*—a quarterly publication covering patient care, diagnostic techniques, medical treatment, research, and technical advances in the field of kidney transplantation. Each issue runs approximately 30 pages and the annual subscription fee is $10.

National Heart, Lung, and Blood Institute

Public Inquiries and Reports Branch (PIRB)   301-496-4236
National Institutes of Health
Building 31, Room 5A-52
9000 Rockville Pike
Bethesda, MD 20205

The Institute, which is part of the federal government, conducts and funds research pertaining to heart and lung transplantation. Staff in the PIRB office can answer questions and refer you to experts in the field, government programs, and research centers. The office has prepared a 30-page information packet about transplantation, which consists of research reports, papers, and articles collected from medical literature, as well as publications geared toward the general public. This packet and other transplantation publications are available free of charge from the above office.

National Institute of Arthritis, Diabetes, Digestive, and Kidney Diseases

Information and Health Research Reports Office   301-496-3583
Building 31, Room 9A-04
9000 Rockville Pike
Bethesda, MD 20205

The Institute, which is part of the federal government, conducts and funds research pertaining to transplantation. The staff can provide information, direct you to government reports, and send you publications. The office's information specialists, listed below, can refer you to research centers, experts, government programs, and literature in their assigned area:

Bone—Barbara Weldon   **Liver—Bill Hall**

Kidney—Jim Fordham   **Pancreas—Charlotte Armstrong**

National Kidney Foundation

2 Park Avenue   212-889-2210
New York, NY 10016

The Foundation sponsors research, provides educational materials and seminars for both the health professional and the general public, and compiles statistics relating to kidney disease and treatment. Its affiliates and local groups nationwide provide patient services and eduational programs. It has several professional councils, one of which is Clinical Nephrology Dialysis and Transplantation. This Council publishes a quarterly newletter about developments in the field. In addition to educational pamphlets about kidney dialysis and transplantation, it also publishes:

● *The American Journal of Kidney Diseases*—a monthly publication geared toward the physician. It covers all aspects of kidney dialysis, transplantation, and disease. The *Journal* is available for $70 a year from Grune and Stratton, Inc., 111 Fifth Avenue, New York, NY 10003, 212-614-3000.

● *The Kidney*—a bimonthly publication written for physicians. It contains medical articles about the kidney. Annual subscription rate is $20.

North America Transplant Coordinator Organization

Ohio Valley Organ Procurement Center   513-872-4156
University of Cincinnati Medical Center
Department of Surgery
231 Bethesda Avenue
Cincinnati, OH 45267

This organization is dedicated to improving the quality of life for people with endstage organ failures. It provides a forum for the exchange of information about all phases of transplantation. Members are clinical transplantation coordinators and procurement coordinators who deal with kidney, heart, liver, heart/lung, and tissue transplants. Many members work with high-tech immunosuppressive drugs and drugs for organ preservation. The organization sponsors meetings, workshops, and training courses, and publishes a quarterly newsletter for members. It has committees on legal/ethical issues, scientific studies, and education.

The organization operates a 24-Hour Alert System to help match donors with patients needing transplants. This voice-activated computer system has facilitated hundreds of organ placements at 41 transplant centers nationwide. It also operates the United Network for Organ Sharing (UNOS) System, an on-line data base designed to facilitate kidney placement.

Transplantation Society

New England Deaconness Hospital   617-732-8549—Dr. Mary Woods
185 Pilgrim Road
Boston, MA 02215

A professional society of physicians and scientists involved in transplantation biology and medicine, the organization keeps its members informed about the latest advances in transplantation technology. It sponsors a yearly conference and publishes:

● *Transplantation*—a monthly collection of papers about clinical and experimental transplantations. Annual subscription fee to the 100-page publication is $65 for members and $95 for nonmembers. It can be ordered from Williams and Wilkins Company, 428 East Preston Street, Baltimore, MD 21202, 301-528-4255.

● *Transplantation Proceedings*—a quarterly publication containing the proceedings of conferences on transplantations. Annual subscription is $89. It is available from Grune and Stratton, Inc., 111 Fifth Avenue, New York, NY 10003, 212-614-3000.

Publications

Dialysis and Transplantation Journal

Dialysis and Transplantation, Inc.
7628 Densmore Avenue
Van Nuys, CA 91406

Published monthly, the *Journal* contains clinical and scientific papers about transplant studies, treatment techniques, nursing and technical skills, and the legal aspects of dialysis and transplantation. Profiles of new products and book reviews are also included. The 60-page publication is available for $35 a year.

Additional Experts and Resources

Listed below are the Directors of several major transplant centers in the United States. Specialty areas are noted when applicable.

Dr. Bartley P. Griffith (Heart)
Director, Department of Surgery
University of Pittsburgh School of Medicine
Scaife Hall, Room 1084
Pittsburgh, PA 15261
412-624-2677

Dr. Thomas Hakala (Kidney)
Director, Renal Transplant Office
University of Pittsburgh School of Medicine
Presbyterian University Hospital
230 Lothrop Street, Room 4414
Pittsburgh, PA 15213
412-624-2680

Dr. Thomas Starzl (Liver)
Director, Falk Clinic
University of Pittsburgh School of Medicine
3601 Fifth Avenue, Room 103
Pittsburgh, PA 15213
412-624-1042

Dr. John Najarian
Chairman, Department of Surgery and
    Transplant
University of Minnesota
420 Delaware Street, SE
Box 195 MAYO
Minneapolis, MN 55455
612-373-8808

Dr. Paul Terasaki
Director, Tissue Typing Laboratory
University of California
1000 Veterans Avenue, Room 15-22
Los Angeles, CA 90024
213-825-7651

Transportation

See also: Marine Technology

Organizations and Government Agencies

American Society of Mechanical Engineers (ASME)
345 East 47th Street   212-705-7722
New York, NY 10017

A nonprofit educational and technical organization, ASME encourages the development of new technologies, including developments in transportation. ASME has several divisions that deal with different aspects of transportation and with the technology involved in the production of vehicles and parts. Its Public Information Office can provide technical information and give referrals. The Society sponsors several technical conferences on transportation annually, and its proceedings are published. A free brochure describing ASME's activities and publications can be obtained from its office.

Institute for Transportation

American Public Works Association   312-667-2200
1313 East 60th Street
Chicago, IL 60637

The Institute is a division of the American Public Works Association, a nonprofit public service organization with more than 20,000 members, mostly employees of federal, state, or local governments. The Institute is concerned primarily with the development and management of the infrastructure required for all modes of transportation. It maintains liaison with related interest groups, evaluates and analyzes laws and regulations related to transportation, undertakes studies, develops research projects, and sponsors educational programs.

Technical inquiries should be directed to Office of Transportation Research. Free brochures describing the Institute, its publications, and the Association are available upon request.

Institute of Transportation Studies

University of California   415-642-3593—Public Information Office
109 McLaughlin Hall   415-642-3604—Transportation Library
Berkeley, CA 94720

Specialists at the Institute conduct technical research in all aspects of surface and air transportation, excluding marine transportation. Its library staff will perform bibliographic data base searches of transportation literature, and a fee may be assessed depending upon the extent of the search. The Public Information Office will answer technical inquiries and/or make referrals. The Institute publishes:

• *ITS Review*—a quarterly publication that features technical articles about transportation topics. It is available free of charge from the Institute's library.

Society of Automotive Engineers (SAE)

400 Commonwealth Drive   412-776-4841
Warrendale, VA 15096

SAE is a nonprofit organization devoted to every field of self-propelled vehicles. It has more than 40,000 members who are engineers or professionals in jobs related to automation engineering. The Society sponsors ten annual technical meetings, all of which are open to the public, and the proceedings from the conference are published. Staff will respond to technical inquiries. SAE maintains a data base of information about the automotive field, and you can request a search (a fee may be assessed for this service). SAE's library services can be used by members and nonmembers. Staff will make copies of articles dating back to 1905 for a cost-recovery fee. Publications include:

• *Publications Catalog*—an annual list of the most recent papers and books issued by SAE. Available free of charge.

• *Handbook SAE*—an annual publication that contains all the standards set by SAE for a wide variety of vehicles and their parts. The four-volume set is $65 for members, and $125 for nonmembers. A separate index is $10.

Transportation Laboratory

Department of Transportation
Office of Engineering Services
PO Box 19128
Sacramento, CA 95819

916-739-2400—General Information
916-739-2452—Library and Technical Information

The Laboratory has transportation specialists who develop research in areas such as transportation engineering; highway materials; testing, research, and development; highway safety systems; and pavement design. If you are seeking technical information about a particular topic, you can contact the Laboratory's library at the phone number listed above. Its staff can answer inquiries and make referrals. The library staff will send you free listings of publications issued by specialists at the Laboratory, as well as abstracts of ongoing Laboratory research.

Transportation Research Board

National Academy of Science
2101 Constitution Avenue, NW
Washington, DC 20418

202-334-2934—General Information
202-334-3213—Publications Department

This unit of the National Research Council serves the National Academy of Sciences and the National Academy of Engineering. The Board is involved with the nature and performance of transportation systems and publishes its research findings. It has over 250 committees, as well as task forces and panels that involve engineers, educators, social scientists, and other transportation professionals. Staff can provide technical information about transportation and give referrals. A free publications catalog can be obtained from the Publications Department.

Transportation Systems Center

Department of Transportation
Kendall Square
Cambridge, MA 02142

617-494-2486—Office of Technology Sharing

A research and development arm of the Department of Transportation, the Center is involved in all aspects of transportation, as well as vehicles materials. Technical staff in the Technology Sharing Office can answer inquiries and give referrals. Abstracts of the technical reports produced by the Center can be searched on the data base TRIS, or be obtained from NTIS, 5885 Port Royal Road, Springfield, VA 22161, 703-487-4807.

Data Bases

TRIS-ON-LINE (Transportation Research Information Services On-line)

Transportation Research Board (TRB)
2100 Constitution Avenue, NW
Washington, DC 20418

202-334-3250

This data base contains summaries of ongoing transportation research projects; abstracts of recently completed research projects, as well as journals, technical papers, and articles; and bibliographic announcements pertaining to railroad, highway, maritime, aviation, pipeline, and urban transportation research. TRIS covers more than 190,000 records and several thousand abstracts of transportation research projects. It is updated annually. TRIS-ON-LINE is accessible through Lockheed DIALOG file #63. If you do not have access to DIALOG, TRB staff will search their data base for you on a cost-recovery basis.

TU

See: Technology Transfer

Ultrasound in Medicine

Organizations and Government Agencies

American Institute of Ultrasound in Medicine (AIUM)
4405 East-West Highway, Suite 504   301-656-6117
Bethesda, MD 20814

The Institute is for those interested in the use of ultrasonic radiation. Its purpose is to promote the use of ultrasonic radiation in clinical medicine, diagnostically and therapeutically, and in research. AIUM maintains a small library and holds an annual convention, after which the *Scientific Proceedings* are published. It also produces the *Journal of Ultrasound in Medicine*, which is published by W. B. Saunders in Hanover, Pennsylvania.

American Society for Ophthalmologic Standardized Echography
Department of Ophthalmology   219-356-2869—Dr. Karl C. Ossoinig, President
University of Iowa Hospital
Iowa City, IA 52242

This professional organization is dedicated to studying the use of ultrasound in ophthalmology. It gives courses on the technology, holds scientific meetings, and publishes the proceedings of the convention, which is held every two years.

International Society for Ophthalmic Ultrasound
c/o B. L. Hodes, M.D., Secretary/Treasurer   717-534-8783
Hershey Medical Center
Hershey, PA 17033

The Society's members are ophthalmologists and physicists interested in the study of the applications of ultrasonography in ophthalmic diagnosis, therapeutics, and biometrics. It holds an annual convention and publishes the proceedings.

National Foundation for Noninvasive Diagnostics

419 North Harrison Street, Suite 229   609-921-2939
Princeton, NJ 08540

The Foundation provides continuing education for physicians and technologists in the field of echocardiology (the use of ultrasound in examining the heart and diagnosing abnormalities). Its main function is to hold three educational seminars annually. Brochures that describe the Foundation and the seminars are available.

Society of Diagnostic Medical Sonographers

PO Box 31782   214-369-4332
Dallas, TX 75231

This nonprofit organization aims to advance the science of ultrasound techniques. Each spring a symposium is held, and the resulting papers are published. Its *Journal of Medical Ultrasound* is published by John Wiley & Sons in New York City.

Publications

John Wiley & Sons

605 Third Avenue   212-692-6034
New York, NY 10158

John Wiley & Sons publishes the following dealing with the use of ultrasound in medicine:

• *Journal of Clinical Ultrasound*—devoted to the clinical application of ultrasound in medicine. Annual subscription rate is $62 for nine issues.

• *Journal of Medical Ultrasound*—a quarterly journal that reports on innovative new techniques of ultrasound in medicine. Annual subscription rate is $33 for four issues.

Journal of Ultrasound in Medicine

W. B. Saunders, Publisher   717-632-3535
Box 465
Hanover, PA 17331

This monthly periodical for physicians and sonographers includes scientific articles on many aspects of ultrasound as used in medicine. Besides articles, it offers literature reviews, news for members of the American Institute of Ultrasound in Medicine, "Case of the Month" features, and information about meetings and conferences. Subscription rate is $75 a month for institutions, and $60 a year for individuals.

Institute of Electrical and Electronics Engineers (IEEE)

345 East 47th Street   212-705-7900
New York, NY 10017

The IEEE publishes the following dealing with ultrasound and medical imaging:

● *Sonics and Ultrasound*—a bimonthly, the journal covers the theory, design, and applications on generation, transmission, and detection of bulk and surface mechanical wave; and fundamental studies in physical acoustics, design of sonic and ultrasonic devices, and their applications in industry, biomedicine, and signal processing. Annual subscription rate is $80.

● *Transactions on Medical Imaging*—this quarterly publication covers medical imaging relating to ultrasonics; x-ray imaging and tomography; nuclear isotope imaging systems; image processing by computers; microwave imaging; nuclear magnetic imaging; radiation sensors and detectors; mathematical tools and analysis of image formation; and perception, display, and pattern recognition. Annual subscription rate is $110.

University/Government/Industry Partnerships

See: Cooperative Programs

Used Computers

Organizations and Government Agencies

Computer Dealers and Lessors Association

1212 Potomac Street, NW   202-333-0102
Washington, DC 20007

The Association offers a free membership directory of over 200 companies that deal in new or used computers. It is a good resource for referral to appropriate brokers.

International Computer Orphanage (ICO)

6711 Mississauga Road, Suite 103   416-826-7955
Mississauga, Ontario L5N 2W3
Canada

A brokerage system for used computers, ICO is a network of "computer orphan agents" who bring "homeless computers into computerless homes." ICO agents can help you lease, buy, or sell used computers. The company estimates clients save 30% to 40% off the computer list price.

Publications

Computer Shopper

PO Box F
407 South Washington Avenue
Titusville, FL 32780

305-269-3211

This monthly newsletter offers the largest single source of classified advertising for used personal computers and equipment. Each issue presents approximately 500 used computers, as well as display advertising for new equipment at discount prices. Features and editorials about how to buy a used computer also appear in each issue.

Computer Classified Blue Book

PO Box 3395
Reno, NV 89505

702-322-8811

This is a quarterly publication presenting specifications and current used market values of 390 microcomputers. It contains one-page profiles of 180 PC models, with information on standard features, such as memory and disc storage capabilities. The prices are revised with each printing. The 100-page publication is available for $70 a year, or $20 a copy with a $10 discount if the order is prepaid.

Additional Experts and Resources

Trenton Computer Festival

Trenton State College
Trenton, NJ 08625

609-771-2666—Allen Katz

This is the oldest and largest computer festival and flea market in the U.S. Held every April, the festival features both new and used computers. Contact Allen Katz for further information.

Computer User Groups

Many computer user groups hold annual flea markets where participants buy, sell, or trade hardware, software, programs, and so forth. For further information contact an appropriate user group in your geographical area (see the next section of this book).

Computer Magazines

Most computer magazines have classified advertising that offers used computers for sale. For locating relevant magazines, see listings under the sections *Computers (General)* or *Microcomputers*.

Computer Dealers

Local computer stores, as well as regional sales offices of computer manufacturers, are knowledgeable about the used computer market in your area. See the Yellow Pages for computer stores and contact them to identify regional sales offices.

Newspapers

Your local newspaper is likely to list used computers in its classified advertising section under "Office and Business Equipment" and/or "Merchandise."

User Groups, Computer

See also: Computers (General)

Organizations and Government Agencies

The best way to locate an appropriate user group is to contact a local dealer who sells the equipment you use. If you cannot find such a dealer in the Yellow Pages of your telephone book, contact the manufacturer of your equipment. Most manufacturers keep listings of their user groups located throughout the country. Often manufacturers have an 800 toll-free number to assist consumers. Contact your 800 operator to obtain this number if it is not listed here. Major manufacturers include:

Apple
International Apple Core
9108 George Street
Santa Clara, CA 95050
408-727-7652

Atari
c/o Joe Waters
122 North Johnson Road
Sterling, VA 22170
703-430-1215

Commodore Business Machines
1200 Wilson Drive
West Chester, PA 19380
215-431-9100

Corona
31324 Via Colinas, Suite 110
West Lake Village, CA 91361
800-621-6746

Digital
DICUS
249 North Boro
Marlboro, MA 08752
617-480-3419
617-480-3463

Epson
3415 Kaslina Street
Torrance, CA 90505
800-421-5426
213-539-9140

Hewlett-Packard
Interex
2570 El Camino West, 4th Floor
Mountain View, CA 94040
415-941-9960

IBM
IBM PC Computer User Group
PO Box 3022
Department 2L9/218
Boca Raton, FL 33432
800-426-3333
305-998-7284

Kaypro
Kaypro User Group
c/o Richard Coude
533 Stevens Avenue
Sonoma Beach, CA 92075
619-481-3941

NEC
1401 Estes Avenue
Elk Grove Village, IL 60007
312-228-5900
800-323-1728

Radio Shack/Tandy
Tandy Computer Business Users Group
PO Box 17580
Fort Worth, TX 76102
817-390-2197
404-676-4318

Sanyo
51 Joseph Street
Moonchild, NJ 07074
800-221-2097

Sony Corp.
Sony Drive
Parkridge, NJ 07656
800-222-7669
201-930-1000

Texas Instruments
PO Box 402430
Department DCA-00000
Dallas, TX 75240
800-842-2737

Xerox
530 NCC
6401 Security Boulevard
Baltimore, MD 21235
301-594-0128

Zenith Data Systems Corp.
1000 Milwaukee Avenue
Glenview, IL 60025
312-391-8860

VCR

See: Videocassette Players

Venture Capital

Organizations and Government Agencies

American Association of Minority Enterprise Small Business Investment Companies (AAMESBIC)

915 15th Street, NW, Suite 700   202-347-8600
Washington, DC 20005

This trade association is composed of minority enterprise small business investment companies (MESBICs). The Association only represents investment companies that provide venture capital for minority-owned small businesses, many of which are high-technology oriented, particularly in the field of telecommunications. A convention is sponsored each year, which includes workshops, meetings, and guest speakers. The staff can answer your questions about minority investment companies and, in particular, can inform you about regulations and legislation affecting MESBICs. Several publications are available:

• *AAMESBIC Directory*—a listing of the approximately 150 MESBICS across the country including names, addresses, and telephone numbers. It also describes each company's investment preferences and investment policy. The 20-page publication costs $3.

• *Perspective*—a monthly newsletter that is geared toward minority small business investors. It contains articles about legislation and regulations affecting MESBICs, including those regarding venture capital investment. The subscription to the two- to six-page newsletter is $36 a year.

• *Small Business Perspective*—a monthly newsletter intended for the general small business investment company. It focuses on legislative issues and regulations, including those that affect venture capital investment. Subscription rate to the two- to four-page newsletter is $25 a year.

• *Journal of Minority Small Businesses Enterprise*—an annual periodical containing five to ten articles about specific areas of interest. Recent issues have focused on small businesses and high technology, accessing capital, and start-up companies. The journal averages 40 pages in length and costs $5.50.

BankAmerica Capital Corporation

Bank of America   415-953-3001
Department 3908
555 California Street
San Francisco, CA 94104

BankAmerica Capital Corporation invests in second- and third-stage development companies, which are, for the most part, high-tech oriented. Financing ranges between $500,000 and $2 million. A publication of particular interest is:

● *Financing Small Business*—a booklet in the Small Business Reporter series prepared by the Bank of America. It provides direct, practical information on basic financing situations, types of financing, including venture capital investment, and preparation of a business plan. The 33-page publication is available for $3, and can be ordered from Small Business Reporter, Bank of America, Dept. 3401, PO Box 37000, San Francisco, CA 94137, 415-622-2491.

Center for Entrepreneurial Management (CEM)

83 Spring Street   212-925-7304
New York, NY 10012

CEM, a nonprofit educational and membership association for entrepreneurial managers, provides information for small growing businesses. Special in-depth reports are prepared by CEM on specific topics such as venture capital. Among CEM's publications are:

● *The Entrepreneurial Manager's Newsletter*—a 12-page monthly source of information for those who advise entrepreneurial managers and professionals. Articles cover such topics as finance. The subscription price is $71 for nonmembers and free to members.

● *How to Prepare and Present a Business Plan*—the 300-page guide by Joseph Mancusco describes writing a business plan that will attract venture capital investors. It includes sample business plans, and costs $20 a copy.

Coopers and Lybrand (C&L)

1251 Avenue of the Americas   212-536-2000
New York, NY 10020

C&L provides accounting, tax, and business advisory services to its clients, many of whom are growing businesses and high-potential start-up companies. Helping entrepreneurs obtain venture capital financing is a primary service. C&L can provide you with several publications, including:

● *The Emerging Business*—gives advice to business owners on how to effectively manage a growth-oriented company. Includes information about available sources of capital, taxes, techniques for improving cash flow, preparing a business plan, and going public. The 425-page book is $35 a copy.

● *Three Keys to Obtaining Venture Capital*—a free three-page booklet designed to provide entrepreneurs with insights into obtaining the venture capital needed to launch or expand a business. A step-by-step guide is included on how to prepare a business plan.

● *Meeting the Business Challenges in High Technology*—a free 10-page booklet that overviews the services provided by C&L that are necessary to meet the needs and objectives of high-technology companies.

Deloitte Haskins and Sells

1114 Avenue of the Americas   800-842-9526
New York, NY 10036

This international firm of independent CPAs provides business consulting services to various companies. Through some of its individual offices, it sponsors local seminars and workshops on financial strategies for high-tech and other entrepreneurs. Staff at the local offices can assist you with questions about venture capital as well as send you the following complimentary publications:

• *Raising Venture Capital*—a 150-page entrepreneur's guidebook prepared by the firm's High-Technology Industry Group. It contains advice and suggestions for competing successfully for venture capital. It shows how to organize a business, write a business plan, describe the product, convince a prospective investor that you have a market, determine financial expectations, and negotiate a final deal.

• *Strategies for Going Public*—a 109-page guidebook prepared for the firm's High-Technology Industry Group. It provides information necessary to understand how to go public successfully. An appendix surveys alternatives to going public.

• *Forming R&D Partnerships*—the 109-page guidebook's main text provides general information about the workings of R&D partnerships, as well as explaining how to form one.

Howard & Company

1528 Walnut Street   215-735-2815
Philadelphia, PA 19102

This corporate financial consulting firm assists companies in financing growth plans. It has a venture capital division which includes venture capital associates who offer limited partnerships to companies. The firm also cosponsors seminars on how to raise venture capital. Some publications of interest are:

• *Going Public: The IPO Reporter*—an eight- to ten-page weekly newsletter for IPO professionals and investors. It provides current information on IPO trends and developments, and provides newly registered companies with details on filing, revenues, incomes, earnings, and assets. Each subscription, $600 a year, includes a glossary of terms and a bibliography of current books and articles.

• *Private Placements*—a six- to eight-page bimonthly newsletter on venture capital, with comprehensive detailed profiles of companies that have recently completed a private placement. It monitors the venture capital community and identifies new funds, new areas of investment, and the latest trends and developments. An annual subscription is $450.

• *Growth Capital*—an eight-page monthly newsletter of entrepreneurial finance that looks at all the sources of business capital. It offers insight into techniques for financing a business on the best terms at the lowest cost. Each subscription, at $95 a year, includes an annual index referencing all the companies, individuals, and techniques cited during the previous year.

National Association of Small Business Investment Companies (NASBIC)
1156 15th Street, NW, Suite 1101   202-833-8230
Washington, DC 20005

NASBIC is a national trade association for the small business investment company industry that serves as an information clearinghouse on venture capital. Staff can direct you to venture capital sources, experts, and literature. NASBIC publishes several industry guides and research studies on a regular basis. Of particular interest are:

• *The SBIC Venture Incentive Compensation Survey*—updated approximately every other year, this publication is a tabulation of salaries, bonuses, and fringe benefits paid by companies in the industry. It is broken down by type and size of firm. The price is $25 a copy.

• *SBIC Audit Fee Survey*—updated annually, this publication provides data on audit costs for various sizes and types of SBICs. Total assets of private capital are broken down into four categories, and venture firms are classified into six categories. Price is $50 a copy.

• *Venture Capital: Where to Find It*—published every spring, this 28-page guide lists more than 350 small business investment companies (SBICs) and minority enterprise SBICs (MESBICs) representing approximately 90% of the industry's resources. It gives the name of the manager, address, phone number, investment policy, industry preference, and preferred dollar limits on loans and investments. Known also as the *NASBIC Membership Directory*, it enables members to contact each other. The *Directory* is available for $1.

National Venture Capital Association (NVCA)
1655 North Fort Meyer Drive   703-528-4370
Arlington, VA 22203

The Association is open to venture capitalist organizations, financiers, and individuals who are responsible for investing private monies in young companies on a professional basis. The purpose of NVCA is to improve the government's knowledge and understanding of the venture capital process. The staff can answer questions about federal legislation and regulations, as well as provide several pages of statistics on venture capital. Some of their publications include:

• *NVCA Fact Sheet*—gives an overview of the industry from 1969 to the present.

• *NVCA Membership Directory*—a free 40-page brochure that describes the association and its activities, including a listing of the membership with addresses, phone numbers, and contacts.

Price Waterhouse
Smaller Business Services   202-296-0800
1801 K Street, NW
Washington, DC 20006

This certified public accounting firm provides auditing, taxation, management consulting, and a variety of other business services to public and private businesses and organizations, as well as to the government. It sponsors seminars that sometimes cover venture capital. Several publications are available, including:

• *Business Review*—a four-page newsletter that is published eight times a year. It deals with management topics, especially those affecting small businesses and entrepreneurs, and often features current statistical information and legislative issues. Available free of charge to companies and entrepreneurs.

• *Financing Your Business*—this guide for smaller businesses and entrepreneurs deals with getting funds for a start-up situation. It directs the user to finding sources including venture capital. The 65-page guide costs $5.

Venture Capital Network, Inc.

23 School Street 603-224-5388
Concord, NH 03301

This private nonprofit corporation is organized by the Business and Industry Association of New Hampshire in cooperation with the University System of New Hampshire's Office of Small Business Programs. Its purpose is to introduce entrepreneurs to active, informal, independent investors. It maintains a computerized data base of entrepreneurs and investors, and the data are compared regularly to determine if potential matches exist. The staff can answer your questions about smaller-scale equity investments, and can send you information.

Venture Economics, Inc.

PO Box 348 617-347-8600
Wellesley Hills, MA 02181

This company provides information, research, and consulting services for corporations, major institutions, and state and local government agencies on venture capital trends. The firm's data base contains information on about 6,000 companies that have received venture capital. The information includes business descriptions, financing resources, and specific investors. The data base forms part of the firm's Venture Intelligence Service, which is available by subscription. Venture Economics sponsors approximately 25 regional seminars a year that deal with how to raise venture capital. Staff can answer questions as well as refer you to other information sources. Brochures describing the company and the services available, as well as the following publications, can be sent upon request:

• *Venture Capital Journal*—a monthly periodical that sights new issues and trends in venture capital investments. It also provides the Venture Capital 100 Index in each issue, which tracks the market performance of each company. The subscription rate to the 55-page publication is $495.

• *Pratt's Guide to Venture Capital Sources*—an annual directory that lists 500 venture capital firms in the U.S. and Canada. Also included are 20 articles recommending ways to raise venture capital. The 530-page publication is $95.

• *Venture Capital Journal Yearbook*—an annual publication that summarizes the investment activities of the previous year. It includes statistics and data about capital commitments and investment activities in specific industries. The 50-page publication costs $150.

• *Venture Capital Investments*—an annual review of the industry with overviews of the industry by categories. It also sights technological trends, identifies newly backed companies, and looks at sources of venture capital. The cost of the 150-page publication is $925 to nonsubscribers of the Venture Intelligence Service.

• *The Venture Capital Industry—Opportunities and Considerations*—an annual study and comprehensive analysis of venture capital for prospective corporations and institutional investors. It has information about the history of the industry, legal considerations, industry performance and trends. The 100-page publication costs $2,500.

Office of Investment

U.S. Small Business Administration   202-653-6672—John Werner
1441 L Street, NW
Washington, DC 20416

This government office licenses, regulates, and funds about 500 Small Business Investment Companies (SBICs) nationwide. It is the only federal source of venture capital in the U.S. Staff can direct you to your nearest SBIC and provide information about approaching the SBIC. The office can also send you a list of licensed small business investment companies, an information kit designed for people interested in forming an SBIC, and a directory of the 500 SBICs supported by SBA. These are available free of charge.

Arthur Young and Company

National Headquarters   212-407-1715—Director of Marketing
277 Park Avenue   800-3HI-TECH—National High-Tech Group
New York, NY 10172   415-393-2731—in California

Arthur Young's High-Tech Group works with companies through every stage of development, offering professional business services. The Group helps companies develop business plans and determine sources of financing, including venture capital funding. Arthur Young offers many publications and newsletters directed to the interests of high-tech companies, and it will send a publications list upon request. These publications can be ordered from High-Tech Center, Arthur Young and Company, 1 Post Road, San Francisco, CA 94104, 800-344-8324, in CA 415-393-2731. Some of these publications are:

• *Helping High Technology Companies Grow*—a free 16-page booklet that describes Arthur Young's services to high-tech clients. Several other information sources are recommended in it.

• *Research and Development Partnerships: Maximizing the Advantages*—a booklet that offers an overview of the business aspects of research and development partnerships. It reports on the advantages and disadvantages of these partnerships with respect to cost, risk, control, benefits for the investor and the company, and different sources of investment. The 26-page publication is available free of charge.

• *Outline for a New High-Technology Business Plan*—a booklet describing how to develop a business plan for a growing business. It covers topics such as marketing, analysis, research and development, manufacturing, management and ownership, and financial backing. The 12-page booklet is available free of charge.

Publications

A Businessman's Guide to Capital Raising Under Securities Laws

c/o Packard Press   215-236-2000
Tenth and Spring Garden
Philadelphia, PA 19123

This 55-page, nontechnical pamphlet familiarizes the businessperson with the current regulatory environment for raising capital by selling securities. It assists businesspeople in anticipating and understanding the decisions that must be made, and the process to follow in the securities offering. It is updated every year and can be ordered for $10.

How to Raise Capital: Preparing and Presenting the Business Plan

c/o Dow Jones-Irwin   312-798-6000
1818 Ridge Road
Homewood, IL 60430

The book is a complete guide to writing a proposal that will attract potential investors and lenders. It offers detailed explanations on technical business terminology and concepts. The 250-page publication is available for $24.95.

Inc. Magazine

Inc. Publishing Corporation   617-227-4700
38 Commercial Wharf
Boston, MA 02110

Inc. Magazine is a monthly publication that provides information about trends, investment companies, capital, and company profiles. Sometimes special features relate to venture capital. Subscription to the 150- to 200-page publication is $24 a year.

Venture Capital Handbook

Prentice-Hall, Inc.   201-592-2000
200 Old Tappan Road
Old Tappan, NJ 07675

The *Handbook* covers the entire relationship between the venture capitalist and the entreprenuer. It explains how to prepare a proposal, what negotiations are like, what the investor is looking for, and what occurs at closing. The 402-page publication costs $35.

Venture Capital Investment Trends: 1981-82

Science Indicators Unit   202-634-4682
National Science Foundation
1800 G Street, NW (L-611)
Washington, DC 20550

The report is included in *Science Indicators—1986*, a biannual publication. The report aggregates confidential information on investments of venture capital firms. It covers total transactions and divides them into 40 technological categories. The publication is available free of charge.

Venture Magazine

521 Fifth Avenue   212-682-7373
New York, NY 10175

This monthly publication is for entrepreneurs who want to invest, and includes articles about start-up companies. It has information about financing, such as new trends and different ways to finance public offerings, and profiles venture capital firms. In each June issue the Venture Capital 100 are listed, which are the largest 100 venture firms. Each December a directory of venture capital firms is included. Subscription rate to the 150-page magazine is $18.

Additional Experts and Resources

William E. Wetzel, Jr., Professor of Finance

Whittemore School of Business and Economics   603-862-2771
University of New Hampshire
Durham, NH 03824

Professor Wetzel's special interest is seed capital financing of entrepreneurial ventures. He is knowledgeable about risk capital from individual investors, and can tell you what circumstances are necessary for an entrepreneur to seek funding from individuals as opposed to a professional venture capital firm. He has had articles published in journals, and has conducted a research effort that focused on individual investors. The resulting 120-page report, *Informal Risk Capital in New England* (PB81-196149), costs $13 and is available from NTIS, 5285 Port Royal Road, Springfield, VA 22161.

VHSIC

See: Chips and Integrated Circuits

Video Games, Coin-Operated

Organizations and Government Agencies

Amusement and Music Operators Association (AMOA)
2000 Spring Road, Suite 220   312-654-2662
Oak Brook, IL 60521

AMDA is a trade association that covers coin-operated machines in the entertainment industry, including video games. Members are mostly operators, but some are manufacturers and distributors. Its monthly newsletter, *The Locator*, covers the coin-operated industry, with each issue containing special feature articles, business information about the industry, technological developments, innovative techniques, legislative matters, and news about the industry. The newsletter is currently available only to members, but in the future it will be sold to the public. Inquiries about the association and the newsletter are handled by their personnel.

Amusement Game Manufacturers Association (AGMA)
205 The Strand, Suite 3   703-548-8044
Alexandria, VA 22314

This trade association represents 45 manufacturers in the coin-operated amusement industry, which includes video games. It holds an annual meeting, as well as an annual trade show. The staff will answer questions and refer you to other sources of information. AGMA publishes a monthly newsletter, *The Coin Gram*, that covers developments in the industry, especially in the area of copyrights.

National Coin Machine Institute (NCMI)
2455 East Sunrise Boulevard, #311   305-561-0886
Ft. Lauderdale, FL 33304

This 140-member trade association covers developments in all areas that affect amusement vending operation, including coin-operated video games. They hold an annual seminar, as well as regional meetings. Bulletins about the industry are published on an irregular basis. They will answer questions and make referrals to other sources of information.

Publications

Play Meter
PO Box 24170   213-347-3820
Woodland Hills, CA 70184

> *Play Meter* is a bimonthly magazine that covers the coin-operated games industry. Each issue presents technical topics, lists of top-earning games, feature articles, and editorials. The magazine is $50 a year.

Replay
PO Box 2550   213-347-3820
Woodland Hills, CA 70184

> Published biweekly, this trade publication covers the coin-operated equipment industry, including video games. Each issue includes general news, technology advances, and the latest developments in the industry. Subscription is $40 a year.

StarTech Journal
PO Box 1065   609-662-3432
Merchantsville, NJ 08109

> *StarTech Journal* is a monthly journal that presents information on troubleshooting tips and technology, equipment symptoms and modifications, circuit analysis, system checkout procedures, theories of operation, field services reports, and videodisc maintenance and operations. Available for $56 a year.

Additional Experts and Resources

Michael J. Maasen
U.S. Department of Commerce   202-377-1140
International Trade Administration (ITA)
Room 4312
Washington, DC 20230

> Mr. Maasen is the industry specialist at ITA for toys and the toy industry. He will answer questions and can refer you to other good sources of information.

Video Games, Home and Computer

Organizations and Government Agencies

Consumer Electronics Group
Electronic Industries Association (EIA)   202-457-4919
2001 Eye Street, NW
Washington, DC 20006

The Consumer Electronics Group represents many consumer electronics manufacturers, including those in the video game industry. The group holds an annual consumer electronics show, and makes available many brochures about what it does. Staff members in the Public Relations department are always available to answer questions. The group publishes the *Consumer Electronics Annual Review*, with statistical and technical information; and *Consumer Electronics: U.S. Sales by Product Category*, a statistical booklet.

Publications

Reese Communications
460 West 34th Street   212-947-6500
New York, NY 10001

Reese Communications publishes the following that deal with video games:
• *Electronic Games*—this consumer monthly reports on the electronic gaming field, covering all aspects of video games, computer games, hand-held games, and coin-operated games. Subscription is $20 a year.
• *Hotline*—a biweekly newsletter that covers the latest developments in the industry. Annual subscription rate is $45.

Link News Briefs
Link Resources Corporation   212-473-5600
215 Park Avenue South
New York, NY 10003

A monthly electronic newsletter covering developments in all new electronic media. Topics covered include PCs, cable television, on-line data bases, videocassettes, STV, teletext, video games (software, hardware, producers, and so forth), and satellites. Available only online through NewsNet (800-345-1301; 215-527-8030 in PA).

Additional Experts and Resources

The annual reports of the largest manufacturers of video games are excellent sources of industry and technology information. Two of the biggest manufacturers are Warner Communications and Mattel.

Ralph Watkins
International Trade Analyst   202-724-0976
U.S. International Trade Commission
Washington, DC 20436

Mr. Watkins, an expert in the video game industry, prepared a onetime study on the industry, *Comparative Assessment of the U.S. Video Game Industry*. The report is available free of charge. Mr. Watkins will answer questions and make referrals to other sources of information.

Michael J. Maasen
U.S. Department of Commerce   202-377-1140
International Trade Administration (ITA)
Room 4312
Washington, DC 20230

Mr. Maasen is the industry specialist at ITA for toys and the toy industry. He will answer questions and can refer you to other sources of information.

Videocassette Players

Organizations and Government Agencies

American Video Association
557 East Juanita Road   602-892-8553
Mesa, AZ 85204

The Association for video dealers provides technical information about equipment and the videocassette industry. Questions are answered and referrals can be made to other sources. It publishes a weekly update on new equipment available in the market, and a monthly newsletter providing broad coverage of this specific industry. The Association holds an annual meeting in the winter. Membership cost is $600 a year.

Electronic Industries Association (EIA)

2001 Eye Street, NW   202-457-4900
Washington, DC 20006

 EIA members include manufacturers and distributors of videocassette players. It provides services to its members in engineering, marketing, trade shows, government relations, and communications affairs. EIA has experts in videocassette systems among its staff of engineers, and it will furnish orientation and make referrals to other sources of information. *Consumer Electronics Review* is a free annual publication by the EIA, and it has statistical information on videocassette players.

Publications

Consumer Electronics Monthly

CES Publishing   212-686-7744
345 Park Avenue South
New York, NY 10010

 This trade magazine is oriented toward retail dealers. The publication deals with videocassette players among other items of the consumer electronics sector. The annual cost is $50.

Television Digest, Inc.

1836 Jefferson Place, NW   202-872-9200
Washington, DC 20036

 Television Digest has the following publications dealing with videocassette players:

 • *Video Week*—a weekly publication devoted to videocassette players. Annual subscription rate is $378.

 • *The VCR Explosion*—a special report that covers the VCR boom of 1982-83, detailing what happened and why. It includes complete market-share ranking of all U.S. VCR brands, product reports, technology updates, and so forth. Price is $95 a copy.

 • *Consumer Electronics Video Data Book*—this annual compilation of key statistics covers videocassettes, including tables of weekly sales to dealers, data on market supply, import-export summaries, analyses, and other information. Subscribers pay $95, while others pay $125.

Video

Reese Communications   212-947-6500
460 West 34th Street
New York, NY 10001

 Published monthly, the magazine is edited for the serious home video user as well as for the prospective purchasers of home video products and services. It has columns, departments, and feature articles aimed to educate and instruct readers on the many forms and functions of videocassette players and accessories. Annual subscription is $15.

Additional Experts and Resources

U.S. Department of Commerce
International Trade Administration
14th Street and Constitution Avenue, NW, Room 1015B
Washington, DC 20230

202-377-2872—Arthur Pleasants
202-377-0337—E. MacDonald Nyher

ITA specialists monitor and analyze the videocassette player industry. They can make referrals and provide information about equipment available. Mr. Pleasants is an expert on the products from the industrial point of view, and Mr. Nyhen deals with the consumer-oriented equipment.

Videodisc Technology

Organizations and Government Agencies

Society for Applied Learning Technology (SALT)
50 Culpepper Street
Warrenton, VA 22186

703-347-0055

SALT holds an annual conference on the current state of the art of videodiscs and then publishes the proceedings. A special interest group on interactive video learning has been formed. It holds training conferences and publishes material on the subject. For answers to questions about SALT and what it does in the field of videodisc technology, contact Mr. Fox at the Society, or Bob Bellinger (chairman of the special interest group) at 401-841-4763. Publications include:
- *Interactive Videodisc in Education and Training* (1983)—$35.
- *Videodisc for Military Training and Simulation* (1982)—$35.
- *Videodisc, Microcomputer, and System Technologies in Education and Training* (1982)—$40.

Publications

Video Play Report
Winslow Information
923 Sixth Street, SW
Washington, DC 20024

202-484-3853—Ken Winslow

Issued biweekly, this newsletter covers the technology and techniques of video, concentrating on the key issues and events of the day in the production,distribution, and use of video programming. Emphasis of the reports is on the crossover market, the home video equipment and programming made to be sold to home users. The subscription is $135 a year for 27 issues.

Interactive Video Technology
Heartland Communications   216-567-3732
223 Sunrise Drive
Sheve, OH 44676

    Interactive Video Technology is a monthly newsletter covering interactive video, presenting information on new products, legal issues, changes in the industry, projects in interactive video, and a calendar of events. The newsletter is also available electronically through NewsNet (800-345-1301; call 215-527-8030 in Pennsylvania). The subscription rate is $45 a year.

The Home Video and Cable Report
Knowledge Industry Publications, Inc.   914-328-9157
701 Westchester Avenue   800-431-1800
White Plains, NY 10604

    This weekly newsletter covers the news in every segment of the home video industry from the latest about discs and interactive systems to availability of satellite space to the latest figures on basic and pay cable services. Annual subscription rate is $295 for 52 issues.

Video Marketing
1680 Vine Street, Suite 820   213-462-6530
Hollywood, CA 90028

    Video Marketing publishes the following:

    • *Video Marketing Newsletter*—this news service offers news and analysis of home video software and hardware. Subscription rate is $277 for 24 issues.

    • *Video Marketing Game Letter*—a biweekly that covers home electronic games. It provides information on sales statistics, market forecasts, financial news of key hardware and software companies, teledelivery, piracy and counterfeiting, videodisc interface developments, and so forth. Subscription is $395 a year for 24 issues.

Video Discs
Knowledge Industry Publications, Inc.   800-431-1800
701 Westchester Avenue   914-328-9157
White Plains, NY 10604

    This book by Efrem Sigel examines the applications (actual and projected) and the future of videodisc technology. Among the areas covered are programming, production, and cost considerations; education and training; the consumer market; the international market production; information storage; and competing technologies. The 200-page book is available for $34.95.

The Video Source Book

National Video Clearinghouse, Inc.   516-364-3686
100 Lafayette Drive
Syosset, NY 11791

The book is a complete annual reference guide to 35,000 videotapes and video-discs in every field, cataloged by title, with a detailed indexing system pinpointing programs in 405 precise subject areas. The entries describe each video program in detail, and cover eight subject areas: business/industry; children/juvenile; fine arts; general interest/education; health/science; how-to/instruction; movies/entertainment; and sports/recreation. It is available for $125. The Clearinghouse maintains a specially-designed computer data base that is regularly updated to include new titles and distributions.

Video Disc Book: A Guide & Directory

John Wiley & Sons   212-850-6000
605 Third Avenue
New York, NY 10058

Written by Rod Daynes, this is a yellow pages approach to the videodisc industry. The first section includes articles on various videodisc topics, from analyses of the job to be done to design of discs, implementations, and the future of the videodisc industry. The second section is a listing of all people and organizations involved in the industry, as well as all videodisc players and discs developed on the market.

Additional Experts and Resources

Manufacturers

Many companies that manufacture videodisc equipment are good sources of information. They usually print brochures describing their product and the technology involved. Their Public Affairs departments will answer questions about their products. Some of the bigger manufacturers are:

Pioneer Video

200 West Grand Avenue
Montvale, NJ 07645
201-573-1122, ext. 205

Brochures are available describing the individual products, as well as a publication that explains the technology behind many of their products.

Sony Communications Products Company

1 Sony Drive
Park Ridge, NJ 07656
201-930-6106—Marketing

Sony regularly conducts workshops on different topics in interactive video. Brochures on videodisc products and their technology are also available.

RCA

David Sarnoff Research Center
Public Affairs Office
Princeton, NJ 08540
609-734-2507—Julie Maddocks

RCA has brochures describing its videodisc products and its technology. Their bimonthly magazine occasionally dedicates an issue to the topic of videodiscs.

3M Broadcast Products
3M Center Building
St. Paul, MN 55144
612-733-9073

3M has developed a directory called the *Videodisc Connection*, which lists suppliers and services involved in the interactive laser videodisc business. The alphabetically listed directory is designed to match customer needs with the appropriate vendors.

Universities

Several universities do research in videodisc technology, and they can be very good sources of information. The staff will answer questions on the technology of videodiscs and can recommend other good sources of information. Some of these are:

Nebraska Videodisc Design/Production Group
University of Nebraska
PO Box 83111
Lincoln, NE 68501
402-472-3611

The staff disseminates information on videodisc technology and, for a small fee, will send an information packet covering many aspects of videodiscs. The packet includes *Investigating Interactive Video*, a brochure describing the Videodisc Group; six project papers; several manuscript papers on interactive video; and *Videodisc Design Production Group News*, a newsletter of the Nebraska Videodisc Design Group.

MIT Architecture Machine Group
9-516 MIT
105 Massachusetts Avenue
Cambridge, MA 02139
617-253-5981—Andy Lippman

University of Delaware
Instructional Resource Department
East Hall
Newark, DE 19716
302-451-2685—Ed Schwartz

The university has a grant to produce videodiscs for music instruction.

Rochester Institute of Technology
AVI/RIT
PO Box 9887
1 Lomb Memorial Drive
Rochester, NY 14623
716-475-6625—Lamont Seckman

Videotex/Teletext

Organizations and Government Agencies

Videotex Industry Association (VIA)
1901 North Fort Meyer Drive, Suite 200
Rosslyn, VA 22209

703-522-0883

This national trade association is devoted to the development of videotex in the U.S. Activities include researching technical and public policy issues; distributing information on videotex and teletext to the public; and informing members about current state and federal legislation and regulatory action. The Association monitors and gathers the latest information on the industry, and the staff will answer questions. A brochure on videotex/teletext is available, as well as a bibliography of resource materials.

Institute for the Future (IFTF)

2740 Sand Hill Road   415-854-6322
Menlo Park, CA 94025-7097

The Institute is involved in three general areas of research: forecasting and planning; teleconferencing; teletext and videotex. Publications in the area of videotex and teletext are:

- *Teletext and Videotex: Tomorrow's Technology Today*—$3.
- *Videotex in the U.S.—Toward Information Diversity*—$3.
- *Forecasting a New Hybrid Technology—Teletext and Videotex*—$3.

Publications

International Videotex and Teletext News

Arlen Communications, Inc.   301-656-7940
PO Box 40871
Washington, DC 20016

Each monthly newsletter reports on new companies and products, regulations, marketing and financial affairs, as well as summarizing fundamental business, marketing, and technical issues. Available for $225 a year for 12 issues.

Teletext and Videotex in the United States: Market Potential, Technology, and Public Policy Issues (1982)

Data Communications   212-997-2015
McGraw-Hill, Inc.
1221 Avenue of the Americas
New York, NY 10020

This book describes the origins of these technologies in Europe and summarizes the current state of teletext and videotex in the U.S. It forecasts their technological uses and possible markets, then focuses on a range of policy questions expected to arise during their further development. Contents include definitions and overview, experience outside the U.S., current state-of-the-art, future applications and alternatives, technology forecasts, policy issues, case studies, and societal impacts. The 341-page publication is $30.

Telidon Programme
Department of Communications   613-996-2623
300 Slater Street
Ottawa, Ontario K1A OC8
Canada

The Department of Communications of the Canadian government has done extensive research on the topic of videotex. The Canadian teletext system, Telidon, is sponsored by the government, and many reports exist describing the system and its operation. The Informatics Applications Management Branch of the Department publishes, on an irregular basis, *Telidon Reports*, a newsletter covering the Canadian videotex system. Free subscriptions are available from Room 1706, Journal Tower South, Department of Communications, at the above address.

Videodisc/Videotex
Meckler Publishing   203-226-6967
520 Riverside Avenue
Westport, CT 06880

This bimonthly magazine is devoted to the reporting and critical analysis of research and development in videodisc, videotex, and related video technologies. It features articles on the applications of videodisc and videotex systems in business, industry, government, education, libraries, and the home. The subscription rate is $75 a year.

VideoPrint
30 High Street   203-866-6914
Norwalk, CT 06851

Issued bimonthly, the newsletter covers teledelivery of information, computer software, music, and video, including videotex and teletext systems and services. It covers such as areas as major trials and innovations, market research, new ventures, development of new terminals, and advertising and new media. Available for $180 a year for 24 issues.

Videotex Directory
Arlen Communications, Inc.   301-656-7940
PO Box 40871
Washington, DC 20016

This industry directory, published annually, concentrates exclusively on the organizations in the videotex industry in the U.S. and worldwide. It lists dozens of projects underway or planned, with a complete profile of each system operator. Also included is information about hardware and software providers, specialists serving the industry, and sources of equipment and programming for videotex and teletext projects. Seven sections provide information on system operators, hardware and software vendors, associations, government agencies, consultants, publications, and the international arena. Available for $150 a copy or free when you subscribe to *International Videotex and Teletext News*.

Videotex Products

PO Box 138   617-449-1603
Babson Park Branch
Boston, MA 02157

The company publishes:

● *Worldwide Videotex Update*—this monthly newsletter covers many aspects of the videotex field such as new developments and products in the U.S. and abroad, as well as videotex projects and services. Also available through NewsNet (800-345-1301; 215-527-8030 in PA).

● *Videotex Products*—a quarterly newspaper oriented toward qualified professionals in videotex/teletext in the U.S. and Canada, providing all the important facts and news about products and services necessary to make a decision to buy or rent. Free to qualified persons, or available to others for $6 a year.

Videotex/Viewdata Report

Link Resources Corporation   212-473-5600
215 Park Avenue South
New York, NY 10003

A monthly, this industry newsletter presents a synopsis of current events in the videotex/teletext industry, with a detailed discussion of what's going on in the field. It keeps track of new trends in technology, marketing, and so forth. The newsletter is available on-line in data base form through NewsNet (800-345-1301). Subscription rate is $300 a year in printed form.

Voice Recognition

See also: Recognition Technology

Publications

IEEE Transactions on Acoustics, Speech, and Signal Processing

IEEE Service Center   201-781-0060
445 Hoes Lane
Piscataway, NJ 08854

Issued bimonthly, the journal deals with transmission, recording, reproduction, processing, and measurement of speech and other signals by digital, electrical, acoustic, mechanical, and optical means; the components and systems to accomplish these and related aims; and the environmental, psychological, and physiological factors.

Journal of the Accoustical Society of America
American Institute of Physics   212-349-7800
335 East 45th Street
New York, NY 10017

    This monthly publication covers all phases of acoustics with articles and papers on speech in almost every issue. It is a technical journal directed toward scientists. The cost is $195 a year, and $19.50 a single copy.

Media Dimensions, Inc.
525 East 82nd Street   212-680-6451
New York, NY 10028

    This publishing company sponsors an annual voice input/output (I/O) applications show and conference each spring, featuring professional sessions on the application of voice I/O and exhibitors. Official proceedings of "Speech Tech" are available for $54.40, as is an audiocassette with ten different companies demonstrating voice input/output systems, available for $12.50. A book club and information service offer books and equipment at a discount price. The following is published:

    • *Speech Technology—Man/Machine Voice Communications*—this quarterly is devoted to voice synthesis and recognition and features articles on applications, technology, new products, news items, and other features about the voice processing industry. An annual directory issue lists over 130 companies, consultants, and so forth. The periodical is directed toward design engineers, corporate and management engineers, researchers, and financial analysts. The subscription price is $50 a year.

Voice Recognition
Stoneridge Technical Services   301-424-0114
PO Box 1891
Rockville, MD 20850

    This newsletter, published ten times a year, covers primarily current events in the fields of voice recognition, voice synthesis, and related computer-voice technologies. It reports on products, companies, conferences, and other information sources, including books, articles, and reprints. It is directed toward a broad cross-section of people, including researchers, engineers, marketers, executives, and venture capital firms. The subscription price is $95 a year.

Additional Experts and Resources

Dr. Janet Baker
Dragon Systems, Inc.   617-965-5200
Chapel Bridge Park
55 Chapel Street
Newton, MA 02158

    Dr. Baker is chairman of the Institute of Electrical and Electronics Engineers' working group for speech input/output performance evaluation. She has a doctorate in computer science from Carnegie-Mellon University and has worked for over a decade on the highest performance speech recognition-understanding systems. She is available to answer questions related to this technology and performance evaluation and can make referrals.

Dr. David F. Pallot
Speech Input/Output Technology   301-921-3427
Institute for Computer Sciences and Technology
National Bureau of Standards
Washington, DC 20234

   Dr. Pallot and his colleagues are involved in the development of test methods for speech recognition systems. They are available to answer questions and/or make referrals.

Video, Interactive

See: Interactive Video Technology

Warfare

See: Military Technology; Star Wars: Antimissile Systems in Space

Waste Utilization

See also: Fuels, Alternative; Renewable Energy Technology; Solar Technology

Organizations and Government Agencies

Energy From Municipal Waste Division
Renewable Technology   202-252-6104
Conservation and Renewable Energy
Department of Energy
Forrestal Building, CE-323, Room 5F081
1000 Independence Avenue, SW
Washington, DC 20558

This office conducts research in the area of utilizing municipal waste as an energy source, and it is responsible for DOE programs in waste-to-energy research and development.

Energy Cascading Branch

Waste Energy Reduction Division   202-252-2084
Office of Industrial Programs
Conservation and Renewable Energy
Department of Energy
Forrestal Building, CE-121, Room 5F035
1000 Independence Avenue, SW
Washington, DC 20558

This division conducts technological development in cogeneration, waste heat recovery and utilization, and combustion efficiency improvement, especially in industries that use energy intensive processes.

Waste Management Division

Office of Environmental Engineering and Technology   202-382-2583
Environmental Protection Agency
RD-681, West Tower
401 M Street, SW
Washington, DC 20460

This office plans, develops, and manages a comprehensive research program to (1) develop and demonstrate methods to prevent, manage, or control the discharge of pollutants and the disposal of wastes from municipal, recreational, and other domestic sources; (2) assess the environmental and socioeconomic impact of such methods; (3) develop methods to reduce the production of wastes, including recycling; and (4) provide technical expertise and management assistance in the foregoing areas.

Waste Products Utilization Branch

Conservation Research and Development   202-252-2369
Industrial Programs
Conservation and Renewable Energy
Department of Energy
Forrestal Building, CE-12, Room 5F035
1000 Independence Avenue, SW
Washington, DC 20558

The Branch supports research and technological developments in industrial waste utilization, and develops technologies, processes, and systems to convert materials to feedstocks, chemicals, and fuels. Projects include research in mechanical and biochemical separation and conversion, combustion-pyrolysis, controls, and catalysis development with all forms of industrial waste materials.

Water Pollution

See: Marine Technology

Weapons Technology

See: Military Technology

Weight Reduction

See: Nutrition

Who's Who in Technology

See: People in Technology

Wind Energy

See also: Fuels, Alternative; Renewable Energy Technology; Solar Technology

Organizations and Government Agencies

American Wind Energy Association (AWEA)
1516 King Street   703-684-5196
Alexandria, VA 22314

AWEA is a trade association and lobbying group for the wind energy industry. Its members are manufacturers of wind energy conversion systems (WECS), retailers and distributors of the systems, wind farm developers, and investors. The Association develops standards for the industry and holds an annual conference. Conference proceedings cover many technical subjects within the wind energy field. Publications include:

● *Wind Energy Weekly*—covers business opportunities, legislative and regulatory developments, international issues, research developments, technical findings, market forecasts, and wind farm projects. Annual subscription rate is $125.

Geophysical Fluid Dynamics Section

Pacific Northwest Laboratory   509-375-3870
PO Box 999
Richland, WA 99352

Research on wind characteristics as they apply to design, performance, and operation of wind turbines is conducted by this office. Recent work includes wind turbine wake research, microscale turbulence analyses for dynamic stress load studies, and wind forecasting research for energy production studies. The laboratory has performed wind energy resource assessments for the U.S. and has developed an international wind energy resource assessment. It also has done research on siting methods for large and small wind turbines.

National Climatic Data Center

National Environmental Satellite, Data and Information Service   703-259-0682
National Oceanic and Atmospheric Administration
Federal Building
Asheville, NC 28801-2696

The office provides, for the cost of reproduction, climatic summaries for sites in or near a locality. Climatic summaries for major U.S. weather stations, including wind data, are available in different forms, some more detailed than others. Detailed summaries include wind variation by hour of day and month of year. Wind tabulations that present the percentage frequency of the directions and wind-speed groups have been constructed for many of the weather stations. In addition, the office has available raw, unedited insolation data for 26 cities in the U.S.

Small Wind Energy Conversion Systems Program

Rocky Flats Area Office   303-497-2647
Department of Energy (DOE)
PO Box 928
Golden, CO 80401

DOE's small wind energy conversion systems program is managed by this office. It operates Rocky Flats National Test Center for test and evaluation of small wind systems. It also performs and directs research and development directed toward providing information to manufacturers to improve performance of commercially available wind systems. The office answers technical inquiries from system manufacturers and utilities.

Wind Energy Technology

Solar Electric Technologies   202-252-1776
Conservation and Renewable Energy
Department of Energy
Forrestal Building, CE-331.1, Room 5H048
1000 Independence Avenue, SW
Washington, DC 20558

This office coordinates government research and development programs on large wind energy conversion systems. Most research and development for large wind energy systems is done through NASA (use this office for contact).

Large Wind Technology Branch

Forrestal Building, CE-331.1, Room 5H048
202-252-1995

Darrieus vertical axis wind turbine research and technology program is managed by this office.

Small Wind Technology Branch

Forrestal Building, CE-331.2, Room 5H048
202-252-6268

This Branch is responsible for the small wind energy (100 kilowatts or less) technology program.

Publications

Wind Industry News Digest

Alternative Sources of Energy, Inc.   612-983-6892
107 South Central Avenue
Milaca, MN 56353

This bimonthly newsletter is for those involved in the wind energy industry, including manufacturers, consultants, investors, suppliers, and researchers. It contains financial, legislative, and technical news. Annual subscription is $54.

Data Bases

National Climatic Data Center

National Oceanic and Atmospheric Administration (NOAA)   704-259-0682
User Services Branch
Federal Building
Asheville, NC 28801

The Center collects statistical information on solar energy and wind, and other meteorological data and information. It can supply data tapes and publications, and frequently prepares custom tapes to meet requester needs. Just specify the type of data you need, the time period, and the geographic location. The following tape is available:

• *Wind Energy Resource Information System*—the Pacific Northwest Laboratory, Battelle Memorial Institute, Richland, Washington, integrated regional wind resource assessment data and atlas maps into this data base, with approximately 975 stations. Standard information given with each table includes the table number, station name, station WBAN number, period of record, number of valid observations, and anamometer height and reference location. The time period covered is variable through 1978.

Additional Experts and Resources

Alternative Sources of Energy, Inc. is an excellent resource for wind. Its Energy Information Referral Service (EIRS) can answer questions and/or make referrals. Contact: Alternative Sources of Energy, Inc., 107 South Central Avenue, Milaca, MN 56353, 612-983-6892.

Windmills and Wind Turbines

See: Wind Energy

Word Processing

See: Minicomputers; Office Automation

Worker Displacement

See: Labor and Technological Change

X-Ray Technology

See: Medical Imaging

Appendix

New Tech Definition Sources

Publications

The Computer Language Company

140 West 30th Street

New York, NY 10001

212-736-8364

The company offers the following books providing definitions of computer terms:

- *The Computer Glossary*—a paperback that reduces technical computer concepts into words that can be understood by everyone. Geared toward computer beginners, businesspeople, and students, it contains illustrations and more than 1,000 definitions, each of which is cross-referenced. The 289-page publication is available for $14.95.

- *The Computer Coloring Book*—this paperback is oriented toward those wanting to understand computer basics. It contains 25 full-page illustrations as well as 50 terms and explanations. The 50-page publication is available for $3.95.

Dictionary of Occupational Titles

Superintendent of Documents

U.S. Government Printing Office

Washington, DC 20402

202-783-3238

This comprehensive directory lists, defines, and describes all the occupations currently in existence in the United States. Each occupational entry also includes the educational requirements and work experience generally needed to obtain a job. The publication was prepared for the Department of Labor's Employment and Training Administration. The cost is $23 and the GPO stock number is 029-013-0079.

IEEE Service Center

445 Hoes Lane

Piscataway, NJ 08854

201-981-0060

The following sources defining electrical and electronics terms are available from the IEEE:

- *IEEE Standard Dictionary of Electrical and Electronics Terms*—this volume contains 23,521 entries, with the terminology drawn from IEEE standards publications. Updated every five years, the 1,173-page publication is available for $49.95. The combination of this dictionary and the *IEC Multilingual Dictionary of Electricity* (described below) is available for $79.95.

- *IEC Multilingual Dictionary of Electricity*—this dictionary is based on the current 40 chapters of the International Electrotechnical Vocabulary. It defines more than 7,500 terms. This edition consists only of the English text, giving terms and definitions in English and equivalent terms in eight languages. Updated irregularly, the publication is available for $49.95.

McGraw-Hill Book Company

Princeton-Hightstown Road   609-426-5254

Hightstown, NJ 08520

This company publishes several books that are good sources of high-technology terms and definitions:

- *McGraw-Hill Dictionary of Scientific and Technical Terms*—the volume contains 98,500 technical and scientific terms and 115,000 definitions. It is a comprehensive dictionary, covering 100 different fields of science and engineering. Updated approximately every five years, this 1,846-page book is available for $70.

- *McGraw-Hill Dictionary of Science and Engineering*—derived from the work described above, this volume represents about 30 percent of that dictionary. Covering the same fields, it consists largely of basic terms and eliminates more esoteric terminology. Updated approximately every five years, it contains 942 pages and is available for $32.50.

- *Electronics Dictionary*—a guide to the current language of electronics which contains more than 17,000 entries and 1,100 illustrations. It features new meanings for older terms plus explanations of the hundreds of new words that have appeared in the electronics literature since the publication of the previous edition. Another feature is the *Electronic Style Manual,* which summarizes in convenient form the most troublesome spelling and grammatical problems encountered in writing about electronics. Updated irregularly, the 745-page publication is available for $39.50. The manual is also sold separately for $6.95.

Saunders Dictionary and Encyclopedia of Laboratory Medicine and Technology

210 West Washington Square   215-574-4878

Philadelphia, PA 19105

The publication contains more than 20,000 terms relevant to all aspects of laboratory medicine. It provides the terms explaining physics, chemistry, and biology underlying laboratory tests, as well as those pertaining to instrumentation, reference values, units of measurement, and lab safety. The 1,700-page publication is available for $47.95.

Telephony's Dictionary

Telephony Publishing Corporation   312-922-2435

Books Department

55 East Jackson Boulevard

Chicago, IL 60604

More than 14,500 terms, primarily in the areas of telecommunications and computers are defined in this volume. Some of the fields it covers include telephony, computers, satellites, radio, and special transmission media and techniques. Updated irregularly, the approximately 315-page publication is available for $35.

Van Nostrand's Scientific Encylopedia

Van Nostrand Reinhold Company   606-525-6600
7625 Empire Drive
Florence, KY 41042

A basic general reference, the Encyclopedia covers many technical topics and emphasizes the physical sciences. Its articles range from 30 lines to a couple of pages and are directed to scientifically oriented individuals at the undergraduate and graduate levels. Updated approximately every five years, the 3,500-page publication is available either in a one-volume edition for $107.50 or in two volumes for $139.50. Both editions contain exactly the same content.

Additional Experts and Resources

Many associations publish glossaries of terms used in the fields they represent. In addition, association staff members, particularly the librarian, may be able to give you definitions of terms.

New Tech General Resources

Organizations and Government Agencies

Association for Computing Machinery (ACM)

11 West 42nd Street   212-689-7440
New York, NY 10036

A large educational and scientific society, ACM is concerned with all aspects of computer science and its applications. The Association's 32 Special Interest Groups (SIGs) and numerous local chapters throughout the United States provide members with a variety of forums and continuing education programs. The SIGs focus on subjects ranging from automata and computability theory (ACT) to software engineering (SIGSOFT). Each group holds an annual conference, the proceedings of which are published by the Association, and produces its own newsletter. A listing of these groups and contact information are available from the Association's main office in New York. These local chapters sponsor meetings and workshops, including professional development seminars, which are open to the public. Contact the parent organization at the above address to be directed to the nearest local chapter. Each chapter can assist you by answering questions and making referrals. In addition to special interest group materials, the Association also publishes 11 journals covering different aspects of research, development, and applications in the computer field. A free publications catalogue is available. Examples of publications are:

• *Communications of the ACM*—this monthly contains technical papers, representing original contributions to the computing field. Reports, articles, professional activities, and features of interest to its members are also included. The "Computing Practices" section presents articles of current use to practitioners and periodically includes "Self-Assessment Procedures" so you can test yourself in key areas of computing. The publication is available free to members and costs nonmembers $78 a year.

● *Journal of the Association for Computing Machinery*—this quarterly is the pioneer research publication in the computer sciences. The *Journal* presents scholarly papers in all areas of the field and is the primary publication of theoretical computer science, treating the subject in its broadest terms and covering the whole spectrum of disciplines. The publication is available to nonmembers for $60 a year and to members for $12 a year.

● *Computing Review*—a monthly publication that covers the literature on computing and its pertinent applications. Over a thousand volunteer reviewers evaluate books, papers, articles, and so on. More than 200 serial publications that discuss the latest advances and developments are scanned. The Association's *Guide to Computing Literature,* an annual compendium, serves as a guide to the entire body of computing literature; it lists all work reviewed and abstracted in *Computing Reviews* and other references. The publication is available to nonmembers for $60 a year and to members for $19.

Congressional Caucus for Science and Technology
House Annex Building #2, H2-226   202-226-7788
2nd and D Streets, SW
Washington, DC 20515

Established to serve the U.S. Congress, the Caucus has a staff that works to provide elected representatives with accurate and timely information about the impact of science and technology on economic development in the United States. In addition to serving as a forum for its members, the Caucus has three goals: to become a clearinghouse on the impact of science and technology on American economic development and employment trends, to sponsor national and regional seminars and meetings designed to increase the scientific literacy in the United States, and to identify scientists, technologists, educators, and others who can advise Congress about scientific and technological issues. The staff can answer questions about high-tech-related federal legislation, and they will send you copies of proposed laws and statutes under consideration by Congress. They refer scientific and technological inquiries to the Caucus's research arm, the Research Institute for Space, Science and Technology. The Institute conducts and publishes results of studies about public policy issues arising from the advancement of technology. Topics they have examined include: expanding United States exports of high-tech products, protecting patents of high-tech products, and what can be done to assure minorities of equal access to employment in high-tech fields.

The Caucus publishes newsletters, summaries, and other materials. Of particular interest is:

● *Sci-Tech Briefs*—Published irregularly, these fact sheets cover a wide range of topics. Briefs have been published about education, high technology, research and development, technological innovations, and technology transfer. The briefs can be purchased from the Research Institute for Space, Science and Technology, 1245 4th Street, SW, Washington, DC 20024.

Federal Laboratory Consortium (FLC)

U.S. Department of Agriculture   202-447-7185—Ted Maher, FLC Director
3865 South Building
Washington, DC 20250

FLC is a national network of 300 individuals from federal laboratories and centers across the country. Members are responsible for assessing the technologies developed at their facility and then passing that knowledge on to industry, government, and the general public. Through the FLC the public can gain access to all unclassified research conducted by the federal government. Ted Maher can refer you to an FLC member in your specialty or geographical area. The following directory can also be used to locate members:

● *Federal Lab Directory 1985*—provides information about 388 federal laboratories with 10 or more full-time professionals engaged in research and development. Information is provided about staff size, mission, and major scientific or testing equipment. The directory is available free of charge from: Office of Research and Technology Applications, National Bureau of Standards, Room 402, Administration Building, Washington, DC 20234, 202-921-38144.

Information Analysis Centers (IACs)

National Referral Center   202-287-5670
Library of Congress
Washington, DC 20540

The federal government supports approximately 130 Information Analysis Centers involved in science and technology. These IACs provide a variety of services such as publication of critical data compilations, state-of-the-art reviews, bibliographies, current awareness services, and newsletters. Many centers answer inquiries and provide consulting and literature-search services. The National Referral Center can direct you to an appropriate IAC or send you a listing of all the centers.

Institute of Electrical and Electronics Engineers (IEEE)

IEEE Headquarters   212-705-7890—Technical Activities Information
345 East 47th Street   212-705-7866—Public Information
New York, NY 10017   202-785-0017—Washington Office
                                        201-981-1393—Service Center (to order publications)

This large engineering society focuses on advancing the theory and practice of electrical engineering, electronics, computer engineering, and computer science as well as the standing of members of the profession. IEEE consists of 30 technical societies corresponding to essentially every recognized discipline or interest area related to the fields it represents. There are, for example, societies for biomedical engineering, control systems, communications, and power engineering. The largest of these societies is the IEEE Computer Society. Staff at IEEE can answer questions and refer you to members in regional chapters throughout the United States.

One of IEEE's principal functions is publishing technical literature, and currently it is credited with publishing 15 percent of the world's technical papers in the electrical and electronics field. Each society publishes one or more technical periodicals, usually called *Transactions* or *Journals*, that record and disseminate new information in its respective field. Examples of areas covered by the *Transactions* and *Journals* include aerospace and electronic systems, computers, electron devices, lightwave technology, microwave theory and techniques, and quantum electronics. One primary periodical is:

● *IEEE Spectrum*—a monthly publication, it is the core magazine of the Institute. It contains state-of-the-art news, review, and applications articles of interest to a wide range of engineers and scientists in the electrical and electronics field; book reviews and listings; letters; and news of IEEE activities, the industry, and the profession. Free to members, the publication is available to nonmembers for $92 a year.

National Bureau of Standards (NBS)

Gaithersburg, MD 20899   301-921-1000—General Information

NBS provides a wide variety of measurement services and plays an advisory and consultant role to both government and industry in the development of standards. As the national reference for physical measurement, NBS produces measurement standards data necessary to create, make and sell American products and services in the United States and abroad. NBS offices provide special services and programs for the public. Its laboratories and centers can provide scientific and technological services as well as measurement, instrumentation, and standards information. As staff members deal directly with industries and users in their specialty areas, they are a good resource for information about the latest high-tech developments in their respective fields. Generally, a staff member can refer you to major companies, appropriate literature, research centers, experts, and government agencies.

National Coalition for Science and Technology (NCST)

1201 16th Street, NW, Suite 503   202-822-7312—Marc Rosenberg
Washington, DC 20036

A nonpartisan organization, NCST is composed of scientists, educators, businesspeople, engineers, high-tech corporations, and educational institutions concerned about the growing crisis in science education, scientific research, and productivity in the United States. Since its founding in 1982, the Coalition has advocated strong, consistent support for education, training, and research in science, technology, and engineering. It lobbies in those areas and sponsors conferences. Other activities include changing antitrust laws to facilitate joint ventures in research and working to continue research and development tax credits. Staff members can answer questions and provide referrals to the Coalition's Advisory Board and members.

National Referral Center (NRC)

Library of Congress   202-287-5670
Washington, DC 20540

NRC maintains an on-line data bank of more than 12,000 qualified organizations and individuals willing to provide information to the general public on topics in science, technology, and the social sciences. The Center can refer you to appropriate associations, government agencies, literature, and data bases. Staff members will perform computer searches and send you a printout free of charge. A typical citation contains the name of the resource, mailing address, telephone number, areas of interest, special collections, data bases, publications, and special services.

National Science Foundation (NSF)

1800 G Street, NW   202-357-9498—Office of Legislative and Public Affairs
Washington, DC 20550

An independent federal agency, NSF supports research in science and technology, educational activities in science and engineering, and surveys about the general state of science and technology in the United States. The Public Affairs staff will refer you to an NSF Program Officer who can tell you about the latest research and development efforts in your field of interest. Program Officers generally are experts in their field, and they can direct you to professionals and institutions performing research as well as available research reports. Upon request NSF will send its Organizational Directory, which can lead you to staff members who can make referrals. The Foundation also publishes a guide to its programs, which is a helpful tool for identifying the various subject areas for which it funds research projects. Described below are several important NSF publications. The last four are available by calling the Division of Science Resource Studies at 202-634-4622. All publications are available free of charge from NSF as long as the supply lasts. When the NSF supply is exhausted, the publications must be purchased from the Superintendent of Documents, U.S. Government Printing Office, Washington, DC 20402, 202-783-3238.

• *Grants and Award for Fiscal Year*—this 200-page publication lists grants and awards for the previous year. It breaks the information down by discipline, listing each state and providing information such as name of recipient and amount of funding.

• *Grants for Scientific and Engineering Research*—this publication contains instructions and applications for submitting a proposal. It goes hand-in-hand with the programs guide mentioned above.

• *Science Indicators*—a biannual publication of the National Science Board, it reports on the overall state of United States scientific endeavors. The report presents quantitative measures for all facets of science and technology. It also provides comparisons of the scientific and technical activities conducted in the United States and other major industrialized areas, primarily Japan, Western Europe, and the Soviet Union.

• *National Patterns of Science and Technology*—published each spring, this report provides statistical information. It assembles and analyzes data collected in the geographic studies and provides a concise and comprehensive summary as well as estimates for various national totals. These totals include the number of scientists, their median salaries, and more within many different categories of scientists and/or scientific fields. The publication complements the National Science Board's *Science Indicators* and NSF's *Science Engineering Personnel: A National Overview.*

• *Science and Engineering Personnel: A National Overview*—updated every two years, this 66-page publication provides a comprehensive overview of the status of American scientific and technological efforts as they relate to the employment of science and engineering personnel. It summarizes with text and appendix tables and provides a framework for analyzing issues relating to personnel.

• *Women and Minorities in Science and Engineering*—updated every two years, this 182-page report looks at current employment of women and minorities in science and engineering. Consisting of text and tables, it includes labor market indicators, comparison salary data, and information on experience levels and unemployment rates. This publication contains a large section on education and training of women and minorities, including precollege preparation.

National Technical Information Service (NTIS)

U.S. Department of Commerce   703-487-4600
5285 Port Royal Road
Springfield, VA 22161

NTIS is the central source for the public sale of United States government–sponsored research, development, and engineering reports as well as computer software and data files. The NTIS collection exceeds one million titles, with 60,000 new reports being released yearly. NTIS provides a variety of information, awareness products, and services to keep the public abreast of the technologies developed by the United States government's multibillion-dollar annual research and engineering effort. For additional information about NTIS, see *Technology Transfer.*

Office of Technology Assessment (OTA)

U.S. Congress   202-224-9241—Public Affairs Office
Washington, DC 20510

A nonpartisan analytical agency, OTA prepares objective analysis of major policy issues related to scientific and technological change. While these reports are prepared primarily for the United States Congress, OTA has published more than 100 reports that are available to the public. Its staff consists of experts in a wide range of areas: international security, health and life sciences, communication and information technologies, space, and productivity and innovation. OTA's staff members are available to answer questions and provide information about the subjects they have researched.

Upon request, OTA will send a free catalogue of its publications as well as descriptions of new projects under way. Summaries of most OTA reports are available, at no cost, directly from their office. Full text of the reports must be purchased from the Superintendent of Documents, U.S. Government Printing Office, Washington, DC 20402, 202-783-3238. Described below are OTA reports covering high tech in general.

• *Census of State Government Initiatives for High Technology Industrial Development, Background Paper* (Stock no. 052-003-00912-9, Publication code OTA-BP-STI-21)—is a directory of 153 state government programs to encourage high-technology development. It includes names and phone numbers of program directors as well as descriptions of program services. Cost is $4.50. (For an excerpt, see *State High-Tech Initiatives* section in this directory.)

• *Encouraging High Technology Development, Background Paper #2* (Publication Code OTA-BP-STI-25)—this publication provides a discussion of high-technology development initiatives by state and local governments, universities, and private-sector groups. It includes a directory of 54 local high-technology initiatives with names and phone numbers of program directors. Cost is $4.75.

• *Technology, Innovation, and Regional Economic Development* (Publication Code OTA-STI-238)—this report assesses the potential for local economic growth offered by high-technology industries, the types of programs used by state and local groups to encourage development of these industries, and the implications of these programs for federal policy. Cost is $4.50.

Republican Task Force on High Technology Initiatives

429 Cannon House Office Building   202-225-5411—Representative Ed Zschau, Chairman
Washington, DC 20515

The Task Force develops and implements legislation proposals related to technology and competitiveness for the Republican leadership in the U.S. House of Representatives. Staff members can provide information about high tech legislation under consideration by the U.S. Congress and they will send you free copies of bills and reports pertaining to the subject. Upon request they will also add your name to the Task Force mailing list for press releases and other material.

Science and Technology Division

Library of Congress   202-287-5639—Reference Section
2nd Street and Independence Avenue, SE   202-287-5580—*Tracer Bullets*
Washington, DC 20540

This division offers reference services based on its collection of 3 million scientific and technical books and pamphlets and 2.5 million technical reports. Reference librarians will assist researchers in using the Library's book and on-line catalogs as well as its abstracting and indexing services. The division publishes:

• *LC Science Tracer Bullets*—a series of bibliographic guides to selected scientific and technical topics such as high technology, computer science, and robotics. Each bullet is designed to help a reader begin to locate published material on a subject about which he or she has only general knowledge. The bullets are excellent information resources, as they contain reference listings of basic texts, handbooks, manuals, dictionaries, abstracting and indexing services, periodicals, selected journal articles, and technical papers. The bullets also list places to contact for additional information. As they are not updated regularly, it is advisable to check with the Science and Technology Division to get recommendations about new material in the Library's file.

Technology Utilization Division (NASA TU)

Code I, NASA Headquarters   202-453-8415
National Aeronautics and Space Administration
400 Maryland Avenue, SW
Washington, DC 20546

Publications:
Scientific and Technical Information Facility
Baltimore-Washington International Airport   301-621-0241
PO Box 8757
BWI Airport, MD 21240

The Technology Utilization Division is involved in transferring NASA-developed technology to industry for the development of useful commercial products. A variety of services and publications are available, and staff members will answer questions and make referrals. All NASA TU information and services are restricted to United States citizens and United States-based industry. For additional information about the program, see the entry for NASA Technology Utilization and Industry Affairs Division under *Technology Transfer*. Available publications include:

• *NASA Tech Brief Journal*—a free quarterly publication covering 125 new technologies in each issue. A current awareness medium, the *Journal* has information about potential products, industrial processes, basic and applied research, shop and laboratory techniques, computer software, and new sources of technical data contacts.

• *Special Publications Series*—NASA TU publishes technology utilization notes, handbooks and bibliographies on specific technologies, conference proceedings, technology utilization surveys on anything from advanced value technology to measurement of blood pressure in the human body, and technology utilization reports and compilations that include all types of documents about products, processes, and other topics in the tech utilization field.

• *NASA Annual Report*—produced to make the public aware of practical uses of aerospace research, this document contains a section profiling at least 35 companies that have used NASA technologies to improve their products. It is issued in June or July.

Publications

Directory of Federal Technology Resources
National Technical Information Service (NTIS)   703-487-4650
5285 Port Royal Road
Springfield, VA 22161

The directory, order no. PB 84-100015/AAW, describes 100 federal laboratories, agencies, and engineering centers willing to share expertise, equipment, and, in some cases, facilities. It also includes descriptions of 70 federal technical information centers. The directory is indexed by subject, geographical location, and resource name. It was prepared by the Center for the Utilization of Federal Technology, a division of NTIS. The cost is $25.

Gale Research Co.

Penobscot Building 645 Griswold   313-961-2242
Detroit, MI 48226

This company provides several directories describing science and technology information sources:

• *Scientific and Technical Organization and Agency Directory*—includes the scientific and technical organizations listed in other directories Gale publishes as well as some additional organizations. It covers government and private research centers, federal and state government agencies, engineering consultants, data bases and computerized information sources, sci-tech book publishers, technical libraries and information centers, and so on. Updated annually, the directory contains more than 12,000 entries. The 1,000-page publication is available for $125.

• *Special Libraries and Information Centers in the United States and Canada* —contains some 16,600 entries offering full descriptions of special libraries, information centers, and similar units. It includes information about computerized services and a listing of over 700 networks and consortia. This 1,640-page volume is Volume I of the three-volume *Directory of Special Libraries* and is available for $280.

• *Subject Directory of Special Libraries*—these five volumes contain the same information found in *Special Libraries and Information Centers in the United States and Canada.* They differ in that material is rearranged under 27 subject sections. The five volumes are: Volume 1, *Business and Law Libraries;* Volume 2, *Education and Information Science Libraries;* Volume 3, *Health Sciences Libraries;* Volume 4, *Social Sciences and Humanities Libraries;* Volume 5, *Science and Technology Libraries.* The cost is $125 per volume and $550 for the five-volume set.

High Technology

38 Commercial Wharf   617-227-4700
Boston, MA 02110

A monthly publication devoted entirely to high technology, this magazine covers the commercial application of new technologies, who is investing in them, and why. Articles also explain, in easy-to-understand terms, the concepts behind a particular technology as well as current research in the field and who is conducting it. Each month, the magazine's Update section announces recent developments in new products and technologies, its Insight Reports department forecasts future directions in technology development, and its Investments section provides a narrative about a particular segment of industry such as medical sales, telecommunications, or computer software products. The magazine is a helpful tool for learning about high technology and getting leads to information sources about specific topics. The annual subscription fee is $21.

High Technology Growth Stocks

14 Nason Street   617-897-9422
Maynard, MA 01754

Published monthly, this publication focuses on emerging high-tech companies estimated to grow in excess of 30 percent per year over the next three years. It features weekly profile markets and follows 75 fast-growing companies. The 40-page publication is available for $165 per year.

McGraw-Hill Book Co.

Princeton-Hightstown Road   609-426-5254
Hightstown, NJ 08520

This company offers three publications providing overall coverage of high-technology topics.

• *McGraw-Hill Concise Encyclopedia of Science and Technology*—updated every five years, this one-volume encyclopedia covers approximately 75 scientific and engineering disciplines. A condensation of the 15-volume set described below, it consists of 7,300 articles, index, and appendix. The 2,065-page publication is available for $89.50.

• *McGraw-Hill Encyclopedia of Science and Technology*—updated every five years, this 15-volume set covers approximately 75 scientific and engineering disciplines. Consisting of more than 7,800 articles and an index, the 13,000-page publication is available for $975.

• *McGraw-Hill Yearbook of Science and Technology*—this annual publication updates the *McGraw-Hill Encyclopedia of Science and Technology,* focusing on recent events and research. Its coverage varies from year to year. It generally contains five to ten feature articles and 150 articles on standard subjects. The 450-page yearbook is available for $50.

Technology Review

Room 10-140   617-253-8292—Subscriptions
Massachusetts Institute of Technology
Cambridge, MA 02139

Edited at the Massachusetts Institute of Technology, *Technology Review* is directed toward leaders in the fields of science, engineering, business, and policy-making. Written for both the professional and nonprofessional, this magazine is devoted to current and future developments in technology, with emphasis on the implications for human affairs and the environment. The annual subscription price for 8 issues is $24.

U.S. Industrial Outlook

Department of Commerce   202-377-4356
International Trade Administration
Industry Publications
14th Street and Constitution Avenue, NW, Room 442
Washington, DC 20230

Published each January, this book provides an overview and prospectus for more than 300 American industries. High-tech industries covered include telecommunications, computing and office equipment, electronic components, aerospace, chemicals, photographic equipment, information services, medical and dental instruments, and more. For each topic, the book provides statistics, information about new technologies and trends, and import/export data. It also provides the name, telephone number, and address of an expert for nearly every industry in the country. These specialists can tell you about the latest developments in the field and refer you to other experts and literature. The 1,000-page book costs $14 and is available at any Government Printing Office (GPO) bookstore. (The office listed above can direct you to a GPO bookstore in your area.)

Who's Who in Technology Today

Research Publication   203-397-2600
12 Lunar Drive
Woodbridge, CT 06525

A five-volume set, this provides bibliographic information about 32,000 leading scientists and technologists. Information given about each individual includes: business and home address, current position, previous position, honors, awards, technical achievements, and so on. The publication is organized according to 50 disciplines and divided into five volumes: Volume 1—*Electronics and Computer Science;* Volume 2 —*Physics and Optics;* Volume 3—*Chemistry and Biotechnology;* Volume 4—*Mechanical Engineering, Civil Engineering and Earth Science;* and Volume 5—*Index.* The 4,086-page publication is updated every two years. The entire set can be purchased for $425, or you can purchase individual volumes for $85.90 each.

Data Bases

Corporate Technology Database

Venture Economics
PO Box 348
Wellesley Hills, MA 02181

The computerized data base contains information about more than 15,000 high-tech companies, including name, address, telephone number, business product descriptions, chief executive officer's name, status of company, venture economic code, number of employees, and year of incorporation. It can be searched by any of the above variables and is available as a subscription service for $7,500 a year. Subscribers receive the entire data base as well as quarterly updates.

Additional Experts and Resources

Industry Specialists

U.S. Department of Commerce   202-377-4356—Publications Staff
Industry Publications
14th Street and Constitution Avenue, NW, Room 442
Washington, DC 20230

The publications staff can refer you to an industry specialist for nearly every industry in the United States. These specialists can tell you about the latest technological developments, provide statistics, give you data on imports and exports, tell you how American industry compares with industry abroad, and refer you to appropriate experts and literature.

Trade Reference Room Libraries

International Trade Administration   202-377-3808
14th Street and Constitution Avenue, NW
Washington, DC 20230

Trade reference room libraries located in ITA district offices across the country have U.S. import/export data by product, country, or both. These libraries have all kinds of trade statistics, but you must obtain the information by visiting the library yourself. Foreign data are only available at the ITA trade reference room in Washington, D.C. Check your telephone directory under "U.S. Government, Department of Commerce" and ask to be referred to International Trade Administration offices. If you can not locate your district office, call ITA's main office, listed above.

Index